Chemistry, Biology and Potential Applications of Honeybee Plant-Derived Products

Editor
Susana M. Cardoso
University of Aveiro,
Portugal

Co-Editor
Artur M.S. Silva
University of Aveiro,
Portugal

advertisements or ideas contained in the Work.

Limitation of Liability:

In no event will Bentham Science Publishers, its staff, editors and/or authors, be liable for any damages, including, without limitation, special, incidental and/or consequential damages and/or damages for lost data and/or profits arising out of (whether directly or indirectly) the use or inability to use the Work. The entire liability of Bentham Science Publishers shall be limited to the amount actually paid by you for the Work.

General:

1. Any dispute or claim arising out of or in connection with this License Agreement or the Work (including non-contractual disputes or claims) will be governed by and construed in accordance with the laws of the U.A.E. as applied in the Emirate of Dubai. Each party agrees that the courts of the Emirate of Dubai shall have exclusive jurisdiction to settle any dispute or claim arising out of or in connection with this License Agreement or the Work (including non-contractual disputes or claims).
2. Your rights under this License Agreement will automatically terminate without notice and without the need for a court order if at any point you breach any terms of this License Agreement. In no event will any delay or failure by Bentham Science Publishers in enforcing your compliance with this License Agreement constitute a waiver of any of its rights.
3. You acknowledge that you have read this License Agreement, and agree to be bound by its terms and conditions. To the extent that any other terms and conditions presented on any website of Bentham Science Publishers conflict with, or are inconsistent with, the terms and conditions set out in this License Agreement, you acknowledge that the terms and conditions set out in this License Agreement shall prevail.

Bentham Science Publishers Ltd.
Executive Suite Y - 2
PO Box 7917, Saif Zone
Sharjah, U.A.E.
Email: subscriptions@benthamscience.org

BENTHAM SCIENCE

CONTENTS

FOREWORD

This book is about relationships: complicated, sophisticated, intriguing, sometimes mysterious but always fascinating and rewarding. These are the relationships between the society of honeybees, the plant world and the human society.

The book is dedicated to bee products of plant origin: honey, bee pollen and propolis (bee glue). The plant-derived bee products are a result of the co-evolution of honeybees and flowering plants, of the activities of two very different types of genetic inheritance, leading to mutual benefits. Much later, a third party appeared in this relationship: humanity, and contributed further to the benefits by beekeeping. Also, curiosity and scientific research have become a part of the relationship, and so the present work is also among the results.

Plant-derived bee products have specific characteristics which set them apart and make their study different, if not more demanding, compared to the study of the bee-synthesized products. Bee products of plant origin deserve such special attention because their study requires a multidisciplinary approach. First of all, there is the phytochemical aspect, revealing the chemical features of the source plant and the product; including the specific phytochemical methods and approaches to bee products. The study of their pharmacological properties, aimed to prove or disprove the numerous anecdotal data about health benefits and healing properties of honey, pollen and propolis, is another important aspect. The combination of the knowledge acquired by these two lines of research leads to the possibility to standardize plant-derived bee products for different purposes. And last but not least, although the use of these products might seem traditional, finding new potential for their innovative applications requires imagination, ingenuity and skills. This book is an attempt to cover the most recent advances in all those aspects.

Being invited to write this Foreword is a great honor. I have spent my professional life studying bee glue and fascinated all these years by the ability of bees to find the chemicals that suit them best in the complex environment they inhabit, in almost every geographic region of the Earth. The authors that have contributed to this book share this fascination; they are devoted and well known experts in their respective fields. The combination of distinct approaches and view points is a merit of the work. All this makes the book particularly valuable to researchers: bee scientists, pharmacologists, phytochemists, but also to anyone interested in bees, bee-plant relationship and apitherapy.

Vassya Bankova
Institute of Organic Chemistry with Centre of Phytochemistry,
Bulgarian Academy of Sciences,
Sofia,
Bulgaria

PREFACE

Domestic bees produce several products including honey, propolis, pollen, royal jelly, beeswax and bee venom, which are essential to their survival and development. Honeybee products have also been used in folk medicine since ancient times and presently constitute one of the most applicable groups of natural products for Humans. While beeswax, bee venom, and royal jelly are chemically synthesized by the bees themselves, the remaining products result from bee´s engineering modification action on plant-derived samples.

The primary usefulness of honeybee plant-derived products to Man is largely based on their utilities in the hive, *i.e.* honey and pollen are stable food sources for bees, with the first being particular enriched in sugars and the latter being an excellent supplement of minerals, vitamins and proteins, whereas propolis is used for the hive protection.

The specific composition of each product is rather variable, depending on the plants found around the hive, as well as on the geographic and climatic characteristics of the site, thus affecting specific biological properties and possible applications. Still, worldwide the usage of such products has been increasing exponentially because of the believed health-benefits of those products. The investigation of the chemical composition and associated biological properties of honey, pollen and propolis has been imperative in elucidating their specific composition and respective potential health benefits, being as well a open door for standardization of the products for distinct usages.

This eBook comprises of comprehensive review on the chemical composition of honey, pollen and propolis of worldwide, complemented with the contribution of distinct analytical techniques for this topic. It also summarizes the current information of relevant biological properties and applications of honey, pollen and propolis, which overall contribute for added-value to these bee plant-derived products. We deeply believe that the present volume is not only important for scientific community, but also for beekepers and readers in general.

Susana M. Cardoso & Artur M.S. Silva
University of Aveiro,
Portugal

List of Contributors

Marta Quicazán	Instituto de Ciencia y Tecnología de Alimentos – ICTA, Universidad Nacional de Colombia. Ciudad Universitaria, Av. Kr. 30 # 45-03 Ed. 500C, 111321, Bogotá, Colombia
Carlos Zuluaga	Instituto de Ciencia y Tecnología de Alimentos – ICTA, Universidad Nacional de Colombia. Ciudad Universitaria, Av. Kr. 30 # 45-03 Ed. 500C, 111321, Bogotá, Colombia
Vassya Bankova	Institute of Organic Chemistry with Centre of Phytochemistry, Bulgarian Academy of Sciences, 1113 Sofia, Bulgaria
Boryana Trusheva	Institute of Organic Chemistry with Centre of Phytochemistry, Bulgarian Academy of Sciences, 1113 Sofia, Bulgaria
Milena Popova	Institute of Organic Chemistry with Centre of Phytochemistry, Bulgarian Academy of Sciences, 1113 Sofia, Bulgaria
Maria G. Campos	Drug Discovery Group, Center for Pharmaceutical Studies, Faculty of Pharmacy & Chemistry Center FCT, University of Coimbra, Health Sciences Campus, Azinhaga de Santa Comba, 3000-548 Coimbra, Portugal
Lokutova Olena	National University of Life and Environmental Sciences of Ukraine, Kiev, Ukraine
Ofélia Anjos	IPCB/ESA-Instituto Politécnico de Castelo Branco, Escola Superior Agrária, Castelo Branco, Portugal CEF/ISA/UTL – Centro de Estudos Florestais, Instituto Superior de Agronomia, Universidade Técnica de Lisboa, Lisboa, Portugal
Marcelo D. Catarino	CERNAS, School of Agriculture, Polytechnic Institute of Coimbra, Bencanta, 3045-601 Coimbra, Portugal
Jorge M. Alves-Silva	CERNAS, School of Agriculture, Polytechnic Institute of Coimbra, Bencanta, 3045-601 Coimbra, Portugal
Soraia I. Falcão	CIMO, School of Agriculture, Polytechnic Institute of Bragança, Campus de St^a Apolónia, Apartado 1172, 5301-855 Bragança, Portugal
Miguel Vilas-Boas	CIMO, School of Agriculture, Polytechnic Institute of Bragança, Campus de St^a Apolónia, Apartado 1172, 5301-855 Bragança, Portugal
Micaela Jordão	CERNAS, School of Agriculture, Polytechnic Institute of Coimbra, Bencanta, 3045-601 Coimbra, Portugal
Susana M. Cardoso	CERNAS, School of Agriculture, Polytechnic Institute of Coimbra, Bencanta, 3045-601 Coimbra, Portugal QOPNA, Department of Chemistry, University of Aveiro, Campus Universitário de Santiago, 3810-193 Aveiro, Portugal

Clementina M.M. Santos	School of Agriculture, Polytechnic Institute of Bragança, 5301-855 Bragança, Portugal
Artur M.S. Silva	Department of Chemistry & QOPNA, University of Aveiro, Campus Universitário de Santiago, 3810-193 Aveiro, Portugal
António M. Peres	LSRE- Laboratory of Separation and Reaction Engineering - Associate Laboratory LSRE-LCM, School of Agriculture, Polytechnic Institute of Bragança, Campus de St[a] Apolónia, Apartado 1172, 5301-855 Bragança, Portugal
Mara E.B. Sousa	CIMO - Mountain Research Centre, School of Agriculture, Polytechnic Institute of Bragança, Campus de St[a] Apolónia, Apartado 1172, 5301-855 Bragança, Portugal
Ana C.A. Veloso	Polytechnic Institute of Coimbra, ISEC, DEQB, Rua Pedro Nunes, Quinta da Nora, 3030-199 Coimbra, Portugal CEB - Centre of Biological Engineering, University of Minho, Campus de Gualtar, 4710-057 Braga, Portugal
Letícia Estevinho	School of Agriculture, Polytechnic Institute of Bragança, Campus de St[a] Apolónia, Apartado 1172, 5301-855 Bragança, Portugal
Luís G. Dias	School of Agriculture, Polytechnic Institute of Bragança, Campus de St[a] Apolónia, Apartado 1172, 5301-855 Bragança, Portugal CQ-VR, Center of Chemistry – Vila Real, University of Trás-os-Montes e Alto Douro, Apartado 1013, 5001-801 Vila Real, Portugal
Daniel Cozzolino	School of Medical and Applied Sciences, Central Queensland University, Rockhampton, QLD 4701, Australia
Marta Quicazán	Instituto de Ciencia y Tecnología de Alimentos – ICTA, Universidad Nacional de Colombia. Ciudad Universitaria, Av. Kr. 30 # 45-03 Ed. 500C, 111321. Bogotá, Colombia
Carlos Zuluaga	Instituto de Ciencia y Tecnología de Alimentos – ICTA, Universidad Nacional de Colombia. Ciudad Universitaria, Av. Kr. 30 # 45-03 Ed. 500C, 111321. Bogotá, Colombia
Pedro A.R. Fernandes	QOPNA, Department of Chemistry, University of Aveiro, Campus Universitário de Santiago, 3810-193 Aveiro, Portugal School of Agriculture, Polytechnic Institute of Viseu, 3500-606 Viseu, Portugal
Sónia S. Ferreira	QOPNA, Department of Chemistry, University of Aveiro, Campus Universitário de Santiago, 3810-193 Aveiro, Portugal School of Agriculture, Polytechnic Institute of Viseu, 3500-606 Viseu, Portugal
Alice Fonte	School of Agriculture, Polytechnic Institute of Viseu, 3500-606 Viseu, Portugal

Dulcineia F. Wessel	School of Agriculture, Polytechnic Institute of Viseu, 3500-606 Viseu, Portugal CI&DETS, Polytechnic Institute of Viseu, 3500-606 Viseu, Portugal
Joana Liberal	Center for Neuroscience and Cell Biology, University of Coimbra 3004-517, Coimbra, Portugal Faculty of Pharmacy, University of Coimbra 3000-548, Coimbra, Portugal
Isabel V. Ferreira	Center for Neuroscience and Cell Biology, University of Coimbra 3004-517, Coimbra, Portugal
Eliza O. Cardoso	Biosciences Institute, UNESP, 18618-000, Botucatu, Brazil
Ana Silva	Center for Neuroscience and Cell Biology, University of Coimbra 3004-517, Coimbra, Portugal
Ariane R. Bartolomeu	Biosciences Institute, UNESP, 18618-000, Botucatu, Brazil
João Martins	Center for Neuroscience and Cell Biology, University of Coimbra 3004-517, Coimbra, Portugal Faculty of Pharmacy, University of Coimbra 3000-548, Coimbra, Portugal
Karina B. Santiago	Biosciences Institute, UNESP, 18618-000, Botucatu, Brazil
Bruno J. Conti	Biosciences Institute, UNESP, 18618-000, Botucatu, Brazil
Bruno M. Neves	Center for Neuroscience and Cell Biology, University of Coimbra 3004-517, Coimbra, Portugal Faculty of Pharmacy, University of Coimbra 3000-548, Coimbra, Portugal Department of Chemistry, Mass Spectrometry Center, QOPNA, University of Aveiro, Campus Universitário de Santiago, 3810-193 Aveiro, Portugal
Maria T. Batista	Center for Neuroscience and Cell Biology, University of Coimbra 3004-517, Coimbra, Portugal Faculty of Pharmacy, University of Coimbra 3000-548, Coimbra, Portugal
José M. Sforcin	Biosciences Institute, UNESP, 18618-000, Botucatu, Brazil
Maria T. Cruz	Center for Neuroscience and Cell Biology, University of Coimbra 3004-517, Coimbra, Portugal Faculty of Pharmacy, University of Coimbra 3000-548, Coimbra, Portugal
Cristina Almeida-Aguiar	CITAB, Centre for the Research and Technology of Agro-Environmental and Biological Sciences, AgroBioPlant Group, University of Minho, Braga, Portugal Biology Department, School of Sciences, University of Minho, Campus de Gualtar, 4710-057 Braga, Portugal
Ricardo Silva-Carvalho	Life and Health Sciences Research Institute (ICVS), School of Health Sciences, University of Minho, Braga, Portugal ICVS/3B's - PT Government Associate Laboratory, Braga/Guimarães, Portugal

Fátima Baltazar	Life and Health Sciences Research Institute (ICVS), School of Health Sciences, University of Minho, Braga, Portugal ICVS/3B's - PT Government Associate Laboratory, Braga/Guimarães, Portugal
Miroslava Kačániová	Department of Microbiology, Faculty of Biotechnology and Food Sciences, Slovak University of Agriculture in Nitra, Tr. A. Hlinku 2, 949 76 Nitra, Slovakia
Cristina Almeida-Aguiar	CITAB, Centre for the Research and Technology of Agro-Environmental and Biological Sciences, AgroBioPlant Group, University of Minho, Braga, Portugal Biology Department, School of Sciences, University of Minho, Campus de Gualtar, 4710-057 Braga, Portugal
Silvia C.F. Iop	Multidisciplinary Department , Federal University of Santa Maria, Rua Francisco Guerino, 407 – Silveira Martins – RS, Brasil
Elsa C.D. Ramalhosa	Mountain Research Centre (CIMO) - School of Agriculture, Polytechnic Institute of Bragança, Campus de Stª Apolónia, Apartado 1172, 5301-855 Bragança, Portugal
Dalva M.R. Dotto	Multidisciplinary Department , Federal University of Santa Maria, Rua Francisco Guerino, 407 – Silveira Martins – RS, Brasil
Andreia Cirolini	Multidisciplinary Department , Federal University of Santa Maria, Rua Francisco Guerino, 407 – Silveira Martins – RS, Brasil
Naiane Beltrami	Multidisciplinary Department , Federal University of Santa Maria, Rua Francisco Guerino, 407 – Silveira Martins – RS, Brasil

CHAPTER 1

Chemical Characterization of Honey

Marta Quicazán*, Carlos Zuluaga

Instituto de Ciencia y Tecnología de Alimentos – ICTA, Universidad Nacional de Colombia. Ciudad Universitaria, Av. Kr. 30 # 45-03 Ed. 500C, 111321. Bogotá, Colombia

Abstract: The chemical composition of honey has been widely investigated around the world. Studies have shown how geographical and botanical origin of honey influences its physical-chemical properties, being a useful tool for evaluation of authenticity and differentiation. In this chapter, a review has been performed in order to describe the chemical composition and the most important nutritional properties of honeys from several countries and diverse botanical origins. Reported data of honeys for water, sugars, ash, minerals, color conductivity, aminoacids and quality indicators, such free acidity, enzymes and hydroxymethylfurfural (HMF) are mentioned and also compared to the established limits given by the Codex Alimentarius and the International Honey Commission.

Keywords: Apiculture, Authenticity, Bioactive compounds, Biological activity, Carbohydrates, Chemical markers, Classification, Color, Differentiation, Electrical conductivity, Flavonoids, HMF, Honey enzymes, Minerals, Moisture, Nutrition, Origin, Phenolic compounds, Quality, Volatile compounds.

1. INTRODUCTION

The Codex Alimentarius defines honey as "the natural sweet substance produced by honey bees from the nectar of plants or from secretions of living parts of plants or excretions of plant sucking insects on the living parts of plants, which the bees collect, transform by combining with specific substances of their own, deposit, dehydrate, store and leave in the honey comb to ripen and mature" [1].

* **Address correspondence to Marta Quicazán:** Instituto de Ciencia y Tecnología de Alimentos – ICTA, Universidad Nacional de Colombia. Ciudad Universitaria, Av. Kr. 30 # 45-03 Ed. 500C, 111321. Bogotá, Colombia; Tel/Fax: +57-1-3165300; Email: mcquicazand@unal.edu.co

Honey market is currently growing worldwide, mainly because consumers have recognized this food as a natural product that provides health benefits. In 2012, honey production was higher than 1.5 million tons and its production in the world is distributed as following: Asia (38.5%), Americas (24.1%), Europe (23.9%), Africa (11.3%) and Oceania (2.2%). The main producers are China, United States, Argentina and Turkey [2].

Honey is not a new product. Since very early in human history, honey was the only source of concentrated sugar and, even, it was already recognized for its therapeutic significance. Many of the myths of the traditional medicinal use of honey have continued even today [3]. Due to the recognition and value-added of honey, unfortunately it is common to find the existence of counterfeit or adulterated products, which have become a constant concern among beekeepers. As a strategy to overcome this problem, several studies have been conducted in order to evaluate, from different points of view, the quality and authenticity of honey [4 - 6].

Bees collect nectars from different flowers, and thus, honey comes from diverse floral sources [7]. According to this, honey can be associated to its botanical or geographical origin, where different kinds of characteristics, such climate or soil, determine the abundance of meliferous flora. Baroni *et al.* and Persano *et al.* [8, 9] suggested that floral origin have an important influence in honey quality. The main componets in honey are carbohydrates, but minor substances such organic acids, proteins, minerals or vitamins can also be found [3].

Several authors reported a moisture content in honey below 20%, reducing sugar content between 60-65% and a content of 1-10% sucrose [10 - 13], characteristics for which no differences can be drawn. Generally, those components present in low concentrations in honey are used to discriminate and detect potential fraud [14, 15].

The regulations of the European Community provides general definitions related to honey, including general and specific compositional characteristics such as hydroxymethylfurfural content, humidity and levels of pesticides, but these parameters have no relation to botanical or geographical origin [9, 16]. Other studies have found that laboratory markers such as volatile compounds, profile of sugars, flavonoids, phenolic compounds, minerals and electrical conductivity are useful to classify and differentiate honeys according to origin [17].

This chapter summarizes existing information regarding the physical-chemical properties, nutritional information and quality of honeys worldwide. Profiles of main components, used for honey characterization as recommended by the International Honey Commission, are described. In addition, the main bioactive compounds are presented. Due to lack of space and the vast amount of information published around the world regarding the composition of honey, it is impossible to report all available articles. Instead, this chapter focuses on the research performed mainly in the past three decades (1985–2014).

2. MAIN COMPONENTS OF HONEY

2.1. Water and Sugars

Honey is composed primarily of water and simple reducing sugars (mainly fructose and glucose), and non-reducing sugars (mainly sucrose and maltose). These parameters depend on many factors, such as the maturity achieved in the bee hive during the harvesting season, climatic and geographic factors, and other elements affecting floral abundance [18].

Water content for worldwide varieties of honey is given in (Table **1**). Bogdanov *et al.* [16] stated that "honey moisture is a quality criterion that determines the capability of honey to remain stable and to resist spoilage by yeast fermentation". Regulation from several countries, European Union and Codex Alimentarius agree to limit the water content in honey to not more than 20%, with some exceptions such as Heather honey for which moisture is allowed up to 23%. In general terms, honeys from around the world fulfill the requirement for moisture content no matter their floral origin.

For instance, Eucalyptus Uruguayan honey is reported to have low levels of water of near 6%; nevertheless, Nigerian honeys can present up to 30%, being more susceptible to spoilage. It is clear, then, this parameter more than a differentiation variable, is a primarily quality criterion.

Table 1. Water content of selected honeys worldwide.

Country	Floral origin	Value ranges (%)	Ref.
Algeria	Multifloral	14.0 - 20.2	[19]
	Eucalyptus	14.0 - 19.8	
Argentina	Multifloral	15.8 - 20.2	[8, 23]
	Eucalyptus	14.4 - 20.0	
Bosnia-Herzegovina	Black Locust	15.6 - 18.8	[26]
Brazil	*Eucalyptus*	19.3 - 22.4	[27]
	Wild flower	16.0 - 23.4	
Burkina Faso	Multifloral	14.7 - 18.7	[31]
Cameroun	Not specified	15.3 - 17.7	[33]
Canada	Canola	16.5	[38]
Colombia	Multifloral	17.1 - 19.8	[39]
	Multifloral	16.8 - 19.0	
Cuba	Linen vine	18.2 - 20.6	[42]
	Morning glory	16.9 - 18.7	
Germany	False acacia	15.02 - 16.92	[44]
	Heather	17.44 - 19.94	
Ireland	White clover and blackberry	15.6 - 18.8	[47]
India	*Citrus*	18.2 - 18.8	[51, 52]
	Multifloral	13.5 - 14.5	
Italy	*Eucalyptus*	15.5 - 16.0	[54]
	Wildflower	15.4 - 17.4	
Kenya	Not specified	16.6 - 16.8	[56]
Libya	*Eucalyptus*	18	[57]
Malaysia	Acacia	15.1 - 15.3	[59, 60]
	Pineapple	14.7 - 15.1	

Country	Floral origin	Value ranges (%)	Ref.
Maroc	Sunflower	16.9 - 18.5	[20 - 22]
	Carob tree	18.3 - 19.3	
	Heather	16.9 - 18.4	
Mexico	Tajonal	16.3 - 20.6	[24, 25]
	Multifloral	18.9 - 24.3	
Nepal	Not specified	14.5 - 19.7	[28]
Nigeria	Not specified	16.3 - 30.8	[29, 30]
Poland	Buckwheat	17.7 - 20.0	[32]
Portugal	Multifloral	13.5 - 19.7	[34 - 37]
	Erica	17.0 - 18.1	
Romania	Polyfloral	16.9 - 18.7	[40]
Saudi Arabia	Sidir Aseer	15.4 - 15.4	[41]
	Sidir Albaha	14.0 - 14.1	
Serbia	Acacia	13.9 - 20.6	[43]
Slovakia	Multifloral	17.0 - 19.5	[45]
Slovenia	Multifloral	14.4 - 18.0	[46]
Spain	Thyme	14.2 - 19.8	[48 - 50]
	Multifloral	13.0 - 18.7	
Sudan	*Cucurbita maxima*	16.0 - 17.2	[53]
	Acacia nilotica	15.7 - 16.8	
Turkey	Flower	16.5 - 20.0	[55]
	Chestnut	16.4 - 20.0	
Uruguay	*Eucalyptus* spp.	5.6 - 29.6	[58]
	Pastures	11.7 - 21.7	
USA	Multifloral	13.4 - 22.9	[61]

On the other hand, because of their floral origin, the main sugars present in honey are glucose, fructose, maltose and sucrose; other disaccharides and oligosaccharides occur in lower proportion and in trace quantities [62]. The sugar composition shown in Table **2** includes the most important sugars.

Table 2. Sugar composition of selected honeys worldwide.

Americas

Country	Botanical origin	Glucose (%)	Fructose (%)	Sucrose (%)	Ref.
Argentina	Multifloral	19.4 - 38.0	26.1 - 48.3	N.R.	[7]
Canada	Canola	38.2 - 40.2	39.3 - 40.7	N.R.	[38]
	Clover	37.5 - 38.9	39.3 - 40.7	N.R.	
Colombia	Multifloral	28.9 - 39.3	36.0 - 44.0	4.5 - 10.7	N.P.
Mexico	Tajonal	35.2 - 36.8	40.4 - 41.4	0.0 - 0.3	[24, 25]
	Multifloral	35.0 - 40.2	37.7 - 44.0	2.8 - 3.1	
USA	Multifloral	22.0 - 40.8	27.3 - 44.3	0.3 - 7.6	[61]
Venezuela	Multifloral	29.2 - 38.7	38.5 - 44.5	2.2 - 5.5	[15]

Africa

Country	Botanical origin	Glucose (%)	Fructose (%)	Sucrose (%)	Ref.
Algeria	Multifloral	21.2 - 35.2	34.0 - 49.1	0 - 4.3	[19]
	Eucalyptus	24.2 - 34.0	34.2 - 47.6	0.4 - 2.1	
	Citrus	25.5 - 28.4	39.5 - 42.8	0.0 -1.0	
Libya	*Eucalyptus*	32	40	1.5	[30]
Maroc	Heather	30.6 - 30.8	36.2 - 38.7	2.9 - 4.5	[20 - 22]
	Eucalyptus	27.3 - 36.2	32.1 - 43.1	0.0 - 0.2	
	Thyme	26.1 - 33.8	37.2 - 41.0	0.6 - 2.6	
	Multifloral	25.1 - 39.3	36.6 - 46.3	0.2 - 2.8	

Europe

Country	Botanical origin	Glucose (%)	Fructose (%)	Sucrose (%)	Ref.
Estonia	Polyfloral	28.8 - 40.3	33.1 - 38.3	N.R.	[63]
Germany	Heather	29.0 - 33.4	40.0 - 42.2	0.0 - 0.2	[44]
Italy	Citrus	29.8 - 35.7	33.5 - 45.1	0.0 - 4.5	[64]
	Thymus	26.8 - 33.8	38.0 - 49.2	0.0 - 0.6	
Poland	Buckwheat	46.6	51.6	0.3 - 1.8	[32]

Asia

Country	Botanical origin	Glucose (%)	Fructose (%)	Sucrose (%)	Ref.
Nepal	Not specified	36.3 - 46.3	42.3 - 50.5	0.0 - 7.8	[28]
Saudi Arabia	Sidir Aseer	24.1 - 25.5	37.6 - 40.2	6.2 - 6.2	[41]
	Sidir Albaha	28.5 - 29.7	42.1 - 43.5	0.0 - 0.1	
	Talh Tehamh	32.4 - 35.2	36.2 - 38.8	0.9 - 1.1	
	Samra Taif	27.4 - 29	33.2 - 35.2	0.7 - 0.8	

(Table 2) contd.....

Americas						Africa					
Country	**Botanical origin**	**Glucose (%)**	**Fructose (%)**	**Sucrose (%)**	**Ref.**	**Country**	**Botanical origin**	**Glucose (%)**	**Fructose (%)**	**Sucrose (%)**	**Ref.**
Romania	Polyfloral	30.8 - 34.0	34.2 - 36.6	1.6 - 1.8	[40]						
Spain	Multifloral	19.3 - 31.2	23.2 - 39.9	N.R.	[48, 50]						
	French Lavender	28.0 - 36.8	36.5 - 39.5	0.0 - 1.3							

N.R.: Not reported. N.P.: Non published data

Glucose content varied between 19.4 and 40.8 g/100 g for honey from Americas, between 21.2 and 39.3 g/100 g for Africa, 19.3 and 40.3 g/100 g for Europe, and between 24.1 and 46.3 g/100 g for Asia/Oceania. In the case of fructose, content varied between 26.1 and 48.3 g/100 g for Americas, 32.1 and 49.1 g/100 g for Africa, 23.2 and 51.6 g/100 g for Europe and 33.2 and 50.5/100 g for Asia/Oceania. Finally, reported data for sucrose showed values ranging from 0.0 to 7.6 g/100 g in Americas and from 0.0 to 4.5 g/100 g in Africa and Europe. In Asia/Oceania the values of sucrose were between 0.0 and 7.8 g/100 g.

Regulation sets the maximum content of glucose and fructose in honeys. The sum of those two monosaccharides should be less than 60 g/100 g, except for honeydew honey which cannot have less than 45 g/100 g [1]. Such divergent values have high variability and probably too few samples analyzed, and thus, further characterization must be performed in order to obtain more reliable data of sugar concentration values as an origin denomination criterion. In the case of sucrose, honey should have no more than 5 g/100 g, except for Alfalfa (*Medicago sativa*), *Citrus* spp., False Acacia (*Robinia pseudoacacia*), French Honeysuckle (*Hedysarum*), Menzies Banksia *(Banksia menziesii*), Red Gum *(Eucalyptus camaldulensis*), Leatherwood (*Eucryphia lucida*), *Eucryphia milligani* which should have no more than 10 g/ 100 g. Another exception for sucrose occurs for Lavender (*Lavandula spp*) and Borage *(Borago officinalis*) honeys, which can have up to 15 g /100 g [1]. From Table **2**, it is clear that some authentic honeys have a higher content of sucrose than regulated, such as honey from USA or Nepal (7.8 g sucrose /100 g honey).

Different oligosaccharides have been quantified in honeys, although they are not present in significant concentrations (usually not more than 3 g/100 g honey). Despite the low amount, their presence is useful for differentiating honeys according to origin. Examples of oligosaccharides present in honey include: maltose, raffinose, kojibiose, meliobiose, turanose, erlose, nigerose, maltotrealose, palatinose, gentiobiose, maltotriose, isomaltose, cellobiose, panose, trehalose [28, 65, 66]. Quantitative data for oligosaccharides from New Zealand, Nepal and Spain are presented in (Table **3**).

Table 3. Amounts of some oligosaccharides present in bee-honey.

Oligosaccharide	**Content (g/100 g honey)**		
	New Zealand	**Nepal**	**Spain**
Trehalose	0.042 - 0.051	2.1	0.51 - 0.86
Turanose	1.2 - 1.8	2.3	N.R
Maltose	1.9 - 2.6	2.2	3.2 - 7.1
Isomaltose	0.2 - 0.4	1.9	0.6 - 1.6
Erlose	0.9 - 1.8	2.2	0.00 - 0.30
Melezitose	0.07 - 0.09	2.2	0.00 - 0.30
Panose	0.27 - 0.73	2.9	N.R
Maltotetraose	0.18 - 0.85	3.2	N.R
Melibiose	N.R	2.01	0.12 - 0.54
Raffinose	N.R	2.2	0.00 - 0.10
Reference	[65]	[28]	[67]

N.R.: Not reported

2.2. Ash and Minerals

The ash and mineral contents depend strongly on botanical and geographical origin [68 - 70]. The contents of ash and of some minerals (Na, K, Ca, Mg, Fe, Cu, Mn and Zn) for worldwide honeys are shown in Tables **4** and **5**, respectively. Despite ash content is not included among composition parameter of Codex Alimentarius, several authors and regulations from different countries establish a maximum of 0.5 g/100 g, which is met by most of the honeys described in (Table **4**). Exceptions, mainly in honeys from America and Europe, with values higher than 1.2 g/100 g, are found. This difference implies that some samples would not meet the suggested value for ash content, in spite of authenticity. Therefore, this limit needs to be clarified.

Table 4. Ash content in selected honeys worldwide.

Origin	Botanical origin	Range value (%)	Ref.	Origin	Botanical origin	Range value (%)	Ref.
Argentina	Multifloral	0.06 - 0.21	[7]	Mexico	Tajonal	0.15 - 0.26	[24, 25]
Bosnia-Herzegovina	Black locust	0.08 - 0.54	[26]		Multifloral	0.06 - 0.08	
Brazil	*Eucalyptus*	0.07 - 0.32	[27]	Nigeria	Not specified	0.10 - 0.52	[29, 30]
	Wild flower	0.02 - 0.92		Pakistan	Not specified	0.22 - 0.38	[73]
Burkina Faso	Multifloral	0.10 - 0.50	[31]	Portugal	Multifloral	0.09 - 0.53	[34 - 37]
Cameroun	Not specified	0.05 - 0.67	[33]	Romania	Acacia	0.03 - 0.28	[40]
Colombia	Multifloral	0.00 - 0.47	N.P.		Polyfloral	0.40 - 0.60	
Cuba	Linen vine	0.47 - 0.53	[42]	Saudi Arabia	Sidir Aseer	0.05	[41]
	Morning glory	0.41 - 0.47			Sidir Albaha	0.09	
India	*Citrus*	0.00 - 0.32	[51, 52]	Slovenia	Lime	0.25 - 0.30	[74]
	Multifloral	0.00 - 0.53			Chestnut	0.66 - 0.72	
	Litchi	0.12 - 0.14		Spain	Thyme	0.16 - 0.60	[49, 50, 67]
Ireland	White clover and blackberry	0.07 - 0.36	[47]		Multifloral	0.00 - 0.99	
Italy	*Citrus*	0.03 - 0.10	[64]	Sudan	*Cucurbita maxima*	0.11 - 0.13	[53]
	Thymus	0.07 - 0.19			*Acacia nilotica*	0.12 - 0.14	
Libya	*Eucalyptus*	0.17	[57]	Turkey	Blossom	0.02 - 0.43	[75]
Maroc	Sunflower	0.12 - 0.20	[20 - 22]	USA	Multifloral	0.02 - 1.03	[61]
	Heather	0.36 - 0.40		Venezuela	Multifloral	0.19 - 0.64	[15]
	Mint	0.18 - 0.23					
	Eucalyptus	0.08 - 0.44					

N.P.: Non published data

Table 5. Mineral composition of selected honeys worldwide.

Origin	Botanical origin	Content (mg/kg)								Ref.
		Potassium	Sodium	Calcium	Magnesium	Iron	Zinc	Manganese	Copper	
Argentina	Multifloral	91 - 1956	6 - 90	19 - 136	6.0 - 47.0	1.1 - 10.3	0.1 - 3.9	0.1 - 0.7	0.05 - 0.68	[76]
Bosnia-Herzegovina	Black locust	63 - 130	55 - 66	N.R.	0.5 - 33.9	2.5 - 5.5	6.0 - 17.9	0.3 - 8.5	0.40 - 6.60	[26]
Cameroun	Not specified	111 - 289	4 - 8	42 - 411	5.8 - 12.6	N.R.	N.R.	0.1 - 0.5	0.10 - 0.30	[33]
Colombia	Multifloral	259 - 701	41 - 113	82 - 172	18.2 - 45.0	0.7 - 9.3	5.3 - 21.1	N.R.	N.R.	N.P.
Estonia	Polyfloral	126 -1382	5 - 19	20 -64	5.5 - 25.5	N.R.	N.R.	N.R.	N.R.	[63]
India	*Citrus*	743 - 744	98	33	N.R.	8.8 - 8.9	2.5 - 2.6	N.R.	1.86 - 1.90	[51, 52]
	Helianthus annus	689 -690	176	64	N.R.	10.4	4.2	N.R.	1.87 - 1.91	
	Multifloral	932 - 933	247	73	N.R.	10.1	11.2	N.R.	1.82 - 1.86	
Ireland	White clover and blackberry	410 - 693	60 - 158	79 - 152	18.9 - 53.3	1.7 - 13.2	1.6 - 8.5	1.7 - 10.2	1.40 - 2.30	[47]
Israel	Avocado	189 -3768	27 -133	58 -137	18.5 - 204.6	0.9 - 9.3	0.8 - 11.5	N.R.	0.00 - 3.20	[77]
Libya	*Eucalyptus*	414	381	3	32.9	7.4	1.6	3.1	0.10	[57]
Maroc	Sunflower	103 - 150	N.R.	103 - 117	19.1 -35.5	6.6 - 12.1	0.9 - 1.9	0.2 - 0.4	0.28 - 1.08	[20, 78]
	Crucifer	50 -58	N.R.	70 - 92	18.9 - 23.2	3.2 - 5.2	0.4	0.2 - 0.4	0.24 - 0.26	
	Carob tree	47 -69	N.R.	210 - 280	30.2 - 42.6	4.0 - 5.2	0.5 - 0.9	0.8 - 1.0	0.39 - 0.45	
	Heather	741 - 890	N.R.	131 - 175	67.9 - 90.3	20.1 - 28.8	1.1 - 1.7	0.6 -2.0	1.40 - 1.50	
	Loeflingia	42 - 200	N.R.	62 - 92	10.8 - 14.2	4.6 - 6.9	0.7 - 1.3	0.8 - 1.8	0.38 - 0.40	
Pakistan	Not specified	340 -1480	71 - 89	51 - 71	18.0 -39.0	4.4 - 7.5	2.0 - 2.9	N.R.	1.50 - 1.90	[73]
Portugal	Multifloral	118 -2591	90 - 728	6 - 134	10.6 - 70.4	N.R.	N.R.	N.R.		[34]
Saudi Arabia	Sidir Aseer	479 - 487	34 -40	N.R.	N.R.	1.0 - 1.1	2.0 - 2.5	0.1	0.45 - 0.49	[41]
	Sidir Albaha	88 -98	28	N.R.	N.R.	1.4 - 1.9	1.9 - 4.2	0.2	0.57 - 0.63	
	Talh Tehamh	49 -53	78	N.R.	N.R.	1.2 - 1.3	0.9 - 5.1	0.1	0.43 - 0.49	

(Table 5) contd.....

Origin	Botanical origin	Content (mg/kg)								Ref.
		Potassium	Sodium	Calcium	Magnesium	Iron	Zinc	Manganese	Copper	
Slovenia	Lime	1610 - 1940	N.R.	60 -72	N.R.	N.R.	N.R.	2.9 -4.4		[74]
	Chestnut	3290 - 3940	N.R.	132 - 1654	N.R.	N.R.	N.R.	1.7 - 25.4		
	Black locust	261 - 313	N.R.	19 -132	N.R.	N.R.	N.R.	3.1 -22.7		
Spain	French Lavender	85 - 1080	98 - 381	41 - 91	17.0 -75.0	3.2 - 10.7	1.5 - 7.2	0.5 - 1.9	0.20 - 3.20	[67, 79]
	Rubus	328 -2579	9 - 151	47 - 159	29.0 -220.0	0.0 - 26.0	1.0 - 24.0	N.R.	1.00 - 6.00	

N.R.: Not reported. N.P.: Non published data.

On the other hand, for all types of honey, the most concentrated mineral element quantified is potassium (47-3940 mg/kg), and the least concentrated element is copper (0.0–6.6 mg/kg). Other minerals, in increasing order of concentration, are zinc (0.1–24.0 ppm), manganese (0.1-25.4 mg/kg), iron (0.0-28.8 mg/kg), magnesium (0.5–220.0 mg/kg), sodium (4–728 mg/kg) and calcium (3-1654 mg/kg). In general, the honey that exhibits higher ash concentration has also a higher concentration of each mineral element, as may be expected. This high variability indicates the reason why this parameter is used as a differentiation criterion, since it has been widely used for *A. mellifera* honey, and other apicultural products [14, 49, 71, 72].

2.3. Color and Conductivity

Both, color and conductivity are closely related to the honey′s type and environmental factors [69]. It has been found that honeys with higher mineral content are darker and have a greater value in electrical conductivity. The measurement of the electrical conductivity has now become standardized in many countries and for the Codex Alimentarius, setting a limit at not more than 0.8 mS/cm, with some exceptions: honeydew, chestnut, strawberry tree (*Arbutus unedo*), bell heather (*Erica*), eucalyptus, lime (*Tilia* spp), ling heather (*Calluna vulgaris*) manuka or jelly bush (*Leptospermum*), tea tree (*Melaleuca* spp), which cannot have less than 0.8 mS/cm [1]. On the other hand, minerals present in honey come almost exclusively from nectar and they react with organic matter to form brown colored compounds. Hence, the greater the amount of mineral matter, the darker the honey will be [80]. Tables **6** and **7** show some results for color and electrical conductivity in honeys, respectively.

Table 6. Color measurement as mmPfund for selected honeys.

Origin	Botanical origin	Color (mmPfund)	Ref.	Origin	Botanical origin	Color (mmPfund)	Ref.
Argentina	*Lotus*	1 - 56	[23]	Italy	*Hedysarum*	11 - 35	[54, 64]
	Melilotus	1 - 47			*Helianthus*	51 - 71	
	Brassicaceae	2 - 66			*Rhododendron*	11 - 27	
	Helianthus	22 - 61			*Robinia*	11 - 27	
	Eucalyptus	1 - 55			*Taraxacum*	41 - 71	
	Clover	7 - 61			*Thymus*	27 - 83	
Canada	Canola	16	[38]		*Tilia*	11 - 71	
	Clover	14		Maroc	Thyme	111 - 119	[22]
Colombia	Multifloral	55 - 81	N.P.		*Euphorbia resinífera*	30 - 146	
Cuba	Linen vine	72 - 106	[42]		*Euphorbia echinus*	83 - 119	
	Morning glory	24 - 46			*Ziziphus*	51 - 110	
	Christmas vine	4 - 20			Rosemary	28 - 51	
	Black mangrove	24 - 46			Multifloral	18 - 96	
	Singing bean	24 - 40		Romania	Acacia	11 - 45	[81]
Italy	Honeydew	97 - 99	[54, 64]		Sunflower	79 - 83	
	Eucalyptus	54 - 56			Lime	36 - 54	
	Chestnut	87 - 91			Honeydew	92 - 103	
	Sulla	17 - 19		Spain	*Rubus*	39 - 150	[79]
	Heather	96 - 100		Uruguay	*Eucalyptus* spp.	59 - 121	[58]
	Wildflower	86 - 102			Pastures	12 - 64	
	Arbutus	55 - 83			*Citrus* spp.	14 - 70	
	Citrus	11 - 35			*Baccharis* spp.	40 - 86	

N.P.: Non published data

Honey electrical conductivity is highly dispersed and dependent not only on botanical but also on geographical factors. Low values of conductivity have been described for honeys from Pakistan (40-60 μS/cm), Nigeria (17-63 μS/cm), while the opposite has been found for Brazilian *Eucalyptus* honeys (331–2865 μS/cm).

Table 7. Electrical conductivity for selected honeys worldwide.

Origin	Botanical origin	E (μS/cm)	Ref.	Origin	Botanical origin	E (μS/cm)	Ref.
Algeria	Not specified	635 - 637	[19]	Maroc	Sunflower	430 - 520	[20 - 22]
	Multifloral	110 - 930			*Eucalyptus*	336 - 889	
	Eucalyptus	100 - 900			Multifloral	150 - 1142	
Argentina	*Lotus*	120 - 340	[8, 23]	Mexico	Tajonal	190 - 220	[24, 25]
	Eucalyptus	160 - 460			Multifloral	240 - 280	
	Clover	220 - 320		Nepal	Not specified	170 - 450	[28]
Bosnia-Herzegovina	Black locust	80 - 560	[26]	Nigeria	Not specified	9 - 173	[29, 30]
Brazil	*Eucalyptus*	331 - 2865	[27]	Pakistan	Not specified	40 - 60	[73]
	Wild flower	161 - 2200		Poland	Buckwheat	312 - 480	[32]
	Orange	212 - 1090		Portugal	*Erica*	520 - 940	[34 - 37]
Colombia	Multifloral	315 - 575	N.P.		Multifloral	460 - 880	
Cuba	Linen vine	200 - 400	[42]	Spain	Thyme	288 - 559	[48, 49, 79, 82]
	Morning glory	100 - 300			Multifloral	119 - 1515	
Estonia	Polyfloral	110 - 400	[63]		*Rubus*	220 - 1070	
Germany	False acacia	180 - 220	[44]	Serbia	Acacia	100 - 690	[43]
	Heather	700 - 900			Sunflower	190 - 550	
	Sunflower	100 - 320		Slovakia	Blossom	135 - 433	[83]
Ireland	White clover and blackberry	170 - 400	[47]		Multifloral	265 - 657	
Italy	Honeydew	1410 - 1594	[54, 64]		Acacia	46 - 548	
	Eucalyptus	462 - 584		Slovenia	Lime	727 - 836	[46, 74]
	Chestnut	1460 - 1500			Chestnut	1348 - 1619	
Libya	*Eucalyptus*	440	[57]		Multifloral	330 - 840	

(Table 7) contd.....

Origin	Botanical origin	E (µS/cm)	Ref.	Origin	Botanical origin	E (µS/cm)	Ref.
Malaysia	Acacia	755 - 765	[59, 60]	Sudan	*Cucurbita maxima*	198 - 202	[53]
	Pineapple	348 - 352		Uruguay	*Eucalyptus* spp.	600 - 800	[58]
	Longan tree	475 - 485			*Citrus* spp.	330 - 490	
					Baccharis spp.	230 - 750	

N.P.: Non published data

Nevertheless, the bibliographic reports included in Table **7** suggests that some authentic honeys can even have higher values of conductivity as limited by the Codex Alimentarius regulation (800 µS/cm), being the majority of them multifloral. Considering this, it would be of importance to update this regulation to the present available information.

On the basis of color, it is possible to make a preliminary classification of honey. Locust, *citrus, rhododendron* and French honeysuckle are light, while honey and honeydew are very dark brown. Some honeys have a particular color such as *Erica* which is orange and *Helianthus*, which is bright yellow [9]. According to data provided in Table **6**, it can be seen that in general, most of unifloral honeys tend to have lighter colors, ranging from extra white to extra white ambar, as specified by Pfund scale. In multifloral honeys it is also found colors light ambar and ambar. Some exceptions can be made with selected unifloral honeys which can present darker colors. *Eucalyptus, Rubus* or *Euphorbia* honeys present mostly dark ambar color.

It is important to consider that the Pfund scale does not take into account the tone of the color of the samples and is not enough to characterize the color of honeys. Recent bibliography can be found with remarkable results on distinguishing honeys by CIE*Lab tristimulus, defined as the representation in rectangular coordinates of color space, with three coordinates: L* - Clarity or luminic intensity, ranging from white to black, and two chromatic coordinates a* and b*, representing the different hue scales, ranging from green to red for a* and from blue to yellow to b* [84]. Some reports for honeys are presented in (Table **8**). As can be inferred, CIE*Lab parameters could be an useful tool for differentiating honeys. González-Miret *et al.* [85], classified honey samples into two groups from the point of view of lightness (L* parameter in the CIE*Lab scale): light honeys had L* values lower than 50 and dark honeys had L* values higher than 50. Only for this description, a notorious classification by geographical or botanical origin of honeys can be performed.

Table 8. Color measurement in honeys by CIE*Lab tristimulus.

Country	Botanical origin	Scale range (adimensional)			Reference
		L*	a*	b*	
Colombia	Multifloral	40.8 - 65.5	-0.09 - 18.72	30.23 - 56.18	N.P.
Mexico	Multifloral	23.9 - 37.4	0.45 - 3.32	2.21 - 11.09	[24]
Slovenia	Acacia	63.5 - 65.7	-3.51 - (-2.40)	13.49 - 22.67	[86]
	Lime	61.4 - 65.6	-3.79 - (-3.01)	20.93 - 33.36	
	Chestnut	42.3 - 53.4	3.79 - 10.82	33.32 - 49.36	
	Fir	38.9 - 46.4	5.67 - 11.12	29.10 - 39.77	
	Spruce	39.6 - 47.1	7.10 - 12.20	29.80 - 40.90	
	Multifloral	50.3 - 57.3	-0.90 - 5.77	43.10 - 49.22	
	Forest	38.0 - 45.6	7.59 - 12.31	27.81 - 37.86	
Spain	Avocado	36.9 - 46.8	24.69 - 29.77	17.09 - 30.78	[85]
	Chestnut	37.6 - 44.0	21.89 -25.60	17.28 - 25.65	
	Citrus	65.6 - 87.4	0.81 - 12.86	38.09 - 61.42	
	Eucalyptus	56.0 - 66.9	20.29 - 29.19	45.89 - 57.82	
	Heather	28.7 - 46.4	1.77 - 31.14	1.59 - 31.37	
	Honeydew	31.0 - 62.4	5.61 - 29.52	5.11 - 52.99	
	Lavender	59.8 - 73.0	16.60 - 26.07	53.17 - 66.12	
	Rosemary	44.3 - 86.5	0.66 - 27.77	24.90 - 59.81	
	Thymus	51.0 - 59.5	25.65 - 29.67	36.46 - 50.30	

N.P.: Non published data

2.4. Amino Acids

The nitrogen content of honey is low and quite variable. Anklam [4] stated that "honey contains about 0.2% of protein, which is of both bee and plant origin". Naturally, the amino acid profile is markedly influenced by the geographical and floral origins and hence, these components can be regard as markers of differentiation. The amino acid proline dominates in honey, however Anklam [4] and Fattori [80] mention that arginine, tryptophan, and cystine have been shown to be characteristic for some floral honey types. In turn, sulfur-containing amino acids (methionine and cysteine) can only be found as a minor component of specific honeys [87].

Pirini and Conte [88] mention that "arginine was present in chestnut honey (mean value of 0.35 mg/100 g) but not in orange blossom, acacia, rosemary and lime tree honeys; tryptophan (mean value of 0.43 mg/100 g) was found only in acacia honey". In addition, Bouseta *et al.* could detect arginine in eucalyptus and lavender honeys [89], while Hermosín *et al.* [87] described it as a major amino acid in honeys samples of *Rosemary*, *Lavender*, *Thyme* and *Orange Blossom* [87]. As can be observed in Table **9**, the proline content of honeys is also highly variable.

Table 9. Proline content in selected honeys.

Country	Botanical origin	Content (mg/kg)	Reference
Argentina	Not specified	166 - 708	[8]
Spain	*Eucalyptus*	112 - 986	[87, 90]
	Citrus	37 - 417	
	Rosemary	10 - 56	
	Lavender	27 - 88	
	Thyme	27 - 40	
	Orange blossom	12 - 48	
Turkey	Not specified	21 - 75	[91]

N.P.: Non published data

2.5. Quality Physicochemical Parameters

Physicochemical analyses are important for establishing the identity of each variety of honey, according to both botanical and geographical origin, and to provide regulatory organizations with objective tools for preventing honey falsification in commerce. Some quality parameters of honey produced by *A. mellifera* are not directly related to nutritive value but instead to authenticity, designation of origin, and safety [pH, acidity, content of hydroxymethylfurfural (HMF), diastase activity, among others]. Such characterizations, together with the need to avoid adulteration and falsification, have led food regulation agencies to set standards. The average international standards (Codex Alimentarius and the International Honey Commission) set the following limits to some physico-chemical parameters: acidity max. 50 meq/kg, HMF max 40 mg/kg, diastase activity min. 8 Schade units. Table **10** resumes worldwide reported data for pH and acidity of honeys.

Honey acidity is strongly dependent on the floral source parameter. Singh and Bath [92] indicated "the presence of organic acids, particularly gluconic acid, in equilibrium with their lactones or esters and inorganic ions such as phosphate and chloride". Honeys *locust* as others, such as those of *rhododendron* and *citrus* origin, have naturally low acidity (less than 20 meq/kg), while this parameter in honeys of *Arbutus, Erica* and *Thymus* has been shown to amount for more than 30 meq/kg [80]. Reported data also point that free acidity in some honeys is above that allowed by the codex regulation. This applies to multifloral honey from Argentina, to Brazilian honeys and to French lavender honey from Spain, for which values of free acidity were above 50 meq/kg (Table **10**).

Table 10. pH, free and lactonic acidity in selected honeys worldwide.

Origin	Botanical origin	pH	Free acidity (meq/kg)	Lact. (meq/kg)	Ref.	Origin	Botanical origin	pH	Free acidity (meq/ Kg)	Lact. (meq/ Kg)	Ref.
Argentina	Multifloral	3.1 - 5.8	10.1 - 52.2	0.1 - 8.2	[8, 23]	Kenya	Not specified	3.8 - 4.1	24.0 - 31.0	6.0 - 15.5	[56]
	Lotus	3.3 - 4.7	10.0 - 30.3	3.4 - 38.5		Maroc	Sunflower	3.5 - 3.8	15.3 - 36.7	10.0 - 12.0	[20 - 22]
	Eucalyptus	3.4 - 3.8	13.3 - 22.9	5.7 - 37.1			Heather	4.2 - 4.5	22.2 - 30.2	2.3 - 2.9	
Brazil	*Eucalyptus*	2.9 - 5.1	12.5 - 55.0	N.R.	[27]		Multifloral	3.8 - 4.7	10.4 - 44.1	5.4 - 14.2	
	Wild flower	2.3 - 5.0	14.0 - 75.5	N.R.		Mexico	Tajonal	3.6 - 3.8	18.8 - 24.8	N.R.	[25]
	Orange	2.7 - 4.6	14.0 - 57.0	N.R.		Poland	Buckwheat	N.R.	38.3 - 44.0	4.5 - 7.4	[32]

(Table 10) contd.....

Origin	Botanical origin	pH	Free acidity (meq/kg)	Lact. (meq/kg)	Ref.
Burkina Faso	Multifloral	3.6 - 4.4	27.1 - 46.1	N.R.	[31]
Colombia	Multifloral	3.8 - 4.2	32.2 - 42.6	0.0 - 9.9	N.P.
Cuba	Linen vine	3.7 - 4.1	21.0 - 34.4	N.R.	[42]
	Morning glory	3.8 - 4.6	13.3 - 21.5	N.R.	
	Christmas vine	3.7 - 4.5	12.2 - 21.4	N.R.	
	Black mangrove	4.7 - 5.1	16.1 - 22.7	N.R.	
Estonia	Polyfloral	3.5 - 5.1	14.0 - 25.0	N.R.	[63]
Germany	False acacia	3.93 - 4.43	0.33 - 6.33	N.R.	[44]
	Heather	4.21 - 4.87	15.5 - 21.5	N.R.	
	Sunflower	3.49 - 3.73	13.64 - 20.7	N.R.	
India	*Citrus*	N.R.	32.3 - 33.1	14.4 - 15	[51, 52]
	Multifloral	N.R.	16.4 - 17	14.4 - 15	
Ireland	White clover and blackberry	3.9 - 4.3	23.8 - 42.1	0.2 - 14.9	[47]
Italy	*Eucalyptus*	3.8 - 4.2	18.8 - 20.0	3.0 - 3.8	[54, 64]
	Chestnut	4.9 - 5.5	13.1 - 15.3	1.4 - 2.4	
	Heather	3.7 - 4.3	34.0 - 36.4	5.1 - 6.9	

Origin	Botanical origin	pH	Free acidity (meq/ Kg)	Lact. (meq/ Kg)	Ref.
Portugal	*Erica*	3.5 - 4.2	21.9 - 45.2	N.R.	[34 - 37]
	Multifloral	3.5 - 4.2	17.1 - 48.9	4.2 - 16.5	
Romania	Multifloral	3.9 - 4.5	17.5 - 27.2	N.R.	[40, 81]
	Acacia	3.9 - 4.8	8.4 - 13.3	N.R.	
	Forest	4.2 - 5.2	18.6 - 37.1	N.R.	
Serbia	Acacia	3.5 - 5.6	7.8 - 29.6	N.R.	[43]
	Linden	4.0 - 5.4	8.2 - 26.2	N.R.	
	Sunflower	3.2 - 4.1	11.0 - 42.7	N.R.	
Slovenia	Black locust	4,0 - 4.2	10.5 - 14.0	3.0 - 5.2	[46, 74]
	Multifloral	4.0 - 4.8	12.3 - 42.7	0.1 - 5.2	
Spain	Thyme	3.6 - 4.8	17.6 - 39.8	4.3 - 11.3	[48 - 50]
	Multifloral	3.6 - 5.0	13.1 - 51.2	0.0 - 13.9	
	French Lavender	3.3 - 4.0	24.7 - 59.8	0.2 - 8.5	
Turkey	Blossom	3.2 - 4.3	14.0 - 30.5	4.0 - 10.5	[55]
USA	Multifloral	3.4 - 6.1	6.8 - 47.2	0 - 18.8	[61]
Venezuela	Multifloral	3.3 - 4.3	24.4 - 53.3	N.R.	[15]

N.R.: Not reported. N.P.: Non published data.

Lactones are considered to be a reserve of acidity, and their value is lower than free acidity [93]. As can concluded from Table **10**, ranges of lactone are highly variable, with the highest amounts being found in some monofloral Argentineans honeys (values up to 38.5 meq/kg). On the other hand, the pH of honey is usually between 3.5 and 4.5 unlike honeydew honey whose value can be between 4.5 and 5.5 [80]. Still, these values are only a reference, as according to Table **10**, low values of pH (close to 3.0) or higher (up to 6.0) can be found.

On the other hand, diastase and HMF are quality parameters widely employed for checking authenticity and freshness of honeys, and they are very useful to note changes during long-term storage and heating [94 - 96]. Diastase (amylase), decomposes starch or glycogen into smaller sugar units [80], meanwhile Zappalà *et al.* [97] mentioned that "HMF is formed during acid-catalysed dehydration of hexoses and, it is connected to the chemical properties of honey, like pH, total acidity and mineral content". Time and heat promotes the loss of thermolabile, aromatic substances, but must important, the decrease of diastase and the increase of HMF. According to the International Honey Commission: "the diastase activity must not be less than or equal to 8, expressed as diastase number (DN). DN in Schade scale, which corresponds to the Gothe scale number and it is defined as g of starch hydrolysed in 1 h at 40°C per 100 g honey. Diastase activity should be determined after processing and blending (Codex draft) or for all retail honey" (USDA draft). Moreover, Codex Alimentarius established that "the HMF content of honey after processing and/or blending must not be higher than 80 mg/kg" [1]. Tosi *et al.* [96] stated: "these properties are used together as their values are indicative of the heating intensity to which honey has been subjected". Some reports of diastase and HMF contents of different types of honeys are presented in (Table **11**).

In general, honeys reported in Table **11** presented values of diastase above 8, fulfilling the Codex regulation and presenting an extraordinary enzymatic activity. In contrast, HMF values for some honeys were higher than maximum allowed. For instance, Brazilian honeys had contents even closer to 250 mg/kg, and Algerian honeys reached up to 120 mg/kg.

Yilmaz and Küfrevioglu [75] showed that the average HMF content of Turkish honey increased from 3.3 to 19.1 mg/kg, and the average diastase number decreased from 14.6 to 10.7 after one year of storage at 20°C. Turhan [98] determined HMF level and formation kinetics of honey after heating process, finding that temperatures higher than 135°C, even for short times, increased HMF values above the accepted by international standards. Khalil *et al.* [99] found that the HMF content of fresh Malaysian honey samples stored for 3–6 months was within the internationally recommended value (2.80–24.87 mg/kg), while honey samples stored for longer periods (12–24 months) contained much higher HMF concentrations (128.19–1131.76 mg/kg). On the other hand, Thrasyvoulou [100] found that the average decrease in diastase number for 20 samples of Greek honey stored for one year at 25°C was 40%, while Sancho *et al.* [101] reported a decrease of 33% in 115 Spanish honey samples after one year storage at 15–25 °C. Moreira *et al.* [102] found a reduction in the diastase activity lower than established by regulations after storing Brazilian honey for 3 months at a temperature range between 35-40°C.

Table 11. Diastase and HMF of selected honeys worldwide.

Origin	Botanical origin	Diastase DN	HMF (mg/kg)	Ref.	Origin	Botanical origin	Diastase DN	HMF (mg/kg)	Ref.
Algeria	Multifloral	4.0 - 40.0	0.5 - 124.0	[19]	Malaysia	Acacia	N.R.	0.1 - 0.5	[59, 60]
	Eucalyptus	8.0 - 30.0	5.8 - 110.7			Pineapple	N.R.	68.6 - 69.4	
	Citrus	7.0 - 9.0	2.2 - 4.0		Maroc	*Eucalyptus*	9.5 - 158.0	7.5 - 39.1	[20 - 22]
Argentina	Multifloral	8.9 - 41.0	0.1 - 34.1	[7]		*Citrus*	1.63 - 290	5.0 - 43.3	
Brazil	*Eucalyptus*	5.0 - 23.8	0.3 - 207.2	[27]		Multifloral	5.9 - 21.8	0.6 - 52.8	
	Wild flower	5.0 - 38.5	1.0 - 122.0		Mexico	Tajonal	9.0 - 31.7	2.8 - 6.8	[50]
	Orange	1.1 - 17.9	7.9 - 254.7		Portugal	Multifloral	3.0 -38.0	1.8 - 32.8	[34 - 37]
Burkina Faso	Multifloral	11.2 - 27.9	0.0 - 37.3	[31]		*Erica*	10.0 -30.0	0.9 - 22.8	
Canada	Canola	14.7	2.1	[38]	Romania	Polyfloral	N.R.	1.2 - 28.4	[40]
	Clover	16.5	1.9			*Tilia*	N.R.	2.4 - 12.4	

(Table 11) contd.....

Origin	Botanical origin	Diastase DN	HMF (mg/kg)	Ref.
Colombia	Multifloral	0.0 - 59.4	0.0 - 36.5	N.P.
Cuba	Linen vine	18.1 - 28.3	1.5 - 8.3	[42]
	Morning glory	15.3 - 26.7	3.8 - 10.4	
	Christmas vine	8.5 - 18.3	3.3 - 17.9	
Estonia	Polyfloral	16.2 - 29.1	N.R.	[63]
Germany	False acacia	N.R.	4.8 - 17.0	[44]
	Heather	N.R.	6.6 - 17.2	
India	Sweet orange	8.1 - 11.1	0.4 - 1.6	[51, 52]
	Ber	10.0 - 12.6	0.9 - 1.3	
	Peach	5.6 - 7.8	1.0 - 1.8	
Italy	Honeydew	23.3 - 24.7	1.6 - 18	[54]
	Eucalyptus	25.8 - 26.2	2.8 - 3.0	
	Chestnut	23.6 - 25.4	1.7 - 2.1	
Kenia	Not specified	15.5 -31.1	2.7 - 34.6	[56]
Saudi Arabia	Sidir Aseer	3.9 - 4.1	1.4 - 1.7	[41]
	Sidir Albaha	9.2 - 9.4	5.6 - 5.7	
	Talh Tehamh	3.4 - 3.5	12.8 - 14.5	
Slovakia	Honeydew	17.3 - 39.9	N.R.	[83]
	Multifloral	N.R.	0.0 - 32.2	
	Acacia	N.R.	2.2 - 20.0	
Spain	Multifloral	10.2 - 63.7	0.0 - 15.7	[48, 50, 79]
	French Lavender	11.8 - 40.0	4.1 - 11.9	
	Rubus	8.5 - 31.9	0.0 - 0.8	
Turkey	Blossom	9.0 - 26.1	0.0 - 11.5	[75]
	Flower	17.9 - 29.4	1.9 - 30.7	
	Chestnut	17.9 - 29.4	2.1 - 8.0	
USA	Multifloral	2.1 - 61.2	N.R.	[61]

N.R.: Not reported. N.P.: Non published data.

Other enzymatic assays, in particular that of invertase activity evaluation (*i.e.* β-fructosidase, which decomposes sucrose into fructose and glucose), has been employed for quality evaluation. At a much lesser extension, glucose oxidase, *i.e.* the enzyme that catalyses the oxidation of glucose to gluconic acid and hydrogen peroxide, has also been used for this purpose [80].

Although there is no regulation about the minimum value of invertase activity, some reports mention the importance of measuring this enzyme for honey quality evaluation [62, 102, 103]. The invertase activity of some selected honeys is shown in (Table **12**).

Table 12. Invertase activity in some selected honeys.

Country	Botanical origin	Invertsase Number*	Reference
Bulgaria	Polyfloral	24.2 - 33.6	[104]
Czech Republic	Floral	0.8 - 20.4	[106]
	Compound	4.0 - 25.9	
	Honeydew	10.8 - 24.6	
Spain	*Castanea sativa*	19.2 - 32.6	[107, 108]
	Rosmarinus officinalis	6.0 - 24.0	
	Dorycnium pentaphyllum	6.7 - 13.2	
	Erica arborea	5.3 - 18.7	
	Erica cinerea	14.0 - 29.0	
	Erica vagans	6.9 - 10.3	
	*Citrus*spp	7.0 - 12.3	
	Eucalyptus spp	6.3 - 16.0	
	Hedysarum coronarium	0.9 - 8.4	
	*Quercus*spp	11.9 - 22.8	
	Robinia pseudoacacia	3.7 - 4.7	
	Lavandula stoechas	11.4 - 17.1	
	Polyfloral	10.6 - 46.2	
	Multifloral	4.8 - 25.1	
Italy	*Robinia*	0.4 - 7.7	[56, 105]
	Arbutus	0.5 - 8.9	
	Citrus	1.0 - 9.5	
	Erica	1.6 - 10.7	
	Rosmarinus	3.9 - 11.8	
	Hedysarum	5.0 - 14.0	
	Taraxacum	5.4 - 14.5	
	Rhododendron	7.3 - 17.7	
	Carduus	6.5 - 16.4	
	Tilia	6.7 - 19.8	
	Helianthus	9.0 - 16.3	
	Multifloral	7.5 - 28.4	
	Thymus	13.8 - 26.8	
	Eucalyptus	13.5 - 28.3	
	Castanea	14.7 - 29.2	

* Invertase number: g sucrose hydrolysed per 100 g of honey per h

2.6. Bioactive Compounds

Recently, there has been an increased interest in bioactive compounds from foodstuff, and their application to the treatment of different diseases caused by oxidative stress. In this context, it is important to highlight that honey is recognized as a rich source of bioactive compounds, which mainly comprise phenolic compounds, in particular flavonoids and phenolic acid [81, 109 - 111].

Several authors [109, 112 - 114] reported anticarcinogenic, anti-inflammatory, antiatherogenic, antithrombotic, immunomodulating and analgesic activities, among others, and mostly exert these functions because of their antioxidant abilities.

The content of phenolic compounds in honey samples has been assessed by distinct techniques. Most commonly, the authors quantify the global content of these bioactive contents, through the Folin-Ciocalteu method (mean values in Table **13**) and/or those of total flavonoids, as determined by the Dowd method, since these are extraordinary approaches to establish in simple manners the presence of phenolic compounds [115, 116].

Table 13. Total phenolics measured in honey.

Country	Botanical origin	Total phenolics (mg GAE/ 100 g)	Ref	Country	Botanical origin	Total phenolics (mg GAE/ 100 g)	Ref
Brazil	Multifloral	42.8 - 78.2	[118]	Poland	Heather	59.9 - 76.2	[119, 120]
	Orange blossom	34.0 - 53.2			Buckwheat	98.3 - 121.4	
Croatia	Multifloral	14.1 - 24.8	[121]		Black locust	12.2 - 15.6	
Czech	Floral	8.4 - 14.7	[122]	Romania	Acacia	2.0-39.0	[81]
	Lime	8.3 - 9.8			Lime	16.0-38.0	
	Raspberry	9.6 - 10.2			Sunflower	20.0-45.0	
Eslovenia	Acacia	2.6 - 6.8	[86]	Saudi Arabia	Multifloral	85.0 - 317.4	[123]
	Lime	9.0 - 15.9		Turkey	Multifloral	0.2 - 141.8	[124, 125]
	Multifloral	12.7 - 19.5			Chestnut	19.1 - 108.2	
Italy	Multifloral	15.1 - 82.5	[126]		Honeydew	22.4 - 25.7	

GAE: Gallic acid equivalents

Honeys′ enrichment in bioactive compounds also depend on the floral source, environmental factors and processing methods [109, 117] and hence overall, the total content of phenolic compounds in honey samples worldwide is extremely variable (Table **13**). *E.g.* multifloral honey from Turkey has been reported to have 0.2 mg GAE/100 g while the phenolic content in honeys from Saudi Arabia were described to reach up to 317 mg GAE/100 g. In a similar way, flavonoids′ content from Brazilian honeys were described to be poor in flavonoids (0.2 mg QE/ 100 g), in opposition to that of sunflower honey from Romania (28.3 mg QE/100 g).

In addition to the content monitoring of total phenolic compounds and/or flavonoids in honey samples, authors also frequently quantified specific flavonoids and/or phenolic acids (see values in Tables **14** and **15**, respectively).

Table 14. Contents of specific flavonoids that are commonly found in honey.

Country	Botanical origin	Myr	Tri	Pin	Api	Que	Lut	Kae	Reference
Australia	*Eucalyptus*	0.11 - 3.57	0.36 - 2.65	N.R	N.R	0.19 - 1.30	0.14 - 3.19	0.02 - 0.16	[127, 128, 129]
	Jelly bush	0.08 - 1.15	0.08 - 0.68	0.04 - 0.37	N.R	0.06 - 0.36	0.03 - 0.57	0.02 - 0.13	
	Malaleuca quinquenervia	2.07 - 4.17	1.07 - 1.10	N.R	N.R	0.47 - 1.10	0.25 - 0.55	0.08 - 0.20	
	Guioa semiglauca	0.12 - 0.46	0.18 - 0.76	N.R	N.R	0.28 - 0.49	0.15 - 0.56	0.10 - 0.17	
	Lophosteron conferta	0.23 - 0.35	1.30 - 1.46	N.R	N.R	0.60 - 0.72	0.94 - 1.41	0.09 - 0.26	
	Banksia ericifolia	0.63 - 0.82	0.41 - 0.54	N.R	N.R	0.30 - 0.36	0.19 - 0.22	0.08 - 0.09	
New Zealand	Manuka	0.07 - 0.14	N.R	N.R	N.R	0.38 - 0.39	0.33 - 0.43	0.13 - 0.17	[128]
Slovenia	Acacia	N.R	N.R	0.06 - 0.14	0.004 - 0.010	0.01 - 0.03	N.R	0.02 - 0.03	[130]
	Linden	N.R	N.R	0.03 - 0.08	0.005 - 0.009	0.02 - 0.05	N.R	0.02 - 0.03	
	Floral	N.R	N.R	0.05 - 0.14	0.012 - 0.040	0.035 - 0.040	N.R	0.02 - 0.05	
	Chsestnut	N.R	N.R	0.01 - 0.04	0.004 - 0.055	0.02 - 0.04	N.R	0.01 - 0.03	
	Spruce	N.R	N.R	0.03 - 0.06	0.007 - 0.025	0.03 - 0.07	N.R	0.02 - 0.04	
	Fir	N.R	N.R	0.02 - 0.05	0.005 - 0.016	0.05 - 0.07	N.R	0.02 - 0.03	
	Forest	N.R	N.R	0.02 - 0.09	0.007 - 0.028	0.02 - 0.03	N.R	0.01 - 0.03	

(Table 14) contd.....

Country	Botanical origin	Myr	Tri	Pin	Api	Que	Lut	Kae	Reference
Spain	Lemon blossom	N.R	N.R	N.R	N.R	0.17 - 0.61	0.00 - 0.61	0.30 - 0.86	[131]
	Orange blossom	N.R	N.R	N.R	N.R	0.07 - 0.35	0.00 - 0.15	0.05 - 0.76	

Data is expressed as mg/100 g honey. Myr: Myricetin, Tri: Tricetin, Pin: Pinobanksin, Api: Apigenin, Que: Quercetin, Lut: Luteolin, Kae: Kaempferol. N.R.: not reported.

Table 15. Contents of specific phenolic acids that are commonly found in honey.

Country	Botanical origin	Gallic acid	Chlorogenic acid	Caffeic acid	Coumaric acid	Ferulic acid	Ellagic acid	Reference
Australia	*Eucalyptus*	0.34 - 6.62	0.10 - 4.26	0.17 - 1.52	0.18 - 1.52	0.15 - 1.08	0.22 - 2.66	[132]
China	Polyfloral	0.07 - 0.14	0.01 -0.12	0.01 - 0.03	0.01 -0.02	0.004 - 0.015	N.R.	[133]
Serbia	Acacia	N.R.	N.R.	0.00 - 0.06	0.00 -0.03	N.R.	N.R.	[134]
	Linden	N.R.	N.R.	0.01 - 0.09	0.00 -0.02	N.R.	N.R.	
	Sunflower	0.00 - 0.15	N.R.	0.01 - 0.08	0.00 -0.02	N.R.	N.R.	
	Basil	N.R.	N.R.	0.03 - 0.08	N.R.	N.R.	N.R.	
Spain	Lemon blossom	N.R.	N.R.	N.R.	0.37 - 1.77	0.30 - 0.81	N.R.	[131]
	Orange blossom	N.R.	N.R.	N.R.	0.20 - 0.97	0.12 - 0.55	N.R.	
	Rubus	N.R.	N.R.	0.00 - 0.60	0.07 -2.08	N.R.	0.00 - 1.70	[135]
Turkey	Multifloral	0.01 - 0.30	0.00 -2.83	0.00 - 0.27	0.04 -1.20	0.01 - 0.05	N.R.	[136]

Data are expressed as mg/100 g honey

3. HONEY CHARACTERIZATION FOR DIFFERENTIATION

Besides its importance for health benefits correlation, the quantification of specific phenolic compounds in honey samples is also useful for geographical and/or botanical differentiation. Indeed, distinct authors have suggested that specific phenolic compounds can serve as floral markers [127, 128, 130, 131, 137,

138]. *E.g.* the use of the flavonoid hesperetin in studies for the identification of the botanical origin of citrus honey [131], kaempferol for rosemary honey, as well as quercetin for sunflower honey [139, 140], has recently been demonstrated. Some phenolic acids (*e.g.*, ellagic acid) in heather honey [141] and hydroxycinnamates (caffeic, *p*-coumaric and ferulic acids) in chestnut honey [142] have also been used as floral markers [138].

Besides phenolic compounds, other compounds have been suggested as honey′s botanical markers, such norisoprenoids, terpenes, aldehydes, among others [143 - 145]. In this field, volatile compounds, which are normally estimated by gas chromatography coupled to mass spectrometry (GC-MS) techniques, are most commonly cited [146, 147]. Selected examples of the main volatile compounds suggested as botanical markers are resumed in (Table **16**).

Table 16. Different instrumental techniques employed for honey characterization and differentiation.

Geographical origin	Botanical origin	Study	Employed technique	Findings	Reference
Australia	Blue gum (*Eucalyptus leucoxylon*) Yellow box (*Eucalyptus melliodora*)	Volatile compounds	GC-MS	Norisoprenoids, terpenes, benzene derivatives	[148]
New Zealand	Heather	Volatile compounds	GC-MS	Aliphatic compounds	[149, 150]
France, Belgium, United-Kingdom, Norway, Germany, Greece and Italy	*Calluna vulgaris*and *Erica arborea*	Volatile compounds	GC-MS	Products obtained from non-enzymatic browning reaction	[144]
Italy	Strawberry-tree (*Arbutus unedo L.)*	Volatile compounds	GC-MS	Norisoprenoid compounds such as α-isophorone, β-isophorone and 4-oxoisophorone	[151]
Spain	*Eucalyptus*	Volatile compounds	GC-MS	Two components were present in all samples, assigned as 2-hydroxy-5-methyl-3-hexanone and 3-hydroxy-5-meth-l-2-hexanone	[145, 152, 153]
	Citrus			Terpenes and derivatives, such as linalool, (*Z*) (*E*)-linalool oxide, α-terpineol, terpineal and isomers of lilac aldehyde and lilac alcohol were found	
	Lavender			Hexanal, nerolidol oxide, coumarin, important concentrations of hexanol and hotrienol	

(Table 16) contd.....

Geographical origin	Botanical origin	Study	Employed technique	Findings	Reference
Greece	Thyme	Volatile compounds	GC-MS	Phenylacetaldehyde, 1-phenylbutane-2,3-dione, 3-hydroxy-4-phenylbutan-2-one, 3-hydroxy-1-phenylbutan-2-one, phenylacetonitrile, and carvacrol.	[154]
France	Different botanical origins	Volatile compounds	GC-MS	Octanal, phenylacetaldehyde, octan-1-ol, 2-methoxyphenol, nonanal, and 2*H*-1-benzopyr-n-2-one	[155]
Palestina	*Thymus capitatus, Thymelaea hirsuta,* and *Tolpis virgata*	Volatile compounds	GC-MS	Phenols, aldehydes, ketones, acids, and alcohols	[156]
Slovakia	Different botanical origins	Volatile compounds	GC-MS	Hydrocarbons, alcohols, aldehydes and ketones, terpenes, benzene derivatives, and compounds containing heteroatoms	[157]
Colombia	Different botanical origins	Volatile compounds	Electronic nose	Geographical classification of honeys achieved by chemometrics	[158]
Croatia	Acacia, chestnut and honeydew	Non-volatile compounds	Electronic tongue	Botanical classification achieved by chemometrics	[159]
Switzerland	Acacia, alpine rose, chestnut, rape, honeydew, alpine polyfloral and lowland polyfloral	Classification by botanical origin	Fluorescence spectroscopy	Correct classification of 90% according to botanical origin	[160]
Italy	Acacia, *rhododendron*, polyfloral produced over 1000 m of altitude, chestnut and high mountain polyfloral	Carbohydrates identification	^{1}H NMR and ^{13}C NMR spectroscopy	Fructose, glucose, gentiobiose, isomaltose, kojibiose, maltose, maltulose, melibiose, nigerose, palatinose, sucrose, turanose, erlose, isomaltotriose, kestose, maltotriose, melezitose, raffinose, and maltotetraose	[161]

(Table 16) contd.....

Geographical origin	Botanical origin	Study	Employed technique	Findings	Reference
Brazil	Wildflower, *citrus*, eucalyptus and Assa-peixe	Compounds identification	^{1}H NMR spectroscopy	Phenylalanine, tyrosine, sucrose, lactic acid	[162]

Some other techniques have been employed recently in order to characterize honeys in a simple, quick and efficient manner. These methods have been also used for classification and differentiation according to geographical and botanical origin. On the other hand, two instrumental techniques: electronic nose and electronic tongue have been recently introduced for aroma and taste objective measurement [147, 163, 164]. Both instruments are intended to mimic sensory human organs. These are non-destructive techniques designed to analyze, recognize and identify very low levels of chemical volatile and non-volatile substances, provided with arrays of electrochemical sensors capable of creating a unique digital fingerprint of the food. The electronic tongue is comprised of potentiometric sensors coupled with an Ag/AgCl reference electrode [159], meanwhile the electronic nose is coupled with different kind of sensors depending on the nature of the sample: Metal Oxide Semiconductors (MOS), Metal Oxide Semiconductors Field Electric Transistors, Polimer Conductors (CP) and Bulk Acoustic Wave (BAW)[158, 165].

Finally, some other techniques such fluorescence, ^{1}H and ^{13}C NMR spectroscopy have been used for honey differentiation and classification [160 - 162]. These latter techniques have been also employed for adulteration detection [166, 167]. Fluorescence spectroscopy provides information on the presence of fluorescent molecules and their environment in honey such as polyphenols and amino acids, and therefore, could be helpful for authenticating the botanical origin of honey [132, 143, 168]. In all cases, it is important to denote that multivariate statistical techniques should be employed due to the huge amount of data available and the need of provide useful information with a certain degree of confidence.

CONCLUSION

The particular knowledge of the composition of honey for each geographical and botanical origin guarantees the authenticity of products and creates value in a food of high recognition among consumers. Despite some parameters are more useful than others to differentiate honeys by origin (minerals, color, electrical conductivity), the certainty to discriminate them is only possible through a complete characterization of their properties. The results shown in honeys from different botanical and geographical origins led to conclude that the values of some properties (electrical conductivity, carbohydrates, acidity, etc.) differ from the limits set by the international regulation (Codex Alimentarius). It is clear that at the time of drafting the regulations, any information concerning the characterization of this product from certain places, particularly outside Europe and North America, was scarce and thus not taken into account. For this reason, it would be appropriate to revise such regulations and adjust them based on the new findings for honeys worldwide. Moreover, novel techniques seek to reduce the use of reagents, highly specialized personal and times of analysis, providing relevant information for a rapid characterization of honeys through specific chemical markers. No matter if traditional or novel techniques for the characterization of honeys are used, the large amount of data so obtained has become one of the main consequences, which requires of multivariate statistical analyses to discriminate honeys by origin in a properly manner. These methods have already demonstrated their efficiency based on the ability to differentiate honeys from a complete analysis of the entire composition information.

CONFLICT OF INTEREST

The author confirms that author has no conflict of interest to declare for this publication.

ACKNOWLEDGEMENTS

The authors thank the colombian Ministry of Agriculture and Rural Development, the Colombian Science, Technology and Innovation Department (COLCIENCIAS), and the following beekeeping associations: Asociación de Apicultores de Boyacá (ASOAPIBOY), the Asociación de Apicultores de la región del Sumapaz (ASOAPIS), the Asociación de Apicultores Conservacionistas de la Sierra Nevada de Santa Marta (APISIERRA), the Red de Productores Ecológicos de la Sierra Nevada de Santa Marta (Red Ecolsierra), the

Asociación Apícola Comunera (ASOAPICOM), Apicultura Los Cerezos and Apiarios Los Cítricos.

REFERENCES

[1] FAO. *Codex Standards for Honey*; Codex Alimentarius Commission Standards: Rome, **2001**.

[2] FAOSTAT. *Production quantities by country. Honey, natural. Food and Agriculture Organization, Available: http://faostat3.fao.org/browse/Q/QL/E,* **2014**. Accessed: 01-Oct-2014

[3] Krell, R. Value-added products from beekeepings. *FAO Agric. Serv,* **1996**, *124*, 87-113.

[4] Anklam, E. A review of the analytical methods to determine the geographical and botanical origin of honey. *Food Chem.,* **1998**, *63*, 549-562. [http://dx.doi.org/10.1016/S0308-8146(98)00057-0]

[5] Bogdanov, S.; Martin, P. *Honey Authenticity: a Review, Swiss Bee Res. Cent*; Mühlheim: Cent., **2002**.

[6] Bogdanov, S. Authenticity of honey and other bee products: state of the art. *Anim. Sci. Biotechnol.,* **2008**, *64*, 1-2.

[7] Wang, J.; Li, Q. *Advances in Food Nutrition and Research*; Waltham, Taylor , Ed.; Elsevier: New York, **2011**, 62, pp. 89-138.

[8] Baroni, M.V.; Arrua, C.; Nores, M.L.; Fayé, P.; Díaz, M.; Chiabrando, G.; Wunderlin, D. Composition of honey from Córdoba (Argentina): Assessment of North/South provenance by chemometrics. *Food Chem.,* **2009**, *114*, 727-733. [http://dx.doi.org/10.1016/j.foodchem.2008.10.018]

[9] Persano, L.; Piro, R. Main European unifloral honeys: descriptive sheets. *Apidologie (Celle),* **2004**, *35*, S38-S81. [http://dx.doi.org/10.1051/apido:2004049]

[10] Jean-Prost, P.; Medori, O. *Apiculture : connaître l'abeille, conduire le rucher*; Lavoisier: Paris, **1987**.

[11] Nagai, T.; Inoue, R.; Inoue, H.; Suzuki, N. Preparation and antioxidant properties of water extract of propolis. *Food Chem.,* **2003**, *80*, 29-33. [http://dx.doi.org/10.1016/S0308-8146(02)00231-5]

[12] Ouchemoukh, S.; Louaileche, H.; Schweitzer, P. Physicochemical characteristics and pollen spectrum of some Algerian honeys. *Food Contr.,* **2005**, *18*, 52-58. [http://dx.doi.org/10.1016/j.foodcont.2005.08.007]

[13] Terrab, A.; Vega-Pérez, J.; Diez, M.; Heredia, F. Characterisation of northwest Moroccan honeys by gas chromatographic-mass spectrometric analysis of their sugar components. *J. Sci. Food Agric.,* **2001**, *82*, 179-185. [http://dx.doi.org/10.1002/jsfa.1011]

[14] Hernández, O.; Fraga, J.; Jiménez, A.; Jiménez, J.; Arias, J. Characterization of honey from the Canary Islands: determination of the mineral content by atomic absorption spectrophotometry. *Food Chem.,* **2005**, *93*, 449-458. [http://dx.doi.org/10.1016/j.foodchem.2004.10.036]

[15] Ojeda, G.; Sulbarán, B.; Ferrer, A.; Rodríguez, B. Characterization of Honey produced in Venezuela. *Food Chem.,* **2004**, *84*, 499-502. [http://dx.doi.org/10.1016/S0308-8146(02)00517-4]

[16] Bogdanov, S.; Martin, P.; Lüllmann, C. Harmonized Methods of the European Honey Commission. *Apidologie.,* **1997**, (Extra issue), 1-59.

[17] Fernández-Torres, R.; Pérez-Bernal, J.L.; Bello-López, M.A.; Callejón-Mochón, M.; Jiménez-Sánchez, J.C.; Guiraúm-Pérez, A. Mineral content and botanical origin of Spanish honeys. *Talanta,* **2005**, *65*(3), 686-691. [http://dx.doi.org/10.1016/j.talanta.2004.07.030] [PMID: 18969853]

[18] Fuenmayor, C.; Díaz, C.; Zuluaga, C.; Quicazán, M. *Pot-Honey. A legacy of stingless bees*; Roubik, Vit Pedro, Ed.; Springer: New York, **2013**, 27, pp. 383-394.

[19] Makhloufi, C.; Kerkvliet, J.; D'albore, R.; Choukri, A.; Samar, R. Characterization of Algerian honeys by palynological and physico-chemical methods. *Apidologie (Celle),* **2010**, *41*, 509-521. [http://dx.doi.org/10.1051/apido/2010002]

[20] Terrab, A.; Diez, M.; Heredia, F. Palynological, physicochemical and colour characterization of Moroccan honeys: III Other unifloral honey types. *Int. J. Food Sci. Technol.,* **2003**, *38*, 395-402. [http://dx.doi.org/10.1046/j.1365-2621.2003.00713.x]

[21] Terrab, A.; Diez, M.; Heredia, F. Palynological, physicochemical and colour characterization of Moroccan honeys: I. River red gum (*Eucalyptus camaldulensis Dehnh*) honey. *Int. J. Food Sci. Technol.,* **2003**, *38*, 379-386. [http://dx.doi.org/10.1046/j.1365-2621.2003.00715.x]

[22] Chakir, A.; Romane, A.; Marcazzan, G.; Ferrazzi, P. Physicochemical properties of some honeys produced from different plants in Morocco. *Arab. J. Chem.,* **2011**, *46* [http://dx.doi.org/10.1016/j.arabjc.2011.10.013]

[23] Malacalza, N.; Caccavari, M.; Fagúndez, G.; Lupano, C. Unifloral honeys of the province of Buenos Aires, Argentine. *J. Sci. Food Agric.,* **2005**, *85*, 1389-1396. [http://dx.doi.org/10.1002/jsfa.2105]

[24] Viuda-Martos, M.; Ruiz-Navajas, Y.; Zaldivar-Cruz, J.; Kuri, V.; Fernández-López, J.; Carbonell-Barrachina, A.; Pérez-Álvarez, J. Aroma profile and physico-chemical properties of artisanal honey from Tabasco, Mexico. *Int. J. Food Sci. Technol.,* **2010**, *45*, 1111-1118. [http://dx.doi.org/10.1111/j.1365-2621.2010.02243.x]

[25] Mora-Escobedo, R.; Moguel-Ordoñez, Y.; Jaramillo-Florez, M.; Gutiérrez-López, G. The composition, rheological and thermal properties of Tajonal (*Viguiera dentata*) Mexican honey. *Int. J. Food Prop,* **2006**, *9*, 299-316. [http://dx.doi.org/10.1080/10942910600596159]

[26] Grujić, S.; Grujić, R.; Popov-Raljić, J.; Komić, J. *The 7th International Congress of Food Technologists, Biotechnologists and Nutritionists,* **2011**, pp. 274-278.

[27] Marchini, L.; Moreti, A.; Otsuk, I.; Sodré, G. Physicochemical composition of Apis mellifera honey samples from São Paulo State, Brazil. *Quim. Nova,* **2007**, *30*, 1653-1657. [http://dx.doi.org/10.1590/S0100-40422007000700029]

[28] Joshi, S.; Pechhacker, H.; William, A.; von der Ohe, W. Physico-chemical characteristics of Apis dorsata, A. cerana and A. mellifera honey from Chitwan district, central Nepal. *Apidologie (Celle),* **2000**, *31*, 367-375. [http://dx.doi.org/10.1051/apido:2000128]

[29] Omafuvbe, B.; Akanbi, O. Microbiological and physico-chemical properties of some commercial Nigerian honey. *Afr. J. Microbiol. Res.,* **2009**, *3*, 891-896.

[30] Adebiyi, F.; Akpan, I.; Obiajunwa, E.; Olaniyi, H. Chemical/physical characterization of nigerian honey. *Pakistan J. Nutr.,* **2004**, *3*, 278-281. [http://dx.doi.org/10.3923/pjn.2004.278.281]

[31] Meda, A.; Lamien, C.; Millogo, J.; Romito, M.; Nacoulma, O. Physicochemical Analyses of Burkina Fasan Honey. *Acta Vet. Brno,* **2005**, *74*, 147-152. [http://dx.doi.org/10.2754/avb200574010147]

[32] Rybak-Chmielewska, H.; Szczesna, T. Composition and properties of polish buckwheat honey. *Curr. Adv. Buckwheat Res.,* **1995**, *114*, 793-799.

[33] Joseph, T.; Julius, A.; Florence, F.; Delphine, D.; Jonnas, P.; Antoine, M. Physico-chemical and microbiological characteristics of honey from the sudano-guinean zone of West Cameroon. *Afr. J. Biotechnol.,* **2007**, *6*, 908-913.

[34] Silva, L.; Videira, R.; Monteiro, A.; Valentão, P.; Andrade, A. Honey from Luso region (Portugal): Physicochemical characteristics and mineral contents. *Microchem. J.,* **2009**, *93*, 73-77. [http://dx.doi.org/10.1016/j.microc.2009.05.005]

[35] Feás, X.; Pires, J.; Iglesias, A.; Estevinho, M.L. Characterization of artisanal honey produced on the Northwest of Portugal by melissopalynological and physico-chemical data. *Food Chem. Toxicol.,* **2010**, *48*(12), 3462-3470. [http://dx.doi.org/10.1016/j.fct.2010.09.024] [PMID: 20870005]

[36] Feás, X.; Pires, J.; Estevinho, M.; Iglesias, A.; De Araujo, J. Palynological and physicochemical data characterisation of honeys produced in the Entre-Douro e Minho region of Portugal. *Int. J. Food Sci. Technol.,* **2010**, *45*, 1255-1262. [http://dx.doi.org/10.1111/j.1365-2621.2010.02268.x]

[37] Estevinho, L.; Rodrigues, S.; Pereira, A.; Feás, X. Portuguese bee pollen: palynological study, nutritional and microbiological evaluation. *Int. J. Food Sci. Technol.,* **2012**, *47*, 429-435. [http://dx.doi.org/10.1111/j.1365-2621.2011.02859.x]

[38] Abell, D.; Friebe, H.; Schweger, S.; Kwok, A.; Sporns, P. Comparison of processed unifloral clover and canola honey. *Apidologie (Celle),* **1996**, *27*, 451-460. [http://dx.doi.org/10.1051/apido:19960603]

[39] Velásquez-Giraldo, A.; Vélez-Acosta, L.; Zuluaga-Gallego, R. Physicochemical and Microbiological Characterization of *Apis mellifera* sp. Honey from Southwest of Antioquia in Colombia. *Ing. y Cienc.,* **2013**, *9*, 61-74.

[40] Oroian, M. Physicochemical and rheological properties of romanian honeys. *Food Biophys.,* **2012**, *7*, 296-307. [http://dx.doi.org/10.1007/s11483-012-9268-x]

[41] Al-Khalifa, A.; Al-Arify, I. Physicochemical characteristics and pollen spectrum of some Saudi honeys. *Food Chem.,* **1999**, *67*, 21-25. [http://dx.doi.org/10.1016/S0308-8146(99)00096-5]

[42] Alvarez-Suarez, J.M.; Tulipani, S.; Díaz, D.; Estevez, Y.; Romandini, S.; Giampieri, F.; Damiani, E.; Astolfi, P.; Bompadre, S.; Battino, M. Antioxidant and antimicrobial capacity of several monofloral Cuban honeys and their correlation with color, polyphenol content and other chemical compounds. *Food Chem. Toxicol.,* **2010**, *48*(8-9), 2490-2499. [http://dx.doi.org/10.1016/j.fct.2010.06.021] [PMID: 20558231]

[43] Lazarević, K.; Andrić, F.; Trifković, J.; Tešić, Z.; Milojković-Opsenica, D. Characterisation of Serbian unifloral honeys according to their physicochemical parameters. *Food Chem.,* **2012**, *132*, 2060-2064. [http://dx.doi.org/10.1016/j.foodchem.2011.12.048]

[44] Smanalieva, J.; Senge, B. Analytical and rheological investigations into selected unifloral German honey. *Eur. Food Res. Technol.,* **2009**, *229*, 107-113. [http://dx.doi.org/10.1007/s00217-009-1031-2]

[45] Kasperová, J.; Nagy, J.; Popelka, P.; Dičáková, Z.; Nagyová, A.; Maľa, P. Physico-chemical indicators and identification of selected Slovak honeys based on colour measurement. *Acta Vet. Brno,* **2012**, *81*, 57-61. [http://dx.doi.org/10.2754/avb201281010057]

[46] Bertoncelj, J.; Golob, T.; Kropf, U.; Korošec, M. Characterisation of Slovenian honeys on the basis of sensory and physicochemical analysis with a chemometric approach. *Int. J. Food Sci. Technol.,* **2011**, *46*, 1661-1671. [http://dx.doi.org/10.1111/j.1365-2621.2011.02664.x]

[47] Downey, G.; Hussey, K.; Daniel-Kelly, J.; Walshe, T.; Martin, P. Preliminary contribution to the characterisation of artisanal honey produced on the island of Ireland by palynological and physico-chemical data. *Food Chem.,* **2005**, *91*, 347-354. [http://dx.doi.org/10.1016/j.foodchem.2004.06.020]

[48] Soria, A.; González, M.; de Lorenzo, C.; Martínez-Castro, I.; Sanz, J. Characterization of artisanal honeys from Madrid (Central Spain) on the basis of their melissopalynological, physicochemical and volatile composition data. *Food Chem.,* **2004**, *85*, 121-130. [http://dx.doi.org/10.1016/j.foodchem.2003.06.012]

[49] Terrab, A.; Recamales, A.; Hernanz, D.; Heredia, F. Characterisation of Spanish thyme honeys by their physicochemical characteristics and mineral contents. *Food Chem.,* **2004**, *88*, 537-542. [http://dx.doi.org/10.1016/j.foodchem.2004.01.068]

[50] Nozal Nalda, M.J.; Bernal Yagüe, J.L.; Diego Calva, J.C.; Martín Gómez, M.T. Classifying honeys from the Soria Province of Spain *via* multivariate analysis. *Anal. Bioanal. Chem.,* **2005**, *382*(2), 311-319. [http://dx.doi.org/10.1007/s00216-005-3161-0] [PMID: 15856192]

[51] Nanda, V.; Singh, B.; Kukreja, V.; Bawa, A. Characterisation of honey produced from different fruit plants of northern India. *Int. J. Food Sci. Technol.,* **2009**, *44*, 2629-2636. [http://dx.doi.org/10.1111/j.1365-2621.2009.02094.x]

[52] Nanda, V.; Sarkar, B.; Sharma, H.; Bawa, A. Physico-chemical properties and estimation of mineral content in honey produced from different plants in Northern India. *J. Food Compos. Anal.,* **2003**, *16*, 613-619.
[http://dx.doi.org/10.1016/S0889-1575(03)00062-0]

[53] Idris, Y.; Mariod, A.; Hamad, S. Physicochemical properties, phenolic contents and antioxidant activity of sudanese honey. *Int. J. Food Prop,* **2011**, *14*, 450-458.
[http://dx.doi.org/10.1080/10942910903243673]

[54] Marini, F.; Magrì, A.; Balestrieri, F.; Fabretti, F.; Marini, D. Supervised pattern recognition applied to the discrimination of the floral origin of six types of Italian honey samples. *Anal. Chim. Acta,* **2004**, *515*, 117-125.
[http://dx.doi.org/10.1016/j.aca.2004.01.013]

[55] Rasgele, P.; Kekecoglu, M. Physico-chemical properties of Rhododendron honey produced in Turkey. *Herba Pol,* **2013**, *59*, 88-97.

[56] Muli, E.; Munguti, A.; Raina, S. Quality of honey harvested and processed using traditional methods in rural areas of kenya. *Acta Vet. Brno,* **2007**, *76*, 315-320.
[http://dx.doi.org/10.2754/avb200776020315]

[57] Owayss, A. Physicochemical analysis for standardizing quality criteria of libyan eucalyptus (*Eucalyptus* sp.) honey. *Egypt J. Appl. Sci,* **2005**, *20*, 247-255.

[58] Corbella, E.; Cozzolino, D. Classification of the floral origin of Uruguayan honeys by chemical and physical characteristics combined with chemometrics. *LWT -. Food Sci. Technol. (Campinas.),* **2006**, *39*, 534-539.

[59] Moniruzzaman, M.; Sulaiman, S.A.; Khalil, M.I.; Gan, S.H. Evaluation of physicochemical and antioxidant properties of sourwood and other Malaysian honeys: a comparison with manuka honey. *Chem. Cent. J.,* **2013**, *7*, 138-149.
[http://dx.doi.org/10.1186/1752-153X-7-138] [PMID: 23938192]

[60] Moniruzzaman, M.; Khalil, M.I.; Sulaiman, S.A.; Gan, S.H. Physicochemical and antioxidant properties of Malaysian honeys produced by *Apis cerana, Apis dorsata* and *Apis mellifera. BMC Complement. Altern. Med.,* **2013**, *13*, 43-54.
[http://dx.doi.org/10.1186/1472-6882-13-43] [PMID: 23433009]

[61] White, J., Jr; Doner, L. Honey composition and properties. *Agric. Handb,* **1980**, *335*, 82-91.

[62] Bogdanov, S.; Jurendic, T.; Sieber, R.; Gallmann, P. Honey for nutrition and health: a review. *J. Am. Coll. Nutr.,* **2008**, *27*(6), 677-689.
[http://dx.doi.org/10.1080/07315724.2008.10719745] [PMID: 19155427]

[63] Kirs, E.; Pall, R.; Martverk, E.; Laos, K. Physicochemical and melissopalynological characterization of Estonian summer honeys. *Procedia Food Sci.,* **2011**, *1*, 616-624.
[http://dx.doi.org/10.1016/j.profoo.2011.09.093]

[64] Oddo, L. Characterization of unifloral honeys. *Apidologie (Celle),* **1995**, *26*, 453-465.
[http://dx.doi.org/10.1051/apido:19950602]

[65] Astwood, K.; Lee, B.; Manley, M. Oligosaccharides in New zealand honeydew honey. *J. Agric. Food Chem.,* **1998**, *46*, 4958-4962.

[http://dx.doi.org/10.1021/jf980720d]

[66] Mateo, R.; Bosch-Reig, F. Classification of Spanish unifloral honeys by discriminant analysis of electrical conductivity, color, water content, sugars and pH. *J. Agric. Food Chem.,* **1998**, *46*(2), 393-400.
[http://dx.doi.org/10.1021/jf970574w] [PMID: 10554252]

[67] Bonvehi, J.; Coll, F. Physico-chemical properties, composition and pollen spectrum of french lavender (Lavandula stoechas L.) honey produced in Spain. *Leb. und-Technologie,* **1993**, *196*, 511-517.

[68] Bogdanov, S.; Haldimann, M.; Luginwühl, W.; Gallmann, P. Minerals in honey: environmental, geographical and botanical aspects. *J. Apic. Res. Bee World,* **2007**, *46*, 269-275.

[69] Terrab, A.; González, A.; Diez, M.; Heredia, F.; Gonzalez, G. Mineral content and electrical conductivity of the honeys produced in north-west Morocco and their contribution to the characterisation of unifloral honeys. *J. Sci. Food Agric.,* **2003**, *83*, 637-643.
[http://dx.doi.org/10.1002/jsfa.1341]

[70] Gonzalez, A.; Gomez, J.; Garcia-Villanova, R.; Rivas, T.; Ardanuy, R.; Sanchez, J. Geographical discrimination of honeys by using mineral composition and common chemical quality parameters. *J. Sci. Food Agric.,* **2000**, *80*, 157-165.
[http://dx.doi.org/10.1002/(SICI)1097-0010(20000101)80:1<157::AID-JSFA506>3.0.CO;2-B]

[71] de Alda-Garcilope, C.; Gallego-Picó, A.; Bravo-Yagüe, J.C.; Garcinuño-Martínez, R.M.; Fernández-Hernando, P. Characterization of Spanish honeys with protected designation of origin "Miel de Granada" according to their mineral content. *Food Chem.,* **2012**, *135*(3), 1785-1788.
[http://dx.doi.org/10.1016/j.foodchem.2012.06.057] [PMID: 22953923]

[72] Vanhanen, L.P.; Emmertz, A.; Savage, G.P. Mineral analysis of mono-floral New Zealand honey. *Food Chem.,* **2011**, *128*(1), 236-240.
[http://dx.doi.org/10.1016/j.foodchem.2011.02.064] [PMID: 25214355]

[73] Rehman, S.; Khan, Z.; Maqbool, T. Physical and spectroscopic characterization of Pakistani honey. *Cienc. Investig. Agrar.,* **2008**, *35*, 199-204.

[74] Kropf, U.; Korošec, M.; Bertoncelj, J.; Ogrinc, N.; Nečemer, M.; Kump, P.; Golob, T. Determination of the geographical origin of Slovenian black locust, lime and chestnut honey. *Food Chem.,* **2010**, *121*, 839-846.
[http://dx.doi.org/10.1016/j.foodchem.2009.12.094]

[75] Yilmaz, H.; Küfrevioglu, I. Composition of honeys collected from eastern and South-eastern anatolia and effect of storage on hydroxymethylfurfural content and diastase activity. *Turk. J. Agric. For.,* **2001**, *25*, 347-349.

[76] Cantarelli, M.; Pellerano, R.; Marchevsky, E.; Camiña, J. Quality of honey from Argentina: Study of chemical composition and trace elements. *J. Argentine Chem. Soc.,* **2008**, *96*, 33-41.

[77] Dag, A.; Afik, O.; Yeselson, Y.; Schaffer, A.; Shafir, S. Physical, chemical and palynological characterization of avocado (*Persea americana* Mill.) honey in Israel. *Int. J. Food Sci. Technol.,* **2006**, *41*, 387-394.
[http://dx.doi.org/10.1111/j.1365-2621.2005.01081.x]

[78] Terrab, A.; Díez, M.; Heredia, F. Characterisation of Moroccan unifloral honeys by their

physicochemical characteristics. *Food Chem.,* **2002**, *79*, 373-379.
[http://dx.doi.org/10.1016/S0308-8146(02)00189-9]

[79] Escuredo, O.; Seijo, M.; Fernández-González, M. Descriptive analysis of Rubus honey from the north-west of Spain. *Int. J. Food Sci. Technol.,* **2011**, *46*, 2329-2336.
[http://dx.doi.org/10.1111/j.1365-2621.2011.02753.x]

[80] Fattori, B. *'La miel' Propiedades, Composición y Análisis Fisicoquímico*; Apimondia: Buenos Aires, **2004**.

[81] Al, M.; Daniel, D.; Moise, A.; Bobis, O.; Laslo, L.; Bogdanov, S. Physico-chemical and bioactive properties of different floral origin honeys from Romania. *Food Chem.,* **2009**, *112*, 863-867.
[http://dx.doi.org/10.1016/j.foodchem.2008.06.055]

[82] Terrab, A.; Heredia, F. Characterisation of avocado (*Persea americana* Mill) honeys by their physicochemical characteristics. *J. Sci. Food Agric.,* **2004**, *84*, 1801-1805.
[http://dx.doi.org/10.1002/jsfa.1888]

[83] Kačániová, M.; Sudzina, M.; Sudzinová, J.; Fikselová, M.; Čuboň, J.; Haščík, P. Microbiological and physico – chemical quality of honey collected from different Slovak habitats. *Slovak J. Anim. Sci.,* **2007**, *40*, 38-43.

[84] Gambaro, A.; Ares, G.; Gimenez, G.; Pahor, S. Preference mapping of color of Uruguayan honeys. *J. Sens. Stud.,* **2007**, *22*, 507-519.
[http://dx.doi.org/10.1111/j.1745-459X.2007.00125.x]

[85] González-Miret, M.L.; Terrab, A.; Hernanz, D.; Fernández-Recamales, M.A.; Heredia, F.J. Multivariate correlation between color and mineral composition of honeys and by their botanical origin. *J. Agric. Food Chem.,* **2005**, *53*(7), 2574-2580.
[http://dx.doi.org/10.1021/jf048207p] [PMID: 15796597]

[86] Bertoncelj, J.; Dobersek, U.; Jamnik, M.; Golob, T. Evaluation of the phenolic content, antioxidant activity and colour of Slovenian honey. *Food Chem.,* **2007**, *105*, 822-828.
[http://dx.doi.org/10.1016/j.foodchem.2007.01.060]

[87] Hermosin, I.; Chicon, R.; Cabezudo, D. Free amino acid composition and botanical origin of honey. *Food Chem.,* **2003**, *83*, 263-268.
[http://dx.doi.org/10.1016/S0308-8146(03)00089-X]

[88] Pirini, A.; Conte, L. Capillary gas chromatography determination of free amino acids in honey as a mean of discrimination between different botanical sources. *J. High Resolut. Chromatogr.,* **1992**, *15*, 165-170.
[http://dx.doi.org/10.1002/jhrc.1240150306]

[89] Bouseta, A.; Scheirman, V.; Collin, S. Flavor and free amino acid composition of lavender and eucalyptus honeys. *Food Sci,* **1996**, *61*, 683-694.
[http://dx.doi.org/10.1111/j.1365-2621.1996.tb12181.x]

[90] Serrano, S.; Villarejo, M.; Espejo, R.; Jodral, M. Chemical and physical parameters of Andalusian honey: classification of Citrus and Eucalyptus honeys by discriminant analysis. *Food Chem.,* **2004**, *87*, 619-625.
[http://dx.doi.org/10.1016/j.foodchem.2004.01.031]

[91] Yilmaz, H.; Yamuz, O. Content of some trace metals in honey from south-eastern Anatolia. *Food Chem.,* **1999**, *65*, 475-476. [http://dx.doi.org/10.1016/S0308-8146(98)00205-2]

[92] Singh, N.; Bath, P. Quality evaluation of different types of Indian honey. *Food Chem.,* **1997**, *58*, 129-133. [http://dx.doi.org/10.1016/S0308-8146(96)00231-2]

[93] White, J., Jr *Composición y propiedades de la miel. La apicultura en Estados Unidos*; Ciudad de México: Limusa, **1979**.

[94] Sahinler, N.; Gul, A. Effect of heating and storage on honey hydroxy Methylfurfural and Diastase Activity. *J. Food Technol.,* **2005**, *3*, 152-157.

[95] Laura, S.; Muste, S.; Tofana, M.; Suharioschi, R. Changes of hydroxymethylfurfural in heated honey. *Agric. -. Sci. Pract,* **2013**, *3*, 87-88.

[96] Tosi, E.; Martinet, R.; Ortega, M.; Lucero, H.; Ré, E. Honey diastase activity modified by heating. *Food Chem.,* **2008**, *106*, 883-887. [http://dx.doi.org/10.1016/j.foodchem.2007.04.025]

[97] Zappalà, M.; Fallico, B.; Arena, E.; Verzera, A. Methods for the determination of HMF in honey: a comparison. *Food Contr.,* **2005**, *16*, 273-277. [http://dx.doi.org/10.1016/j.foodcont.2004.03.006]

[98] Turhan, K. *Innovations in Chemical Biology*; Sener, , Ed.; Springer: Berlin, **2009**, 23, pp. 233-239.

[99] Khalil, M.I.; Sulaiman, S.A.; Gan, S.H. High 5-hydroxymethylfurfural concentrations are found in Malaysian honey samples stored for more than one year. *Food Chem. Toxicol.,* **2010**, *48*(8-9), 2388-2392. [http://dx.doi.org/10.1016/j.fct.2010.05.076] [PMID: 20595027]

[100] Thrasyvoulou, A. The use of HMF and diastase as criteria of quality of Greek honey. *J. Apic. Res.,* **1986**, *25*, 186-195.

[101] Sancho, M.; Muniategui, S.; Huidobro, J.; Simal-Lozano, J. Aging of honey. *J. Agric. Food Chem.,* **1992**, *40*, 134-138. [http://dx.doi.org/10.1021/jf00013a026]

[102] Moreira, R.; De Maria, C.; Pietrolungo, M.; Trugo, L. Chemical changes in the non-volatile fraction of Brazilian honeys. *Food Chem.,* **2007**, *104*, 1236-1241. [http://dx.doi.org/10.1016/j.foodchem.2007.01.055]

[103] Alvarez-Suarez, J.; Tulipani, S.; Romandini, S.; Bertoli, E.; Battino, M. Contribution of honey in nutrition and human health: a review. *Med. J. Nutrition Metab.,* **2010**, *3*, 15-23. [http://dx.doi.org/10.1007/s12349-009-0051-6]

[104] Parvanov, P.; Dinkov, D.; Tananaki, C. Invertase activity and carbohydrate spectrum of organic acacia and polyfloral honey after one-year storage. *Bulg. J. Vet. Med,* **2012**, *15*, 198-205.

[105] Oddo, L. Invertase activity in honey. *Apidologie (Celle),* **1999**, *30*, 57-65. [http://dx.doi.org/10.1051/apido:19990107]

[106] Vorlov, L.; Pridal, A. Invertase and diastase activity in honeys of Czech provenience. *Acta Univ.*

Agric. Silvic. Mendel. Brun., **2002**, *8*, 57-65.

[107] Serra-Bonvehi, J.; Torrento, M.; Raich, J. Invertase activity in fresh and processed honeys. *J. Sci. Food Agric.,* **2000**, *80*, 507-512. [http://dx.doi.org/10.1002/(SICI)1097-0010(200003)80:4<507::AID-JSFA558>3.0.CO;2-5]

[108] Sánchez, M.P.; Huidobro, J.F.; Mato, I.; Muniategui, S.; Sancho, M.T. Evolution of invertase activity in honey over two years. *J. Agric. Food Chem.,* **2001**, *49*(1), 416-422. [http://dx.doi.org/10.1021/jf0003350] [PMID: 11170607]

[109] Kassim, M. Mustafa, M.; Mohd, M.; Kamaruddin, M. Ellagic acid, phenolic acids and flavonoids in Malaysian honey extracts demostrate *in vitro* anti-inflammatory. *Nutr. Res.,* **2010**, *30*, 650-659. [http://dx.doi.org/10.1016/j.nutres.2010.08.008] [PMID: 20934607]

[110] Gheldof, N.; Engeseth, N.J. Antioxidant capacity of honeys from various floral sources based on the determination of oxygen radical absorbance capacity and inhibition of *in vitro* lipoprotein oxidation in human serum samples. *J. Agric. Food Chem.,* **2002**, *50*(10), 3050-3055. [http://dx.doi.org/10.1021/jf0114637] [PMID: 11982440]

[111] Aljadi, A.; Kamaruddin, M. Evaluation of the phenolic contents and antioxidant capacities of two Malaysian floral honeys. *Food Chem.,* **2004**, *85*, 513-518. [http://dx.doi.org/10.1016/S0308-8146(02)00596-4]

[112] Viuda-Martos, M.; Ruiz-Navajas, Y.; Fernández-López, J.; Pérez-Alvarez, J.A. Functional properties of honey, propolis, and royal jelly. *J. Food Sci.,* **2008**, *73*(9), R117-R124. [http://dx.doi.org/10.1111/j.1750-3841.2008.00966.x] [PMID: 19021816]

[113] Nasuti, C.; Gabbianelli, R.; Falcioni, G.; Cantalamessa, F. Antioxidative and gastroprotective activities of anti-inflammatory formulations derived from chestnut honey in rats. *Nutr. Res.,* **2006**, *26*, 130-137. [http://dx.doi.org/10.1016/j.nutres.2006.02.007]

[114] Vinson, J.; Hao, Y.; Su, X.; Zubik, L. Phenol antioxidant quantity and quality in foods: vegetables. *J. Agric. Food Chem.,* **1998**, *46*, 3630-3634. [http://dx.doi.org/10.1021/jf980295o]

[115] Beretta, G.; Granata, P.; Ferrero, M.; Orioli, M.; Maffei, R. Standardization of antioxidant properties of honey by a combination of spectrophotometric / fluorometric assays and chemometrics. *Anal. Chim. Acta,* **2005**, *533*, 185-191. [http://dx.doi.org/10.1016/j.aca.2004.11.010]

[116] Arvouet-Grand, A.; Vennat, B.; Pourrat, A.; Legret, P. Standardisation d'un extrait de propolis et identification des principaux constituants. *J. Pharm. Belg.,* **1994**, *49*(6), 462-468. [PMID: 7884635]

[117] Martos, I.; Cossentini, M.; Ferreres, F.; Tomás-Barberán, F. Flavonoid composition of Tunisian honeys and propolis. *J. Agric. Food Chem.,* **1997**, *45*, 2824-2829. [http://dx.doi.org/10.1021/jf9609284]

[118] Lianda, R.; Sant'Ana, L.; Echevarria, L.; Castro, R. Antioxidant activity and phenolic composition of brazilian honeys and their extracts. *J. Braz. Chem. Soc.,* **2012**, *23*, 618-627.

[119] Jasicka-Misiak, I.; Poliwoda, A.; Dereń, M.; Kafarski, P. Phenolic compounds and abscisic acid as potential markers for the floral origin of two Polish unifloral honeys. *Food Chem.,* **2012**, *131*, 1149-

1156.
[http://dx.doi.org/10.1016/j.foodchem.2011.09.083]

[120] Kuś, P.; Congiu, F.; Teper, D.; Sroka, Z.; Jerković, I.; Tuberoso, C. Antioxidant activity, color characteristics, total phenol content and general HPLC fingerprints of six Polish unifloral honey types. *LWT -. Food Sci. Technol. (Campinas.),* **2014**, *55*, 124-130.

[121] Marković, K.; Škaričić, Z.; Šarić, G.; Krpan, M.; Major, N.; Hruškar, M.; Vahčić, N. *7th International Congress of Food Technologists, Biotechnologists and Nutritionists,* **2011**, , pp. 167-172.

[122] Lachman, J.; Orsák, M.; Hejtmánková, A.; Kovárová, E. Evaluation of antioxidant activity and total phenolics of selected Czech honeys. *LWT -. Food Sci. Technol. (Campinas.),* **2010**, *43*, 52-58.

[123] Al-Hindi, R.R.; Bin-Masalam, M.S.; El-Shahawi, M.S. Antioxidant and antibacterial characteristics of phenolic extracts of locally produced honey in Saudi Arabia. *Int. J. Food Sci. Nutr.,* **2011**, *62*(5), 513-517.
[http://dx.doi.org/10.3109/09637486.2010.550276] [PMID: 21391794]

[124] Silici, S.; Sagdic, O.; Ekici, L. Total phenolic content, antiradical, antioxidant and antimicrobial activities of *Rhododendron* honeys. *Food Chem.,* **2010**, *121*, 238-243.
[http://dx.doi.org/10.1016/j.foodchem.2009.11.078]

[125] Sagdic, O.; Silici, S.; Ekici, L. Evaluation of the phenolic content, antiradical, antioxidant, and qntimicrobial activity of different floral sources of honey. *Int. J. Food Prop,* **2013**, *16*, 658-666.
[http://dx.doi.org/10.1080/10942912.2011.561463]

[126] Giorgi, A.; Madeo, M.; Baumgartner, J.; Lozzia, G.C. The relationships between phenolic content, pollen diversity, physicochemical information and radical scavenging activity in honey. *Molecules,* **2011**, *16*(1), 336-347.
[http://dx.doi.org/10.3390/molecules16010336] [PMID: 21217600]

[127] Yao, L.; Jiang, Y.; Singanusong, R.; D'Arcy, B.; Datta, N.; Caffin, N.; Raymont, K. Flavonoids in australian melaleuca, guioa, lophostemon, banksia and helianthus honeys and their potential for floral authentication. *Food Res. Int.,* **2004**, *37*, 166-174.
[http://dx.doi.org/10.1016/j.foodres.2003.11.004]

[128] Yao, L.; Datta, N.; Toma, F.; Ferreres, F.; Martos, I.; Singanusong, R. Flavonoids, phenolic acids and abscisic acid in Australian and New Zealand *Leptospermum* honeys. *Food Chem.,* **2003**, *81*, 159-168.
[http://dx.doi.org/10.1016/S0308-8146(02)00388-6]

[129] Yao, L.; Jiang, Y.; D'Arcy, B.; Singanusong, R.; Datta, N.; Caffin, N.; Raymont, K. Quantitative high-performance liquid chromatography analyses of flavonoids in Australian Eucalyptus honeys. *J. Agric. Food Chem.,* **2004**, *52*(2), 210-214.
[http://dx.doi.org/10.1021/jf034990u] [PMID: 14733497]

[130] Bertoncelj, J.; Polak, T.; Kropf, U.; Korošec, K.; Golob, T. LC-DAD-ESI/MS analysis of flavonoids and abscisic acid with chemometric approach for the classification of Slovenian honey. *Food Chem.,* **2011**, *127*, 296-302.
[http://dx.doi.org/10.1016/j.foodchem.2011.01.003]

[131] Escriche, I.; Kadar, M.; Juan-Borrás, M.; Domenech, E. Using flavonoids, phenolic compounds and headspace volatile profile for botanical authentication of lemon and orange honeys. *Food Res. Int.,*

2011, *44*, 1504-1513.
[http://dx.doi.org/10.1016/j.foodres.2011.03.049]

[132] Yao, L.; Jiang, Y.; Singanusong, R.; Datta, N.; Raymont, K. Phenolic acids and abscisic acid in Australian Eucalyptus honeys and their potential for floral authentication. *Food Chem.,* **2004**, *86*, 169-177.
[http://dx.doi.org/10.1016/j.foodchem.2003.08.013]

[133] Zhang, X.H.; Wu, H.L.; Wang, J.Y.; Tu, D.Z.; Kang, C.; Zhao, J.; Chen, Y.; Miu, X.X.; Yu, R.Q. Fast HPLC-DAD quantification of nine polyphenols in honey by using second-order calibration method based on trilinear decomposition algorithm. *Food Chem.,* **2013**, *138*(1), 62-69.
[http://dx.doi.org/10.1016/j.foodchem.2012.10.033] [PMID: 23265456]

[134] Kečkeš, S.; Gašić, U.; Veličković, T.Ć.; Milojković-Opsenica, D.; Natić, M.; Tešić, Ž. The determination of phenolic profiles of Serbian unifloral honeys using ultra-high-performance liquid chromatography/high resolution accurate mass spectrometry. *Food Chem.,* **2013**, *138*(1), 32-40.
[http://dx.doi.org/10.1016/j.foodchem.2012.10.025] [PMID: 23265452]

[135] Escuredo, O.; Silva, L.; Valentão, P.; Seijo, M.; Andrade, P. Assessing Rubus honey value: Pollen and phenolic compounds content and antibacterial capacity. *Food Chem.,* **2012**, *130*, 671-678.
[http://dx.doi.org/10.1016/j.foodchem.2011.07.107]

[136] Silici, S.; Sarioglu, K.; Karaman, K. Determination of polyphenols of some Turkish honeydew and nectar honeys using HPLC-DAD. *J. Liquid Chromatogr. Relat. Technol.,* **2013**, *36*, 2330-2341.

[137] Ferreres, F.; Andrade, P.; Tomas-Barberan, F. Flavonoids from Portuguese heather honey. *Leb. und-Technologie,* **1994**, *1991*, 32-37.

[138] Pyrzynska, K.; Biesaga, M. Analysis of phenolic acids and flavonoids in honey. *TrAC. Trends Analyt. Chem.,* **2009**, *28*, 893-902.
[http://dx.doi.org/10.1016/j.trac.2009.03.015]

[139] Petrus, K.; Schwartz, H.; Sontag, G. Analysis of flavonoids in honey by HPLC coupled with coulometric electrode array detection and electrospray ionization mass spectrometry. *Anal. Bioanal. Chem.,* **2011**, *400*(8), 2555-2563.
[http://dx.doi.org/10.1007/s00216-010-4614-7] [PMID: 21229237]

[140] Ranjbari, E.; Biparva, P.; Hadjmohammadi, M.R. Utilization of inverted dispersive liquid-liquid microextraction followed by HPLC-UV as a sensitive and efficient method for the extraction and determination of quercetin in honey and biological samples. *Talanta,* **2012**, *89*, 117-123.
[http://dx.doi.org/10.1016/j.talanta.2011.11.079] [PMID: 22284468]

[141] Antony, S.M.; Han, I.Y.; Rieck, J.R.; Dawson, P.L. Antioxidative effect of maillard reaction products formed from honey at different reaction times. *J. Agric. Food Chem.,* **2000**, *48*(9), 3985-3989.
[http://dx.doi.org/10.1021/jf000305x] [PMID: 10995301]

[142] Merken, H.M.; Beecher, G.R. Measurement of food flavonoids by high-performance liquid chromatography: A review. *J. Agric. Food Chem.,* **2000**, *48*(3), 577-599.
[http://dx.doi.org/10.1021/jf990872o] [PMID: 10725120]

[143] Ruoff, K.; Luginbühl, W.; Künzli, R.; Bogdanov, S.; Bosset, J.O.; von der Ohe, K.; von der Ohe, W.; Amado, R. Authentication of the botanical and geographical origin of honey by front-face fluorescence

spectroscopy. *J. Agric. Food Chem.,* **2006**, *54*(18), 6858-6866.
[http://dx.doi.org/10.1021/jf060697t] [PMID: 16939350]

[144] Guyot-Declerck, C.; Chevance, F.; Lermusieau, G.; Collin, S. Optimized extraction procedure for quantifying norisoprenoids in honey and honey food products. *J. Agric. Food Chem.,* **2000**, *48*(12), 5850-5855.
[http://dx.doi.org/10.1021/jf000504g] [PMID: 11141257]

[145] Castro-Vázquez, L.; Díaz-Maroto, M.; González-Viñas, M.; Pérez-Coello, M. Differentiation of monofloral citrus, rosemary, eucalyptus, lavender, thyme and heather honeys based on volatile composition and sensory descriptive analysis. *Food Chem.,* **2009**, *112*, 1022-1030.
[http://dx.doi.org/10.1016/j.foodchem.2008.06.036]

[146] Overton, S.; Manura, J. *Flavor and aroma in natural bee honey*; American Laboratory: New Brunswick, **1994**.

[147] Cuevas-Glory, L.; Pino, J.; Santiago, L.; Sauri-Duch, E. A review of volatile analytical methods for determining the botanical origin of honey. *Food Chem.,* **2007**, *103*, 1032-1043.
[http://dx.doi.org/10.1016/j.foodchem.2006.07.068]

[148] D'Arcy, B.; Rintoul, G.; Rowland, C.; Blackman, A. Composition of Australian honey extractives. 1. Norisoprenoids, monoterpenes, and other natural volatiles from blue gum (*Eucalyptus leucoxylon*) and yellow box (*Eucalyptus melliodora*). *J. Agric. Food Chem.,* **1997**, *45*, 1834-1843.
[http://dx.doi.org/10.1021/jf960625+]

[149] Tan, S.; Wilkins, A.; Holland, P.; McGhie, T. Extractives from New Zealand honeys. 3. Unifloral thyme and willow honey constituents. *J. Agric. Food Chem.,* **1990**, *38*, 1833-1838.
[http://dx.doi.org/10.1021/jf00099a010]

[150] Tan, S.; Wilkins, A.; Holland, P.; McGhie, T. Extractives from New Zealand honeys. 2. Degraded carotenoids and other substances from heather honey. *J. Agric. Food Chem.,* **1989**, *37*, 1217-1221.
[http://dx.doi.org/10.1021/jf00089a004]

[151] Bianchi, F.; Careri, M.; Musci, M. Volatile norisoprenoids as markers of botanical origin of Sardinian strawberry-tree (*Arbutus unedo* L.) honey: Characterisation of aroma compounds by dynamic headspace extraction and gas chromatography–mass spectrometry. *Food Chem.,* **2005**, *89*, 527-532.
[http://dx.doi.org/10.1016/j.foodchem.2004.03.009]

[152] de la Fuente, E.; Valencia-Barrera, R.; Martínez-Castro, I.; Sanz, J. Occurrence of 2-hydroxy-5-methyl-3-hexanone and 3-hydroxy-5-methyl-2-hexanone as indicators of botanic origin in eucalyptus honeys. *Food Chem.,* **2007**, *103*, 1176-1180.
[http://dx.doi.org/10.1016/j.foodchem.2006.10.020]

[153] Castro-Vásquez, L.; Díaz-Maroto, M.; Pérez-Coello, M. Aroma Composition and new chemical markers of Spanish citrus honeys. *Food Chem.,* **2007**, *103*, 601-606.
[http://dx.doi.org/10.1016/j.foodchem.2006.08.031]

[154] Alissandrakis, E.; Tarantilis, P.; Harizanis, P.; Polissiou, M. Aroma investigation of unifloral Greek citrus honey using solid-phase microextraction coupled to gas chromatographic-mass spectrometric analysis. *Food Chem.,* **2007**, *100*, 396-404.
[http://dx.doi.org/10.1016/j.foodchem.2005.09.015]

[155] Baroni, M.V.; Nores, M.L.; Díaz, Mdel.P.; Chiabrando, G.A.; Fassano, J.P.; Costa, C.; Wunderlin, D.A. Determination of volatile organic compound patterns characteristic of five unifloral honey by solid-phase microextraction-gas chromatography-mass spectrometry coupled to chemometrics. *J. Agric. Food Chem.,* **2006**, *54*(19), 7235-7241. [http://dx.doi.org/10.1021/jf061080e] [PMID: 16968088]

[156] Odeh, I.; Abulafi, S.; Dewik, H.; Alnajjar, I.; Imam, A.; Dembitsky, V.; Hanus, L. A variety of volatile compounds as markers in Palestinian honey from *Thymus capitatus, Thymelaea hirsuta*, and *Tolpis virgata. Food Chem.,* **2007**, *101*, 1393-1397. [http://dx.doi.org/10.1016/j.foodchem.2006.03.046]

[157] Špánik, I.; Janáčová, A.; Šusterová, Z.; Jakubík, T.; Jánošková, N.; Novák, P.; Chlebo, R. Characterisation of VOC composition of Slovak monofloral honeys by GC×GC-TOF-MS. *Chem. Pap.,* **2012**, *67*, 127-134.

[158] Zuluaga, C.; Díaz, C.; Quicazán, M. *Nariz Electrónica. Fundamentos, manejo de datos y aplicación en productos apícolas*; Universidad Nacional de Colombia: Bogotá, **2014**.

[159] Major, N.; Marković, K.; Krpan, M.; Sarić, G.; Hruškar, M.; Vahčić, N. Rapid honey characterization and botanical classification by an electronic tongue. *Talanta,* **2011**, *85*(1), 569-574. [http://dx.doi.org/10.1016/j.talanta.2011.04.025] [PMID: 21645743]

[160] Karoui, R.; Dufour, E.; Bosset, J.; De Baerdemaeker, J. The use of front face fluorescence spectroscopy to classify the botanical origin of honey samples produced in Switzerland. *Food Chem.,* **2007**, *101*, 314-323. [http://dx.doi.org/10.1016/j.foodchem.2006.01.039]

[161] Consonni, R.; Cagliani, L.R.; Cogliati, C. NMR characterization of saccharides in Italian honeys of different floral sources. *J. Agric. Food Chem.,* **2012**, *60*(18), 4526-4534. [http://dx.doi.org/10.1021/jf3008713] [PMID: 22509771]

[162] Boffo, E.; Tavares, L.; Tobias, A.; Ferreira, M.; Ferreira, A. Identification of components of Brazilian honey by 1H NMR and classification of its botanical origin by chemometric methods. *LWT -. Food Sci. Technol. (Campinas.),* **2012**, *49*, 55-63.

[163] Lammertyn, J.; Veraverbeke, E.; Irudayaraj, J. zNoseTM technology for the classification of honey based on rapid aroma profiling. *Sens. Act. B - Chem.,* **2004**, *98*, 54-62.

[164] Ampuero, S.; Bosset, J. The electronic nose applied to dairy products: a review. *Sens. Act. B,* **2003**, *94*, 1-12. [http://dx.doi.org/10.1016/S0925-4005(03)00321-6]

[165] Zuluaga, C.; Díaz, C.; Quicazán, M. La nariz electrónica una novedosa herramienta para el control de procesos y calidad en la industria agroalimentaria. *Vitae,* **2011**, *18*, 209-217.

[166] Simsek, A.; Bilsel, M.; Goren, A. 13C/12C pattern of honey from Turkey and determination of adulteration in commercially available honey samples using EA-IRMS. *Food Chem.,* **2011**, *130*, 1115-1121. [http://dx.doi.org/10.1016/j.foodchem.2011.08.017]

[167] Padovan, G.; De Jong, D.; Rodrigues, L.; Marchini, J. Detection of adulteration of commercial honey samples by the 13C/12C isotopic ratio. *Food Chem.,* **2003**, *82*, 633-636.

[http://dx.doi.org/10.1016/S0308-8146(02)00504-6]

[168] Martos, I.; Ferreres, F.; Yao, L.; D'Arcy, B.; Caffin, N.; Tomás-Barberán, F.A. Flavonoids in monospecific eucalyptus honeys from Australia. *J. Agric. Food Chem.,* **2000**, *48*(10), 4744-4748. [http://dx.doi.org/10.1021/jf000277i] [PMID: 11052728]

CHAPTER 2

Latest Developments in Propolis Research: Chemistry and Biology

Vassya Bankova*, Milena Popova, Boryana Trusheva

Institute of Organic Chemistry with Centre of Phytochemistry, Bulgarian Academy of Sciences, 1113 Sofia, Bulgaria

Abstract: Propolis is a plant derived bee product which serves dual purposes in the honeybee colony: building material and protective substance. Propolis has been used as remedy in the traditional medicine of numerous nations, because of its antimicrobial, antioxidant, and many other beneficial pharmacological actions. In this chapter, the results of the newest (in the last 5 years) chemical studies of propolis from different geographic and plant origin are reviewed, together with the new identified source plants: 152 new constituents of propolis, being 57 new chemical entities, and 12 new chemical types of propolis are listed. The importance of propolis for the bee colony is discussed, with special attention to the activity of propolis and its constituents against bee pathogens and parasites. The review of recent propolis literature demonstrates its potential to serve as a source of new chemical structures and new bioactive compounds due to its chemical diversity. It also reveals the potential of propolis to be used for development of innovative products, mainly in the field of food industries, animal husbandry, and beekeeping. For this to happen, the combined efforts of researchers and technologists from different areas are necessary, in order to make better use of bee glue.

Keywords: *Apis mellifera*, *Ascophaera apis*, Bee health, New constituents, New sources, *Paenibacillus*, Plant sources, Propolis, Social immunity, *Varroa destructor*.

* **Address correspondence to V. Bankova:** Institute of Organic Chemistry with Centre of Phytochemistry, Bulgarian Academy of Sciences, Acad. G Bonchev str., bl.9, 1113 Sofia, Bulgaria. Tel: ++359 2 9606 149; Fax: ++359 2 8700 225; Email: bankova@orgchm.bas.bg

Susana M. Cardoso (Ed)

1. INTRODUCTION

Propolis (bee glue) is a plant derived bee product that serves dual purposes in the honeybee colony. It is a building material, used by worker bees to cover the internal surface of the cavity they inhabit (the so-called "propolis envelope" in the nests of feral bees [1]), to block holes and cracks, to tighten the hive entrance in cold season, etc. On the other hand, propolis is a protective material, guarding the colony against infections: it contains putrefaction of insect, small animals – intruders killed in the hive and too large to be carried out. It is an important element of the social immunity of honeybees [2].

Bees collect resinous materials from different plant parts (lipophilic materials on leaves and leaf buds, mucilages, gums, resins, latices, etc.) [3], and mix these materials with wax to produce propolis. In order to come to the resinous materials, in some cases bees cut fragments of vegetative tissues to release the resin [4]. Bees incorporate in propolis protective plant excretions, which prevent vulnerable plant tissues from infection by harmful microorganisms. They make use of the biosynthetic potential of plants and apply the secondary plant metabolites in the resins for the same purpose as the source plant: for protection. Thus, the action against microorganisms is an essential property of propolis, and this fact has been recognized by human beings since times immemorial. Propolis has been used a remedy in the traditional medicine of numerous nations, mainly to treat burns and wounds, sore throat, stomach ulcer, *etc.* [5]. Modern science has revealed many other beneficial pharmacological properties of bee glue [6, 7].

The chemical composition of propolis varies significantly depending on the particular resin used by the honeybees in propolis production; at different locations with specific climatic and phyto-geographic conditions the chemistry of propolis differs dramatically. However, it took some time for researchers to realize this peculiarity of propolis. The paradigm shift occurred about the turn of the 21st century. It became clear that unlike beeswax or bee venom, from chemical point of view there is no single product that is "just propolis". This resulted in a new approach to propolis research, and especially in studying the biological activity of propolis. It was understood that it was not enough to report that experiments were performed with propolis, but that it was necessary to chemically characterise the particular propolis used in the experiments [8].

Although of different chemical composition, propolis from different locations always demonstrates considerable biological activity [9, 10]. For this reason, the chemical diversity of different propolis samples has the potential to provide valuable leads to active components. The study of bee glue in unexplored regions offers possibilities to uncover new biologically active compounds with important pharmacological effects, especially antibacterial, antioxidant and anticancer substances. In this chapter, the recently discovered new propolis constituents (since 2009), including such coming from newly found source plants will be discussed.

2. GENERAL PHYSICO-CHEMICAL PROPERTIES OF PROPOLIS

Propolis is a sticky material; its colour is variable and depends on the plant source: brown, brown-yellow, yellow, green, brown-orange, red. Its smell also varies with the botanical origin but is in general pleasant, balmy or resinous; the taste is bitter and pungent. At temperatures lower than 15°C propolis is hard and brittle, while over 30°C it is soft, flexible and very sticky. Important characteristic of propolis is the amount of substances soluble in 70% ethanol, called "balsam". Its content can be between 20 and 80%. The balsam contains mainly the plant derived bioactive compounds [5]. Besides the resins collected from plants, propolis contains also variable amounts of beeswax, pollen, mechanical impurities, up to 2% minerals.

The percentage of beeswax varies significantly: values between 5 and 49 % have been reported in the literature [11, 12]. Moreover, the wax amount depends on the harvesting method: the use of a propolis collector usually reduced the wax content below 20% [13].

Mechanical impurities consist of wood particles, remains of dead bees, moth cocoons, plant parts, etc. Their amount is variable too; differences in plant sources could explain the variations. In general, green Brazilian propolis is characterized by much higher percentage of mechanical impurities, up to 40%, compared to poplar type propolis (usually less than 10%). This could be due to the way bees collect resins from *Baccharis dracunculifolia* plants to produce green propolis:

cutting young leaves and buds in order to liberate the resinous material from them [14]. As a result, secretory hairs belonging to the genus *Baccharis* have been found in green Brazilian propolis in high quantities [15].

The density of propolis is about 1.11 – 1.14, its melting point between 80 – 105°C [16]. Propolis has very low solubility in water, even by boiling; it is better soluble in organic solvents such as ethanol, acetone, or ethanol-chloroform mixtures [16]. Further properties, such as antioxidant capacity, total phenolics and flavonoids content, different types of biological activities, depend to a large extent on the specific chemical composition of the particular propolis sample.

3. NEW PROPOLIS CONSTITUENTS AND NEW PROPOLIS SOURCES, IDENTIFIED SINCE 2009

In the 5 year period 2009 – 2014, a significant number of new propolis constituents have been reported. Besides the well known propolis types, originating from "conventional" source plants, such as *Populus nigra, P. tremula, Betula* spp., *Baccharis dracunculifolia, Macaranga tanarius, Mangifera indica,Dalbergia* spp., new propolis types have been reported. The new constituents are listed in (Table **1**). In this table, only isolated compounds, properly characterized by spectral methods, are included.

From all the 152 new constituents of propolis, 57 are new chemical entities. This large number is a substantial confirmation of the chemical diversity of propolis and its potential as source of new biologically active compounds.

There are several further articles reporting new propolis constituents, which are not listed in (Table **1**). The reason is that these compounds have not been isolated and identified by spectral methods. They have been only tentatively identified by LC-UV-MS and ESI-MS-MS spectra. Among them, a number of flavonoid glycosides, including one *C*-glycoside, have been detected by LC-UV-ESI-MS in Brazilian propolis from different regions [48, 49]. In propolis from Portugal, over 10 glycosides of quercetin and kaempferol were found using LC-DAD-ESI-MS^n [50]. Nevertheless, their positive identification after isolation as individual compounds should be the ultimate proof for their presence in propolis.

Table 1. Compounds found for the first time in propolis since 2009.

Geographic origin	Plant source of the compound	Structure type	Compound	Major/minor compound	References
Africa					
Kenya, Mwingi	Unknown	Arylnaphtalene lignane	Tetrahydrojusticidin B*	Minor	[17]
		Arylnaphtalene lignane	6-methoxydiphyllin*	Minor	
		Arylnaphtalene lignane	Phyllamyricin C	Minor	
Kenya, Voi	*Macaranga schweinfurthii*	Geranyl flavonol	Macarangin	Major	[17]
		Geranyl stilbene	Schweinfurthin A	Minor	
		Geranyl stilbene	Schweinfurthin B	Minor	
Egypt	Unknown	Prenyl flavonoid	Isonymphaeol D*	Minor	[18]
Algeria	Unknown	Flavone	Ladanein	Minor	[19]
		Flavone	Pilosin	Minor	
Ghana	Unknown	Prenylated stilbene	(*E*)-5-{2-[8-hydroxy-2-methyl-2-(4-methylpent-3-en-1-yl)-2*H*-chromen-6-yl]vinyl}-2-(3-methylbut-2-en-1-yl)benzene-1,3-diol*	Minor	[20]
		Prenylated stilbene	5-[(*E*)-3,5-dihydroxystyryl]-3-[(*E*)-3,7-dimethylocta-2,6-dien-1-yl]benzene-1,2-diol*	Major	
Cameroon	Unknown	Prenylated phloroglucinone	Deperoxidised derivative of plukenetione C*	Minor	[20]

(Table 1) contd.....

Geographic origin	Plant source of the compound	Structure type	Compound	Major/minor compound	References
Cameroon	*Mangifera indica*	Alkyl phenol	3-Undecylphenol	Minor	[21]
		Alkyl phenol	3-Tetradecylphenol	Minor	
		Alkyl phenol	3-Pentadecylphenol	Minor	
		Alkyl phenol	3-Hexadecylphenol	Minor	
		Alkyl phenol	3-Heptadecylphenol	Minor	
		Alkyl phenol	3-Nonadecylphenol	Minor	
		Alkyl phenol	3-(10'*Z*-Pentadecenyl)phenol	Minor	
		Alkyl phenol	3-(12'*Z*-Pentadecenyl)phenol	Minor	
		Alkyl phenol	3-(8'*Z*-Heptadecenyl)phenol	Minor	
		Alkyl phenol	3-(12'*Z*-Heptadecenyl)phenol	Minor	
		Alkyl phenol	3-(14'*Z*-Heptadecenyl)phenol	Minor	
		Alkyl phenol	3-(13'*Z*-Nonadecenyl)phenol*	Minor	
		Alkyl phenol	3-(14'*Z*-Nonadecenyl)phenol*	Minor	
		Alkyl resorcinol	5-Hexadecylresorcinol	Minor	
		Alkyl resorcinol	5-(10'*Z*-Pentadecenyl)resorcinol	Minor	
		Alkyl resorcinol	5-(12'*Z*-Heptadecenyl)resorcinol	Minor	
		Alkyl resorcinol	5-(14'*Z*-Heptadecenyl)resorcinol	Minor	
		Alkyl resorcinol	5-(14'*Z*-Nonadecenyl)resorcinol*	Minor	
Algeria, North	*Populus* spp.	Dihydroflavonol	Pinobanksin 3-(*E*)-caffeate	Major	[22]
Algeria	*Cistus* spp.	Clerodane diterpene	Cistadiol	Major	[22]
		Clerodane diterpene	18-Hydroxy-*cis*-clerodane-3-ene-15-oic acid	Major	
Cameroon	Unknown	Oleanane triterpene	Olean-12-ene-3в,28-diol	Minor	[23]
Asia					

(Table 1) contd.....

Geographic origin	Plant source of the compound	Structure type	Compound	Major/minor compound	References
Korea, Jeju island	*Angelica keiskei*	Chalcone	Jejuchalcone A*	Minor	[25]
		Chalcone	Jejuchalcone B*	Minor	
		Chalcone	Jejuchalcone C*	Minor	
		Chalcone	Jejuchalcone D*	Minor	
		Chalcone	Jejuchalcone E*	Minor	
		Chalcone	(-)-Jejuchalcone F*	Minor	
		Chalcone	(+)-Jejuchalcone G*	Minor	
		Chalcone	(-)-Jejuchalcone H*	Minor	
		Chalcone	4-Hydroxyderricin	Major	
		Chalcone	2',4,4'-Trihydroxy-3'-(6,7-dihydroxy-3, 7-dimethyl-2-octaenyl)chalcone ((±)-TB5)	Minor	
		Chalcone	(±)-Xanthoangelol B	Minor	
		Chalcone	Xanthoangelol	Major	
		Chalcone	Xanthoangelol F	Minor	
		Chalcone	(±)-Bavachromanol	Minor	
		Chalcone	2',3'-[2-Methyl-2-(4-methylpenten-3-yl)-3-hydroxy-dihydropyrano]-4,4'-dihydroxychalcone ((+)-TB2)	Minor	
		Chalcone	(+)-Lespeol	Minor	
		Chalcone	(+)-TB1	Minor	
		Coumarin	(*S,S*)-(+)-Laserpitin	Major	
		Coumarin	(*S,S*)-(-)-Isolaserpitin	Major	
		Coumarin	(*S*)-(-)-Selinidin	Minor	
		Coumarin	(*R,R*)-(+)-Khellactone	Major	
		Coumarin	(*R,R*)-(+)-3'-Senecioylkhellactone	Minor	
		Coumarin	(2'*S*,3'*R*)-(+)-Vaginidiol	Minor	
		Coumarin	(2'*S*,3'*R*)-(+)-Daucoidin A	Minor	
		Coumarin	(*S*)-(-)-Oxypeucedanin hydrate	Minor	
		Coumarin	(*S*)-(+)-Marmesin	Minor	

(Table 1) contd.....

Geographic origin	Plant source of the compound	Structure type	Compound	Major/minor compound	References
China, Henan Province	Unknown	Flavanol	8-[(*E*)-4-Phenylprop-2-en-1-one]-(2*R*,3*S*)-2-(3,5-dihydroxyphenyl)-3,4-dihydro-2*H*-2-benzopyra--5-methoxyl-3,7-diol*	Major	[25]
		Flavanol	8-[(*E*)-4-Phenylprop-2-en-1-one]-(2*S*,3*R*)-2-(3,5-dihydroxyphenyl)-3,4-dihydro-2*H*-2-benzopyran-5-methxyl-3,7-diol*	Major	
		Flavanol	8-[(*E*)-4-Phenylprop-2-en-1-one]-(2*R*,3*S*)-2-(3-methox-l-4-hydroxyphenyl)-3,4-dihydro-2*H*-2-benzopyran-5-methoxy--3,7-diol	Minor	
		Flavanol	8-[(*E*)-4-*p*-Henylprop-2-en-1-one]-(2*S*,3*R*)-2-(3-methoxyl-4-hydroxyphenyl)-3,4-dihydro-2*H*-2-benzopyra--5-methoxyl-3,7-diol*	Minor	
China, Hubei province	*Populus* spp.	Phenolic glyceride	2-Acetyl-1-coumaroyl-3-cinnamoylglycerol*	Major	[26]
		Phenolic glyceride	(+)-2-Acetyl-1-feruloyl-3-cinnamoylglycerol*	Major	
		Phenolic glyceride	(-)-2-Acetyl-1-feruloyl-3-cinnamoylglycerol*	Minor	
		Phenolic glyceride	2-Acetyl-1,3-dicinnamoylglycerol*	Major	
		Phenolic glyceride	(-)-2-Acetyl-1-(*E*)-feruloyl-3-(3''(ς),16'')-dihydroxypalmitoylglycerol*	Minor	
Thailand	*Styrax* spp.	Flavanone	(7''*S*)-8-[1-(4'-Hydroxy-3'-methoxyphenyl)prop-2-en-1-yl]-(2*S*)-pinocembrin	Minor	[27]
		Cinnamic acid derivative	(*E*)-Cinnamyl (*E*)-cinnamylidenate	Minor	

(Table 1) contd.....

Geographic origin	Plant source of the compound	Structure type	Compound	Major/minor compound	References
Myanmar	*Mangifera indica*	Triterpene	(22*Z*,24*E*)-3-Oxocycloart-22,24-dien-26-oic acid*	Minor	[28]
		Triterpene	(24*E*)-3-Oxo-27,28-dihydroxycycloart-24-en-26-oic acid*	Minor	
		Triterpene	28-Hydroxymangiferonic acid	Minor	
		Triterpene	27-Hydroxymangiferonic acid	Major	
		Triterpene	(24*E*)-3-Oxo-23-hydroxycycloart-24-en-26-oic acid	Minor	
		Triterpene	(24*E*)-3*β*-Hydroxycycloart-24-en-26-al	Minor	
		Triterpene	(24*E*)-3*α*,27-Dihydroxycycloart-24-en-26-oic acid	Minor	
		Triterpene	(24*E*)-3*β*,27-Dihydroxycycloart-24-en-26-oic acid	Major	
		Triterpene	(24*E*)-3*α*,22-Dihydroxycycloart-24-en-26-oic acid	Minor	
		Triterpene	(24*E*)-3*β*,23-Dihydroxycycloart-24-en-26-oic acid	Minor	
		Prenylated flavanone	(2*S*)-5,7-Dihydroxy-4'-methoxy-8,3-diprenylflavanone	Minor	
		Prenylated flavanone	(2*S*)-5,7,4'-Trihydroxy-8,3'-diprenylflavanone	Minor	
		Prenylated flavanone	(2*S*)-5,7-Dihydroxy-4'-methoxy-8-prenylflavanone	Minor	
		Prenylated flavanone	(2*S*)-5,7,4'-Trihydroxy-8-prenylflavanone	Minor	
Jordania	Unknown	Triterpenic acid	24(*Z*)-1*β*-3*β*-Dihydroxyeupha-7,24-dien-26-oic acid*	Major	[29]
		Triterpenic acid	24(*Z*)-3*β*-Dihydroxyeupha-7,24-dien-26-oic acid	Major	
Saudi Arabia	*Psiadia* spp.	Kaurane diterpene	Propsiadin [(*ent*)-2-oxokaur-16-en-6,18-diol]*	Minor	[30]
		Kaurane diterpene	Psiadin	Major	
		Flavone	Psiadiarabin	Minor	
Iran	*Ferula* spp.	Prenylated coumarin	Suberosin	Minor	[31]
		Terpene ester of aromatic acid	Tschimgin (bornyl *p*-hydroxybenzoate)	Minor	
		Terpene ester of aromatic acid	Tschimganin (bornyl vanillate)	Minor	
		Terpene ester of aromatic acid	Ferutinin (ferutinol *p*-hydroxybenzoate)	Minor	
		Terpene ester of aromatic acid	Teferin (ferutinol vanillate)	Minor	
Oman	*Azadiracta indica*	Prenylated flavanone	7-*O*-methyl-8-prenylnaringenin	Minor	[32]
		Prenylated flavanone	3',8-Diprenylnaringenin	Major	
Oman	*Acacia nilotica*	Flavanols	Mixture of 2,3-*trans*-3,4-*trans* and 2,3-*trans*-3,4-*cis*-mollisacacidins	Minor	[32]

(Table 1) contd.....

Geographic origin	Plant source of the compound	Structure type	Compound	Major/minor compound	References
Australia and Oceania					
Solomon islands	Unknown	Prenylated flavonol	Solophenol A*	Minor	[33]
		Prenylated flavanone	Sophoraflavanone A	Minor	
Solomon islands	*Macaranga* spp.	Prenylated flavonol	Solophenol B*	Minor	[34]
		Prenylated flavonol	Solophenol C*	Minor	
		Prenylated flavonol	Solophenol D*	Minor	
		Prenyl stilbene	Solomonin*	Minor	
Kangaroo island	*Lepidosperma viscidum*	Cinnamic acid ester	Methyl (*E*)-4-(4'-hydroxy-3'-methylbut-(*E*)-2'-enyloxy)cinnamate*	Major	[35]
		Cinnamic acid ester	Methyl 4-(3'-methylbut-(*E*)-2'-enyloxy)cinnamate	Minor	
		Prenyl stilbene	5,4'-Dihydroxy-3'-methoxy-3-prenyloxy-(*E*)-stilbene*	Minor	
		Prenyl stilbene	3',4',5-Trihydroxy-3-methoxy-3-prenyloxy-(*E*)-stilbene*	Minor	
		Prenyl stilbene	3,5,3',4'-Tetrahydroxy-2-prenyl-(*E*)-stilbene*	Minor	
		Prenyl stilbene	3,5,4'-Trihydroxy-3'-methoxy-2-prenyl-(*E*)-stilbene*	Minor	
		Prenyl stilbene	5,3',4'-Trihydroxy-3-methoxy-2-prenyl-(*E*)-stilbene*	Minor	
		Prenyl stilbene	5,4'-Dihydroxy-3,3'-dimethoxy-2-prenyl-(*E*)-stilbene*	Minor	
		Prenyl stilbene	3,5,4'-Trihydroxy-4-prenyldihydrostilbene*	Minor	
		Prenyl stilbene	5,4'-Dihydroxy-3-prenyloxy-(*E*)-stilbene*	Minor	
		Prenyl stilbene	3',4'-Dihydroxy-3,5-dimethoxy-(*E*)-stilbene	Minor	
		Prenyl stilbene	3,5,3',4'-Tetrahydroxy-2,4-diprenyldihydrostilbene	Minor	
		Prenyl stilbene	3,5-Dihydroxy-2-prenyl-(*E*)-stilbene	Minor	
		C-methylflavanone	Farrerol	Minor	
		C-methylflavanone	Matteucinol	Minor	
Kangaroo islands	*Acacia paradoxa*	Chalcone	2',3',4'-Trimethoxychalcone*	Major	[36]
		Chalcone	2'-Hydroxy-3',4'-dimethoxychalcone*	Major	
		Chalcone	2',4'-Dihydroxy-3'-methoxychalcone	Major	
		Dihydroflavonol	5,7-Dihydroxy-6-methoxy-2,3-dihydroflavonol 3-acetate*	Major	
Europe					

(Table 1) contd.....

Geographic origin	Plant source of the compound	Structure type	Compound	Major/minor compound	References
Greece (Crete)	*Cupressus sempervirens*	Diterpenic acid	14,15-*Dinor*-13-oxo-8(17)-labden-19-oic acid*	Minor	[37]
		Diterpenic acid ester	Labda-8(17),13*E*-dien-19-carboxy-15-yl oleate*	Minor	
		Diterpenic acid ester	Labda-8(17),13*E*-dien-19-carboxy-15-yl palmitate*	Minor	
		Triterpenic acid	3,4-*Seco*-cycloart-12-hydroxy-4(28),24-dien-3-oic acid*	Minor	
		Triterpenic acid	Cycloart-3,7-dihydroxy-24-en-28-oic acid*	Minor	
		Diterpene	*Trans*-communal	Minor	
		Diterpenic acid	Pimaric acid	Minor	
		Diterpene	Totarolone (3-oxototarol)	Minor	
		Diterpenic acid	15-Oxolabda-8(17),13*E*-dien-19-oic acid	Minor	
		Diterpenic acid	15-Oxolabda-8(17),13*Z*-dien-19-oic acid	Minor	
		Diterpenic acid	Junicedric acid	Minor	
		Flavonoid	Isorhamnetin-3-*O*-rutinoside	Minor	
Malta	*Ferula communis*	Terpenyl ester of substituted benzoic acid	2-Acetoxy-6-*p*-methoxybenzoyl jaeschkeanadiol	Minor	[38]
		Terpenyl ester of substituted benzoic acid	2-Acetoxy-6-*p*-hydroxybenzoyl jaeschkeanadiol	Minor	
Bulgaria	*Populus* spp.	Oxo-fatty acid	9-Oxo-10(*E*)-12(*Z*)-octadecadienoic acid	Minor	[39]
America					
Mexico, Sonora State	Unknown	Phenylallylflavanone	(2*R*,3*R*)-6-[1-(4'-Hydroxy-3'-methoxyphenyl)pr-p-2-en-1-yl]pinobanksin;	Minor	[40]
		Phenylallylflavanone	(2*R*,3*R*)-6-[1-(4'-Hydroxy-3'-methoxyphenyl)pr-p-2-en-1-yl]pinobanksin 3-acetate	Minor	
Mexico, Sonora State	Unknown *Populus* spp. (?)	Phenylallylflavanone	(2*R*,3*R*)-3,5-Dihydroxy-7-methoxyflavanone-3-(2-methyl)butyrate	Minor	[41]
		Phenylallylflavone	(7''*R*)-8-[1-(4'-Hydroxy-3'-methoxyphenyl)prop-2-en-1-yl]chrysin	Minor	
		Phenylallylflavonole	(7''*R*)-8-[1-(4'-Hydroxy-3'-methoxyphenyl)prop-2-en-1-yl]galangin	Minor	
Mexico	*Dalbergia* spp.	Phenol	1-(3',4'-Dihydroxy-2'-methoxyphenyl)-3-phenylpropane	Minor	[42]
		Phenol	(*Z*)-1-(2'-Methoxy-4',5'-dihydroxyphenyl)-3-phenylprop-2-ene	Minor	
		Flavan	3-Hydroxy-5,6-dimethoxyflavan	Minor	
Brazil, Bahia	Unknown	Prenylated benzophenone	Hyperibone A*	Minor	[43]
Brazil	Unknown	-	Tannins	Minor	[44]

(Table 1) contd.....

Geographic origin	Plant source of the compound	Structure type	Compound	Major/minor compound	References
Argentina, Andean region	*Larrea nitida*	Lignane	3'-Methylnordihydroguaiaredic acid*	Major	[45]
		Lignane	Nordihydroguaiaredic acid*	Minor	
		Lignane	4-[4-(4-Hydroxyphenyl)-2,3-dimethyl-butyl]benzene-1,2-diol*	Minor	
		Lignane	*Meso*-(rel 7*S*,8*S*,7'*R*,8'*R*)-3,4,3',4'-tetrahydroxy-7,7'-epoxylignan*	Minor	
		Lignane	(7*S*,8*S*,7'*S*,8'*S*)-3,3',4'-Trihydroxy-4-methoxy-7,7'-epoxylignan*	Minor	
Argentina, Tucuman	*Zuccagnia punctata*	Flavone	3-Hydroxy-7,8-dimethoxyflavone	Minor	[46]
Honduras	*Liguidambar styraciflua*	Ester of cinnamic acid	(*E*)-Cinnamyl (*Z*)-cinnamate	Minor	[47]

* New natural compound

Another new propolis constituent identified in the last 5 years is β-glucosidase. The β-glucosidase activity of propolis has been studied and it was found that glucose derivatives were rapidly hydrolyzed but not conjugates with other sugars [51,52].

Not only new chemical constituents but new chemical types of propolis have been found and respectively their sources have been identified as new propolis-bearing plants. They are listed in (Table **2**).

Table 2. Newly identified propolis plant sources.

Species	Plant family	Geographic location	Reference
Acacia nilotica, (L.) Delile	Fabaceae	Oman	[32]
Acacia paradoxa DC	Fabaceae	Australia	[36]
Angelica keiskei Ito	Apiaceae	Korea, Jeju Island	[24]
Azadiracta indica A.Juss.	Meliaceae	Oman	[32]
Cupressus sempervirens L.	Cupressaceae	Greece, Malta	[37, 53]
Ferula spp., incl. *Ferula communis* L.	Apiaceae	Iran, Malta	[31, 38]
Larrea nitida Cav.	Zygophylaceae	Argentina	[45]
Lepidosperma viscidum R.Br.	Cyperaceae	Australia	[35]
Liquidambar styraciflua L.	Hamamelidaceae	Honduras	[47]
Macaranga schweinfurthii Pax	Euphorbiaceae	Kenya	[17]
Psiadia spp.	Asteraceae	Saudi Arabia	[30]
Zuccania punctata Cav.	Fabaceae	Argentina	[46]

The remarkable chemical variability of propolis is, of course, due to its plant origin and to the fact that at different geographic locations the source plants vary with respect to the local flora at the site of collection. However, there is another important, but often neglected factor - the choices made by bees. It is obvious that bees choose sticky resinous materials because of their physical properties. On the other hand, this material is also their chemical defence against microorganisms, but it is yet not clear how exactly bees recognise these properties in the collected material. The new knowledge on propolis sources brings new insight into the complex relationship between honeybees and plants. Among the new sources, some belong to families already known to include propolis-bearing plants, *e.g.*

Fabaceae (*Dalgbergia*, the source of red South American propolis) and Asteraceae (*Baccharis dracunculifolia*), Euphorbiaceae. Plants of a number of the listed plant families have not been found to be sources of propolis till now: Apiaceae, Hamamelidaceae, Meliaceae. Obviously, bees do not limit their choices but have adapted to make the best of the available plants in any habitat.

4. PROPOLIS AND BEE HEALTH

Propolis has been used since millennia as a remedy in traditional medicine systems all over the world. For this reason, it has been attracting the growing attention of researchers for the last 60 years. In the course of these studies, a number of propolis' diverse and useful pharmacological activities have been revealed but the studies have been directed almost exclusively to the potential health benefits for human beings, and later on for domestic animals such as cattle, pigs, poultry, dogs, etc. Only recently, an interest appeared in the capacity of propolis to act against bee pathogens and in the possibility to apply propolis or its constituents in beekeeping instead of pesticides and the banned in Europe antibiotics.

A number of these studies revealed the important role of propolis as an element of honeybees‘ social immunity. As resin-collecting bees (foraging for propolis) have no direct benefit from this activity, it can be considered a part of the bees‘ social immune system, providing the colony with some defence against infections and parasites [54]. Especially the research of Spivak and her group contributed significantly to this concept. Treating of beehives with propolis extract resulted in lower bacterial loads and in significantly lower expression of two honeybee immune-related genes in the propolis treated colonies [54]. The costs of an elevated immune system have been well-documented across bee species and include reduced life span under stressful conditions, and lowered colony productivity [54, 55]. The study [54] showed that propolis does not suppress the immune system; however it allows for this system to be downregulated and reduces the costs for an elevated immunity in the absence of pathogen challenge. Further, Simone-Finstrom and Spivak [56] observed a case of self-medication of bee colonies by propolis collection: there were an increased number of individuals that collect propolis in response to infection with chalkbrood *Ascophaera apis*.

Several research groups have demonstrated the ability of propolis to suppress the development of *Paenibacillus larvae*, the causal agent of American foulbrood (AFB) disease. *In vitro* inhibition of *P. larvae* was found for propolis samples from different states of Brazil, from Minnesota (USA) [57], Argentina [58], and for European poplar type propolis from Romania and Bulgaria [39,59]. *In vivo* antibacterial effect of propolis was also demonstrated: field assays showed effective treatment of hives affected with AFB by Egyptian propolis in concentrations 0.1 and 0.05% [60]. Antunez *et al.* [58] found significant decrease in the number of *P. larvae* spores/g of honey in naturally infected beehives treated with propolis extract. The authors suggested a mechanism of action including oral ingestion of the extract which facilitates the direct effect on *P. larvae*. They supposed that this mechanism could not prevent the infection of new larvae, but could inhibit the replication of vegetative bacterial cells in larval gut. In addition, they found that propolis extract had low oral toxicity to honeybees. Very recently, a number of propolis samples from different locations in the USA have demonstrated differential activity against the growth of *P. larvae* and *A. apis*. Differential activity was observed both among samples and between pathogens [61].

Of course the chemical constituents of propolis are the active principles that suppress *P. larvae*. Bilikova *et al.* [39] isolated and identified several constituents of poplar propolis with significant activity against *P. larvae* strains: pinocembrin, 3-*O*-acetylpinobanksin, and a well-defined mixture of caffeic acid esters, including CAPE (caffeic acid phenethyl ester). Mihai *et al.* [59] found that the concentration of phenolics and flavonoids in poplar type propolis affects its antimicrobial action against *P. larvae* but also that there is a significant interaction in this respect between flavones/flavonols and flavanones/dihydroflavonols in propolis extract.

In the last 30 years, *Varroa destructor* has become the most significant threat to beekeeping in Europe and all over the world. In the search for natural anti-varroa agents, propolis has also been studied. Garedew *et al.* [62, 63] showed a lethal and narcotic effect of propolis on varrroa mites. Propolis extract with 70% ethanol (10% w/v) resulted in 100% mortality with a brief contact time of 5 s. Even sublethal concentrations resulted in a significant decrease of the heat production rate in the mites (detected by calorimetry, [62]), thus causing weakening of the mites. Damiani *et al.* [64,65] also demonstrated anti-varroa effect of Argentinean propolis from the Pampean region: spraying infested bees with a 10% propolis solution did not affect the bees but was lethal to 78% of the mites. Recently, Popova *et al.* [66] found some relationship between the chemical composition of propolis and the honeybee resistance against *V. destructor*. Propolis of varroa resistant colonies from Avignon, France, contained higher concentrations of caffeic acid and its pentenyl esters: 3-methyl-3-butenyl caffeate, 2-methyl-2-butenyl caffeate and 3-methyl-2-butenyl caffeate, compared to susceptible colonies. It is yet unknown how these compounds affect the ability of bee colonies to resist varroa infestation.

Obviously, the studies on the importance of propolis for the health and wellbeing of honeybees and their social immunity are in their early stage. The next years will bring new and insightful findings in this field which will support the efforts to better understand natural defence mechanisms of bees and help humans to improve honeybee health and, hopefully, stop colony collapse disorder (CCD).

CONCLUSION

The interest in propolis research has been steadily growing since the 1970's. The annually published articles containing "propolis" in title, abstract and/or keywords increased from an average of 20 for the period 1975 – 1985 to 300 for the last five years (Scopus database). At the same time, it has become obvious that propolis is highly variable with respect to its chemical composition, and that many propolis chemical types exist, having distinct chemical profiles determined by their plant sources. For this reason, the chemical profiling and characterization of every propolis sample used in studying its potential applications in any possible field is of crucial importance. It is vital to stress that any research aiming propolis application, done with propolis without chemical characterization is irreproducible and irrelevant, and thus a waste of time and efforts.

The review of recent propolis literature demonstrates the potential of propolis to serve as a source of new chemical structures and new bioactive compounds due to its chemical diversity, resulting from the diversity of plant source used by bees in different geographic regions. It also reveals the potential of propolis to be used for development of innovative products, mainly in the field of food industries, animal husbandry, beekeeping, etc. For this to happen, the combined efforts of researchers and technologists from different areas are necessary, in order to make better use of bee glue.

CONFLICT OF INTEREST

The author confirms that author has no conflict of interest to declare for this publication.

ACKNOWLEDGEMENTS

The authors are grateful to the Institute of Organic Chemistry with Centre of Phytochemistry, BAS, for support.

REFERENCES

[1] Seeley, T.D.; Morse, R.A. The nest of the honey bee (*Apis mellifera* L.). *Insectes Soc.,* **1976**, *23*, 495-512.
[http://dx.doi.org/10.1007/BF02223477]

[2] Simone-Finstrom, M.; Spivak, M. Propolis and bee health: the natural history and significance of resin use by honey bees. *Apidologie (Celle),* **2010**, *41*, 295-311.
[http://dx.doi.org/10.1051/apido/2010016]

[3] Crane, E. *Beekeeping: Science, Practice and World Recourses*; Heinemann: London, **1988**.

[4] Salatino, A.; Teixeira, E.W.; Negri, G.; Message, D. Origin and chemical variation of Brazilian propolis. *eCAM,* **2005**, *2*, 33-38.

[5] Ghisalberti, E.L. Propolis: A review. *Bee World,* **1979**, *60*, 59-84.
[http://dx.doi.org/10.1080/0005772X.1979.11097738]

[6] Burdock, G.A. Review of the biological properties and toxicity of bee propolis (propolis). *Food Chem. Toxicol.,* **1998**, *36*(4), 347-363.
[http://dx.doi.org/10.1016/S0278-6915(97)00145-2] [PMID: 9651052]

[7] Sforcin, J.M.; Bankova, V. Propolis: is there a potential for the development of new drugs? *J. Ethnopharmacol.,* **2011**, *133*(2), 253-260.
[http://dx.doi.org/10.1016/j.jep.2010.10.032] [PMID: 20970490]

[8] Bankova, V. Chemical diversity of propolis and the problem of standardization. *J. Ethnopharmacol.,* **2005**, *100*(1-2), 114-117.
[http://dx.doi.org/10.1016/j.jep.2005.05.004] [PMID: 15993016]

[9] Kujumgiev, A.; Tsvetkova, I.; Serkedjieva, Y.; Bankova, V.; Christov, R.; Popov, S. Antibacterial, antifungal and antiviral activity of propolis of different geographic origin. *J. Ethnopharmacol.,* **1999**,

64(3), 235-240.
[http://dx.doi.org/10.1016/S0378-8741(98)00131-7] [PMID: 10363838]

[10] Seidel, D.J.; Qiang, F.; Randel, W.J.; Reichler, T.J. Widening of the tropical belt in a changing climate. *Nat. Geosci.,* **2008**, *1*, 21-24.

[11] Popova, M.; Chen, C-N.; Chen, P-Y.; Huang, C-Y.; Bankova, V. A validated spectrophotometric method for quantification of prenylated flavanones in pacific propolis from Taiwan. *Phytochem. Anal.,* **2010**, *21*(2), 186-191.
[PMID: 19856482]

[12] Puker, A.; Abot, R.A.; Matias, R.; Rodrigues, S.R.; Pinto, A.M. Propolis produced by africanized honeybees in the Cerrado-Pantanal ecotone: Effects of seasonality in production and physico-chemical characteristics. *Sociobiology,* **2010**, *56*, 149-162.

[13] Stan, L.; Mărghitaş, L.A.; Dezmirean, D. Quality criteria for propolis standardization. *Animal Sci. Biotechnol.,* **2011**, *44*, 137-140.

[14] Weinstein Teixeira, E.; Negri, G.; Meira, R.M.S.A.; Message, D.; Salatino, A. Plant origin of green propolis: bee behavior, plant anatomy and chemistry. *eCAM,* **2005**, *2*, 85-92.

[15] Santos, F.A.; Bastos, E.M.; Maia, A.B.; Uzeda, M.; Carvalho, M.A.; Farias, L.M.; Moreira, E.S. Brazilian propolis: physicochemical properties, plant origin and antibacterial activity on periodontopathogens. *Phytother. Res.,* **2003**, *17*(3), 285-289.
[http://dx.doi.org/10.1002/ptr.1117] [PMID: 12672164]

[16] Bogdanov, S. Propolis. *http://www.apitherapie.at/ATdownloads/Bogdanov_Propolis.pdf,* 1999 [Accessed on 8 January 2015];

[17] Petrova, A.; Popova, M.; Kuzmanova, C.; Tsvetkova, I.; Naydenski, H.; Muli, E.; Bankova, V. New biologically active compounds from Kenyan propolis. *Fitoterapia,* **2010**, *81*(6), 509-514.
[http://dx.doi.org/10.1016/j.fitote.2010.01.007] [PMID: 20079813]

[18] El-Bassuony, A.; AbouZid, S. A new prenylated flavanoid with antibacterial activity from propolis collected in Egypt. *Nat. Prod. Commun.,* **2010**, *5*(1), 43-45.
[PMID: 20184018]

[19] Segueni, N.; Zellagui, A.; Moussaoui, F.; Lahouel, M.; Rhouati, S. Flavonoids from Algerian propolis. *Arab. J. Chem.,* in press
[http://dx.doi.org/10.1016/j.arabjc.2011.05.013]

[20] Almutairi, S.; Eapen, Ben.; Chundi, S.M.; Akhalil, A.; Siheri, W.; Clements, Carol.; Fearnley, J.; Watson, D.G.; Edrada-Ebel, R. New anti-trypanosomal active prenylated compounds from African propolis. *Phytochem. Lett.,* **2014**, *10*, 35-39.
[http://dx.doi.org/10.1016/j.phytol.2014.07.002]

[21] Kardar, M.N.; Zhang, T.; Coxon, G.D.; Watson, D.G.; Fearnley, J.; Seidel, V. Characterisation of triterpenes and new phenolic lipids in Cameroonian propolis. *Phytochemistry,* **2014**, *106*, 156-163.
[http://dx.doi.org/10.1016/j.phytochem.2014.07.016] [PMID: 25104230]

[22] Piccinelli, A.L.; Mencherini, T.; Celano, R.; Mouhoubi, Z.; Tamendjari, A.; Aquino, R.P.; Rastrelli, L. Chemical composition and antioxidant activity of Algerian propolis. *J. Agric. Food Chem.,* **2013**, *61*(21), 5080-5088.

[http://dx.doi.org/10.1021/jf400779w] [PMID: 23650897]

[23] Paul, S.; Emmanuel, T.; Matchawe, C.; Alembert, T.T.; Elisabeth, Z.M.; Sophie, L.; Luce, V.E.; Maurice, T.F.; Joel, Y.G.A.; Alex, A.D.T.; Joseph, M.T. Pentacyclic triterpenes and crude extracts with antimicrobial activity from Cameroonian brown propolis samples. *J. App. Pharm. Sci.,* **2014**, *4*, 1-9.

[24] Shimomura, K.; Sugiyama, Y.; Nakamura, J.; Ahn, M-R.; Kumazawa, S. Component analysis of propolis collected on Jeju Island, Korea. *Phytochemistry,* **2013**, *93*, 222-229. [http://dx.doi.org/10.1016/j.phytochem.2012.02.018] [PMID: 22483235]

[25] Sha, N.; Guan, S.H.; Lu, Z.Q.; Chen, G.T.; Huang, H.L.; Xie, F.B.; Yue, Q.X.; Liu, X.; Guo, D.A. Cytotoxic constituents of chinese propolis. *J. Nat. Prod.,* **2009**, *72*(4), 799-801. [http://dx.doi.org/10.1021/np900118z] [PMID: 19278239]

[26] Shi, H.; Yang, H.; Zhang, X.; Sheng, Y.; Huang, H.; Yu, L. Isolation and characterization of five glycerol esters from Wuhan propolis and their potential anti-inflammatory properties. *J. Agric. Food Chem.,* **2012**, *60*(40), 10041-10047. [http://dx.doi.org/10.1021/jf302601m] [PMID: 22978445]

[27] Athikomkulchai, S.; Awale, S.; Ruangrungsi, N.; Ruchirawat, S.; Kadota, S. Chemical constituents of Thai propolis. *Fitoterapia,* **2013**, *88*, 96-100. [http://dx.doi.org/10.1016/j.fitote.2013.04.008] [PMID: 23660244]

[28] Li, F.; Awale, S.; Zhang, H.; Tezuka, Y.; Esumi, H.; Kadota, S. Chemical constituents of propolis from Myanmar and their preferential cytotoxicity against a human pancreatic cancer cell line. *J. Nat. Prod.,* **2009**, *72*(7), 1283-1287. [http://dx.doi.org/10.1021/np9002433] [PMID: 19572611]

[29] Shaheen, S.A.; Zarga, M.H.; Nazer, I.K.; Darwish, R.M.; Al-Jaber, H.I. Chemical constituents of Jordanian propolis. *Nat. Prod. Res.,* **2011**, *25*(14), 1312-1318. [http://dx.doi.org/10.1080/14786419.2010.509060] [PMID: 21732907]

[30] Almutairi, S.; Edrada-Ebel, R.; Fearnley, J.; Igoli, J.O.; Alotaibi, W.; Clements, C.J.; Gray, A.I.; Watson, D.J. Isolation of diterpenes and flavonoids from a new type of propolis from Saudi Arabia. *Phytochem. Lett.,* **2014**, *10*, 160-163. [http://dx.doi.org/10.1016/j.phytol.2014.08.022]

[31] Trusheva, B.; Todorov, I.; Ninova, M.; Najdenski, H.; Daneshmand, A.; Bankova, V. Antibacterial mono- and sesquiterpene esters of benzoic acids from Iranian propolis. *Chem. Cent. J.,* **2010**, *4*, 8. [http://dx.doi.org/10.1186/1752-153X-4-8] [PMID: 20350297]

[32] Popova, M.; Dimitrova, R.; Al-Lawati, H.T.; Tsvetkova, I.; Najdenski, H.; Bankova, V. Omani propolis: chemical profiling, antibacterial activity and new propolis plant sources. *Chem. Cent. J.,* **2013**, *7*(1), 158. [http://dx.doi.org/10.1186/1752-153X-7-158] [PMID: 24053750]

[33] Inui, S.; Shimamura, Y.; Masuda, S.; Shirafuji, K.; Moli, R.T.; Kumazawa, S. A new prenylflavonoid isolated from propolis collected in the Solomon Islands. *Biosci. Biotechnol. Biochem.,* **2012**, *76*(5), 1038-1040. [http://dx.doi.org/10.1271/bbb.120021] [PMID: 22738984]

[34] Inui, S.; Hosoya, T.; Shimamura, Y.; Masuda, S.; Ogawa, T.; Kobayashi, H.; Shirafuji, K.; Moli, R.T.; Kozone, I.; Shin-ya, K.; Kumazawa, S. Solophenols B-D and solomonin: new prenylated polyphenols isolated from propolis collected from the Solomon Islands and their antibacterial activity. *J. Agric. Food Chem.,* **2012**, *60*(47), 11765-11770. [http://dx.doi.org/10.1021/jf303516w] [PMID: 23067056]

[35] Abu-Mellal, A.; Koolaji, N.; Duke, R.K.; Tran, V.H.; Duke, C.C. Prenylated cinnamate and stilbenes from Kangaroo Island propolis and their antioxidant activity. *Phytochemistry,* **2012**, *77*, 251-259. [http://dx.doi.org/10.1016/j.phytochem.2012.01.012] [PMID: 22321386]

[36] Tran, V.H.; Duke, R.K.; Abu-Mellal, A.; Duke, C.C. Propolis with high flavonoid content collected by honey bees from *Acacia paradoxa. Phytochemistry,* **2012**, *81*, 126-132. [http://dx.doi.org/10.1016/j.phytochem.2012.06.002] [PMID: 22784552]

[37] Popova, M.P.; Chinou, I.B.; Marekov, I.N.; Bankova, V.S. Terpenes with antimicrobial activity from Cretan propolis. *Phytochemistry,* **2009**, *70*(10), 1262-1271. [http://dx.doi.org/10.1016/j.phytochem.2009.07.025] [PMID: 19698962]

[38] Popova, M.; Boryana, T.; Daniela, A.; Cutajar, S.; Mifsud, D.; Farrugia, C.; Tsvetkova, I.; Najdenski, H.; Bankova, V. The specific chemical profile of Mediterranean propolis from Malta. *Food Chem.,* **2011**, *126*, 1431-1435. [http://dx.doi.org/10.1016/j.foodchem.2010.11.130]

[39] Bilikova, K.; Popova, M.; Trusheva, B.; Bankova, V. New anti-Paenibacillus larvae substances purified from propolis. *Apidologie (Celle),* **2013**, *44*, 278-285. [http://dx.doi.org/10.1007/s13592-012-0178-1]

[40] Li, F.; He, Y.M.; Awale, S.; Kadota, S.; Tezuka, Y. Two new cytotoxic phenylallylflavanones from Mexican propolis. *Chem. Pharm. Bull. (Tokyo),* **2011**, *59*(9), 1194-1196. [http://dx.doi.org/10.1248/cpb.59.1194] [PMID: 21881271]

[41] Li, F.; Awale, S.; Tezuka, Y.; Esumi, H.; Kadota, S. Study on the constituents of Mexican propolis and their cytotoxic activity against PANC-1 human pancreatic cancer cells. *J. Nat. Prod.,* **2010**, *73*(4), 623-627. [http://dx.doi.org/10.1021/np900772m] [PMID: 20307087]

[42] Lotti, C.; Campo Fernandez, M.; Piccinelli, A.L.; Cuesta-Rubio, O.; Márquez Hernández, I.; Rastrelli, L. Chemical constituents of red Mexican propolis. *J. Agric. Food Chem.,* **2010**, *58*(4), 2209-2213. [http://dx.doi.org/10.1021/jf100070w] [PMID: 20121106]

[43] Castro, M.L.; do Nascimento, A.M.; Ikegaki, M.; Costa-Neto, C.M.; Alencar, S.M.; Rosalen, P.L. Identification of a bioactive compound isolated from Brazilian propolis type 6. *Bioorg. Med. Chem.,* **2009**, *17*(14), 5332-5335. [http://dx.doi.org/10.1016/j.bmc.2009.04.066] [PMID: 19497755]

[44] Mayworm, M.A.S.; Lima, C.A.; Tomba, A.C.B.; Fernandes-Silva, C.C.; Salatino, M.L.F.; Salatino, A. Does Propolis Contain Tannins? *eCAM,* **2014**, *2014*

[45] Agüero, M.B.; Svetaz, L.; Sánchez, M.; Luna, L.; Lima, B.; López, M.L.; Zacchino, S.; Palermo, J.; Wunderlin, D.; Feresin, G.E.; Tapia, A. Argentinean Andean propolis associated with the medicinal plant Larrea nitida Cav. (Zygophyllaceae). HPLC-MS and GC-MS characterization and antifungal

activity. *Food Chem. Toxicol.,* **2011**, *49*(9), 1970-1978.
[http://dx.doi.org/10.1016/j.fct.2011.05.008] [PMID: 21600954]

[46] Agüero, M.B.; Gonzalez, M.; Lima, B.; Svetaz, L.; Sánchez, M.; Zacchino, S.; Feresin, G.E.; Schmeda-Hirschmann, G.; Palermo, J.; Wunderlin, D.; Tapia, A. Argentinean propolis from Zuccagnia punctata Cav. (Caesalpinieae) exudates: phytochemical characterization and antifungal activity. *J. Agric. Food Chem.,* **2010**, *58*(1), 194-201.
[http://dx.doi.org/10.1021/jf902991t] [PMID: 19916546]

[47] Lotti, C.; Piccinelli, A.L.; Arevalo, C.; Ruiz, I.; Migliani, D.E.; Figueira, R. DeSáL.; Tessis, A.C.; Ferreira-Pereira, A.; Rastrelli, L. Constituents of Honduran Propolis with Inhibitory Effects on Saccharomyces cerevisiae Multidrug Resistance Protein Pdr5p. *J. Agric. Food Chem.,* **2012**, *60*, 10540-10545.
[http://dx.doi.org/10.1021/jf302578r] [PMID: 23004023]

[48] Righi, A.A.; Alves, T.R.; Negri, G.; Marques, L.M.; Breyer, H.; Salatino, A. Brazilian red propolis: unreported substances, antioxidant and antimicrobial activities. *J. Sci. Food Agric.,* **2011**, *91*(13), 2363-2370.
[http://dx.doi.org/10.1002/jsfa.4468] [PMID: 21590778]

[49] Righi, A.A.; Negri, G.; Salatino, A. Comparative Chemistry of Propolis from Eight Brazilian Localities. *eCAM,* **2013**, *2013*

[50] Falcão, S.I.; Vale, N.; Gomes, P.; Domingues, M.R.; Freire, C.; Cardoso, S.M.; Vilas-Boas, M. Phenolic profiling of Portuguese propolis by LC-MS spectrometry: uncommon propolis rich in flavonoid glycosides. *Phytochem. Anal.,* **2013**, *24*(4), 309-318.
[http://dx.doi.org/10.1002/pca.2412] [PMID: 23172843]

[51] Zhang, C.P.; Zheng, H.Q.; Hu, F.L. Extraction, partial characterization, and storage stability of β-glucosidase from propolis. *J. Food Sci.,* **2011**, *76*(1), C75-C79.
[http://dx.doi.org/10.1111/j.1750-3841.2010.01941.x] [PMID: 21535657]

[52] Zhang, C.P.; Liu, G.; Hu, F.L. Hydrolysis of flavonoid glycosides by propolis β-glycosidase. *Nat. Prod. Res.,* **2012**, *26*(3), 270-273.
[http://dx.doi.org/10.1080/14786419.2010.541877] [PMID: 21851328]

[53] Popova, M.; Trusheva, B.; Cutajar, S.; Antonova, D.; Mifsud, D.; Farrugia, C.; Bankova, V. Identification of the plant origin of the botanical biomarkers of Mediterranean type propolis. *Nat. Prod. Commun.,* **2012**, *7*(5), 569-570.
[PMID: 22799077]

[54] Simone, M.; Evans, J.D.; Spivak, M. Resin collection and social immunity in honey bees. *Evolution,* **2009**, *63*(11), 3016-3022.
[http://dx.doi.org/10.1111/j.1558-5646.2009.00772.x] [PMID: 19619221]

[55] Evans, J.D.; Pettis, J.S. Colony-level impacts of immune responsiveness in honey bees, *Apis mellifera. Evolution,* **2005**, *59*(10), 2270-2274.
[http://dx.doi.org/10.1111/j.0014-3820.2005.tb00935.x] [PMID: 16405170]

[56] Simone-Finstrom, M.D.; Spivak, M. Increased resin collection after parasite challenge: a case of self-medication in honey bees? *PLoS One,* **2012**, *7*(3), e34601.
[http://dx.doi.org/10.1371/journal.pone.0034601] [PMID: 22479650]

[57] Bastos, E.M.; Simone, M.; Jorge, D.M.; Soares, A.E.; Spivak, M. *In vitro* study of the antimicrobial activity of Brazilian propolis against *Paenibacillus larvae. J. Invertebr. Pathol.,* **2008**, *97*(3), 273-281. [http://dx.doi.org/10.1016/j.jip.2007.10.007] [PMID: 18054037]

[58] Antúnez, K.; Harriet, J.; Gende, L.; Maggi, M.; Eguaras, M.; Zunino, P. Efficacy of natural propolis extract in the control of American Foulbrood. *Vet. Microbiol.,* **2008**, *131*(3-4), 324-331. [http://dx.doi.org/10.1016/j.vetmic.2008.04.011] [PMID: 18508208]

[59] Mihai, C.M.; Mărghitaş, L.A.; Dezmirean, D.S.; Chirilă, F.; Moritz, R.F.; Schlüns, H. Interactions among flavonoids of propolis affect antibacterial activity against the honeybee pathogen *Paenibacillus larvae. J. Invertebr. Pathol.,* **2012**, *110*(1), 68-72. [http://dx.doi.org/10.1016/j.jip.2012.02.009] [PMID: 22386493]

[60] Kamel, A.A.; Moustafa, A.A.; Nafea, E.A. Propolis as a natural antibiotic to control American foulbrood disease in honey bee colonies. *Afr. J. Agric. Res.,* **2013**, *8*, 3047-3062.

[61] Wilson, M.B.; Brinkman, D.; Spivak, M.; Gardner, G.; Cohen, J.D. Regional variation in composition and antimicrobial activity of US propolis against *Paenibacillus larvae* and *Ascosphaera apis. J. Invertebr. Pathol.,* **2015**, *124*, 44-50. [http://dx.doi.org/10.1016/j.jip.2014.10.005] [PMID: 25450740]

[62] Garedew, A.; Lamprecht, I.; Schmolz, E.; Schricker, B. The varroacidal action of popolis: a laboratory assay. *Apidologie (Celle),* **2002**, *33*, 41-50. [http://dx.doi.org/10.1051/apido:2001006]

[63] Garedew, A.; Schmolz, E.; Schricker, B.; Lamprecht, I. Microcalorimetric investigation of the action of propolis on *Varroa destructor* mites. *Thermochim. Acta,* **2002**, *382*, 211-220. [http://dx.doi.org/10.1016/S0040-6031(01)00737-7]

[64] Damiani, N.; Fernández, N.J.; Maldonado, L.M.; Alvarez, A.R.; Eguaras, M.J.; Marcangeli, J.A. Bioactivity of propolis from different geographical origins on *Varroa destructor* (Acari: Varroidae). *Parasitol. Res.,* **2010**, *107*(1), 31-37. [http://dx.doi.org/10.1007/s00436-010-1829-7] [PMID: 20336318]

[65] Damiani, N.; Maggi, M.D.; Gende, L.B.; Faverin, C.; Eguaras, M.J.; Marcangeli, J.A. Evaluation of the toxicity of a propolis extract on *Varroa destructor* (Acari: Varroidae) and *Apis mellifera* (Hymenoptera: Apidae). *J. Apic. Res.,* **2010**, *49*, 257-264. [http://dx.doi.org/10.3896/IBRA.1.49.3.05]

[66] Popova, M.; Reyes, M.; Le Conte, Y.; Bankova, V. Propolis chemical composition and honeybee resistance against *Varroa destructor. Nat. Prod. Res.,* **2014**, *28*(11), 788-794. [http://dx.doi.org/10.1080/14786419.2014.881366] [PMID: 24483289]

CHAPTER 3

Chemical Composition of Bee Pollen

Maria G. Campos[1,*], Lokutova Olena[2], Ofélia Anjos[3,4]

[1] *Drug Discovery Group, Center for Pharmaceutical Studies, Faculty of Pharmacy & Chemistry Center FCT/ University of Coimbra, Health Sciences Campus, Azinhaga de Santa Comba, 3000-548 Coimbra, Portugal*

[2] *National University of Life and Environmental Sciences of Ukraine, Kiev, Ukraine*

[3] *IPCB/ESA – Instituto Politécnico de Castelo Branco, Escola Superior Agrária, Castelo Branco, Portugal*

[4] *CEF/ISA/UTL – Centro de Estudos Florestais, Instituto Superior de Agronomia, Universidade Técnica de Lisboa, Lisboa, Portugal*

Abstract: Bee pollen, usually used as an important source of nutrients and micronutrients for the young bees in the hive, is also an important food for humans. This product is very rich in proteins, lipids, free sugars, carbohydrates, and it contains trace amounts of minerals, phenolic acids, flavonoids and a good range of vitamins. A brief look at bee pollen composition, it is easily recognised that it is a balanced food that can be used as a stand-alone food or as a nutritional supplement or even as a medicinal product. Several bioactivities, due to some of these compounds, were studied in bee pollen samples from different floral sources and the results conduce to important properties. The amount and diversity of micronutrients could induce vast benefits if used for health purposes following a complete risk assessment. Nevertheless, the results pointing towards the encouraged use of bee pollen, the risk assessment of some floral species containing toxic compounds has not been fully studied to insure the safety of consumption for all the gathered flowers, so this will also be discussed in this chapter. Admiration for its goodness and medicinal properties, bee pollen has been consumed for centuries, however, currently the efficacy and safety for all consumed products, foods, supplements or medicines is an important tool to guarantee correct quality control and essential to add value to the product.

[*] **Correspondence to M.G. Campos:** Faculty of Pharmacy, University of Coimbra, Health Sciences Campus, Azinhaga de Santa Comba, 3000-548 Coimbra, Portugal. Tel: + 351 239 488 484; Fax: +351 239 487 362; Email: mgcampos@ff.uc.pt.

Susana M. Cardoso (Ed)

To summarise, in this chapter we will put the situation of gaps in bee pollen research into some kind of perspective, outlining some important points and discussing in more depth the implications of collecting samples, chemical composition and risk assessment.

Keywords: *Apis mellifera*, Chemical composition, Collecting, Dietary product, Food, Gametophyte, Medicinal product, Micronutrients, Nutrients, Nutritional supplement, Pollen, Risk assessment.

1. INTRODUCTION

Bee pollen is flower pollen collected by the honey bee, *Apis mellifera,* for the purpose of feeding its larvae in the early stages of development. Collected flower pollen is accumulated as pellets (corbicular pollen) in pouches on the rear legs of the bee and it is the mixture of these pellets that comprises bee pollen.

Pollen itself is the male gametophyte in flowers. The female gametophyte produces nectar, a sweet liquid gathered and eaten by insects and other animals. This important substance, rich in sucrose and water, is the optimal environment necessary for the germination of the pollen tube and the release of its DNA in the female organ. For the fertilization of the flower minerals are also needed. We also suppose that polyphenolic compounds can be important in this step, perhaps in an allelopatic way, as they are species-specific [1].

Bees are very selective in the flora they choose to collect pollen. In fact some research has pointed out that for each *genera* they prefer only certain *species*, sometimes probably only one [2]. Spontaneous plants are the main floral resources selected. In any case they also pollinate some breeding plants such as the fructiferous trees. For example, bees can be "forced" to pollinate kiwis (*Actinidia deliciosa*), but once these insects have found a better source of pollen from a spontaneous plant, they leave the fields. In fact kiwis are genetically modified and they do not have any mate in nature to do the pollination.

Genetically modified (GM) plants are an important issue associated to bee pollens because their impact on bees, as on Humans is still unclear. Theoretically, bees do not visit breeding plants, but in practice, for example, they collect maize (*Zea mays* L.) pollen that is one of the major genetically modified plants. In these plants the introduction of *Bt* toxin from *Bacillus thuringiensis* is mainly in pollen and the impact on insects and humans is still controversial. *Bt* toxin is one of many microbial pesticides. Its formulated fermentation cultures can be sprayed on foliage to control selected insects because the ubiquitous bacterium synthesizes a toxic protein, known as the delta-endotoxin, every time it stops growing and produces a spore [3, 4]. The insecticidal gene that molecular biologists moved into corn, cotton, or potatoes is actually a truncated version of the natural gene. For the gene to function in plant cells, small snippets of DNA are attached that allow the code to be read. To track the location of the gene and to help select plant cells that have successfully incorporated the gene into their chromosome, marker genes encoding for either antibiotic or herbicide resistance are also spliced onto the toxic protein gene. We need to know how much protein we might be exposed to when eating food made from transgenic corn [4]. This is also true for bee pollen from maize that is commercialized for human consumption. Fifteen years ago researchers studied the sensibility of *Apis mellifera* and in the protocol they concluded that *Bt* toxin did not affect bees [5]. Nevertheless up until now, we have considered all the results in this field as preliminary and a further wait is necessary to see what will happen. Those toxins are the compounds that can be found in pollen, so reporting on these aspects is the new point that needs to be highlight. In the meantime humans continue to use the product in simultaneous to these evaluations.

2. BEE POLLEN COLLECTING FOR HUMAN INTAKE

The definition of "Bee pollen" is the result of the agglutination of flower pollens, made by the worker bees, with nectar (and/or honey) and salivary substances, which is collected at the entrance to the hive [6].

The concept of honeybees is usually associated to honeybees collecting nectar, however there are many other import products collected from the hive. Pollen is a crucial part of the honeybees′ diet, providing a wide range of nutrients namely protein, lipids, carbohydrates, vitamins, and minerals, *e.g.* pollen is the major source of amino acids in their diet, although it does have many other constituents. Before discussing some of them in detail we will give an overview on pollen grains themselves.

Sporopollenin is a biopolymer very resistant, which protects pollen from environmental stressors. However, bees produce some specific enzymes in their digestive system that allows the grains to split. Most of the pollen collected is to be eaten by nurse bees that use the nourishing compounds absorbed from it to secrete royal jelly from their hypopharyngeal glands. The royal jelly will feed the young larvae, including workers and drones (only up to the 3rd day of larvae phase) and the queens that receive a steady diet of royal jelly throughout their development and during their whole life. Then, about three days later, the royal jelly is mixed with bee bread (a mixture of pollen, honey, and enzymes) until they spin their cocoons.

In the colony, the bees usually only specifically collect either pollen or nectar on a single trip, although it is possible that some bees, can collect both products on the same journey, depending on the beekeeper practices. Pollen is stuffed into hairy receptacles on their hind legs called corbiculae.

Once back at the hive, pollen is stored in cells close to the nest [7]. During the bees more active season, pollen is stored for only a few days, however during the winter it is stored for much longer.

The relationship between bees and honey plants is very important because bees can collect 10-1000 more pollen than other insects; individual bees can go in and out of a hive 20-30 times a day and a colony can visit 20 to 30 million flowers during a season [8].

In order to allow the bee pollen to be gathered, a trap is used at the entrance of the hive with small holes [9] where bees can go through but the pellets cannot. While crawling inside the hive, bees push themselves through the holes and the stick pellets on their legs remain at the entrance falling into a draw. The draw containing the pollen loads is removed (Fig. **1**) and these are cleaned, dried and properly conditioned for sale.

When beekeepers trap the pollen loads at the entrance of the hive, the manipulation of this product by bees is near null because they only agglutinated the pollen grains with nectar in order to make pellets. The different bee pollen pellets are a result of the diverse flora visited and they usually have different colours too.

Fig. (1). Pollen pellets.

During individual trips from the hive or nest, bees regularly visit flowers of only one kind, ignoring others, unless the supply is very restricted. The flower constancy of bees is one of the factors restricting interspecific hybridization [10]. When they do not harvest nectar from the flowers, they carry honey from the hive for the same purpose.

The condition of harvest and storage is crucial to have a good quality product at the end of the production chain for human consumption. If bee pollen is not stored properly it loses its nutritional value and becomes susceptible to moulds and bacteria development given its high moisture content. The recommendation to avoid some of these issues is to dry the product under 40 °C, but bee pollen can also be consumed fresh. Note still that the storage of the fresh product requires refrigeration temperatures of at least between 5 and 10 °C. Barajas *et al.* [11], recommend a storage temperature under 0ºC or even lower. Nonetheless, the drying of bee pollen is necessary to extend its shelf life. This however must be done carefully in order to maintain its nutritional value [11].

The health benefits of bee pollen will be discussed in the Section of Health-associated benefits of honeybee plant-derived products. As with everything else, there are risks associated to the intake of this product and these will be discussed at the end of this chapter.

Moving on to questions related to digestibility of bee pollen associated to the release of nutrients, there are still some doubts about the percentage of constituents available for human consumption. As referred to above, the digestibility of pollen exine is almost null because this structure is very tough and as a consequence, some constituents would not be completely absorbed. For some authors these facts will compromise the benefits of the product. However, almost all pollens are operculated and these holes (apertures) allow the release of the content inside the exine. Both arguments have merit and to split the difference this research gap needs to be filled in the near future.

3. CHEMICAL COMPOSITION OF BEE POLLEN

Given the standing of bee pollen as nutrient, there is still a lack of legislation regarding physicochemical, hygienic and sanitary quality control for this product to be accepted worldwide. The possible validation in the near future of certain floral origins of bee pollen for medicinal purposes, will imply a rigorously and deeply monitored regulation of its quality [6]. The majority of the Standard Methods for bee pollen quality control and research will be published in the Coloss Bee Book, vol III in the current year.

Pollen has different natural functions in nature and each of these is associated to a specific macro and micronutrients. For the reproduction of some flora and to feed larvae and insects, pollen is involved in the development of life and the equilibrium of nutrients as glucids (carbohydrates), lipids, proteins and free amino acids (including the entire essential amino acids) is understandable.

The content in macronutrients makes bee pollen a good complement to the daily diet, mainly due to the well balanced proportions of proteins, fats and carbohydrates. According to Villanueva *et al.* [12], bee pollen has a low caloric value, usually averaging about 381.70 ± 14.69 kcal/100 g of pollen, which can serve the purposes of hypocaloric dietetic strategies. Similar results were found by others [13].

The high variability of the bee pollen nutrient composition is given by the great variability of the floral species present in the blends, collection at distinct seasons of the year, as well as with the different methods of analysis [7, 14]. It is known that the major *taxon* on the mix of bee-collected pollen makes a difference in its average constituents; hence being reflected in different reported values.

An overall chemical composition of pollen can be presented as an average of nutrients, namely 10-40% of proteic material (from which approximately half is in the form of free amino acids which can be assimilated immediately by the body); 13-55% of carbohydrates, of which about 40% are simple sugars (*e.g.* fructose and glucose) coming from the added nectar and/or honey used to bind the pollen grains together, while about 21% are polysaccharides that cannot be absorbed by the Human organism, including cellulose, hemicelluloses, lignin, sporopolenin and others[2]; 1-13% of lipids and a variety of secondary plant components such as minerals, vitamins, phenolics including flavonoids, sterols, steroids, organic acids [7, 9, 15, 16]. However, outside values from those indicated above can be found in literature. In particular, Bogdanov [17] reported that carbohydrates, crude fibers, proteins and lipids ranged between 13 and 55%, 0.3 and 20%, 10 and 40%, 1 and 10%, respectively. In addition, according to [14] and depending on the pollen type and region, values of pollen protein content could vary between 5% and 30%, while content ranges for carbohydrate and lipids were between 10% and 40% and 1% to 5% , respectively.

Notably, bee pollen loads contain twenty-two basic amino acids in their composition [18, 19]. A summary of the content of specific amino acids from pollen loads of distinct floral origins collected in Ukraine [20] is presented in (Table **1**), albeit readers must be aware that these values are only average, since they are affected by several factors such as *e.g.* geographic region and the time of collection [14]. In a similar way, values in (Table **2**) intend to be indicative for the lipids content of bee pollen loads of different floral resources.

Table 1. Amino acids content in bee pollen loads of distinct floral origin collected in Ukraine [20].

Amino acids (g/100g)	***Trifolium pratense***	***Sinapis alba***	***Phacelia tanacetifolia***	***Malus domestica***	***Taraxacum officinale***	***Papaver rhoeas***
Total	21.53	17.30	20.81	21.20	18.0	10.25
Irreplaceable amino acids						
Arg	0.94	0.88	1.10	1.00	1.00	0.37
Val	1.30	1.15	1.30	1.40	1.10	0.62
His	0.49	0.40	0.50	0.64	0.47	0.39
Ile	0.86	0.88	1.0	1.00	0.82	0.41
Leu	1.50	1.30	1.70	1.70	1.30	0.74
Lys	1.30	1.30	1.40	1.40	1.50	1.00
Met	0.34	0.26	0.41	0.47	0.39	0.19
Thr	1.00	0.95	0.94	1.20	1.00	0.38
Phe	0.9	0.86	0.94	1.00	0.88	0.47
Total	8.63	7.98	9.29	9.81	8.46	4.57
Replacement amino acids						
Ala	1.20	1.00	1.20	1.20	1.10	0.74
Asp	2.50	1.90	2.30	2.20	1.90	0.98
Gly	0.94	0.97	1.00	1.00	0.96	0.63
Glu	2.94	2.50	2.70	2.90	2.50	1.1
Pro	3.10	1.50	2.20	1.10	1.20	1.00
Ser	1.20	1.00	1.00	1.20	1.10	0.63
Tyr	0.94	0.26	1.00	0.94	0.50	0.44
Cys	0.08	0.20	0.12	0.25	0.28	0.26
Total	12.9	9.33	11.52	11.39	9.54	5.68

In general, the major part (60%) of the fatty acid in bee pollen is in the free form. Bound fatty acids were characterised by a high content of α-linolenic acid (70%), and followed by small amounts of linoleic and oleic acid. Palmitic acid is the most abundant amongst saturated fatty acids [21]. Bee pollen contains the essential n-3 fatty acid, alpha-linolenic acid and good ratios of PUFA/SFA and n-6/n-3 fatty acids [22]. It is important to highlight that these polyunsaturated fatty acids (PUFAs), such as omega-3 and omega-6, are not synthesized by humans and thus, must be included in the food diet. In this context, pollen and beebread could compensate for the imbalance of fatty acids in the diet.

Table 2. Lipids content in bee pollen loads (pellets) for different floral resources (mean standard variation) collected in Ukraine [20].

Total lipids, fatty acids (g/kg)	*Helianthus annuus*	*Scilla bifolia*	*Artemisia*	*Salix caprea*	*Acer negundo*	*Alnus glutinosa*	*Fagopyrum esculentum*	*Papaver rhoeas*	*Taraxacum officinale*
Total lipids	90.84± 0.961	72.82± 0.745	32.78± 0.364	58.59± 0.568	63.53± 0.628	22.08± 0.247	78.29± 0.744	81.17± 0.796	117.72± 1.080
Caprylic, 8:0	-	-	-	-	-	0.19± 0.002	0.19± 0.002	-	-
Capric, 10:0	-	-	-	-	-	0.20± 0.002	0.18± 0.002	-	-
Lauric, 12:0	1.42± 0.015	0.04± 0.001	0.12± 0.002	0.13± 0.002	0.56± 0.005	0.02± 0.001	0.11± 0.002	0.02± 0.001	3.45± 0.024
Myristic, 14:0	0.51± 0.004	0.04± 0.001	0.01± 0.001	0.03± 0.001	0.04± 0.001	0.02± 0.001	0.03± 0.001	0.09± 0.001	0.35± 0.002
Pentadecanoic,15:0	0.30± 0.002	0.11± 0.001	0.10± 0.001	0.16± 0.001	0.72± 0.006	0.02± 0.001	0.04± 0.001	0.03± 0.001	0.02± 0.001
Palmitic, 16:0	11.49± 0.089	6.95± 0.062	2.15± 0.021	4.04± 0.039	5.57± 0.052	2.15± 0.020	4.11± 0.039	6.47± 0.057	6.42± 0.052
Palmitoleic, 16:1	0.93± 0.007	0.56± 0.004	0.13± 0.001	0.32± 0.003	0.44± 0.003	0.17± 0.002	0.32± 0.003	0.44± 0.004	0.04± 0.001
Stearic, 18:0	3.74± 0.030	1.76± 0.011	0.23± 0.002	0.31± 0.003	0.83± 0.007	1.42± 0.010	0.48± 0.004	0.58± 0.005	3.26± 0.028
Oleic, 18:1	10.16± 0.072	9.05± 0.078	2.83± 0.023	3.27± 0.025	3.09± 0.020	0.98± 0.007	1.08± 0.009	5.68± 0.050	5.32± 0.047
Linoleic, 18:2	13.69± 0.084	6.09± 0.052	2.97± 0.025	6.15± 0.056	21.86± 0.168	3.54± 0.031	4.10± 0.036	11.48± 0.092	14.34± 0.104
Linolenic, 18:3	40.92± 0.328	35.76± 0.315-	1.78± 0.014	20.06± 0.165	16.31± 0.122	5.85± 0.042	27.89± 0.228	39.89± 0.316	33.28± 0.268

The composition of fatty acids in pollen, as for other constituents, depends on plant species too [19, 23, 24]. As an example, Avni *et al.* [25] when working with pollen from Israel found a narrow range of 2.3% to 6.6% of total fatty acids in bee pollen mixtures, while Arruda *et al.* [26] reported the range of 4.6-6.1% of total fatty acids for seven mixed bee pollen samples of subtropical origin.

In Portugal Féas *et al.* [27] also found linolenic as the principal fatty acid, followed by linoleic acid, palmitic acid and oleic acid. The phenolic and flavonoid contents varied from 12.9 to 19.8 mg of gallic acid equivalents/g of extract and from 4.5 to7.1 mg of catechin equivalents/g of extract, respectively.

Furthermore pollen is also an important source of vitamins. Several vitamins in bee pollen like B complex vitamins, ascorbic acid, along with vitamins A, D and E were identified in their composition. Still, it is important to note that vitamin B1 activity in biological samples is not only due to itself but also to the mono-, di- and triphosphate derivatives of thiamine, with prevalence of mono- and diphosphate forms. Herbert *et al.* [28] verified that the seasonal thiamine levels in bee-collected pollen fluctuated greatly depending on the floral source and time of year. Nevertheless, this vitamin has been shown to exist for up to 4 years unmated in stored pollen [29].

Bee pollen is also an important source of minerals depending on the type of flower pollen used. As known, minerals are very important for human well-being; however they cannot be synthesized in the human body and must instead be obtained through the diet. These compounds can be found in different kinds of foods and are commonly differentiated in major elements (*e.g.* Ca, Cl, K, Mg, N, Na, P and S), minor and trace elements (*e.g.* Fe, B, Br, I and Si). However, some minerals can be toxic if the intake is high or over a long period, (*e.g.* Zn, Se, Mn and Mo). Others at excessive levels may exhibit high toxicity [30 - 33].

Stanciu *et al.* [34] analysed concentrations of selected macro and microelements in bee pollen with specific floral origin, harvested from Transylvania (Romania), by flame atomic spectrometry (F-AES/AAS). These authors confirmed that the examined bee pollen is a natural source of nutritionally essential minerals and can contribute to a better balanced diet or used in special therapeutic applications. In their work, the authors reported that the main macro elements were potassium (2483-7620 $mg.kg^{-1}$), calcium (553-2798 $mg.kg^{-1}$) and magnesium (205-3555 $mg.kg^{-1}$), while the main microelements were iron (18.8-135 $mg.kg^{-1}$) and zinc 18.8-60.5 $mg.kg^{-1}$. Variations in the mineral levels of the analysed monofloral bee pollens were due to differences in the floral origin.

Amongst the literature data, one can point out that the highest values of potassium were reported by Stanciu *et al.* [34] in *Salix* sp. pollen (8686 mg.kg^{-1}), followed by *Helianthus annuus* L. pollen (5421 mg.kg^{-1}), and for multifloral bee pollen, different authors reported different values for pollen potassium concentration, namely: 4000 mg.kg^{-1} [12], 5530 mg.kg^{-1} [35], 2843-5976 mg.kg^{-1} [36] and 4950-5131 mg.kg^{-1} [37].

Yang *et al.* [38] analysed twelve common varieties of monofloral bee pollen collected from China's main producing regions. On average the values obtained were the following: P 5946, K 5324, Ca 2068, Mg 1449, sodium 483, aluminium 129, iron 119, manganese 70, zinc 45, and cupper 17. Overall, the mean values for the main elements were similar but significant differences were observed for the minor and trace elements, depending of the region.

Differences in mineral composition were also found by Paulo *et al.* [39] when analysing pollen from different botanical origins. The concentrations of micronutrients (Cu, Fe, Mn and Zn) differed between the three analysed plant species (*Cistus ladanifer* L., *Rubus ulmifolius* Schott, and *Calluna vulgaris* (L.) Hull) but were similar in the two geographic areas of the same country. These authors concluded that the determination of micronutrients in pollen could be suitable for the identification of botanical species. The research for element profile, in combination with machine learning techniques, can be a promising approach to help in the identification of botanical sources of pollen.

The contents of Cd, Pb, As and Hg were usually lower or not detected in pollen samples. In Campos *et al.* [18], the limits for these trace elements were: Cd $\leq$ 0.1 mg.kg^{-1}, Pb $\leq$ 0.5 mg.kg^{-1}, As $\leq$ 0.5 mg.kg^{-1}, and Hg $\leq$ 0.03 mg.kg^{-1}.

According to Sager [40], the analysis of materials from different regions of the world (Austria, Bulgaria, Great Britain, Greece, Canada, Poland, Hungary, Croatia, Czech Republic) showed that bee products can be used as biomarkers of secondary pollution, at least for the elements Pb, Cd and Cu.

Moving on to another group of compounds belonging to Bee pollen composition, phenolic compounds are beneficial for human health given the free radical scavenging power that is related to a possible decrease of risk of degenerative diseases [41 - 45]. Polyphenolic compounds are powerful antioxidants and in pollen a vast group of flavonoids like flavones and flavonol glycosides can be found in the majority of the species. Among phenolic acids, gallic and caffeic acids are the most frequently detected in pollens and, most of the time, they appear as polymers in the majority of the *taxa* visited by bees for bee pollen collection [45]. Several studies in pure or mixed pollen reported important data for bee pollen antioxidative activity [43, 46 - 52].

The previous authors [43, 46 - 52] identified various polyphenols such as flavones, flavonols, isoflavones, flavonones, anthocyanins, catechin and isocatechin in different pollen samples. Derivatives of kaempferol and quercitin at the glycosidic forms are the more common structures. Despite being minor constituents of pollen, therapeutic and protective effects of bee pollen have been related to the content of polyphenols in pollen, turning these compounds into crucial components [7].

Furthermore, other minor constituents in certain pollens are toxic, as is the case of pyrrolizidine alkaloids (PAs) that are plant defence compounds. Boppré *et al.* [54, 55] found these pyrrolizidine alkaloids in pollen and/or pollen baskets of bees as rather limited and ranged from 6 to 14 000 $mg.g^{-1}$. Kempf *et al.* [56] found 1.08 to 13.36 ($\mu g.g^{-1}$) of PA in samples of *Echium* spp from different provenances. Pyrrolizidine alkaloids content of pollen must be very well monitored, because this group of compounds occur more frequently (31%) in pollen than in honey (9%) and their concentration is also usually higher (on average 5.17 $mg.g^{-1}$*versus* 0.056 $mg.g^{-1}$, calculated as retronecine equivalents) [53, 57]. More than 350 different PA structures are known, however all of them are ester alkaloids composed of a necine base esterified to one or more necic acids [53].

Bee pollens also frequently contain small quantities of moulds. Most of these moulds produce caprylate esterase–lipase, leucine aminopeptidase, acid phosphatase, phosphoamidase, β–glucosidase and *N*–acetyl–β–glucosamidase. A high percent of the isolates (50%) from all sources gave positive reactions for alkaline phosphatase. The majority of the moulds identified were *Aspergilli* (17%), *Mucorales* (21%) and *Penicillia* (32%) [various authors *in*58]. In general, the number of isolates decreased in pollen as it was collected and stored by bees.

Each type of pollen samples appeared to differ with regard to mould flora and dominant species. *Aureobasidium pullulans, Penicillium corylophilum, Penicillium crustosum and Rhizopus nigricans* were among the moulds that may have been introduced by bees during the collection and storage of pollen.The *Mucor* sp., the dominant mould in floral pollen, was not found in corbicular pollen and bee bread. Thus, as has happened with yeast and *Bacillus* ssp, the old flora of corbicular pollen and bee bread maybe the result of microbial inoculations by bees and chemical changes in pollen resulting from additions by bees from the regurgitation of honey sac contents and secretions of glands as well as microbial fermentation which allow some species, but not others, to survive. Potential microbial spoilage of pollen provisions might be controlled by antibiotic substances produced by the normal microflora, bees, pollen and/or honey [58]. These results could be somehow involved in the antibiotic potential of bee pollen.

4. POSSIBLE RISKS ASSOCIATED TO THE CONSUMPTION OF BEE POLLEN

Bee pollen is consumed directly or processed and its extracts exhibit various pharmacological and microbiological properties, which can be correlated to the content of various compounds essential for life support [58]. However, this product as with other natural substances still needs further investigation and attention in order to be used for human purposes [58]. This field has medicinal requirements, and still there are plenty of research gaps that need to be filled as soon as possible.

Herbal preparations are marketed as natural and safe options in conventional medicines for the prevention and treatment of a variety of diseases. However, consumers may not be fully aware of their potential side effects. Nowadays this issue is controversial because people self–medicate based on many of these products (sold without quality control) and sometimes mix these with medicines for the same diseases, which can cause very complicated herbal–drug interactions (www.oipm.uc.pt). These facts are not new. For example, fifteen years ago Shad *et al.* [59] reported two cases of acute hepatitis after the ingestion of herbal preparations. One of the mixtures included chaparral and bee pollen and another was pure bee pollen. Chaparral (*Larrea tridentate*) itself has been reported to have similar effects in other patients, but for bee pollen the authors did not find reports of acute hepatitis. In the same review the authors pointed out another clinical case involving a 33 year old woman who developed acute hepatitis after taking 1 ounce (approximately 28 grams) of pure bee pollen daily for several months and the case of a 69 year old man who developed acute hepatitis after taking 14 tablets daily of a mixed herb preparation containing bee pollen, chaparral, and 19 other ingredients. Both patients made a full recovery within 6 to 8 weeks of discontinuing the herbal products.

The problem related with the above clinical situations is because bee pollen is consumed without control and for several days, even years, and the possibility remains that a chronic exposure may occur that can conduce to chronic toxicity if toxic compounds which are part of the constituents (as for example, hepatotoxic alkaloids). However many other compounds not yet identified could induce an acute toxicity as may have been the case described by Shad *et al.* [59].

Pyrrolizidine alkaloids (PAs) in pollen found by Boppré *et al.* [54, 55] are hepatotoxic and the maximum allowed for human consumption is not completely clear, especially for bee pollen blends. The Committee on Herbal Medicinal Products (HMPC) from the European Medicinal Agency in a document from the 24th of November 2014 (EMA/HMPC/893108/2011) with the title Public statement on the use of herbal medicinal products containing toxic, unsaturated pyrrolizidine alkaloids (PAs) refers that "It has become apparent during the assessment of *certain plants* that the risk assessment of pyrrolizidine alkaloids (PAs) poses considerable difficulties, with several PAs being regarded as both hepatotoxic and carcinogenic". Considering that PAs are natural constituents of a number of plants used for medicinal purposes and that PAs might be part of the food chain, the HMPC decided to prepare a public statement on the use of herbal preparations, of which 35 contained PAs. As a conclusion of this statement, the following was published on the 24th of November 2014 (EMA/HMPC/893108/2011, Committee on Herbal Medicinal Products) [60] on page 15 in relation to honey and pollen:

"...*The levels of toxic unsaturated PAs and N–oxides found in many honeys could, according to published risk assessments, cause chronic diseases such as liver* *cirrhosis, pulmonary hypertension and cancer if these honeys are regularly consumed at the recommended serving sizes of 15 – 25 g. PA levels up to 3900 μg.kg^{-1} honey were found. In the United Kingdom the highest honey consumers are infants eating up to 32 g/day of honey, school children consuming up to 60 g/day and adults eating as much as 92 g/day [61]. If honey contains 2500 μg.kg^{-1} of PAs with two average serving sizes of 40 g a person would be exposed to 100 μg PAs/day. This would exceed the recommended doses. It has been reported that a woman who consumed 20 – 30 μg of PAs/day during her pregnancy gave birth to a child suffering fatal liver damage [62]. Kempf et [56, 63, 64] reported that 17 (31%) of 55 commercial bee pollen products purchased in Europe have been found to contain 1080 – 16350 μg PA/kg. The authors have calculated, based on a 30% probability of PA occurrence, that consumption of the recommended daily amount of 10 g of bee pollen would expose an average consumer to 15 μg (retronecine equivalents) of PAs*."

Some data can be arguable but the recommendations of the EMA/HMPC, because of their known involvement in human poisoning and their putative carcinogenicity, exposure to toxic unsaturated PAs should be kept as low as practically achievable.

Another important issue that should be considered in this nutritive food and potential medicinal product is the fact that it has been previously found to cause anaphylactic reactions, especially when used as a food supplement by people at risk. Greenberger and Flais [65] described a study with an atopic patient who experienced a non-life–threatening anaphylactic reaction upon her initial ingestion of bee pollen. Microscopic examination of the pollen sample and ELISA inhibition assays were performed. The patient had a 7 mm/28 mm wheal/erythema reaction to bee pollen at 1 mg.mL^{-1} concentration. Bee pollen caused 52% inhibition of IgE binding to short ragweed and 55% to ryegrass. Microscopic analysis of the bee pollen revealed ragweed, *Alternaria, Cladosporium*, honeysuckle (*Lonicera* sp), privet shrub (*Ligustrum* sp), and vetch (*Vicia sativa*). In addition, an unknowingly sensitized atopic patient experienced an anaphylactic reaction after ingesting a small quantity of bee pollen that contained pollens and fungi. Previously administered allergen immunotherapy that had reduced rhinitis symptoms did not prevent this allergic reaction.

Moreover, a 34–year–old Spanish woman with a lifelong history of seasonal rhinoconjunctivitis and honey intolerance (pyrosis and abdominal pain) developed astenia, anorexia, abdominal pain, diarrhoea, peripheral blood hypereosinophilia and elevated serum total IgE levels, 3 weeks after starting ingestion of bee pollen. A duodenal biopsy showed eosinophilic infiltration of the mucosal layer. Other causes of hypereosinophilia were not found. Repeated parasitological stool studies, as well as a duodenal aspirate showed negative results. Symptoms, hypereosinophilia and elevated IgE levels were resolved after bee pollen ingestion was stopped. This is a typical case of eosinophilic gastroenteritis by ingestion of bee pollen in a woman with intolerance to honey bee, because the patient fulfilled the usual diagnostic criteria: gastrointestinal symptoms were present, eosinophilic infiltration of the digestive tract was demonstrated by biopsy, no eosinophilic infiltration of other organs was found and the presence of parasites was excluded. Honey intolerance and/or bee pollen administration should be considered as a cause of eosinophilic gastroenteritis [66].

More recent research and according to some authors [67], bee pollen under controlled and supervised situations, could be used for the treatment of oral desensitization of children who have allergies and some cases of benign prostatitis.

To avoid the risk situations, primary care physicians are remind to ask their patients about herbal use and discuss with them their potential toxicities and patients should not forget to talk about bee pollen too.

CONCLUSION

Bee pollen is an important crude material that can have several applications for human purposes and a lot of research has been carried out in the last 3 decades to validate the product as with good practices of collecting and storage and standard methods for quality control. The main floral sources of bee pollen now have good data about the chemical composition and bioactivity. Nevertheless, some gaps need to be filled, especially concerning efficacy and safety. The World Health Organization has as a strategy for the next years (2014–2023) the total integration of complementary therapies in Health Systems and the use of herbal medicines too. Pollens from selected flora, with medicinal activities could be used under strict legislation but to give this status to bee pollen (monofloral blends) the "General Guidelines of Safety and Efficacy" needs to be followed and gaps in research filled as soon as possible.

CONFLICT OF INTEREST

The authors confirm that authors have no conflict of interest to declare for this publication.

ACKNOWLEDGEMENTS

The authors wish to thank to "Projeto Estratégico – RG–Centro–177–3717 da 434 FCT" and FCT UID/QUI/00313/2013 Center of Chemistry from Faculty of Sciences and Technology of University of Coimbra (Portugal) and Centro de Estudos Florestais that is a research unit funded by Fundação para a Ciência e a Tecnologia (Portugal) within UID/AGR/UI00239/2013.

The authors would like to express their gratitude to Isabele Salavessa (IPCB Languages centre) for the English revision.

REFERENCES

[1] Markham, K.R.; Campos, M. 7- and 8-O–methylherbacetin–3–O–sophorosides from bee pollens and some structure/activity observations. *Phytochemistry,* **1996**, *43*(4), 763-767. [http://dx.doi.org/10.1016/0031-9422(96)00286-5]

[2] Campos, M.; Markham, K.R.; Mitchell, K.A.; da Cunha, A.P. An approach to the characterization of bee pollens *via* their flavonoid/phenolic profiles. *Phytochem. Anal.,* **1997**, *8*(4), 181-185. [http://dx.doi.org/10.1002/(SICI)1099-1565(199707)8:4<181::AID-PCA359>3.0.CO;2-A]

[3] Schmitz, G.; Bartsch, D.; Pretscher, P. Selection of relevant non-target herbivores for monitoring the environmental effects of Bt maize pollen. *Environ. Biosafety Res.,* **2003**, *2*(2), 117-132. [http://dx.doi.org/10.1051/ebr:2003007] [PMID: 15612277]

[4] Felsot, A. Herbicide Tolerant Genes: Part 1: Squaring Up Roundup Ready Crops. *Agrichem Environ News,* **2000**, *173*, 8-15.

[5] Malone, L.A.; Burgess, E.P.; Stefanovic, D. Effects of a Bacillus thuringiensis toxin, two Bacillus thuringiensis biopesticide formulations, and a soybean trypsin inhibitor on honey bee (*Apis mellifera* L.) survival and food consumption. *Apidologie (Celle),* **1999**, *30*(6), 465-473. [http://dx.doi.org/10.1051/apido:19990601]

[6] Campos, M.G.; Bogdanov, S.; de Almeida-Muradian, L.B.; Szczesna, T.; Mancebo, Y.; Frigerio, C.; Ferreira, F. Pollen composition and standardisation of analytical methods. *J. Apic. Res.,* **2008**, *47*(2),

154-161.
[http://dx.doi.org/10.3896/IBRA.1.47.2.12]

[7] Almeida-Muradian, L.B.; Pamplona, L.C.; Coimbra, S.; Barth, O.M. Chemical composition and botanical evaluation of dried bee pollen pellets. *J. Food Compos. Anal.,* **2005**, *18*(1), 105-111. [http://dx.doi.org/10.1016/j.jfca.2003.10.008]

[8] Dongock, D.N.; Tchoumboue, J.; Ricciardelli D'Albore, G.; Youmbi, E.; Pinta, Y.J. Spectrum of melliferous plants used by Apis mellifera adansonii in the Sudano Guinean western highlands of Cameroon. *Grana,* **2007**, *46*(2), 123-128. [http://dx.doi.org/10.1080/00173130701318459]

[9] Krell, R. Value-added products from beekeeping. Rome: FAO Agricultural Services Bulletin 124. In: *Food and Agriculture Organization*; , **1996**; pp. 87-113.

[10] Cane, J.H. Specialist Osmia bees forage indiscriminately among hybridizing Balsamorhiza floral hosts. *Oecologia,* **2011**, *167*(1), 107-116. [http://dx.doi.org/10.1007/s00442-011-1977-1] [PMID: 21468665]

[11] Barajas, J.; Cortes-Rodriguez, M.; Rodríguez-Sandoval, E. Effect of Temperature on the Drying Process of Bee Pollen From Two Zones of Colombia. *J. Food Process Eng.,* **2012**, *35*(1), 134-148. [http://dx.doi.org/10.1111/j.1745-4530.2010.00577.x]

[12] Orzáez Villanueva, M.T.; Díaz Marquina, A.; Bravo Serrano, R.; Blaźquez Abellán, G. Mineral content of commercial pollen. *Int. J. Food Sci. Nutr.,* **2001**, *52*(3), 243-249. [http://dx.doi.org/10.1080/09637480020027000-3-1] [PMID: 11400473]

[13] Colin, L.J.; Jones, C.E. Pollen Energetics and Pollination Modes. *Am. J. Bot.,* **1980**, *67*(2), 210-215. [http://dx.doi.org/10.2307/2442644]

[14] Llnskens, H.F.; Jorde, W. Pollen as food and medicine - A review. *Econ. Bot.,* **1997**, *51*(1), 78-86. [http://dx.doi.org/10.1007/BF02910407]

[15] Loper, G.M.; Standifer, L.N.; Thompson, M.J.; Gilliam, M. Biochemistry and microbiology of bee-collected almond (Prunus dulcis) pollen and bee bread. In: *I- Fatty Acids, Sterols Minerals, Vitamins and Minerals. Apidologie*; Springer Verlag: Germany, **1980**; pp. 63-73.

[16] Orzáez Villanueva, M.T.; Díaz Marquina, A.; Bravo Serrano, R.; Blazquez Abellán, G. The importance of bee-collected pollen in the diet: a study of its composition. *Int. J. Food Sci. Nutr.,* **2002**, *53*(3), 217-224. [http://dx.doi.org/10.1080/09637480220132832] [PMID: 11951585]

[17] Bogdanov, S. Pollen: Nutrition, Functional Properties, Health: A Review. In: *Bee Product Science*; Science, B.P., Ed.; , **2011**.

[18] Szczesna, T. Protein content and amino acid composition of bee-collected pollen from selected botanical origins. *J Apic Sci.,* **2006**, *50*(2), 81-90.

[19] Čeksterytė, V.; Račys, J.; Kaškonienė, V.; Venskutonis, P.R. Fatty acid composition in beebread. *Biologija (Vilnius),* **2008**, *54*(4), 253-257. [http://dx.doi.org/10.2478/v10054-008-0052-2]

[20] Lokutova, O.A. *Evaluation of bee pollen load on species, contents of nutritious substance and morphological characteristics of pollen grains.,* **2006**.

[21] Seppänen, T.; Laakso, I.; Wójcicki, J.; Samochowiec, L. An analytical study on fatty acids in pollen extract. *Phytother. Res.,* **1989**, *3*(3), 115-116. [http://dx.doi.org/10.1002/ptr.2650030311]

[22] Estevinho, L.M.; Rodrigues, S.; Pereira, A.P.; Feás, X. Portuguese bee pollen: palynological study, nutritional and microbiological evaluation. *Int. J. Food Sci. Technol.,* **2012**, *47*(2), 429-435. [http://dx.doi.org/10.1111/j.1365-2621.2011.02859.x]

[23] Saa-Otero, M.P.; Díaz-Losada, E.; Fernández-Gómez, E. Analysis of fatty acids, proteins and ethereal extract in honeybee pollen - Considerations of their floral origin. *Grana,* **2000**, *39*(4), 175-181. [http://dx.doi.org/10.1080/00173130051084287]

[24] Odoux, J-F.; Feuillet, D. Aupinel. P.; Loublier, Y.; Tasei, J-N.; Mateescu, C. Territorial biodiversity and consequences on physico-chemical characteristics of pollen collected by honey bee colonies. *Apidologie (Celle),* **2012**, *43*(5), 561-575. [http://dx.doi.org/10.1007/s13592-012-0125-1]

[25] Avni, D.; Hendriksma, H.P.; Dag, A.; Uni, Z.; Shafir, S. Nutritional aspects of honey bee-collected pollen and constraints on colony development in the eastern Mediterranean. *J. Insect Physiol.,* **2014**, *69*, 65-73. [http://dx.doi.org/10.1016/j.jinsphys.2014.07.001] [PMID: 25038311]

[26] De Arruda, V.A.; Pereira, A.A.; de Freitas, A.S.; Barth, O.M. de Almeida-Muradian, L.B. Dried bee pollen: B complex vitamins, physicochemical and botanical composition. *J. Food Compos. Anal.,* **2013**, *29*(2), 100-105. [http://dx.doi.org/10.1016/j.jfca.2012.11.004]

[27] Feás, X.; Vázquez-Tato, M.P.; Estevinho, L.; Seijas, J.A.; Iglesias, A. Organic bee pollen: botanical origin, nutritional value, bioactive compounds, antioxidant activity and microbiological quality. *Molecules,* **2012**, *17*(7), 8359-8377. [http://dx.doi.org/10.3390/molecules17078359] [PMID: 22785265]

[28] Herbert, E.W.; Vanderslice, J.T.; Huang, M.H.; Higgs, D.J. Levels of thiamine and its esters in bee collected pollen using liquid chromatography and robotics. *Apidologie (Celle),* **1987**, *18*(2), 129-136. [http://dx.doi.org/10.1051/apido:19870203]

[29] Hagedorn, H.H. Effect of the age of pollen used in pollen supplements on their nutritive value for the honeybee. II. Effect of vitamin content on pollens. *Apic Res.,* **1968**, *7*, 97-101.

[30] Ihnat, M. A survey of methods of analysis for minerals in feedstuffs. *J. Anim. Sci.,* **2003**, *81*(12), 3218-3225. [PMID: 14677879]

[31] Orzáez Villanueva, M.T.; Díaz Marquina, A.; Bravo Serrano, R.; Blaźquez Abellán, G. Mineral content of commercial pollen. *Int. J. Food Sci. Nutr.,* **2001**, *52*(3), 243-249. [http://dx.doi.org/10.1080/09637480020027000-3-1] [PMID: 11400473]

[32] Bonvehí, J.S.; Jordà, R.E. Nutrient composition and microbiological quality of honeybee-collected pollen in spain. *J. Agric. Food Chem.,* **1997**, *45*(3), 725-732. [http://dx.doi.org/10.1021/jf960265q]

[33] Yang, K.; Wu, D.; Ye, X.; Liu, D.; Chen, J.; Sun, P. Characterization of chemical composition of bee

pollen in China. *J. Agric. Food Chem.,* **2013**, *61*(3), 708-718. [http://dx.doi.org/10.1021/jf304056b] [PMID: 23265625]

[34] Stanciu, O.G.; Marghitas, L.A.; Dezmirean, D.; Campos, M.G. Specific distribution of minerals in selected unifloral bee pollen. *Food Science and Technology Letters,* **2012**, *3*(1), 27-31.

[35] Somerville, D.C.; Nicol, H.I. Mineral content of honeybee-collected pollen from southern New South Wales. *Aust. J. Exp. Agric.,* **2002**, *42*(8), 1131. [http://dx.doi.org/10.1071/EA01086]

[36] Szczesna, T. Concentration of selected elements in honeybee-collected pollen. *J Apic Sci.,* **2007**, *51*(1), 5-13.

[37] Salamanca, G.G.; Pérez, C.R.; González, V.E. Origen botánico propiedades fisicoquimicas microbiológicas del polen colectado en algunas zonas apícolas de la Campiña de Boyacá. *V Congreso Español de Ingeniería de Alimentos,* **2008**.Barcelona

[38] Yang, K.; Wu, D.; Ye, X.; Liu, D.; Chen, J.; Sun, P. Characterization of chemical composition of bee pollen in China. *J. Agric. Food Chem.,* **2013**, *61*(3), 708-718. [http://dx.doi.org/10.1021/jf304056b] [PMID: 23265625]

[39] Paulo, L.; Antunes, P.; Anjos, O. Mineral composition of pollen using inductively coupled plasma atomic emission spectroscopy. *Planta Med.,* **2014**, *80*(16), 2-P16. [http://dx.doi.org/10.1055/s-0034-1394851]

[40] Sager, M.; Maleviti, E. *Elemental Composition of Honeys from Greece-Possible Use as Environmental Indicators.,* **2014**.

[41] Bogdanov, S. Quality and standards of pollen and beeswax. *Apiacta,* **2004**, *38*, 334-341.

[42] Silva, B.M.; Andrade, P.B.; Valentão, P.; Ferreres, F.; Seabra, R.M.; Ferreira, M.A. Quince (*Cydonia oblonga* Miller) fruit (pulp, peel, and seed) and Jam: antioxidant activity. *J. Agric. Food Chem.,* **2004**, *52*(15), 4705-4712. [http://dx.doi.org/10.1021/jf040057v] [PMID: 15264903]

[43] Campos, M.G.; Frigerio, C.; Lopes, J.; Bogdanov, S. What is the future of Bee-Pollen? *J ApiProduct ApiMedical Sci.,* **2010**, *2*(4), 131-144. [http://dx.doi.org/10.3896/IBRA.4.02.4.01]

[44] Erhardt, A.; Baker, I. Pollen amino acids-an additional diet for a nectar feeding butterfly? *Plant Syst. Evol.,* **1990**, *169*, 111-121. [http://dx.doi.org/10.1007/BF00935989]

[45] Rebiai, A.; Lanez, T.; Belfar, M.L. Determination of caffeic acid and gallic acid in Algerian bee pollen by an HPLC method. *Phytochem BioSub J,* **2014**, *8*(3), 190-197.

[46] Morais, M.; Moreira, L.; Feás, X.; Estevinho, L.M. Honeybee-collected pollen from five Portuguese Natural Parks: palynological origin, phenolic content, antioxidant properties and antimicrobial activity. *Food Chem. Toxicol.,* **2011**, *49*(5), 1096-1101. [http://dx.doi.org/10.1016/j.fct.2011.01.020] [PMID: 21291944]

[47] Almaraz-Abarca, N.; Campos, M.; Ávila-Reyes, J.A.; Naranjo-Jiménez, N.; Corral, J.H.; González-Valdez, L.S. Antioxidant activity of polyphenolic extract of monofloral honeybee-collected pollen from mesquite (*Prosopis juliflora, Leguminosae*). *J. Food Compos. Anal.,* **2007**, *20*(2), 119-124.

[http://dx.doi.org/10.1016/j.jfca.2006.08.001]

[48] Carpes, S.T.; Begnini, R.; Alencar, S.M.; Masson, M.L. Study of preparations of bee pollen extracts, antioxidant and antibacterial activity. *Cienc. Agrotec.,* **2007**, *31*(6), 1818-1825. [http://dx.doi.org/10.1590/S1413-70542007000600032]

[49] Mărghitaş, L.A.; Stanciu, O.G.; Dezmirean, D.S.; Bobis, O.; Popescu, O.; Bogdanov, S.; Campos, M.G. *In vitro* antioxidant capacity of honeybee-collected pollen of selected floral origin harvested from Romania. *Food Chem.,* **2009**, *115*(3), 878-883. [http://dx.doi.org/10.1016/j.foodchem.2009.01.014]

[50] Lopes, J.; Stanciu, O.G.; Campos, M.G.; Muradian, L.B.; Marghitas, L.A. Bee pollen antioxidant activity - a review: Achievements and further challenges. *J Phcog,* **2011**, *2*(2), 25-38.

[51] Anjos, O.; Amâncio, D.; Serrano, M.; Campos, M. **2014**.

[52] Amâcio, D.; Serrano, M.; Anjos, O.; Campos, M. Therapeutic potential of pollen *Planta Med,* **2014**, *80*(16), P2B37. [http://dx.doi.org/10.1055/s-0034-1394914]

[53] Kempf, M.; Heil, S.; Hasslauer, I.; Schmidt, L.; von der Ohe, K.; Theuring, C.; Reinhard, A.; Schreier, P.; Beuerle, T. Pyrrolizidine alkaloids in pollen and pollen products. *Mol. Nutr. Food Res.,* **2010**, *54*(2), 292-300. [http://dx.doi.org/10.1002/mnfr.200900289] [PMID: 20013884]

[54] Boppré, M.; Colegate, S.M.; Edgar, J.A. Pyrrolizidine alkaloids of Echium vulgare honey found in pure pollen. *J. Agric. Food Chem.,* **2005**, *53*(3), 594-600. [http://dx.doi.org/10.1021/jf0484531] [PMID: 15686407]

[55] Boppré, M.; Colegate, S.M.; Edgar, J.A.; Fischer, O.W. Hepatotoxic pyrrolizidine alkaloids in pollen and drying-related implications for commercial processing of bee pollen. *J. Agric. Food Chem.,* **2008**, *56*(14), 5662-5672. [http://dx.doi.org/10.1021/jf800568u] [PMID: 18553916]

[56] Kempf, M.; Reinhard, A.; Beuerle, T. Pyrrolizidine alkaloids (PAs) in honey and pollen-legal regulation of PA levels in food and animal feed required. *Mol. Nutr. Food Res.,* **2010**, *54*(1), 158-168. [http://dx.doi.org/10.1002/mnfr.200900529] [PMID: 20013889]

[57] Kempf, M.; Beuerle, T.; Bühringer, M.; Denner, M.; Trost, D.; von der Ohe, K.; Bhavanam, V.B.; Schreier, P. Pyrrolizidine alkaloids in honey: risk analysis by gas chromatography-mass spectrometry. *Mol. Nutr. Food Res.,* **2008**, *52*(10), 1193-1200. [http://dx.doi.org/10.1002/mnfr.200800051] [PMID: 18792927]

[58] Campos, M.G.; Frigerio, C.; Ferreira, F. Bee-Pollen Therapeutical Value. , **2010**.

[59] Shad, J.A.; Chinn, C.G.; Brann, O.S. Acute hepatitis after ingestion of herbs. *South. Med. J.,* **1999**, *92*(11), 1095-1097. [http://dx.doi.org/10.1097/00007611-199911000-00011] [PMID: 10586838]

[60] HMPC - Committee on Herbal Medicinal Products. Public statement on the use of herbal medicinal products containing toxic , unsaturated pyrrolizidine alkaloids (PAs) Public statement on the use of herbal medicinal products containing toxic, unsaturated pyrrolizidine alkaloids (PAs) Table of conten. **2014**, *44*, 1-22.

[61] Edgar, J.A.; Colegate, S.M.; Boppré, M.; Molyneux, R.J. Pyrrolizidine alkaloids in food: a spectrum of potential health consequences. *Food Addit. Contam. Part A Chem. Anal. Control Expo. Risk Assess.,* **2011**, *28*(3), 308-324. [http://dx.doi.org/10.1080/19440049.2010.547520] [PMID: 21360376]

[62] Rasenack, R.; Müller, C.; Kleinschmidt, M.; Rasenack, J.; Wiedenfeld, H. Veno-occlusive disease in a fetus caused by pyrrolizidine alkaloids of food origin. *Fetal Diagn. Ther.,* **2003**, *18*(4), 223-225. [http://dx.doi.org/10.1159/000070799] [PMID: 12835579]

[63] Kempf, M.; Wittig, M.; Schönfeld, K.; Cramer, L.; Schreier, P.; Beuerle, T. Pyrrolizidine alkaloids in food: downstream contamination in the food chain caused by honey and pollen. *Food Addit. Contam. Part A Chem. Anal. Control Expo. Risk Assess.,* **2011**, *28*(3), 325-331. [http://dx.doi.org/10.1080/19440049.2010.521771] [PMID: 20967664]

[64] Kempf, M. Schreier, P.; Reinhard, A. Beuerle, T. Pyrrolizidinalkaloide in Honig und Pollen. *J für Verbraucherschutz und Leb.,* **2010**, *5*(3-4), 393-406. [http://dx.doi.org/10.1007/s00003-009-0543-9]

[65] Greenberger, P.A.; Flais, M.J. Bee pollen-induced anaphylactic reaction in an unknowingly sensitized subject. *Ann. Allergy Asthma Immunol.,* **2001**, *86*(2), 239-242. [http://dx.doi.org/10.1016/S1081-1206(10)62698-1] [PMID: 11258697]

[66] Puente, S.; Iñíguez, A.; Subirats, M.; Alonso, M.J.; Polo, F.; Moneo, I. [Eosinophilic gastroenteritis caused by bee pollen sensitization]. *Med. Clin. (Barc.),* **1997**, *108*(18), 698-700. [PMID: 9324586]

[67] Morais, M.; Moreira, L.; Feás, X.; Estevinho, L.M. Honeybee-collected pollen from five Portuguese Natural Parks: palynological origin, phenolic content, antioxidant properties and antimicrobial activity. *Food Chem. Toxicol.,* **2011**, *49*(5), 1096-1101. [http://dx.doi.org/10.1016/j.fct.2011.01.020] [PMID: 21291944]

CHAPTER 4

Chromatography as a Tool for Identification of Bioactive Compounds in Honeybee Products of Botanical Origin

Marcelo D. Catarino[1,2], Jorge M. Alves-Silva[1], Soraia I. Falcão[3], Miguel Vilas-Boas[3], Micaela Jordão[1,2], Susana M. Cardoso[1,2,*]

[1] *QOPNA, Department of Chemistry, University of Aveiro, Campus Universitário de Santiago, 3810-193 Aveiro, Portugal*

[2] *CERNAS, School of Agriculture, Polytechnic Institute of Coimbra, Bencanta, 3045-601 Coimbra, Portugal*

[3] *CIMO, School of Agriculture, Polytechnic Institute of Bragança, Campus de St[a] Apolónia, Apartado 1172, 5301-855 Bragança, Portugal*

Abstract: Honey, propolis, and pollen are three important components of the beehive produced by honeybees mixing different plant parts (nectar, resin and pollen) with their own secretions, for further usage with different purposes in the hive. The fact that these natural products have been associated with numerous health benefits has attracted the attention of researchers resulting in a significant raise of scientific studies attesting their biological properties. Among the various constituents of honey, propolis and pollen, the phenolic compounds are the ones most frequently related to the beneficial properties of these products and hence, one of the main investigated groups. Their characterization is important to understand individual contribution(s) and synergistic effects of each compound for the overall biological effects of the bee product. To pursuit this goal, spectrophotometric techniques including HPLC, GC and TLC, alongside with the respective detection methods such as DAD, FLD and MS, have been developed and improved in order to offer better and more accurate separative performances.

* **Address correspondence to S.M. Cardoso:** Department of Chemistry, University of Aveiro, Campus Universitário de Santiago, 3810-193 Aveiro, Portugal; Tel: +351 234 370360; Fax: +351 234 370084; Email: susanacardoso@ua.pt

Susana M. Cardoso (Ed)

The aim of this review is to give an approach on the course that the chromatographic techniques have taken until the most recent trends on this field applied to the separation and characterization of the phenolic constituents of honey, propolis and bee pollen as well as an overall perspective of variability in terms of phenolic composition that can be found in the three bee products mentioned.

Keywords: Bee pollen, Benzoic acids, Bioactive compounds, Caffeic acid derivatives, Chromatography, Cinnamic acids, Coumaric acids, DAD, Flavonoids, FLD, GC, Honey, Honeybee-derived products, HPLC, MS, Phenolic compounds, Propolis, TLC.

1. INTRODUCTION

Honey, propolis, and bee pollen are produced from nectar, resin and floral pollen, respectively, mixed with different bee secretions and further used in the beehive for distinct purposes [1, 2]. Notably, these three bee products have also been used for centuries by Men, for food and medicinal purposes [3 - 5].

More recently, Men´s interest for these natural products have significantly raised, since scientific studies have attested their abundance on nutrients and bioactive compounds, together with their association with beneficial properties, including those of cardio-, neuro-, hepato- and chemo-protective, as well as chemo-preventive, antiseptic, antimicrobial, anti-allergic, antioxidant, anticancer, anti-radiation, anti-inflammatory and wound-healing activities, among many others [6 - 9]. Hence, overall, honey, propolis and pollen are now envised as very tempting and useful for a large spectrum of applications in different industries including foods, cosmetics, perfumes and pharmaceutics [7].

Among the numerous compounds from honey, propolis or bee pollen, the phenolic compounds are undoubtedly more frequently associated with the beneficial properties of these products [4, 5, 10, 11]. These compounds are a class of metabolites that are ubiquitously distributed through plant kingdom and plant-derived products [12], where they are important players in growth and reproduction, providing protection against pathogens and predators, besides contributing towards the color and sensory features of fruits, vegetables and their derived products [13].

Chemically, all the phenolic compounds possess at least one phenyl ring in its structure, and most of them arise from a common origin: the amino acids phenylalanine or tyrosine. These amino acids are deaminated to give cinnamic acids, further entering the phenylpropanoid pathway where one or more hydroxyl groups are added to the aromatic ring(s), ranging from simple phenols to complex compounds generally known as polyphenols or phenolic compounds [14]. The most common examples of phenolic compounds that can be found in foods include the phenolic acids (C_6-C_1), cinnamic acids (C_6-C_3) and flavonoids (C_6-C_3-C_6) [13] (Fig. **1**) and hence, in general these are also important groups of phenolic constituents of honey, propolis and bee pollen.

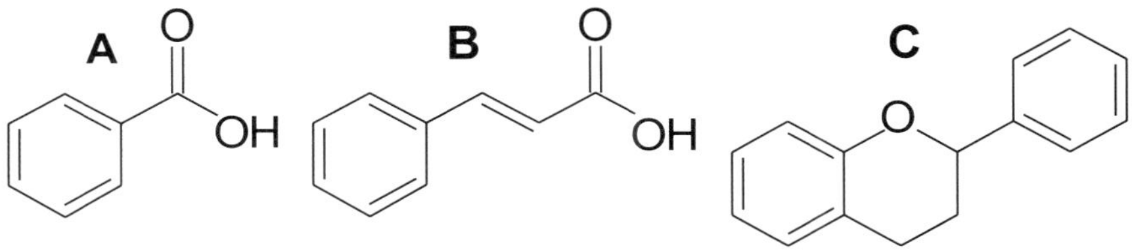

Fig. (1). Basic structure of phenolic acids (**A**), cinnamic acids (**B**) and flavonoids (**C**).

2. CHROMATOGRAPHIC METHODS

The close association between the beneficial properties of honey, propolis and pollen with their phenolic constituents boosted the need for characterizing them so that individual contribution(s) and synergistic effects on their biological activities can be elucidated.

Spectrophotometric assays, including Folin-Ciocalteu and Folin-Denis, for determination of the total phenolic content in plant samples, or reaction with $AlCl_3$ for total flavonoids measurement, are simple and economical, and can be useful for rapid and relatively inexpensive screening of numerous samples. However these techniques only give an estimation of the concentrations of the phenolic compounds over a certain minimum level and do not quantify phenolics individually. Besides, these reagents do not react specifically with phenols, since cross reactions commonly occur in complex samples and hence, unreliable data can be generated [15].

Therefore, it was necessary to replace these traditional methods by more efficient and precise separative equipment. In this field, chromatographic methods are irrefutably the most widely applied techniques due to their excellent separation ability and capability in analyzing multiple compounds in simultaneous [16]. These techniques are characterized by having two distinct phases, *i.e.*, a mobile phase consisting of a sample to be analyzed and a fluid/eluent that carries it out through an immobilized structure, which is known as the stationary phase. Based on differential partitioning between the mobile and stationary phases, the various constituents of the sample travel at different speeds, causing them to separate [17].

Further discussion of chromatographic techniques, namely high-performance liquid chromatography (HPLC), gas chromatography (GC) and thin-layer chromatography (TLC) will be carried out in this section, alongside with the detection systems since they are also important contributors for the characterization of polyphenols.

2.1. High-Performance Liquid Chromatography (HPLC)

Liquid chromatography (LC) is a separation technique that began in the early 1900s´. This technique was initially carried out in a glass cylinder packed with a fine powder and loaded with the sample mixture on top. An eluent was then poured into the column flowing down on it by gravity, dragging the sample compounds through the column at different speeds, causing them to separate [18].

HPLC is an upgraded technique from the traditional low pressure LC, which employ high operational pressures (up to 400 bar) [19]. In this case, smaller sample amounts and solvents are necessary and, therefore, the conventional HPLC typical column dimensions are 250 mm length x 4.6 mm diameter, filled with 5 μm particles, resulting in high plate numbers and consequent faster separations [20, 21]. All these features render this technique a superior resolving power when separating mixtures.

Several different separation modes of HPLC are known. Among them, normal-phase (NP), reverse-phase (RP), ion-exchange (IE) and size-exclusion (SE) are the major four chromatography techniques applied [21]. The NP-HPLC, also known as liquid-solid or adsorption chromatography, is the traditional separation method based on adsorption/desorption of the sample in a polar stationary phase (typically silica or alumina). On the other hand, the RP-HPLC functioning is quite the opposite of the NP-HPLC, *i.e.*, the separation is based on analytes' partition coefficients between a polar mobile phase and an hydrophobic (nonpolar) stationary phase [22]. Regarding to the IE- and SE-chromatography, as the method names imply, the former is based on the exchange of ionic analytes with the counter-ions of the ionic groups attached to the solid support [23], by while the latter is based in the separation of molecules according to their sizes, *i.e.*, smaller molecules migrate slower through the gel pores than the larger ones which are excluded from the pores migrating faster down the column [24].

Despite the previous four methods are the most common, other chromatographic techniques are known, including the affinity chromatography, chiral chromatography, hydrophilic and hydrophobic interaction chromatography, electrochromatography, supercritical fluid chromatography, among others.

The RP-HPLC is however the most popular and consistent analytical technique for characterization of polyphenolic compounds and hence, this prediction applies to honey, propolis and bee pollen samples (Table **1**). The earliest stationary phases used for this method were solid particles coated with nonpolar liquids, which were later replaced by more permanently bonded hydrophobic groups, made up of hydrophobic alkyl chains such as butyl (C4), octyl (C8) or octadecyl (C18) groups bounded on silica support [21, 25].

Table 1. Selected HPLC conditions for determination of phenolic compounds in honeybee-derived products from the last four years.

Stationary Phase	Mobile phase	Detection method	Reference
Honey			
Betasil RP-C18 column (150 mm × 4.6mm, 3 μm p.s.)	Various compositions of water, acetonitrile, methanol, and mixtures of 1% formic acid aqueous solution with methanol or acetonitrile	PDA	[4]
Luna C-18 RP column (250 × 4.6 mm, 5 μm p.s.)	A: acetonitrile B: 1% formic acid in water	UV-Vis	[30]
Zorbax SB-C18 column (250 mm × 4.6 mm, 5 μm p.s,)	A: 0.5% formic acid in water B: methanol	PDA and ESI-MS/MS	[31]

(Table 1) contd.....

Stationary Phase	Mobile phase	Detection method	Reference
Hypersil gold C18 (100 mm × 2.1 mm, 1.9 µm p.s.)	A: water + 0.1% formic acid B: acetonitrile + 0.1% formic acid	ESI-MS/MS	[32]
C18 column (4.6 × 250 mm, 5 µm p.s.)	A: methanol B: 5% formic acid in water	PDA	[33]
Gemini C18 110A column (150 mm × 4.60 mm, 3 µm p.s.)	A: 0.2 M phosphoric acid B: acetonitrile	PDA	[34]
KinetexTM C-18 column (100 × 2.1 mm, 2.6 µm p.s.)	A: 8 mmol.L^{-1} formic acid in water B: acetonitrile	ESI-MS/MS	[35]
Propolis			
Agilent Eclipse XDB-C18 column (4.6 mm × 150 mm, 5 µm p.s.)	A: acetonitrile B: 0.4% acetic acid in water	VWD	[36]
Zorbax Eclipse XDB C18 (50 × 2.1 mm, 1.8µm p.s.)	A: 0.1% formic acid in a water:acetonitrile 98:2 solution B: 0.1% formic acid in an acetonitrile:water 98:2 solution	ESI-MS/MS and NMR	[10]
WondasilTM column C18 (250 × 4.6 mm, 5 µm p.s.)	A: 1% formic acid in water B: methanol	PDA	[16]
Agilent Zorbax SBC18 column (250mm × 4.6mm, 5 µm p.s.)	A: water:methanol:formic acid (93:5:2) B: water:methanol:formic acid (3:95:2)	PDA	[6]
Zorbax Eclipse XDB C18 column (50 mm × 2.1 mm, 1.8 µm, p.s.)	A: 0.1% formic acid in a water:acetonitrile 98:2 solution B: 0.1% formic acid in a acetonitrile:water 98:2 solution	ESI-MS/MS	[37]
Bee Pollen			
RP-18 ODS-A column (250 mm × 4.6 mm, 5 µm p.s.)	A: acetic acid: water (1:20) B: methanol	PDA	[38]
RP Eclipse XDB-C8 column (150 mm × 4.6 mm, 5 µm p.s)	A: 0.1% formic acid in 5% acetonitrile B: 100% acetonitrile	PDA	[39]
RP Zorbax Eclipse XDB-C18 column (4.6 × 150 mm, 5-µm p.s.)	A: 0.5% acetic acid in acetonitrile:water (1:1) B: 2% acetic acid in water	UV-vis	[40]
RP C18 Zorbax 5B-RP-18 (Hewlett-Packard) column (4.6 × 250 mm, 5 µm p.s.)	A: 0.1% acetic acid in water B: methanol	PDA, and ESI-MS/MS	[41]
Thermo-Hypersil GOLD C18 RP column (250 mm × 4.0 mm, 5 µm p.s.)	A: ortho-phosphoric acid B: methanol	PDA	[42]

ESI-MS – Electrospray ionization mass spectrometer; NMR – Nuclear magnetic resonance; p.s. – Particle size; PDA – Photodiode array; RP – Reversed phase; UV-vis – Ultraviolet visible; VWD – Variable wavelength detector.

Due to the chemical complexity and similarity of polyphenols, the identification and quantification of these compounds is usually performed using a gradient elution of numerous mobile phases instead of an isocratic mode. However, the most common method usually consists of a binary system, where the mobile phase is composed of an acidified aqueous solution (with phosphoric, acetic or formic acids) and a less polar organic solvent (normally acetonitrile or methanol) [26]. The small quantity of acid added to the solvent system aims to suppress the ionization of phenolic and carboxylic groups, improving certain parameters such as retention time and resolution [25].

Despite HPLC is a very well established reliable technique, it could still suffer some improvements, especially concerning to the separation times which are usually long in food samples, bee products included. In general, higher efficiencies are obtained from narrower analytical columns diameters (10–150 μm) as well as from smaller particles dimensions (1.5–2 μm) [27]. However, despite the dramatic improvements in the sensitivity, resolution and speed of analysis, there are also some drawback aspects to be considered, being the major one the need for higher pressures in order to turn possible the passage of the mobile phase through the columns. Therefore, the creation of new instruments capable of generating and supporting pressures up to 1000 bar gave rise to what is currently known as ultra-performance LC (UPLC), a term that was firstly used by Waters Corporation [28]. With this new equipment, the same separation that would take over 20 min on RP-HPLC, can now be accomplished under 3 min [29].

Alongside with the separation techniques, the detection systems also play a very important role for the good functioning of these procedures. Particularly, the most common detection system for phenolics employed in HPLC are undoubtedly the ultraviolet–visible (UV–vis), photodiode array (PDA), and UV-fluorescence detectors (FLD) [15] (Table **1**). The existence of conjugated double and aromatic bonds, makes every phenol absorb in the UV or UV–vis region, ranging from 200–290 nm for benzoic acids and 190 to 380 nm for cinnamic acids, while the typical UV–vis spectra for flavonoids comprises two bands: the first with a maximum in the 240–285 nm range, which arises from the A-ring and the second in the 300–550 nm range, that arises from the B-ring. Therefore the most frequent wavelength used for identification of phenolics is 280 nm, although dual monitoring at 254 and 280 nm, or 280 and 320 nm, can be ideal wavelengths [43, 44].

Although less common, detection by FLD offers a higher selective and sensitive quantification of compounds (λ_{ex} = 280; λ_{em} = 310) [45]. This is very useful to determine phenolic compounds in complex sample matrixes such as honey, allowing the quantification of compounds at trace levels. Indeed, it has been reported that the detection limits for *p*-hydroxybenzoic acid and quercetin in honeys from different botanical origins ranged from 25 $ng.kg^{-1}$ to 0.75 $\mu g.kg^{-1}$, respectively [46].

HPLC coupled to mass spectrometry has tremendously improved the analysis of non-volatile species, including those of phenolic compounds [47]. This analytical technique engages the generation of charged molecules in the apparatus ion source that will be sorted by electromagnetic fields according to their mass-to-charge ratio in the mass analyzer and finally measured, usually by a quantitative method in the detector [48]. Although several MS ionization sources can be applied (*e.g.* electrospray ionization (ESI), atmospheric pressure chemical ionization (APCI), fast atom bombardment (FAB) and thermospray (TSP), ESI has been proven to be the most suitable one [49].

The mass spectrometry analysis of phenolic compounds are commonly performed in the negative ionization mode, since this provides the highest sensitivity and results in limited fragmentation of the molecular ion, making it most appropriate to inferring the molecular mass of the separated phenolic compounds, particularly when their concentrations are low [29]. Notably, the use of tandem MS^n in combination with collision-induced dissociation (CID) are crucial for elucidation compounds′ structure or even distinguish between isomers [50], as this technique allows multiple stages precursor ion *m/z* selection followed by product ion detection for successive fragmentation and generation of further product ions.

The coupling of LC to a nuclear magnetic resonance (NMR) spectroscopy is another alternative that has increasingly attracted attention. This technology was initially used in the late 1940s′ to clarify the structure of molecules in organic chemistry [51] and it consists in the submission of certain atomic nuclei to a magnetic field, allowing to exploit their magnetic properties [52, 53]. Notwithstanding, the low sensitivity, long run times and expensive instrumentation are the major weaknesses of this technique [54, 55].

2.2. Gas Chromatography (GC)

GC is also used in some instances for determining phenolic compounds both qualitatively and quantitatively. As observed in Table **2**, various literature reports describe the employment of this technique in its beginnings as an attempt to enable the determination of polyphenolic compounds either in honey, propolis or bee pollen [56 - 60]. The major drawback of GC in the analysis of phenolic compounds is the need for volatility, limiting the range of compounds that can be analyzed. To surpass this gap, compounds usually must pass through a derivatization process, *i.e.*, a chemical modification of the compounds to produce derivatives with properties that are more suitable for GC analysis [50]. Still, derivatization can be a challenge for analytes of interest in complex food matrixes since the presence of glycosides may interfere with their chemical modification [26]. Another alternative is the high-temperature, high-resolution gas chromatography (HT-HRGC) which is an established technique for separating complex mixtures and identifying high-molecular weight compounds that do not elute when analyzed on ordinary GC columns [15].

Earlier GC work was typically performed with flame ionization detection (FID), which is based on the detection of ions formed during combustion of organic compounds in an hydrogen flame [61 - 63], however the combination of GC with MS became widespread, since MS allows the acquisition of molecular mass data and structural information together with the identification of compounds. Notwithstanding, as reported by Araújo *et al.* [64], some of the compounds contained in propolis are not volatile enough for direct GC–MS analysis even upon derivatization or HT-HRGC–MS. Table **2** resumes the main characteristics applied in diverse studies using GC technique for analysis of phenolic compounds from honey, pollen and propolis.

Table 2. Selected GC conditions for the determination of phenolic compounds on bee-derived products from the last ten years.

Derivatization	Detection method	Column conditions	Reference
Honey			
----	MS	SPB-1 fused silica capillary column (25 m × 0.25 mm, 0.25 μm)	[58]
Methylation	FID-MS	HP-5MS capillary column (30 m × 0.25 mm, 0.25 μm c.t.)	[34]
Pyridine + BSTFA	MS	HP-5MS column (30 m × 0.25 mm, 0.25 μm c.t.)	[65]
BF_3 in methanol	FID	CBP1-Shimadzu non-polar column (20 m × 0.2 mm, 0.25 μm c.t.)	[66]
Propolis			
BSTFA	MS	Borosilicate capillary column (20 mm × 0.3 mm, 0.1 μm c.t.)	[15]
BSTFA	FID	SE-54 capillary column (9 mm × 0.25 mm)	Adapted from [44]
BSTFA	MS	Borosilicate capillary column (20 mm × 0.30 mm, 0.1 μm c.t.)	
BSTFA	FID	SE−54 fused-silica capillary column (9 m × 0.25 mm, 0.25 μm c.t.)	Adapted from [26]
Pyridine + BSTFA including 1% TMCS	MS	OV1 capillary column (25 m × 0.25 mm)	[59]
Pyridine + BSTFA	MS	DB1 column (30 m × 0.32 mm)	Adapted from [67]
Pyridine + BSTFA	MS	HP5-MS capillary column (23 m × 0.25 mm, 0.5 μm c.t.)	
Pyridine + BSTFA	MS	HP5-MS capillary column (23 m × 0.25 mm, 0.5 μm c.t.)	
Methylation	MS	HP1 methyl silicone capillary column (25 m × 0.25 mm)	
BSTFA	MS	Borosilicate capillary column (20 mm × 0.3 mm i.d.)	
Methylation	MS	CBP5 column (30 m × 0.25 mm i.d.)	
BSTFA	FID-MS	Glass column (22 m × 0.2 mm)	
----	FID-MS	Fused silica capillary 55 (10 m × 0.3 mm, 0.1 μc.t)	
Pyridine + BSTFA	MS	HP5-MS capillary column (23 m × 0.25 mm, 0.5 μm c.t.)	[68]
TMSi	MS	Zebron ZB-5HT column (30 m × 0.25 mm, 0.25 μm c.t.)	[69]

(Table 2) contd.....

Derivatization	Detection method	Column conditions	Reference
Bee Pollen			
----	MS	Varian Factor Four column (30 m × 0.25 mm)	[60]
Methylation with ethereal diazomethane	MS	HP-5MS capillary column (30 mm × 0.25 mm, 0.25 μm c.t.)	[38]
Pyridine + BSTFA	EI-MS	HP-5 capillary column (30 m × 0.25 mm, 0.25 μm c.t.)	[11]

AED – atomic emission detection; BF_3 – Boron trifluoride; BSTFA – Bis(trimethylsilyl) trifluoroacetamide; c.t. – coating thickness; EI-MS – Electron ionization mass spectrometer; FID – Flame ionization detector; MS- Mass spectrometer; PA – Polyacrylate fiber; TMCS – trimethylchlorosilane; TMSi – trimethylsilylimidazole.

2.3. Thin-Layer Chromatography (TLC)

TLC is another chromatographic technique which is widely used for qualitative analysis of organic compounds, isolation of the individual compounds from multi-component mixtures, quantitative analysis, and preparative-scale isolation [70]. This technique is fast, inexpensive and several samples can be examined at the same time, side by side, providing a chromatographic fingerprint of the sample which is very useful for identification purposes [15, 44]. In fact, with the adequate fractionated multi-component mixtures, video images of the chromatograms can be obtained, making this technique the only chromatographic method that enables presentation of the obtained results in the picture form [71].

Among the many available TLC pre-coated plates (*i.e.*, those with the inorganic adsorbent layers like silica or silica gel and alumina; organic layers like polyamide and cellulose; organic, polar covalently bonded modifications of the silica gel matrix such as diol, cyanopropyl, and aminopropyl; and organic, non-polar bonded modifications namely RP2, RP8, RP18), it is possible to choose a suitable stationary phase according to the sample characteristics [70]. For example, a classical stationary phase of silica gel is widely used to separate more apolar flavonoids such as flavonols and isoflavonoids from propolis [67].

Likewise, as the samples are eluted with different mobile phases, these may also be adapted to the sample of matter. As an example, for the quantification of flavonoids and phenolic acids in propolis, Medic-Saric *et al.* [72] used two-dimensional TLC with *n*-hexane/ethyl acetate/glacial acetic acid (31:14:5, nu/nu) (System A) and chloroform/methanol/formic acid (44:3.5:2.5) (System B) as mobile phases.

In this chromatographic method, visualization is usually read using a common wavelength performed in short- and long-wavelength UV light and in some cases spraying with different reagents, including methanolic diphenylboryloxyethylamine or ethanolic polyethyleneglycole 4000 [68, 73 - 75].

However TLC has some limitations including long development times, relative low reproducibility and moderate sensitivity [76, 77]. Moreover, the volatile mobile phase makes contact with the ambient atmosphere around the chamber, so factors such as humidity and temperature can affect the chromatogram [78]. In Table **3** it is possible to observe the main characteristics applied in diverse studies using TLC technique for analysis of phenolic compounds from propolis.

Table 3. Selected TLC conditions for the determination of phenolic compounds in propolis from the last ten years.

Mobile phase	Stationary phase	Detection method	Reference
n-Hexane-Ethyl acetate-Acetic acid (5:3:1, v/v/v)	Silica gel plates 60, 20 cm x 10 cm, 8 mm band	DART-MS	[79]
Ethyl acetate-Methanol-Water (75:15:0), Ethyl acetate-Formic acid-Water (80:10:10), Ethyl acetate-Formic acid-Acetic acid-Water	Silica	Densitometer	Adapted from [55]
Chloroform-Methanol-Formic acid (various v/v) n-Hexane-Ethyl acetate-Acetic acid (31:14:5)	Silica	UV	Adapted from [55]
n-Hexane-Ethyl acetate-Acetic acid, 31 + 14 + 5 (v/v), (mobile phase 1) or Chloroform-Methanol - Formic acid, 44 + 3.5 + 2.5 (v/v), (mobile phase 2)	20 cm x 20 cm silica gel 60 F254 plates	CAMAG Reprostar 3 densitometer	[72]
Petroleum ether/Ethyl acetate 7:3	20 cm x 20 cm silica gel 60 F254 plates	UV	[68]

3. TYPICAL PHENOLIC COMPOUNDS OF HONEYBEE PRODUCTS

The chemical composition of bee products of botanical origin is greatly dependent on the plants found nearby the hive, as well as on the geographic and climatic characteristics of the place. In this way, distinct samples of honey, propolis or bee pollen can greatly differ from each other with respect to their phenolic profile. Besides differences in the type of compounds, samples of honey, propolis or bee pollen also diverge with respect to the quantity of phenolics, even if they are from a close geographic region. *E.g.* Heather (*Calluna vulgaris*) honeys from different regions of Poland have been shown to contain a divergent total content of phenolic compounds (306 mg GAE/kg [80] and 698 mg GAE/kg [81]). Moreover, the three major propolis extracts available on the market are very distinct between them according to the presence of certain phenolic composition. The poplar type of propolis, also known as CAPE-based propolis and commonly found in European, eastern Asia and New Zealand regions, is typically abundant in caffeic acid phenethyl ester. Another very popular type of propolis is the green propolis from Brazil which is commonly found in areas where plants from *Baccharis* species are abundant and is particular rich in artepillin C. The third is the red propolis from Brazil or China which contains neither CAPE nor artepillin C and it is the less studied up to date [82]. Attending to the mentioned differences, as well as to the huge number of phenolic compounds present in bee products of botanical origin (*e.g.* more than 300 in propolis), this chapter will not be devoted to the deep comprehension of phenolic profiles of specific samples of honey, propolis and bee pollen but instead, it will focus on those more commonly found in these three products, as analysed by chromatographic techniques.

3.1. Non-Flavonoids

Caffeic acid and its derivatives are among the most frequently described non-flavonoids compounds in chromatographic analysis of honey, propolis and pollen samples. The identification of these compounds has been frequently performed by comparison of the retention time in RP-HPLC, together with spectral data gathered by PDA, which typically present wavelength maxima at approximately 324 nm, with a shoulder around 296 nm [83, 84]. These compounds are also commonly identified in ESI-MS^n analysis in the negative mode. The parent ion of caffeic acid in full MS spectra appears at *m/z* 179, which in turn fragments with a common base peak ion in MS^2 spectra at *m/z* 135, due to the loss of the carboxylic group, and other at *m/z* 161, which results from the loss of a water molecule [85 - 91]. These characteristic ions are also commonly observed in MS^n spectra of caffeic acid derivatives.

Table 4. Common hydroxycinnamic acids in honey, propolis and bee pollen worldwide.

Compound	**Honey (BO)**	**Propolis (BO)**	**Pollen (BO)**
Caffeic acid **(1)**	*Ziziphus spina-christi* [4, 121]; *Castanea sativa* [90, 95, 102, 122, 123]; *Robinia pseudoacacia* [35, 85, 88, 95, 107, 108, 122, 123,]; *Tilia* spp. [32, 35, 65, 88, 89, 95, 122]; *Eucalyptus* spp. [90, 95, 97, 102, 122]; *Lavandula* spp. [95, 122]; *Brassica* spp. [35, 88, 89, 95, 107, 122] ; *Helianthus annus* [88, 89, 95, 122]; *Rosmarinus officinalis* [92, 95, 122]; *Citrus* spp. [94 - 96, 102, 108, 122, 123]; *Hedysarum* spp. [102, 103, 108, 122, 123]; *Echium plantagineum* [122]; *Erica* spp. [46, 65, 90, 95, 122, ; *Calluna* spp. [35, 46, 65, 81, 90, 122]; *Rubus* spp. [97, 124]; *Leptospermum scoparium* [90, 93, 99, 106]; *Fagopyrum esculentum* [35, 65, 81, 85, 86, 88, 97]; honeydew [85, 89, 123]; *Melaleuca* spp. [93, 104, 125]; *Ananas comosus* spp. [104]; *Acacia* spp. [93, 97, 98]; *Satureja hortensis, Ailanthus altissima* [123]; *Thymus* spp. [65, 123]; *Turbina corymbosa, Ipomoea triloba, Avicennia germinans, Govania polygama, Lysiloma latisiquum* [101]; *Mimosa scabrella* [100]; *Trifollium* spp. 89; *Solidaga virgaurea* [88]; *Ocimum basilicum* 88; pine, hawthorn, nettle, black chokeberry, aloe 65; milk vetch, wild chrysanthemum, jujube, acacia [98]; multifloral/heterofloral [65, 92, 123]; unknown floral origin [105]	*Baccharis dracunculifolia* [109, 112 - 114, 126, 127]; *Populus* spp. [5, 10, 37, 87, 110, 111, 115, 116, 118, 119, 128]; Black propolis [126]; multifloral origin [117]	*Camelia sinensis*[120]; Multifloral [40]; *Cystus incanus* [129]

(Table 4) contd.....

Compound	Honey (BO)	Propolis (BO)	Pollen (BO)
3-Caffeoyl quinic acid **(2)**	*Ziziphus spina-christi*; *Calluna vulgaris* [81]; *F. esculentum* [81, 85, 88, 97]; *Melaleuca* spp. [104, 125]; *Ananas comosus* spp. [104]; *R. pseudoacacia* [85, 88, 107, 123]; *Brassica* spp. [88, 89, 107]; *Hedysarum* spp. [102, 123]; *Eucalyptus* spp. [97, 102]; milk vetch, wild chrysanthemum, jujube, acacia [98]; *Rubus idaeus, Acacia catechu* [97]; *C. sativa* [102, 123]; *S. hortensis, A. altissima, T. vulgare* [123]; *H. annus* [88, 89]; *Tilia* spp. [35,88, 89]; *Trifollium* [89]; *Ocimum basilicum, Solidaga virgaurea* [88]; honeydew [85, 89, 123]; heterofloral/multifloral [123]	*B. dracunculifolia* [109,126,130]; *Populus* spp. [110, 115] Black propolis [126]	*Camelia sinensis* [120]
4-Caffeoyl quinic acid **(3)**	-	*B. dracunculifolia* [130]	-
5-Caffeoyl quinic acid **(4)**	-	*B. dracunculifolia* [130]	-
3,5-Dicaffeoyl quinic acid **(5)**	-	*B. dracunculifolia* [130]	-
3,4-Dicaffeoyl quinic acid **(6)**	-	*B. dracunculifolia* [131]	-
4,5-Dicaffeyol quinic acid **(7)**	-	*B. dracunculifolia* [130]	-
Dicaffeoyl quinic acid	-	*B. dracunculifolia,* Black propolis [126]	-
3,4,5-Tricaffeoyl quinic acid **(8)**	-	*B. dracunculifolia* [130]	-
Tricaffeoyl quinic acid	-	*B. dracunculifolia,* Black propolis [126]	-
Chlorogenic acid derivatives	-	Multifloral origin [132]	-
Rosmarinic acid **(14)**	*C. vulgaris* [81]; *F. esculentum* [81, 97]; *R. idaeus, A. catechu, E. globules* [97]; *R. pseudoacacia, B. napus* [107]	-	-

(Table 4) contd.....

Compound	Honey (BO)	Propolis (BO)	Pollen (BO)
Prenyl caffeate **(15)**	*R. pseudoacacia, H. annus, T. cordata, O. Basilicum, F. esculentum, Solidago virgaurea, B. napus* [88]; *Mimosa scabrella* [100]	*Populus* spp. [87, 118, 133]	-
Phenethyl caffeate **(16)**	Chaste [134]; *R. pseudoacacia, H. annus, T. cordata, O. Basilicum, F. esculentum, Solidago virgaurea* [88]; *Brassica* spp. [88, 134]	*Populus* spp. [5, 87, 111, 117, 119, 133]	-
Cinnamyl caffeate **(17)**	-	*Populus* spp. [5, 37, 87, 118] Multifloral origin [132]	-
Benzyl caffeate **(18)**	-	*Populus* spp. [87, 118, 133]	-
Caffeic acid derivatives	*R. pseudoacacia, C. sativa, C. sinensis, E. camaldulensis, Erica* spp., *Lavandula* spp., *Tilia europea, Brassica* spp., *R. officinalis, H. annus* [95]	Multifloral origin [132], *B. dracunculifolia* [112]	-
Caffeic acid esters	*Erica* spp., *Brassica* spp., *Aesculus* spp., *Calluna* spp., *Helianthus* spp., *Rosmarinus* spp., *Abies* spp., *Frangula* spp., *Lavandula* spp., *Citrus* spp., *Rhododendron* spp., *Tilia* spp. [135]	-	-

(Table 4) contd.....

Compound	Honey (BO)	Propolis (BO)	Pollen (BO)
Ferulic Acid **(9)**	*Ziziphus spina-christi* [4, 121]; *R. pseudoacacia* [35, 85, 95, 107, 136]; *F. esculentum* [35, 85, 86, 97]; *C. vulgaris* [35]; *Tilia* spp. [35, 95]; *Brassica* spp. [35, 95, 107, 134; *Melaleuca* spp. [104, 125]; *Ananas comosus* spp. [104]; *Citrus* spp. [94, 95, 102, 136]; *R. idaeus* [97]; *Acacia* spp. [97, 121]; *Eucalyptus* spp. [95, 97, 102, 136]; *Hedysarum* spp. [102, 103]; *C. sativa* [95, 102, 136]; *Erica* spp., *R. officinalis, H. annus* [95]; chaste [134]; *Mimosa scabrella* [100]; *Turbinia corymbosa, Ipomoea triloba, Avicennia germinans, Govania polygama, Lysiloma latisiquum* [101]; *Abies alba, Quercus* spp. [136]; *Lavandula* spp. [95, 136]; *Prosopis julifora, L. scoparium* [121]; *Gochnatia* spp., *Croton* spp., *Vernonia* spp. [137]; *Gossypium hirsutum* [138]; *Quillaja saponaria* [139]; milk vetch, wild chrysanthemum, jujube, acacia [98]; multifloral/heterofloral [35, 121]; honeydew [85]	*B. dracunculifolia* [109], [112 - 114]; *Populus* spp. [5, 6, 10, 37, 110, 111, 115, 116, 118, 119, 128]; Multifloral origin [132]	*Camelia sinensis* [120]; multifloral [40, 140]

(Table 4) contd.....

Compound	Honey (BO)	Propolis (BO)	Pollen (BO)
p-Coumaric acid **(10)**	*Ziziphus spina-christi* [4, 121]; milk vetch, wild chrysanthemum, jujube, acacia [98]; *Prosopis julifora,* [121]; *Gochnatia* spp., *Croton* spp., *Vernonia* spp. [137]; *Gossypium hirsutum* [138]; *C. sativa* [95, 102, 122, 123]; *R. pseudoacacia* [35, 85, 88, 95, 107, 108, 122, 123, 141]; *Tilia* spp. [32, 88, 89, 95, 122, 142]; *Eucalyptus* spp. [95, 97, 122]; *Lavandula* spp. [95, 122]; *Brassica* spp. [35, 88, 89, 95, 107, 122 , 142]; *H. annus* [88, 898,122]; *R. officinalis* [92, 95, 122]; *Citrus* spp. [31, 92, 94, 95, 102, 108, 122, 123, 143]; *Hedysarium* spp. [102, 108, 122]; *Echium plantagineum* [122]; Heather (*Erica* spp./*Calluna* spp.) [35, 65, 81, 95, 122]; *F. esculentum* [35, 65, 81, 85, 86, 88, 97, 141]; *Melaleuca* spp. [93, 104, 125]; *Ananas comosus* spp. [104]; *L. scoparium* [31, 93, 99, 121]; *Acacia* spp. [93, 97, 121]; *Rubus* spp. [97, 124]; *S. hortensis, A. altissima, T. vulgaris* [123] ; *Mimosa scabrella* [100]; *Trifollium* spp. [89, 142, 144]; *Turbinia corymbosa, Ipomoea triloba, Avicennia germinans, Govania polygama, Lysiloma latisiquum* [101] ; *O. basilicum, S. virgaurea* [88]; *Rhododendron* spp. [33]; *Salix* spp., fruit tree [142]; *Epilobium angustifolium, Nyssa aquatica, Schinus terebinthinfolius, Melilotus* spp., *Glycine max* [141]; gallberry, palmetto, tupelo [31]; Blueberry [144]; honeydew [85, 89, 145]; heterofloral/multifloral [31, 35, 65, 92, 112, 121, 123, 143]	*B. dracunculifolia* [109, 113, 114, 130] *Populus* spp. [5, 6, 10, 37, 110, 111, 115, 116, 118, 119]	*Camelia sinensis*[120]; *Schisandra chinensis* [146]; multifloral [40, 140]; *Cistus ladaniferus* [147]

(Table 4) contd.....

Compound	Honey (BO)	Propolis (BO)	Pollen (BO)
o-Coumaric acid (**11**)	*C. sativa, R. pseudoacacia, Tilia* spp., *Eucalyptus* spp., *Lavandula* spp., *Brassica napus, H. annus, R. officinalis, C. aurantium, C limon, Hedysarum* spp., *E. plantagineum,* Heather (*Erica* spp./*Calluna* spp.) [122]	*Populus* spp. [115, 116]	Multifloral/heterofloral [40, 140]; *Cistus ladaniferus* [147]
m-Coumaric acid (**12**)	*Gochnatia* spp., *Croton* spp., *Vernonia* spp. [137]; *C. sativa, R. pseudoacacia, Tilia* spp., *Eucalyptus* spp., *Lavandula* spp., *Brassica napus, H. annus, R. officinalis, C. aurantium, C limon, Hedysarum* spp., *E. plantagineum,* Heather (*Erica* spp./*Calluna* spp.) [122]	-	-
Methyl *p*-coumarate	-	*Populus* spp. [87]	-
Prenyl *p*-coumarate	-	*Populus* spp. [87, 118]	-
Cinnamyl *p*-coumarate	-	*Populus* spp. [118]	-
Benzyl *p*-coumarate	-	*Populus* spp. [118]	-
p-Coumaric acid derivatives	-	Multifloral origin [132]; *B. dracunculifolia* [112]	-

(Table 4) contd.....

Compound	Honey (BO)	Propolis (BO)	Pollen (BO)
trans-Cinnamic acid **(13)**	*Gochnatia* spp., *Croton* spp., *Vernonia* spp. [137]; *Ziziphus spina-christi* [4, 121]; *Gossypium hirsutum* [138]; *Epilobium angustifolium, Nyssa aquatica, Schinus terebinthinfolius, Melilotus* spp., *Glycine max* [141]; Blueberry [144]; *F. esculentum* [88, 87, 141]; *Eucalyptus* spp. [97, 122]; *Tilia* spp., *H. annus, Brassica* spp., *R. pseudoacacia* [88, 122]; *Citrus* spp. [122, 143]; Heather (*Erica* spp./*Calluna* spp.) [122]; *C. sativa, R. officinalis, Lavandula* spp., *Hedysarum* spp., *E. plantagineum* [122]; *O. basilicum, S. virgaurea* [88], *Clidemia* spp., *Serjania* spp., *Myrcia* spp. [148]; Heterofloral/multifloral [121, 143, 148]; unknown floral origin [105]	*B. dracunculifolia* [109, 112 - 114]; *Populus* spp. [10, 87, 115]	Multifloral/heterofloral [40, 140]; *Cistus ladaniferus* [147]
3,4-Dimethoxy cinnamic acid	*R.pseudoacacia, C.sativa, Citrus* spp., *Eucalyptus camaldulensis, Lavandula* spp. [95, 136]; *R. officinalis, H. annus, Brassica* spp.,*Tilia europea, Erica* spp. [95]; *Abies alba, Quercus* spp. [136]	*Populus* spp. [5, 87, 118]	-
p-Methoxy cinnamic acid	*Citrus* spp., multifloral [143], *Gossypium hirsutum* [138]	-	-
m-Methoxy cinnamic acid	*Gochnatia* spp., *Croton* spp., *Vernonia* spp. [137]	-	-
3,5-Diprenyl-4-hydroxy cinnamic acid	-	*B. dracunculifolia* [112, 149]	-
3-Prenyl-4-hydroxy cinnamic acid	-	*B. dracunculifolia* [112, 149]	-
Cinnamylideneacetic acid	-	*Populus* spp. [5, 111, 119]	-
Cinnamyl methoxycinnamate	-	*Populus* spp. [118]	*Populus* spp. [118]
Cinnamic acid derivatives	-	*B. dracunculifolia* [150] Multifloral origin [132]	-
Methoxycinnamic acid derivative	-	Multifloral origin [132]	-

(Table 4) contd.....

Compound	Honey (BO)	Propolis (BO)	Pollen (BO)
Ellagic acid	*Citrus*spp. [92, 95, 135]; Heather (*Erica* spp./*Calluna* spp. [81, 95, 135]; *F. esculentum* [81] ; *R. pseudoacacia* [95, 107]; *Brassica* spp. [89, 95, 107, 135]; *Tilia* spp., *H. annus* [89, 95, 135]; *Eucalyptus camaldulensis, R. officinalis, C. sativa* [95]; *Lavandula* spp. [95, 135]; *Melaleuca* spp. [104, 125]; *Ananas comosus* spp. [104]; *Aesculus* spp, *Abies* spp., *Frangula* spp., *Rhododendron* spp. [135]; *Trifollium* spp., honeydew [89]; multifloral/heterofloral [151]	*Populus* spp. [87]	-

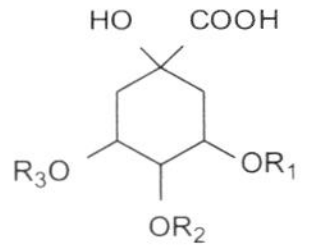

2 R_1=Caf; R_2=R_3=H

3 R_1=R_3=H; R_2=Caf

4 R_1=R_2=H; R_3=Caf

5 R_1=R_2=Caf; R_3=H

6 R_1=H; R_2=R_3=Caf

7 R_1=R_3=Caf; R_2=H

8 R_1=R_2=R_3=Caf

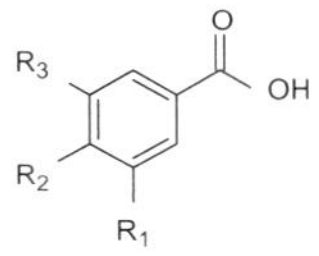

19 R_1=R_2=R_3=H

20 R_1=R_2=R_3=OH

21 R_1=R_2=OH; R_3=H

22 R_1=R_3=OCH_3; R_2=OH

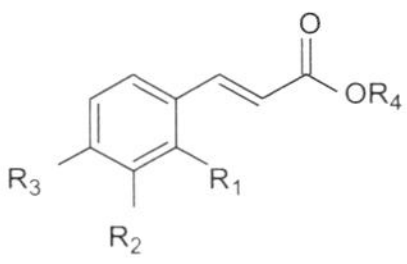

1 R_1=R_4=H; R_2=R_3=OH

9 R_1=R_4=H; R_2=OH; R_3=OCH_3

10 R_1=OH; R_2=R_3=R_4=H

11 R_1=R_3=R_4=H; R_2=OH

12 R_1=R_2=R_4=H; R_3=OH

13 R_1=R_2=R_3=R_4=H

14 R_1=H; R_2=R_3=OH; R_4= 3,4-dihydroxyphenil lactic acid

15 R_1=H; R_2=R_3=OH; R_4= 3-methyl-2butenyl

16 R_1=H; R_2=R_3=OH; R_4= phenetyl alcohol

17 R_1=H; R_2=R_3=OH; R_4= cinnamyl

18 R_1=H; R_2=R_3=OH; R_4= Benzyl

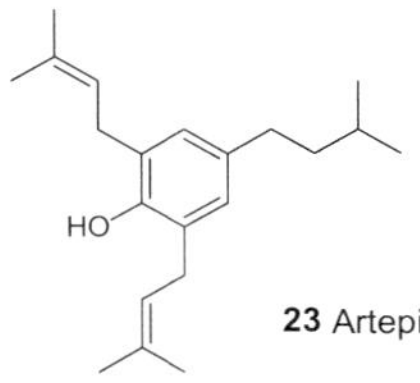

23 Artepilin C

Fig. (2). Structure of the main non-flavonoid components found in honey, propolis and/or pollen. Caf – Caffeic acid.

Caffeic acid (**1** in Fig. **2**) has been described as a phenolic constituent of honey, propolis and bee pollen from many countries (see Table **4**), with concentration levels ranging from 0.001 to 33 μg/g of honey, 0.02 to 32.20 mg/g of propolis extract and 0.446 to 410 μg/g of bee pollen [4, 5, 10, 32, 37, 46, 72, 81, 85 - 90, 92 - 119]. These values show that the composition is highly variable and dependent on the geographic and floral origin of samples. Among the literature, it is clear that the most enriched bee products in caffeic acid are the Brazilian honeys from *Mimosa scrabella* (33 μg/g honey), Chinese *Populus* spp. propolis (32.20 mg/g extract) and Taiwanese *Camellia sinensis* bee pollen (410 μg/g pollen) [5, 100, 120].

Ester derivatives of caffeic acid are quite abundant in honey, propolis and bee pollen. Among these derivatives, 3-*O*-caffeyoilquinic acid (**2** in Fig. **2**), commonly known as chlorogenic acid, is clearly the most abundant in these three bee products, being detected in samples of diverse geographical regions including Brazil, China, Germany, Italy, Lithuania, Malaysia, Poland, Taiwan and Yemen (Table **4**).

In addition to this, several other ester derivatives can be found in honeybee products of botanical origin. Propolis is undoubtedly the most enriched in these compounds typically containing distinct mono-caffeoylquinic acids like 3-*O*-caffeoylquinic acid (3-CQA), 4-*O*-caffeoylquinic acid (4-CQA), 5-*O*-caffeoylquinic acid (5-CQA), together with di-caffeoylquinic acids 3,4-di-*O*-caffeoylquinic acid (3,4-diCQA), 3,5-di-*O*-caffeoylquinic acid (3,5-diCQA) and 4,5-di-*O*-caffeoylquinic acid (4,5-diCQA) and also 3,4,5-tri-*O*-caffeoylquinic acid (3,4,5-triCQA) (**2–8** in Fig. **2** and Table **4**). All these compounds show an identical UV_{max} which is close to that of caffeic acid, but caffeic acid and mono-, di- and tri-CQA acids can be easily distinguishable through full MS detection [126]. Moreover all these isomers can be accurately identified in RP-HPLC coupled to tandem MS^n analysis [130, 152]. Alternatively, identification of individual CQA derivative compounds can be achieved through GC-FID or GC-MS analysis, though a derivatization step is needed first. However this technique have fallen out of favor since the RP-HPLC is more appropriate for the analysis of complex samples [153].

The content of chlorogenic acid in honey has been described to reach 79 μg/g (in Lithuanian multifloral samples), while substantially higher concentrations have been described on *Camelia sinensis* bee pollen samples from Taiwan [81, 97, 120]. However, as mentioned, it is in propolis samples where higher concentrations of chlorogenic acid and of its isomers and/or derivatives have been detected. The values vary from 0.7 to 7.2 mg/g of crude propolis for monoCQAs (3-, 4- and 5-), 5 to 31 mg/g for diCQAs (3,4-, 3,5- and 4,5-) and 1 to 4 mg/g for 3,4,5-triCQA, with 4,5-diCQA clearly predominating [115, 130, 131].

Besides the mono-, di- and triCQAs, other ester derivatives have been frequently detected in honey, propolis and bee pollen samples. These comprise rosmarinic acid, prenyl caffeate, phenethyl caffeate, cinnamyl caffeate and benzyl caffeate (**14-18** in Fig. **2**, Table **4**), which also show an UV_{max} close to that of caffeic acid [118, 154] but are easily identified when HPLC is coupled to MS^n analysis. Honey samples from different botanical and geographical origins are particularly enriched in rosmarinic acid, prenyl caffeate and phenethyl caffeate, with the former being the most predominant, where it varies from 0.012 to 15.85 μg/g [81, 97, 107].

In opposition, caffeates including caffeic acid phenethyl, cinnamyl and/or benzyl esters, prevail in propolis samples, and particularly in CAPE-based propolis. In this regard, CAPE has been found particularly abundant in samples collected from Spain, Argentina and China in concentrations of 19.2, 11.9 and 8.7 mg/g of ethanolic extracts, respectively. Caffeic acid cinnamyl ester is another compound that can also be found abundantly in propolis, being recorded to reach 25.1 mg/g of ethanolic extract in Chinese propolis [5, 111, 119, 155]. On the other hand, prenyl caffeate has been described to be abundant in polar extracts of propolis from Italy (0.01-4.1 mg/g extract) [118] and Macedonia (0.95 and 1.6 mg/g extract, respectively) [133] while the Italian (0.02-1.1 mg/g extract) [118] and Chinese samples are a good source of benzyl caffeate (4.1 and 1.5 mg/g extract, respectively) [133].

Besides caffeic acid and its derivatives, honey products of botanical origin contain considerable amounts of other hydroxycinnamic acids, including ferulic, coumaric and cinnamic acid (**9-13** in Fig. **2**), along with their derivatives (Table **4**). Ferulic acid (typically characterized by UV_{max} at 215, 287 and 312sh nm and $[M - H]^-$ at *m/z* 193, with products ions at *m/z* 149, 178 and 134) is commonly detected in honey and propolis from different geographical and botanical origins, while occurring at a lesser extent in bee pollen. Naturally, the content of this compound in each of the three products is very variable, with reported amounts ranging from 0.004 to 174 µg/100 g honey [4, 85, 97, 121, 156], 0.005 to 12.5 mg/g of propolis extract [111, 112, 115, 132] and 0.37 to 450 µg/g of bee pollen [40, 120, 140].

p-Coumaric acid (**10** in Fig. **2**), which is characterized by having an absorption maxima peak around 310 nm and a $[M - H]^-$ at *m/z* 193 with a corresponding base product ion at *m/z* 119 in ESI-MS analysis in negative mode, has been described in honey samples in concentrations that range from 0.004 to 77.9 µg/g [84, 86, 125, 143, 148, 157] (highest values belonging to the multifloral and *Fagopyrum esculentum* honeys from Brazil) [35, 81, 143]. Typically, *p*-coumaric acid is also found in propolis. Samples from Chinese, Brazilian and Italian beehives, have been described as important sources of this phenolic acid (52.2, 16.0 and 13.5 mg/g extract, respectively) [5, 10, 112]. Instead, its isomer *o*-coumaric acid (**11** in Fig. **2**) has been described to occur in lower concentrations in several other propolis samples, namely in those of Turkish origin in which concentration may vary between 2.1 to 23.3 µg/g raw propolis [115]. Moreover, contrasting to the two other honeybee-derived products, propolis is a source of coumaric acid derivatives, mainly in an ester form such as methyl *p*-coumarate, prenyl *p*-coumarate and cinnamyl *p*-coumarate (Table **4**) [10, 85 - 87]. Quantification of prenyl and cinnamyl *p*-coumarates demonstrates that despite the variations observed between samples from different origins (ranging from 0.02 to 0.65 mg/g extract and 0.01 to 1.12 mg/g extract, respectively) *p*-coumaric acid cinnamyl ester may be found in more abundance than its prenyl counterpart [87, 118].

Cinnamic acid (**13** in Fig. **2**, UV_{max} at approximately 277 nm, typical $[M - H]^-$ at *m/z* 147 with a fragment at *m/z* 103) has been detected in honeys, propolis and bee pollens from several geographic areas and origins (Table **4**), but its major abundance is actually in propolis samples, with concentrations reaching up to 0.1 mg/g raw propolis [87, 115]. Propolis is also the bee product with further diversity in cinnamic acid derivatives, like 3,4-dimethoxycinnamic acid, *p*- and *m*-methoxycinnamic acid, artepillin C, 3-prenyl-4-hydroxycinnamic acid, cinnamylideneacetic acid and cinnamyl methoxycinnamate. Indeed, depending on the geographical and botanical origin, different amounts of these compounds can be found in propolis samples worldwide. As referred before, artepillin C (**23** in Fig. **2**) is a very important compound characteristic of the Brazilian green propolis. In fact, the quantification of this compound has become an important indicative of the Brazilian green propolis quality, and therefore it is generally used as a chemical marker for the quality control of this product [158]. Other cinnamic acids have also been reported as relevant constituents of propolis. *E.g.* 3,4-dimethoxycinnamic acid in propolis from China [5], has been described to amount up to concentrations of 5.8 and 57.4 mg/g. Likewise, concentrations of cinnamylideneacetic acid in the range of 0.8 to 45.4 mg/g extract were detected in propolis of Chinese origin [5, 111, 119].

In addition to hydroxycinnamic acids, honeybee-derived products are also enriched in phenolic acids such as benzoic vanillic , gallic, protocatechuic , syringic acid (**19–22** in Fig. (**2**), respectively), as well as several derivatives of these compounds (Table **5**). Similarly to what was previously mentioned for hydroxycinnamic acids, the identification of these compounds has been mainly carried out by HPLC coupled to DAD and/or MS^n analysis.

Table 5. Common phenolic acids in honey, propolis and bee pollen worldwide.

Compound	Honey (BO)	Pollen (BO)	Propolis (BO)
Benzoic acid (**19**)	*Ziziphus spina-christi* [4]; *C. sativa, Citrus* spp., *Eucalyptus* spp [102, 122, 136]; *R. pseudoacacia* [122, 136]; *Tilia* spp., *B. napus, H. annus, R. officinalis, E. plantagineum* [122]; Heather (*Erica* spp./*Calluna* spp.) [65, 122]; *Lavandula* spp. [122, 136]; *Hedysarum* spp. [102, 103, 122]; *L. scoparium,* Tualang tree, *Melaleuca* spp., *Acacia mangium* [93]; *F. esculentum* [65, 86]; *Gochnatia* spp., *Croton* spp., *Vernonia* spp. [137]; *Mimosa scabrella* [100]; *Gossypium hirsutum* [138]; *Centaurea cyanus* [161]; Pine, hawthorn, nettle, thyme, black chokeberry, aloe [65]; Multifloral/Heterofloral [65, 112]	Multifloral/Heterofloral [40]	*B. dracunculifolia* [112]; *Populus* spp. [87, 115]; Multifloral origin [132]

(Table 5) contd.....

Compound	Honey (BO)	Pollen (BO)	Propolis (BO)
Vanillic acid	*Ziziphus spina-christi* [4, 121]; *C. sativa* [122]; *R. pseudoacacia* [35, 85, 108, 122, 141] ; *Tilia* spp. [35, 46, 89, 122]; *Hedysarium* spp. [108, 122]; *F. esculentum* [35, 81, 85, 141]; *Citrus* spp. [108,122,143]; Heather (*Erica* spp./*Calluna* spp.) [35, 46, 81, 122]; *Eucalyptus* spp. [122]; *Brassica* spp. [35, 89, 122]; *L. scoparium* [93, 121]; *H. annus* [89, 122]; *Hedysarum* spp. [108, 122]; *Acacia* spp. [93, 121]; *Lavandula* spp., *R. officinalis, E. plantagineum* [122]; Tualang tree, *Melaleuca* spp. [93]; *Turbina corymbosa, Ipomoea triloba, Avicennia germinans, Govania polygama, Lysiloma latisiquum* [101]; *Trifolium* spp. [89, 144]; *Gossypium hirsutum* [138]; *Pinus* spp., *Thymus* spp., *Abies cephallonica* [162]; *E. angustifolium, N. aquatic, S. terebinthifolius, Melilotus* spp., *G. max* [141] ; *Clidemia* spp., *Serjania* spp., *Myrcia* spp. [148]; *Prosopis julifora* [121]; Honeydew [85, 89]	*Schisandra chinensis* [146]; Multifloral [40]; *Cistus ladaniferus* [147]	*Populus* spp. [115]; Multifloral origin [132]
Gallic acid	*Clidemia* spp., *Serjania* spp., *Myrcia* spp. [148]; *Pinus* spp., *Thymus* spp., *Abies cephallonica* [162]; *E. angustifolium, N. aquatic, S. terebinthifolius, Melilotus* spp., *G. max* [141] ; *Ziziphus spina-christi* [4, 121]; *C. sativa* [90,102,122]; *Kunzea ericoides, Knightia excelsa* [163]; *Quillaja saponaria* [139]; milk vetch, wild chrysanthemum, jujube, acacia [98]; *R. pseudoacacia* [85, 88, 107, 122, 141]; *F. esculentum* [85, 88, 97, 141]; *Tilia* spp. [32, 46, 88, 89, 122]; *Eucalyptus* spp. [90, 97, 102, 122]; *Trifollium* spp. [89, 163]; *Hedysarum* spp. [102, 103, 122]; *Citrus* spp. [102, 122, 143]; *Brassica* spp. [88, 89, 107, 122]; *H. annus* [88, 89, 122]; Heather (*Erica* spp./*Calluna* spp.) [46, 90, 122]; *E. plantagineum, R. officinalis* [122]; *L. scoparium* [90, 121, 163]; *Melaleuca* spp. [104, 125]; *Ananas comosus* spp. [104]; *Acacia* spp. [97, 121]; *R. idaeus* [97]; *O. basilicum, S. virgaurea* [88]; Honeydew [85, 89]; Multifloral/Heterofloral [121, 143, 148]	*Camelia sinensis* [120]; *Schisandra chinensis* [146]	*Populus* spp. [110, 115]

(Table 5) contd.....

Compound	Honey (BO)	Pollen (BO)	Propolis (BO)
Protocatechuic acid	*Citrus* spp. [122, 143]; *F. esculentum* [81, 86, 88]; *R. pseudoacacia* [88, 122]; *Tilia* spp., *Brassica* spp., *H. annus* [88, 89, 122]; Heather (*Erica* spp./*Calluna* spp.) [81, 122]; *Eucalyptus* spp., *Lavandula* spp., *R. officinalis*, *Hedysarum* spp., *E. plantagineum* [122]; *Trifolium* spp., honeydew [89]; *O. basilicum, S. virgaurea* [88]; multifloral/heterofloral [143]; unknown floral origin [105]	*Schisandra chinensis* [146]; Multifloral [40]; *Cistus ladaniferus* [147]	*Populus* spp. [115]
p-Salicylic acid	*Ziziphus spina-christi* [4]; *Pinus* spp., *Thymus* spp., *Abies cephallonica* [162]; *Gossypium hirsutum* [138]; milk vetch, wild chrysanthemum, jujube, acacia [98]; *C. sativa* [90, 122]; *R. pseudoacacia* [85, 107, 122, 141]; *Tilia* spp. [46, 89, 122]; *Eucalyptus* spp. [90, 122]; *F. esculentum* [65, 81, 85, 86, 122, 141]; *Brassica* spp. [89, 122]; *Citrus* spp. [89, 122, 143]; Heather (*Erica* spp./*Calluna* spp.) [46,65,81,90,122]; *H. annus* [89, 122]; *Lavandula* spp., *R. officinalis, Hedysarum* spp., *E. plantagineum* [122]; *L. scoparium* [90]; Pine, hawthorn, nettle, thyme, black chokeberry, aloe [65]; honeydew [85, 89]; *Trifolium* spp. [89, 144]; multifloral/heterofloral [65, 122, 143]; *E. angustifolium, N. aquatica, S. terebinthifolius, Melilotus* spp., *G. max* [141]	Multifloral/Heterofloral [40]	*Populus* spp. [115]
Syringic acid	*E. angustifolium, N. aquatic, S. terebinthifolius, Melilotus* spp., *G. max* [141]; *Clidemia* spp., *Serjania* spp., *Myrcia* spp. [148]; *Pinus* spp., *Abies cephallonica* [164]; *Thymus capitatus* [164, 165]; *Kunzea ericoides, Knightia excelsa* [163]; milk vetch, wild chrysanthemum, jujube, acacia [98]; *Ziziphus spina-christi* [4, 121]; *L. scoparium* [93, 121, 163]; *F. esculentum* [81, 85, 86, 141]; *Citrus* spp. ; *Turbina corymbosa, Ipomoea triloba, Avicennia germinans, Govania polygama, Lysiloma latisiquum Trifolium* spp. [89, 163]; *Tilia* spp. [46, 89, 122]; *R. pseudoacacia* [85, 107, 122, 141]; Heather (*Erica* spp./*Calluna* spp.) [46, 81, 122]; *Brassica* spp. [89, 107, 122]; *Eucalyptus* spp., *Lavandula* spp., *R. officinalis, Hedysarum* spp., *E. plantagineum* [122]; Tualang tree, *Melaleuca* spp.[93]; honeydew [85, 89]; *Acacia* spp. [93, 121]; multifloral/heterofloral [121, 143, 148]; *Prosopis julifora* [121]	Multifloral [40]; *Cistus ladaniferus* [147]	*Populus* spp. [115]

(Table 5) contd.....

Compound	Honey (BO)	Pollen (BO)	Propolis (BO)
o-Anisic acid	*L. scoparium* [90, 163]; *Kunzea ericoides, Trifollium* spp., *Knightia excelsa* [163]	-	-
o-Salicylic acid	*Gossypium hirsutum* [138]; *C. sativa, R. pseudoacacia, Eucalyptus* spp., *Citrus* spp., *Lavandula* spp. [122, 136]; *Abies alba, Quercus* spp. [136]; *Tilia* spp., *Brassica napus, H. annus, R. officinalis, Hedysarum* spp., *E. plantagineum,* Heather (*Erica* spp./*Calluna* spp.) [122]	-	-
m-Salicylic acid	*C. sativa, R. pseudoacacia, Tilia* spp., *Eucalyptus* spp., *Lavandula* spp., *B. napus, H. annus, R. officinalis, Citrus aurantium, Citrus limon, Hedysarum* spp., *E. plantagenium,* Heather (*Erica* spp./*Calluna* spp.) [122]	-	-
Methyl anisate	-	Heterofloral [38]	-
p-Anisic acid	*L. scoparium, Kunzea ericoids, Knigthia excelsa, Trifollium* spp. [163], *Citrus* spp., Multifloral/heterofloral [143]; *Gochnatia* spp., *Croton* spp., *Vernonia* spp. [137]	-	
Benzoic acid derivatives	-	-	Multifloral origin [132]
Methyl benzoate	-	Heterofloral [38]	-
Methyl gallate	-	*Camelia sinensis* [120]	-
Methyl syringate	*L. scoparium* [90]; Heather (*Erica* spp./*Calluna* spp.) [80, 90]; *C. sativa* [90, 136]; *Eucalyptus* spp. [90, 136]; *Mimosa scabrella* [100]; *F. esculentum* [80, 136]; *R. pseudoacacia* [80, 136]; *A. alba, C. sinensis, Lavandula* spp., *Quercus* spp. [136]; *Tilia* spp., *Solidago* spp., *B. napus* [80]; *C. cyanus* [161]	-	-

BO – Botanical origin

From those, benzoic acid (typical UV_{max} at 229 and 274 nm [86, 87, 90] and deprotonated molecular ion at *m/z* 121 in ESI-MS analysis in negative mode and a typical fragment peak base at *m/z* 77, caused by the loss of the carboxylic acid moiety [159]) is perhaps the mostly widely distributed amongst bee products. This has been described in honeys from different geographical and botanical origins with variable abundance, being particular prevalent in honeys obtained from Brazilian *Gochnatia* spp. and in Yemeni *Ziziphus spina-christi* (141 and 56 μg/g, respectively) [4, 112, 137], as well as in propolis (amounts varying from 5.3 μg to

48.6 mg/g extract) [112, 115, 132] and in pollen (very scarce quantification).

Vanillic acid is also very commonly found in floral honeybee derived products. Typically, the UV analysis of this acid reveals wavelength maxima peaks at 260 and 294 nm [101] and its full MS spectrum in the negative mode is characterized by a deprotonated molecular ion at *m/z* 167 [85, 89] and a typical ion fragment at *m/z* 152, originated by the loss of a CH_3 group [85, 160], along with ions at *m/z* 122 and 108 [160]. As can be observed in Table **5**, this phenolic acid has been described in honeys, propolis and pollen collected worldwide and according to literature, its amounts can be up to 211 μg/g in honey (as described for honey from Brazilian *Citrus* spp. [81, 143, 148]) or to 334 μg/g extract in propolis, as described for Finnish Coniferous Forest propolis samples. Quantification in pollen is scarce, although values of 0.23 to 15.0 μg/g have been reported for Chinese *Schisandra chinensis* and Turkish pollen [115, 146].

Besides benzoic and vanillic acids, gallic acid (MW 170 g/mol and UV_{max} at approximately 272 nm), protocatechuic acid (MW 154 g/mol and UV_{max} at nearly 260 and 290 nm), *p*-salicylic (MW 138 g/mol and UV_{max} around 254 nm), and syringic acid (MW 198 g/mol and UV_{max} close to 276), among others, are often detected in bee-floral products (Table **5**). Considerable amounts of these compounds have been reported in specific samples, *e.g.* the levels of gallic acid in honeys of Italian *Hedysarium* spp. and in a multifloral honey from Brazil accounted for 89.5 and 77.3 μg/g, respectively [103, 143], while the highest levels of protocatechuic acid in honeys were described to occur in those from Brazilian *Citrus* spp. and from Polish *Fagopyrum esculentum* (66.7 and 37.0 μg/g, respectively) [81, 143]. Notably, *p*-salicylic acid levels in Brazilian *Baccharis dracunculifolia* have been reported to account for up to 267 μg/g. Despite lower, the concentrations of this phenolic in multifloral Brazilian honey and Polish *Fagopyrum esculentum* honey were yet significant (98.1 and 62.1 μg/100 g, respectively) [81, 112, 143]; On the other hand, syringic acid was generally found in lower concentrations in honeys, although heterofloral *Clidemia* spp./*Mimosa pudica* and monofloral *Citrus* spp. were described as a potential source of this acid (15.7 and 13.8, 6.4 μg/100 g, respectively) [4, 143, 148].

Amongst propolis worldwide samples, gallic, syringic and *p*-salicylic acids seem to appear in considerable amounts in Turkish temperate propolis (8.7-70.1, 4.4-11.9 and 41.5-139 μg/g raw propolis, respectively). In turn the concentrations of these acids in pollen samples are in general much lower than those found in honey and propolis. From the reported literature, one can ascribe Brazilian, Turkish and Chinese bee pollen as sources of gallic acid (0.08 to 5.6 μg/g) and protocatechuic acid (0.05 and 4.6 μg/g) [40, 146].

3.2. Flavonoids

Flavonoids, whose structures are based on a C_6-C_3-C_6 skeleton (Fig. **1**) are sub-divided into different subclasses differing in the oxidation state of the central heterocyclic ring, including flavonols, flavones, flavanones, dihydroflavonols, anthocyanidins and flavanols [166]. Depending on the matrix, these compounds may be prevalent in the form of aglycones or as glycosides, both varying according to their pattern of hydroxylation and/or methoxylation. Glycosylation can occur as *O*-glycosylation of their hydroxyl groups as well as *C*-glycosylation directly to carbon atom of the flavonoid skeleton. In addition, flavonoid glycosides are frequently acylated with aliphatic or aromatic acid molecules. Such derivatives are thermo-labile and their isolation and further purification without partial degradation is difficult [167].

The structural elucidation of the different subclasses of flavonoids is commonly achieved by comparison of their chromatographic behavior, UV spectra and MS information, with those of reference compounds. In fact, the chromatographic behavior and UV spectroscopy provide particularly wealthy information when applied to flavonoids. Interestingly, a reasonable idea of the structure from many of the most common flavonoid glycosides can be obtained only looking at these two parameters [168].

Notably, one of the most important features of flavonoids is the existence of two main wavelength maxima. In a general perspective, the longer wavelength absortion corresponds to the Band I absortion that ranges between 300 – 380 nm and is considered to be associated to the B-ring cinnamoyl function, while the shorter corresponds to Band II absortion which ranges between 240 – 280 nm and is correlated with the absorption involving the A-ring benzoyl function (Fig. **3**) [169, 170]. However, the exact wavelength and magnitudes of the absorption maximas depend on several factors including the nature of the C-ring, the point of attachment of the B-ring, and the nature of the substituents at A- and B-rings [168].

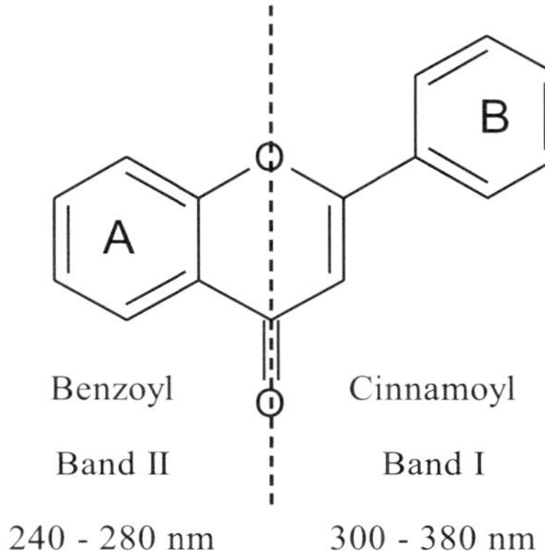

Fig. (3). Representation of the two distint chromophore functions of flavonoids (benzoyl function that comprises the A-ring and C-4 carbonyl group, and cinammoyl function comprising the B-ring and the three carbons of the C-ring) and respective association with the Band I and Band II absorption maxima regions.

Just as important as UV–vis detectors, mass spectrometers are crucial when it comes to the identification of flavonoids. Notably, the fragmentation profile of compounds from distinct subclasses is clearly influenced by their substitution pattern, although several common features can be found. The MS^2 spectrum of many flavonoids reveal the fragments at *m/z* 151 or at *m/z* 165, which are resultant from the retro Diels-Alder mechanism [171]. Besides, neutral losses commonly described to occur in these compounds, such as the small molecules CO (-28 Da), CO_2 (-44 Da), C_2H_2O (-42 Da), as well as the successive losses of these molecules, are also observed [171].

The flavonoid content in honeybee-products is very variable, however it has been reported that their concentration can reach up to 6 $mg.kg^{-1}$ in honey, and about 10% and 0.5% of propolis and bee pollen extracts respectively [172].

Among them, flavonols and flavones are particularly important constituents from bee plant-derived products, with special relevance in honey and bee pollen [173]. In general, flavonols exhibit a Band I and II between 350 – 385 and 250 – 280 nm, respectively, while the absence of the 3-hydroxyl group in flavones causes an hypsochromic shift of the Band I to 310 – 350 nm. Also, the *O*-substitution 3-hydroxyl group in flavonols (*O*-alkyl or *O*-glycosyl), modify the general shape of the Band I absortion wavelength that tends to enlarge to 330 – 360 nm (bathochromic shift), approaching to those of flavones [168].

From flavonols, quercetin (**23** in Fig. (**4**)), kaempferol (**24** in Fig. (**4**)), isorhamnetin (**25** in Fig. (**4**)) and myricetin (**26** in Fig. (**4**)) are the most commonly found, along with several of their methyl and/or glycosidic derivatives (Table **6**).

All these compounds exhibit an UV maxima at 370 nm with exception of kaempferol which reveals a hypsochromic shift of its UV maxima to 366 nm [174]. Despite their similarity in the UV spectra, these compounds can be easily distinguishable through MS spectrometry, because of their distinct MS ($[M-H]^-$ at *m/z* 301, 285, 315 and 317, repectively) and characteristic MS^n data.

Table 6. Relevant flavonols in honey, propolis and bee pollen worldwide.

Flavonol	Honey (BO)	Propolis(BO)	Pollen (BO)
Quer **(23)**	*Eucalyptus* spp. [95,102,176,177]; *R. pseudoacacia, H. annus, Lavandula* spp., *Erica* spp., *Brassica* spp. [95]; *F. esculentum* [35, 86]; *C. sativa* [95, 102, 177]; *Citrus* spp. [31,95,102,164,177]; *Tilia* spp. [32, 95]; *Q. saponaria* [139]; *Mimosa caesaepiniifolia* [178]; *R. officinalis* [95, 177]; *Thymus* spp. [164,165,177]; palmetto berry, *L.scoparium*, tupelo [31]; *Pinus* spp., *Abies* spp. [164]; Heather [35, 177]; *Ziziphus spina-christi* [4]; *Hedysarum* spp. [102]; multifloral [31,102,177]	*Populus* spp. [37,72,87,115,116,133, 177]; *B. dracunculifolia* [73, 126]; *Dalbergia* spp. [179]; *Clusia* spp. [126]; Commercial [16, 118]	Anzer* [40]; *Cystus incanus* [129]; *Cistus ladaniferus* [147]; *Eupatorium* spp., *Ricinus* spp., *Mimosa arenosa, Eucalyptus* spp., *Cecropia* spp., *Mimosa pudica, Elaeis* spp. [180]; *Typha angustifolia* [175]
3-*O*-CH_3Quer	*Eucalyptus* spp. [176], La Alcarria* [181]	*Populus* spp. [182]; Commercial [118]	*Eucalyptus globulus* [183]

(Table 6) contd.....

Flavonol	**Honey (BO)**	**Propolis(BO)**	**Pollen (BO)**
7-*O*-CH_3Quer	-	Commercial [118]	-
3,7-diCH_3Quer	*Citrus* spp., *H. annus*, *Erica* spp. [95]	-	-
3,3'-diCH_3Quer	*H. annus*, *Erica* spp. [95]; *R. officinalis* [2]	-	-
7-3'-diCH_3Quer	*R. officinalis* [2]	-	-
diCH_3Quer	-	*Populus* spp. [87]; Commercial [118]; *B. dracunculifolia*, *Clusia* spp. [126]	-
tetraCH_3Quer	-	*Populus* spp. [87]	-
Quer-3-*O*-Glc	Multifloral (6)	*Populus* spp. [87]; *B. dracunculifolia*, *Clusia* spp. [126]	*Prosopis juliflora* [184]; *Eupatorium* spp., *Ricinus* spp., *Mimosa arenosa*, *Cecropia* spp., *Eucalyptus* spp., *Mimosa pudica*, *Elaeis* spp. [180]
Quer-3-*O*-Rha	Multifloral [185]	*Populus* spp. [87]; *B. dracunculifolia*, *Clusia* spp. [126]	*E. australia* [183]; Multifloral [11]; Multifloral [41]
Quer-3-*O*-Gal			Multifloral [11]
Quer-3-*O*-Rut	Orange blossom [186], *Q. saponaria* [139]; *C. sativa*, *Eucalyptus* spp., *Citrus* spp., *Hedysarum* spp. [102]; *R. pseudoacacia*, *F. esculentum*, Heather, *Brassica* spp., *Tilia* spp. [35]; palmetto berry, *L. scoparium* [31]; *Brassica napus* [186]	*Populus* spp. [87, 115]; Commercial [16]	*Ranunculus petiolaris* [187]; Anzer* [40]; *Cistus ladaniferus* [147]; Multifloral [41]
Quer-3-*O*-Hex(1→2)Hex	*Citrus* spp., *Brassica* spp. [186]	-	-
Quer-3-*O*-Rut-7-*O*-Rha	*Brassica napus* [186]	-	-
Quer-3-*O*-Gluc	-	*Populus* spp. [87]	Multifloral [11]
Quer-3-*O*-Soph	-	-	*E. globulus* [183]; *R. petiolaris* [187]
Quer-3-*O*-Ara	-	*B. dracunculifolia*, *Clusia* spp. [126]	Multifloral [41]
diCH_3Quer-*O*-Rut	-	*Populus* spp. [87]	-
diCH_3Quer-*O*-Gluc	-	*Populus* spp. [87]	-

(Table 6) contd.....

Flavonol	**Honey (BO)**	**Propolis(BO)**	**Pollen (BO)**
Quer-3-O-diGly	-	-	*R. sardous* [183]
Kaemp **(24)**	*Eucalyptus* spp. [176]; *Tilia* spp. [35, 95]; *Citrus* spp., *Brassica* spp., *R. officinalis*, *Erica* spp. [95], *R. pseudoacacia* [35, 95]; *F. esculentum* [86]; *Q. saponaria* [139]; Heather [35]; *Pinus* spp., *Thymus* spp., *Abies* spp., *Citrus* spp. [164]; *Ziziphus spina-christi* [4]; palmetto berry, *L. scoparium* [31]	*Populus* spp. [6, 37, 72, 74, 87, 116, 119, 177, 188]; *B. dracunculifolia* [73,74,109]; Commercial [118, 189]	*Cystus incanus* [129]; *Cistus ladaniferus* [147]; *Eucalyptus* spp., *Mimosa pudica, Elaeis* spp., *Cecropia* spp., *Eupatorium* spp., [180]; *Typha angustifolia* [175]
CH_3Kaemp	*Tilia* spp., *Brassica* spp., *R. officinalis*, *Erica* spp., *Citrus* spp., *Lavandula* spp., *R. pseudoacacia* [95], La Alcarria* [181]	*Populus* spp. [87]	-
OCH_3-CH_3Kaemp	-	*Populus* spp. [87]	-
diCH_3Kaemp	-	*Populus* spp. [87]; Finnish* [132]	-
Kaemp-*O*-*p*-CouRha	-	*Populus* spp. [87]	-
Kaemp-3,4’-di-*O*-Hex	*Tilia* spp., *Brassica* spp. [186]	-	-
CH_3Kaemp-*O*-Glc	-	*Populus* spp. [87]	-
CH_3Kaemp-*O*-Rut		*B. dracunculifolia*, *Clusia* spp. [126]	
8-OCH_3kaemp-3-*O*-Hex(1→2)Hex	*Brassica napus, Prunus avium, Eucalyptus* spp., *Mendicago sativa* [186]	-	-
8- OCH_3kaemp -3-*O*-Neohesp	*Brassica napus, Brássica* spp. [186]	-	-
Kaemp-3-*O*-Hex(1→2)Hex	*Tilia* spp., *Citrus* spp., *Brassica* spp., *R. officinalis* [186]	-	-
Kaemp-3-*O*-Neohesp	*Prunus avium, Eucalyptus* spp., *Rhododendron* spp., *R. officinalis, Taraxacum* spp. [186]	-	*Salix atrocinera* [183]
Kaemp-3-*O*-Rut	*Brassica napus* [186]	*Populus* spp. [87]	
Kaemp-3-*O*-Glc	-	-	Multifloral [41]
Kaemp-3-*O*-Rha	-	-	Multifloral [11]
Kaemp-3-*O*-Rha-Glc	-	-	Multifloral [41]

(Table 6) contd.....

Flavonol	Honey (BO)	Propolis(BO)	Pollen (BO)
Kaemp-3-*O*-Hex(1→2)Hex-7-*O*-Rha	*Brassica napus, Prunus avium, Brassica* spp., *Taraxacum* spp. [186]	-	-
Kaemp-3-*O*-Rut-7-*O* -Rha	*Brassica napus, Brassica* spp. [186]	-	-
Kaemp-3-*O*-(Hex)Rob-7-*O*-Rha	*R. pseudoacacia* [190]	-	-
Kaemp-3-*O*-(Hex)Rob	*R. pseudoacacia* [190]	-	-
Kaemp-3-*O*-Hex-7-*O* -Rha	*R. pseudoacacia* [190]	-	-
Kaemp-3-*O*-Rob-7-*O* -Rha	*R. pseudoacacia* [190]	-	-
Kaemp-7-*O*-Rob	*R. pseudoacacia* [190]	-	-
Kaemp -7-*O*-Rha	*R. pseudoacacia* [190]	-	-
Kaemp-3-*O*-Soph	-	-	*Raphanus raphanistrum* [183]
Isorhm **(25)**	La Alcarria* [181], *Mimosa caesaepiniifolia* [178], *F. esculentum* [86]	*Populus* spp. [87]; *B. dracunculifolia* [73, 126], *Clusia* spp. [126]; Commercial [118, 189]	*Cystus incanus* [129]; *Cistus ladaniferus* [147]; *Eupatorium* spp., *Eucalyptus* spp., *Cecropia* spp., *Mimosa pudica, Elaeis* spp., *Elephantopus* spp. [180]
Isorhm-*O*-Pen	-	*Populus* spp. [87]	-
Isorhm-Glc		*B. dracunculifolia, Clusia* spp. [126]	
Isorhm-3-*O*-(He- -Hex)-7-*O*-Hex	*Taraxacum* ssp. [186]	-	-
Isorhm-3-*O*-(2'', 3''-diRha)Glc	-	-	Multifloral [41]
Isorhm-3-*O*-Rha-Glc	-	-	Multifloral [41]
Isorhm-3-*O*-Hex(1→2)Hex	*Tilia* spp., *Brassica* spp., *Rhododendron* spp. [186]	-	-
Isorhm-3-*O*-Neohesp	*Prunus avium, Brassica* spp., *Taraxacum* spp., *Tilia* spp. [186]	-	-
Isorhm-3-*O*-Xyl(1→6)Glc	*Brassica napus, Brassica* spp., *H. annus* [186]	-	Multifloral [11]
Isorhm-3-*O*-Rut-7-*O*-Hex	*Mendicago sativa* [186]		

(Table 6) contd.....

Flavonol	Honey (BO)	Propolis(BO)	Pollen (BO)
Isorhm-3-O-Rut		*Populus* spp. [87]	
Isorhm-3-O-diGlc			Multifloral [41]
Isorhm-O-Gluc		*Populus* spp. [87]	
Isorhm*O*-AcRut		*Populus* spp. [87]	
Isorhm-3-*O*-Sop--diGly			*Taraxacum* ssp. [183]
Myr **(26)**	*Eucalyptus* spp.; La Alcarria*; *F. esculentum* [35], heather [135]; *Q. saponaria* [139]; *C.sativa, Citrus* spp., *Hedysarum* spp. [102]; *Pinus* spp., *Thymus* spp., *Abies* spp., *Citrus* spp. [164]; *Ziziphus spina-christi* [4];*R. officinalis,* Heather [177]	*Populus* spp. [110]	*E. globulus* [183]; *Cistus ladaniferus* [147]; *Eucalyptus globulus, Elaeis* spp., *Cecropia* spp., *Mimosa pudica, Scoparia* spp. [180]
3-CH_3Myr	Heather [191, 192]	-	-
3,7,4',5'-tetraCH_3Myr	*Eucalyptus* spp., *Thymus algeriensis, R. officinalis, C. sinensis, B. campestris, H. annus,* multifloral [193]	*Dalbergia* spp. [182]	-
Myr-3,7-di-*O*-Glc	*Eucalyptus* spp. [186]	-	-
Myr-3-*O*-Gal	-	-	*R. raphanistrum* [183]
Myr-3-*O*-Rha-Glc	-	-	Multifloral [41]
Rhm	Heather, *R. pseudocacia, F. esculentum, Tilia* spp., *Brassica* spp. [35]	*Populus* spp. [87]; Black propolis [126]; *B. dracunculifolia, Clusia* spp. [126]	-
Fis	*Citrus* spp., *R. officinalis,* Heather, *Eucalyptus* spp., *C. sativa, Thymus* spp [177]	-	-
Kaempf	*R. officinalis* [2]	*Populus* spp. [74, 87,133]; *B. dracunculifolia* [74, 109]	-
Gala	La Alcarria* [181]; *F. esculentum*, heather [86, 7]; *Tilia* spp. [32, 194]; *Citrus* spp. [194]; *Ziziphus spina-christi* [4]; palmetto berry, *L. scoparium* [31]	*Populus* spp. [37, 68, 72, 87, 111, 133, 188, 195, 196]; Commercial [118, 189]	*Cystus incanus* [129]
5-CH_3Gala	-	*Populus* spp. [87]; Commercial [118]	-

(Table 6) contd.....

Flavonol	Honey (BO)	Propolis(BO)	Pollen (BO)
Herb-7-*O*-CH_3-3-*O*-Glc-8-*O*-Gal	-	-	-
Herb-7-*O*-CH_3-3-*O*-diGly	-	-	*R. sardous* [183]; *R. petiolaris* [187]
Herb-8-*O*-CH_3-3-*O*-diGly	-	-	*R. sardous* [183]
Herb-7-*O*-CH_3-3-*O*-Soph	-	-	*R. sardous* [183]; Multifloral [11]
Herb-8-*O*-CH_3-3-*O*-Soph	-	-	*Ulex europeaus, Lotus corniculatus* [183]

BO – Botanical origin; "*" after the name refer to geographic origin since no botanical origin is referred. (Hex)Rob – (Hexosyl)robinoside; AcRut – Acetylrutinoside; Ara – Arabinoside; CH_3 – Methyl; CouRha – Coumaroylrhamnoside; Fis – Fisetin; Gal – Galactoside; Gala – Galangin; Glc – Glucoside; Gluc – Glucoronide; Gly – Glycoside; Herb – Herbacetin; Hex – Hexoside; Hex(1→2)Hex – Hexosyl(1→2)Hexoside; Hex-Hex – hexosyl-hexoside; IsoRhm – Isorhamnetin; Kaemp – Kaempferol; Kaempf – Kaempferide; Myr – Myricetin; Neohesp – Neohesperidoside; OCH3 – Methoxy; OH – Hydroxyl; Pen – Pentoside; Quer – Quercetin; Rhm – Rhamnetin, Rha – Rhamnoside; Rut – Rutinoside: Soph – Sophoroside; Xyl(1→6)Glc – Xylosyl(1→6)glucoside

Naturally, the concentrations of each of these compounds is rather variable according to the botanic and geographical origin of each of the three honey-bee derived products.

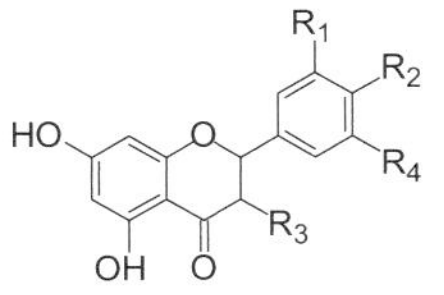

23 $R_1=R_2=R_3=OH$; $R_4=H$

24 $R_1=R_4=H$; $R_2=R_3=OH$

25 $R_1=OCH_3$; $R_2=R_3=OH$; $R_4=H$

26 $R_1=R_2=R_3=R_4=OH$

27 $R_2=OH$; $R_1=R_3=R_4=H$

28 $R_1=R_2=OH$; $R_3=R_4=H$

29 $R_1=R_2=R_3=R_4=H$

30 $R_1=R_2=R_3=H$

31 $R_2=OH$; $R_1=R_3=H$

32 $R_1=R_2=H$; $R_3=OH$

33 $R_1=R_2=R_3=OH$

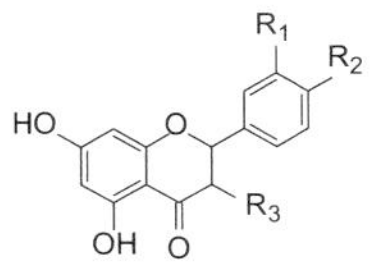

Fig. (4). Structures of relevant flavonoid compounds found on honeybee-derived products.

Bee pollen is specially rich in quercetin when compared to the remaining flavonols (high values registered in *Typha angustifolia* pollen from China (920 μg/100g) [175]). Propolis, on the other hand, is particularly enriched in kaempferol (*e.g.* concentration of 19.5 mg/g of extract from *Populus* spp. origin from Argentina [119]) while myricetin is the most abundant compound in honeys (244.7 μg/g in *Thymus capitatus* honeys from Greece) [165].

As flavones lack hydroxylation in the position C-3, the molecular ion of these compounds exhibit 18 Da less than their respective flavonols [197]. Significant concentrations of these compounds have also been reported in honeybee products. In particular, apigenin (**27** in Fig. (**4**)), luteolin (**28** in (Fig.(**4**)) and chrysin (**29** in Fig. (**4**), together with their methylated derivatives can be found in those products, although with distinct frequency and concentrations [198 - 200] (Table 7).

Table 7. Relevant flavones in honey, propolis and bee pollen worldwide.

Flavone	**Honey (BO)**	**Propolis(BO)**	**Pollen (BO)**
Aca	*R. pseudoacacia* [204]	*Populus* spp. [72, 133]	-
Aca-Gly	-	-	*R. petiolaris* [187]
Api **(27)**	*Erica* spp., *Citrus* spp., *Lavandula* spp. [95]; La Alcarria* [181]; *F. esculentum* [86]; *R. pseudoacacia, Brassica* spp. [35]; *Tilia* spp. [32, 35]; *Ziziphus spina-christi* [4]	*Populus* spp. [6, 37, 72, 74, 87, 116, 119, 133, 188, 195]; Commercial [118, 189]; *B. dracunculifolia* [73]; Finnish* [132]	-
Api-7-*O*-Gly	-	-	*R. petiolaris* [187]
Api-6,8-di-*C*-Gly	-	-	*R. petiolaris* [187]
Baic	*Citrus* spp., *R. officinalis,* Heather, *Eucalyptus* spp., *C. sativa, Thymus* spp. [177]	-	-
Chr **(29)**	*Eucalyptus* spp. [95,176,177,194]; *Tilia* spp. [32,95,194]; *C. sativa, Citrus* spp. [95,177,194]; Heather, *R. officinalis* [95, 177]; *H. annus, Lavandula* spp., *R. pseudoacacia* [95]; *F. esculentum* [86], [7]; *Pinus* spp., *Thymus* spp., *Abies* spp., *Citrus* spp. [164]; *Ziziphus spina-christi* [4]; palmetto berry, *L. scoparium* [31]	*Populus* spp. [6, 37, 68, 72, 74, 87, 111, 119, 133, 177, 188, 195, 196, 205]; African Propolis [206]; Commercial [118, 189]; *B. dracunculifolia* [73]	*Cystus incanus* [129]
5-CH_3Chr	-	*Populus* spp. [87]	-
6-OCH_3Chr	-	*Populus* spp. [87]	-
5,7-diCH_3Chr	-	*Populus* spp. [87]	-
6-CnnChr	-	*Populus* spp.[195]	-
CH_3Chr	-	*Populus* spp. [87]	-
Genk	La Alcarria*[181]	-	-

(Table 7) contd.....

Flavone	Honey (BO)	Propolis(BO)	Pollen (BO)
Lut **(28)**	*Eucalyptus* [176]; *R. officinalis, Citrus* spp., *H. annus, Lavandula* spp. [95]; *R. pseudoacacia, F. esculentum,* Heather, *Tilia* spp., *Brassica* spp. [35]; palmetto berry [31]; *L. scoparium* [31]	*Populus* spp. [87, 110]; *Dalbergia* spp. [179]; Commercial [16]	*E. globulus* [183]; *Eucalyptus globulus, Cecropia* spp., *Mimosa pudica, Elaeis* spp. [180]; *Cistus incanus* [129]
7-CH_3Lut	La Alcarria*[181]	*Populus* spp. [87]	
Sel	-	-	*Eucalyptus* spp., *Cecropia* spp., [180]
Tect	*R. officinalis* [95]	*Populus* spp. [74, 87, 119, 133, 195, 196, 205], *B. dracunculifolia* [74]	-
Tri	*Eucalyptus* spp. [176]; *R. pseudoacacia, F. esculentum,* Heather, *Tilia* spp., *Brassica* spp. [35]	-	*E. globulus* [183]; *Mimosa pudica, Cecropia* spp., *Elaeis* spp., *Eupatorium* spp., *Eucalyptus globulus*

BO – Botanical origin; "*" after the name refer to geographic origin since no botanical origin is referred. Aca – Acacetin; Api – Apigenin; Baic – Baicalein; CH_3 – Methyl; Chr – Chrysin; Cnn – Cinnamyl; Genk – Genkwanin; Gly – Glycoside; OCH_3 – Methoxy; Sel – Selagin; Tect – Tectochrysin; Tri – Tricetin.

With the exception of some Brazilian propolis, chrysin may account up to 4% of the flavonoid content in propolis [200]. The concentration of this flavone in Argentinian *Populus* spp. propolis reach values as high as 68.7 mg/g extract [119]. High concentrations of chrysin have also been reported in Tunisian *H. annuus* honey (1.3 mg/100g) [193].

Apigenin is also widespread in honey and propolis, although with less extent and abundancy than chrysin [119], while luteolin is most of the times absent from the flavonoid profile of the majority of propolis [6, 201 - 203]. Commonly these flavones are detectable in bee pollen samples though in concentrations below the apparatus limit of quantification thus no relative abundance is found [129, 180, 183].

Flavanones and dihydroflavonols (stucturally equivalents of flavones and flavonols respectively, except for the saturated bond between C2-C3 of the C-ring) have also been identified in honeybee-derived products, though with less representativeness (*i.e.*, only few compounds of these groups have been detected).

Both of these groups are characterized by an intense Band II absorption in the range 277 – 295 nm with only a shoulder or low intensity peak representing Band I in the 300 – 330 nm. This happens because the reduction of the C2=C3 double bond eliminates the cinnamoyl chromophore, leaving only the A-ring benzoyl function intact [168, 169]. Interestingly, the UV spectra of compounds belonging to flavanones are almost identical to those obtained for the equivalent dihydroflavanols, indicating that, contrarily to what happens between flavones and flavonols, the presence or absence of the C-3 hydroxyl group in flavonoids lacking a C2-C3 double bond makes little difference in the UV spectra [169].

Despite less represented, flavanones and dihydroflavanols ocupy a place no less important than flavones and flavonols in the composition of bee products. Notably, pinocembrin (**30** in Fig. (**4**)), pinobanksin (**31** in Fig.(**4**)) and their respective derivatives are commonly present in several honeys and propolis samples of different origins, although they are absent in bee pollen (Tables **8** and **9**). In propolis, these two compounds represent, right after chrysin, the main flavonoid constituents of this bee product, reaching up to 4 and 3% of its total flavonoid content. However, concentrations of pinocembrin and pinobanksin may vary from 1 to 85 mg/g extract and 2 to 77 mg/g extract respectively [72, 111, 118, 196], being these associated with their concentrations in honey, *i.e.*, the higher content of these flavonoids in propolis, the greater their concentration in the honey of the hive [192, 200, 207].

Table 8. Relevant flavanones in honey, propolis and bee pollen worldwide.

Flavanone	**Honey (BO)**	**Propolis(BO)**	**Pollen (BO)**
Pinoc **(30)**	*Tilia* spp. [32,95,194]; *Citrus* spp., *Eucalyptus* spp., *C. sativa* [95, 194]; *Erica* spp., *R. officinalis, H. annus, R. pseudoacacia, Lavandula* spp. [95]; *F. esculentum* [86]; palmetto berry, *L. scoparium* [31]	*Populus* spp. [6, 37, 68, 72, 74, 87, 111, 119, 133, 196, 205]; African propolis [206]; Commercial [118]; *Dalbergia* spp. [179]; *B. dracunculifolia* [73]	*Cystus incanus* [129]
5-CH_3Pinoc	-	*Populus* spp. [87]	-
7-CH_3Pinoc	*Eucalyptus* spp., *Thymus* spp., *R. officinalis, Citrus* spp., *B. campestris, H. annuus*, multifloral [193]	*Populus* spp. [133]	-

(Table 8) contd.....

Flavanone	Honey (BO)	Propolis(BO)	Pollen (BO)
Pinoc-5-*O*-3-O- -4-OCH_3PhProp	-	*Populus* spp. [87]	-
Hesp **(32)**	*Citrus* spp., [95,177,194]; *C. sativa, Eucalyptus* spp. [95, 102, 177, 194]; *Tilia* spp. [35,95,177,194]; *R. pseudoacacia, Brassica* spp.,*F. esculentum* [35]; *Hedysarum* spp. [102]; *R. officinalis, Thymus* spp. [177]; Heather [35, 177]	-	-
5,7-diCH_3Hesp	-	*Populus* spp. [210]	-
Hesp-7-*O*-Rut	Multifloral [185]; *Q. saponaria* [139]	-	-
Nar **(33)**	*H. annus* [193]; *Q. saponaria* [139]; *Mimosa caesaepiniifolia* [178]; *Tilia* spp. [32]; *Citrus* spp., *R. officinalis,* Heather, *Eucalyptus* spp., *C. sativa, Thymus* spp. [177]	*Dalbergia* spp. [182]; *Populus* spp. [37,72,116,177]	*Cystus incanus* [129]
Nar-7-*O*-Hesp	Multifloral [185]; *C. sativa, Citrus* spp. [194]; *Ziziphus spina-christi* [4]; *Tilia* spp. [32, 194]	-	-
Liqu	-	*Dalbergia* spp. [182]	-
Sak	-	*Populus* spp. [133]	-
3-OH-5-CH_3 flavanone	-	*Populus* spp. [87]	-
SophoB	-	African propolis [206]	-

BO – Botanical origin; "*" after the name refer to geographic origin since no botanical origin is referred. CH_3 – Methyl; Hesp – Hesperitin; Liqui – Liquiritigenin; Nar – Naringenin; OCH_3 – Methoxy; OH – Hydroxyl; Pinoc – Pinocembrin; Rut – Rutinoside; Sak – Sakuranetin; SophoB – Sophoraflavanone B.

Table 9. Relevant dihydroflavonols in honey, propolis and bee pollen worldwide.

Di-hydroflavonols	Honey (BO)	Propolis(BO)	Pollen (BO)
Pinob **(31)**	*Eucalyptus* spp. [95, 176]; *Tilia* spp., *Citrus* spp., *H. annus, C. sativa, Erica* spp., *R. officinalis, Lavandula* spp., *R. pseudoacacia* [95]; *F. esculentum* [86]	*Populus* spp. [6, 37, 68, 72, 74, 87, 111, 119, 133, 196]; *Dalbergia* spp. [179]; African propolis [206]; Commercial [118]	-
5-CH_3Pinob	-	*Populus* spp. [87, 196]; Commercial [118]; African propolis [206]	-

(Table 9) contd.....

Di-hydroflavonols	**Honey (BO)**	**Propolis(BO)**	**Pollen (BO)**
5-CH_3Pinob-3-*O*-Ac	-	*Populus* spp. [87]; Commercial [118]	-
5,7-diCH_3Pinob	-	*Populus* spp. [200]	-
Pinob-3-*O*-Ac	-	*Populus* spp. [74,87,188]; *Dalbergia* spp. [179]; African propolis [206]	-
Pinob-3-*O*-Ac-5-*O*-*p*-OHPhProp	-	*Populus* spp. [87]	-
Pinob-3-*O*-Prop	-	*Populus* spp. [87]; Commercial [118]	-
5-CH_3Pinob-3-*O*-Pent	-	*Populus* spp. [87]; Commercial [118]	-
7-CH_3Pinob-5-*O*-*p*-OHPhProp	-	*Populus* spp. [87]	-
Pinob-3-*O*-But or IBut	-	*Populus* spp. [87]	-
Pinob-3-*O*-Pente	-	*Populus* spp. [87]	-
Pinob-3-*O*-Pent or 2-CH_3But	-	*Populus* spp. [87]	-
Pinob-*O*-Hexe	-	*Populus* spp. [87]; Commercial [118]	-
Pinob-3-*O*-PhProp	-	*Populus* spp. [87]	-
Pinob-3-*O*-Hex	-	*Populus* spp. [87]	-
Taxifolin	*Clidemia* spp., *Serjania* spp., *Myrcia* spp., *Mimosa pudica, Mora* spp., *Tapirira* spp., *Schefflera* spp. [148]	-	*Cystus incanus* [129]

BO – Botanical origin; "*" after the name refer to geographic origin since no botanical origin is referred. Ac – Acetate; But – Butyrate; CH_3 – Methyl; Hex – Hexoside; Hexe – Hexenoate; IBut – Isobutyrate; OHPhProp – Hydroxyphenylpropionate; Pent – Pentanoate; Pente – Pentenoate; PhProp – Phenylpropionate; Pinob – Pinobanksin; Prop – Propionate.

Although unconstant, the flavanones naringenin (**33** in Fig. (**4**)) and hesperitin (**32** in Fig. (**4**)) also participate in the composition of certain honeys. Notwithstanding, when these compounds are detected, they normally give valuable information about the origin of this bee product. Hesperitin is a good example of this, since its detection is unique to honeys of *Citrus* spp. origin [208, 209]. The concentration of this compound in Italian *Citrus* spp. honey has been pointed to reach 4.09 μg/100g [102], thought in general, its concentrations can reach about 8 μg/100g.

CONCLUSION

In summary, since honeybee-derived products, in particular honey, propolis and bee pollen, have been showing to possess a series of valuable health benefiting compounds, including phenolic compounds, researchers interest have grown in order to identify, characterize and comprehend their individual and/or synergistic contributions for the health benefits claimed in these three bee products. Therefore, advances in separative and chromatographic techniques, together with the improvement of the detection methods, have greatly contributed to the progress and production of very rich data in this field.

Despite the attemps to find patterns in the phenolic composition of honey, propolis and bee pollen, they exhibit such an immense rich and diversified composition wich are very likely to variations according to their botanical and geographical origin. Still, it is possible to distinguish some phenolics that are ubiquitous among these products such as the non-flavonoids caffeic, *p*-coumaric, cinnamic, benzoic, vanillic acids and their derivatives, and four distinct groups of flavonoids including flavonols, flavones, flavanones and dihydroflavonols.

CONFLICT OF INTEREST

The author confirms that author has no conflict of interest to declare for this publication.

ACKNOWLEDGEMENTS

The authors gratefully acknowledge the financial support provided by the Foundation for Science and Technology (FCT) to CERNAS (Project PEst-OE/AGR/UI0681/2011) and of FCT the European Union, QREN, FEDER, COMPETE, for funding the Organic Chemistry Research Unit (QOPNA) (project PEst-C/QUI/UI0062/2013; FCOMP-01-0124-FEDER- 037296).

REFERENCES

[1] Viuda-Martos, M.; Ruiz-Navajas, Y.; Fernández-López, J.; Pérez-Alvarez, J.A. Functional properties of honey, propolis, and royal jelly. *J. Food Sci.,* **2008**, *73*(9), R117-R124. [http://dx.doi.org/10.1111/j.1750-3841.2008.00966.x] [PMID: 19021816]

[2] Arráez-Román, D.; Gómez-Caravaca, A.M.; Gómez-Romero, M.; Segura-Carretero, A.; Fernández-

Gutiérrez, A. Identification of phenolic compounds in rosemary honey using solid-phase extraction by capillary electrophoresis-electrospray ionization-mass spectrometry. *J. Pharm. Biomed. Anal.,* **2006**, *41*(5), 1648-1656.
[http://dx.doi.org/10.1016/j.jpba.2006.02.035] [PMID: 16574364]

[3] Leja, M.; Mareczek, A.; Wyzgolik, G.; Klepacz-Baniak, J.; Czekońska, K. Antioxidative properties of bee pollen in selected plant species. *Food Chem.,* **2007**, *100*(1), 237-240.
[http://dx.doi.org/10.1016/j.foodchem.2005.09.047]

[4] Badjah Hadj Ahmed, A.Y.; Obbed, M.S.; Wabaidur, S.M.; AlOthman, Z.A.; Al-Shaalan, N.H. High-performance liquid chromatography analysis of phenolic acid, flavonoid, and phenol contents in various natural Yemeni honeys using multi-walled carbon nanotubes as a solid-phase extraction adsorbent. *J. Agric. Food Chem.,* **2014**, *62*(24), 5443-5450.
[http://dx.doi.org/10.1021/jf5011758] [PMID: 24871957]

[5] Ahn, M.R.; Kumazawa, S.; Hamasaka, T.; Bang, K.S.; Nakayama, T. Antioxidant activity and constituents of propolis collected in various areas of Korea. *J. Agric. Food Chem.,* **2004**, *52*(24), 7286-7292.
[http://dx.doi.org/10.1021/jf048726s] [PMID: 15563208]

[6] Barbarić, M.; Mišković, K.; Bojić, M.; Lončar, M.B.; Smolčić-Bubalo, A.; Debeljak, Z.; Medić-Šarić, M.; Debeljak, Z.; Medić-Šarić, M.; Debeljak, Ž.; Medić-Šarić, M. Chemical composition of the ethanolic propolis extracts and its effect on HeLa cells. *J. Ethnopharmacol.,* **2011**, *135*(3), 772-778.
[http://dx.doi.org/10.1016/j.jep.2011.04.015] [PMID: 21515353]

[7] Ajibola, A.; Chamunorwa, J.P.; Erlwanger, K.H. Nutraceutical values of natural honey and its contribution to human health and wealth. *Nutr. Metab. (Lond),* **2012**, *9*, 61.
[http://dx.doi.org/10.1186/1743-7075-9-61] [PMID: 22716101]

[8] Shruthi, E.; Suma, B.S. Health from the Hive: potential uses of propolis in general health. *Int. J. Clin. Med.,* **2012**, *03*(3), 159-162.
[http://dx.doi.org/10.4236/ijcm.2012.33033]

[9] Feás, X.; Vázquez-Tato, M.P.; Estevinho, L.; Seijas, J.A.; Iglesias, A. Organic bee pollen: botanical origin, nutritional value, bioactive compounds, antioxidant activity and microbiological quality. *Molecules,* **2012**, *17*(7), 8359-8377.
[http://dx.doi.org/10.3390/molecules17078359] [PMID: 22785265]

[10] Bertelli, D.; Papotti, G.; Bortolotti, L.; Marcazzan, G.L.; Plessi, M. ^{1}H-NMR simultaneous identification of health-relevant compounds in propolis extracts. *Phytochem. Anal.,* **2012**, *23*(3), 260-266.
[http://dx.doi.org/10.1002/pca.1352] [PMID: 21853496]

[11] Graikou, K.; Kapeta, S.; Aligiannis, N.; Sotiroudis, G.; Chondrogianni, N.; Gonos, E.; Chinou, I. Chemical analysis of Greek pollen - Antioxidant, antimicrobial and proteasome activation properties. *Chem. Cent. J.,* **2011**, *5*(1), 33.
[http://dx.doi.org/10.1186/1752-153X-5-33] [PMID: 21699688]

[12] Pereira, D.M.; Valentão, P.; Pereira, J.a.; Andrade, P.B. Phenolics: From chemistry to biology. *Molecules,* **2009**, *14*(6), 2202-2211.
[http://dx.doi.org/10.3390/molecules14062202]

[13] Balasundram, N.; Sundram, K.; Samman, S. Phenolic compounds in plants and agri-industrial by-products: Antioxidant activity, occurrence, and potential uses. *Food Chem.,* **2006**, *99*, 191-203. [http://dx.doi.org/10.1016/j.foodchem.2005.07.042]

[14] Fernandez-Panchon, M.S.; Villano, D.; Troncoso, A.M.; Garcia-Parrilla, M.C. Antioxidant activity of phenolic compounds: from *in vitro* results to *in vivo* evidence. *Crit. Rev. Food Sci. Nutr.,* **2008**, *48*(7), 649-671. [http://dx.doi.org/10.1080/10408390701761845] [PMID: 18663616]

[15] Khoddami, A.; Wilkes, M.A.; Roberts, T.H. Techniques for analysis of plant phenolic compounds. *Molecules,* **2013**, *18*(2), 2328-2375. [http://dx.doi.org/10.3390/molecules18022328] [PMID: 23429347]

[16] Sun, Y.M.; Wu, H.L.; Wang, J.Y.; Liu, Z.; Zhai, M.; Yu, R.Q. Simultaneous determination of eight flavonoids in propolis using chemometrics-assisted high performance liquid chromatography-diode array detection. *J. Chromatogr. B Analyt. Technol. Biomed. Life Sci.,* **2014**, *962*, 59-67. [http://dx.doi.org/10.1016/j.jchromb.2014.05.027] [PMID: 24907544]

[17] Guiochon, G.; Felinger, A.; Shirazi, D.G. *Fundamentals of Preparative and Nonlinear Chromatography,* 2nd ed; Elsevier Academic Press: USA, **2006**.

[18] Snyder, L.R.; Kirkland, J.J.; Dolan, J.W. *Introduction to modern liquid chromatography,* 3rd ed; John Wiley & Sons, Inc.: New Jersey, **2009**. [http://dx.doi.org/10.1002/9780470508183]

[19] Jing, J.; Parekh, H.S.; Wei, M.; Ren, W.C.; Chen, S.B. Advances in analytical technologies to evaluate the quality of traditional Chinese medicines. *Trends Analyt. Chem.,* **2013**, *44*, 39-45. [http://dx.doi.org/10.1016/j.trac.2012.11.006]

[20] Buckenmaier, S.; Miller, C.A.; van de Goor, T.; Dittmann, M.M. Instrument contributions to resolution and sensitivity in ultra high performance liquid chromatography using small bore columns: comparison of diode array and triple quadrupole mass spectrometry detection. *J. Chromatogr. A,* **2015**, *1377*, 64-74. [http://dx.doi.org/10.1016/j.chroma.2014.11.086] [PMID: 25547218]

[21] Dong, M.W. Modern HPLC for practicing scientists. John Wiley & Sons, **2006**. [http://dx.doi.org/10.1002/0471973106]

[22] Fanali, C.; Dugo, P.; Mondello, L.; D'Orazio, G.; Fanali, S. Recent Developments in High-Performance Liquid Chromatography. In: *Food Anal. by HPLC,* 3rd ed; CRC Press, **2012**. [http://dx.doi.org/10.1201/b13024-2]

[23] Lang, K.M.; Kittelmann, J.; Dürr, C.; Osberghaus, A.; Hubbuch, J. A comprehensive molecular dynamics approach to protein retention modeling in ion exchange chromatography. *J. Chromatogr. A,* **2015**, *1381*, 184-193. [http://dx.doi.org/10.1016/j.chroma.2015.01.018] [PMID: 25618359]

[24] Yang, Y.; Li, H.; Li, Z.; Zhang, Y.; Zhang, S.; Chen, Y.; Yu, M.; Ma, G.; Su, Z. Size-exclusion HPLC provides a simple, rapid, and versatile alternative method for quality control of vaccines by characterizing the assembly of antigens. *Vaccine,* **2015**, *33*(9), 1143-1150. [http://dx.doi.org/10.1016/j.vaccine.2015.01.031] [PMID: 25604799]

[25] Haminiuk, C.W.; Maciel, G.M.; Plata-Oviedo, M.S.; Peralta, R.M. Phenolic compounds in fruits - an overview. *Int. J. Food Sci. Technol.,* **2012**, *47*(10), 2023-2044. [http://dx.doi.org/10.1111/j.1365-2621.2012.03067.x]

[26] Robbins, R.J. Phenolic acids in foods: an overview of analytical methodology. *J. Agric. Food Chem.,* **2003**, *51*(10), 2866-2887. [http://dx.doi.org/10.1021/jf026182t] [PMID: 12720366]

[27] Wren, S.; Tchelitcheff, P. Use of ultra-performance liquid chromatography in pharmaceutical development. *J. Chromatogr. A.,* **2006**, *1119*(1-2), 140-146.

[28] Nollet, L.M.; Toldrá, F. *Food analysis by HPLC,* 3rd ed; CRC Press: New York, **2012**.

[29] Pyrzynska, K.; Biesaga, M. Analysis of phenolic acids and flavonoids in honey. *Trends Analyt. Chem.,* **2009**, *28*(7), 893-902. [http://dx.doi.org/10.1016/j.trac.2009.03.015]

[30] Ćetković, G.; Čanadanović-Brunet, J.; Vulić, J.; Djilas, S.; Tumbas Šaponjac, V. Antioxidant and sensorial properties of linden honey with dried apricots. *Chem. Biodivers.,* **2014**, *11*(11), 1861-1870. [http://dx.doi.org/10.1002/cbdv.201400028] [PMID: 25408327]

[31] Marshall, S.M.; Schneider, K.R.; Cisneros, K.V.; Gu, L. Determination of antioxidant capacities, α-dicarbonyls, and phenolic phytochemicals in Florida varietal honeys using HPLC-DAD-ESI-MS(n.). *J. Agric. Food Chem.,* **2014**, *62*(34), 8623-8631. [http://dx.doi.org/10.1021/jf501329y] [PMID: 25102012]

[32] Gašić, U.; Šikoparija, B.; Tosti, T.; Trifković, J.; Milojković-Opsenica, D.; Natić, M.; Tešić, Ž. Phytochemical fingerprints of lime honey collected in serbia. *J. AOAC Int.,* **2014**, *97*(5), 1259-1267. [http://dx.doi.org/10.5740/jaoacint.SGEGasic] [PMID: 25902974]

[33] Silici, S.; Sarioglu, K.; Dogan, M.; Karaman, K. HPLC-DAD Analysis to Identify the Phenolic Profile of Rhododendron Honeys Collected from Different Regions in Turkey. *Int. J. Food Prop.,* **2014**, *17*(5), 1126-1135. [http://dx.doi.org/10.1080/10942912.2012.698441]

[34] Jerković, I.; Kuś, P.M.; Tuberoso, C.I.; Šarolić, M. Phytochemical and physical-chemical analysis of Polish willow (Salix spp.) honey: identification of the marker compounds. *Food Chem.,* **2014**, *145*, 8-14. [http://dx.doi.org/10.1016/j.foodchem.2013.08.004] [PMID: 24128442]

[35] Sergiel, I.; Pohl, P.; Biesaga, M. Characterisation of honeys according to their content of phenolic compounds using high performance liquid chromatography/tandem mass spectrometry. *Food Chem.,* **2014**, *145*, 404-408. [http://dx.doi.org/10.1016/j.foodchem.2013.08.068] [PMID: 24128495]

[36] Zhang, J.; Cao, X.; Ping, S.; Wang, K.; Shi, J.; Zhang, C.; Zheng, H.; Hu, F. Comparisons of Ethanol Extracts of Chinese Propolis (Poplar Type) and Poplar Gums Based on the Antioxidant Activities and Molecular Mechanism. *Evidence-Based Complement. Altern. Med.,* **2015**, *2015*, 15.

[37] Papotti, G.; Bertelli, D.; Bortolotti, L.; Plessi, M. Chemical and functional characterization of Italian propolis obtained by different harvesting methods. *J. Agric. Food Chem.,* **2012**, *60*(11), 2852-2862. [http://dx.doi.org/10.1021/jf205179d] [PMID: 22360702]

[38] Carpes, S.T.; de Alencar, S.M.; Cabral, I.S.; Oldoni, T.L.; Mourão, G.B.; Haminiuk, C.W.; da Luz, C.F.; Masson, M.L. Polyphenols and palynological origin of bee pollen of Apis mellifera L. from Brazil. Characterization of polyphenols of bee pollen. *J. Food.,* **2013**, *11*(2), 150-161.

[39] Lv, H.; Wang, X.; He, Y.; Wang, H.; Suo, Y. Identification and quantification of flavonoid aglycones in rape bee pollen from Qinghai-Tibetan Plateau by HPLC-DAD-APCI / MS. *J. Food Compos. Anal.,* **2015**, *38*, 49-54.
[http://dx.doi.org/10.1016/j.jfca.2014.10.011]

[40] Ulusoy, E.; Kolayli, S. Phenolic composition and antioxidant properties of anzer bee pollen. *J. Food Biochem.,* **2014**, *38*(1), 73-82.
[http://dx.doi.org/10.1111/jfbc.12027]

[41] Negri, G.; Teixeira, E.W.; Alves, M.L.; Moreti, A.C.; Otsuk, I.P.; Borguini, R.G.; Salatino, A. Hydroxycinnamic acid amide derivatives, phenolic compounds and antioxidant activities of extracts of pollen samples from Southeast Brazil. *J. Agric. Food Chem.,* **2011**, *59*(10), 5516-5522.
[http://dx.doi.org/10.1021/jf200602k] [PMID: 21500799]

[42] Ketkar, S.S.; Rathore, A.S.; Lohidasan, S.; Rao, L.; Paradkar, A.R.; Mahadik, K.R. Investigation of the nutraceutical potential of monofloral Indian mustard bee pollen. *J. Integr. Med.,* **2014**, *12*(4), 379-389.
[http://dx.doi.org/10.1016/S2095-4964(14)60033-9] [PMID: 25074888]

[43] Medić-Sarić, M.; Rastija, V.; Bojić, M. Recent advances in the application of high performance liquid chromatography in the analysis of polyphenols in wine and propolis. *J. AOAC Int.,* **2011**, *94*(1), 32-42.
[PMID: 21391479]

[44] Stalikas, C.D. Extraction, separation, and detection methods for phenolic acids and flavonoids. *J. Sep. Sci.,* **2007**, *30*(18), 3268-3295.
[http://dx.doi.org/10.1002/jssc.200700261] [PMID: 18069740]

[45] Delpino-Rius, A.; Eras, J.; Vilaró, F.; Cubero, M.Á.; Balcells, M.; Canela-Garayoa, R. Characterisation of phenolic compounds in processed fibres from the juice industry. *Food Chem.,* **2015**, *172*, 575-584.
[http://dx.doi.org/10.1016/j.foodchem.2014.09.071] [PMID: 25442594]

[46] Michalkiewicz, A.; Biesaga, M.; Pyrzynska, K. Solid-phase extraction procedure for determination of phenolic acids and some flavonols in honey. *J. Chromatogr. A,* **2008**, *1187*(1-2), 18-24.
[http://dx.doi.org/10.1016/j.chroma.2008.02.001] [PMID: 18282581]

[47] Kumazawa, S.; Goto, H.; Hamasaka, T.; Fukumoto, S.; Fujimoto, T.; Nakayama, T. A new prenylated flavonoid from propolis collected in Okinawa, Japan. *Biosci. Biotechnol. Biochem.,* **2004**, *68*(1), 260-262.
[http://dx.doi.org/10.1271/bbb.68.260] [PMID: 14745198]

[48] Stobiecki, M. Application of mass spectrometry for identification and structural studies of flavonoid glycosides. *Phytochemistry,* **2000**, *54*(3), 237-256.
[http://dx.doi.org/10.1016/S0031-9422(00)00091-1] [PMID: 10870178]

[49] Wang, L.; Morris, M.E. Liquid chromatography-tandem mass spectroscopy assay for quercetin and conjugated quercetin metabolites in human plasma and urine. *J. Chromatogr. B Analyt. Technol.*

Biomed. Life Sci., **2005**, *821*(2), 194-201.
[http://dx.doi.org/10.1016/j.jchromb.2005.05.009] [PMID: 15925552]

[50] Capriotti, A.L.; Cavaliere, C.; Foglia, P.; Piovesana, S.; Ventura, S. Chromatographic Methods Coupled to Mass Spectrometry Detection for the Determination of Phenolic Acids in Plants and Fruits. *J. Liquid Chromatogr. Relat. Technol.,* **2014**, *38*, 353-370.
[http://dx.doi.org/10.1080/10826076.2014.941263]

[51] Marcone, M.F.; Wang, S.; Albabish, W.; Nie, S.; Somnarain, D.; Hill, A. Diverse food-based applications of nuclear magnetic resonance (NMR) technology. *Food Res. Int.,* **2013**, *51*(2), 729-747.
[http://dx.doi.org/10.1016/j.foodres.2012.12.046]

[52] Abraham, R.J.; Fisher, J.; Loftus, P. *Introduction to NMR spectroscopy*; John Wiley & Sons, Ltd., **1989**.

[53] Günther, H. *NMR spectroscopy: basic principles, concepts and applications in chemistry*; John Wiley & Sons, **2013**.

[54] Cawley, A.T.; Flenker, U. The application of carbon isotope ratio mass spectrometry to doping control. *J. Mass Spectrom.,* **2008**, *43*(7), 854-864.
[http://dx.doi.org/10.1002/jms.1437] [PMID: 18523972]

[55] de Rijke, E.; Out, P.; Niessen, W.M.; Ariese, F.; Gooijer, C.; Brinkman, U.A. Analytical separation and detection methods for flavonoids. *J. Chromatogr. A,* **2006**, *1112*(1-2), 31-63.
[http://dx.doi.org/10.1016/j.chroma.2006.01.019] [PMID: 16480997]

[56] Horváth, K.; Molnár-Perl, I. Simultaneous GC-MS quantitation of o-phosphoric, aliphatic and aromatic carboxylic acids, proline, hydroxymethylfurfurol and sugars as their TMS derivatives: In honeys. *Chromatographia,* **1998**, *48*(1), 120-126.
[http://dx.doi.org/10.1007/BF02467527]

[57] Wilkins, A.; Lu, Y. Extractives from New Zealand honeys. 5. Aliphatic dicarboxylic acids in New Zealand rewarewa (Knightea excelsa) honey. *J. Agric. Food Chem.,* **1995**, *7*, 3021-3025.
[http://dx.doi.org/10.1021/jf00060a006]

[58] Sanz, M.L.; Gonzalez, M.; De Lorenzo, C.; Sanz, J.; Martínez-Castro, I. A contribution to the differentiation between nectar honey and honeydew honey. *Food Chem.,* **2005**, *91*, 313-317.
[http://dx.doi.org/10.1016/j.foodchem.2004.06.013]

[59] Kartal, M.; Kaya, S.; Kurucu, S. GC-MS analysis of propolis samples from two different regions of Turkey. *Zeitschrift Fur Naturforsch. - Sect. C J. Biosci.,* **2002**, *57*, 905-909.

[60] LeBlanc, B.W.; Davis, O.K.; Boue, S.; DeLucca, A.; Deeby, T. Antioxidant activity of Sonoran Desert bee pollen. *Food Chem.,* **2009**, *115*(4), 1299-1305.
[http://dx.doi.org/10.1016/j.foodchem.2009.01.055]

[61] Vande Casteele, K.; De Pooter, H.; Van Sumere, C.F. Gas chromatographic separation and analysis of trimethylsilyl derivatives of some naturally occurring non-volatile phenolic compounds and related substances. *J. Chromatogr. A,* **1976**, *121*, 49-63.
[http://dx.doi.org/10.1016/S0021-9673(00)82297-9]

[62] Schulz, J.M.; Herrmann, K. Analysis of hydroxybenzoic and hydroxycinnamic acids in plant material. *J. Chromatogr. A,* **1980**, *195*(1), 95-104.

[http://dx.doi.org/10.1016/S0021-9673(00)81546-0]

[63] Grob, R.L.; Barry, E.F. Modern practice of gas chromatography. John Wiley & Sons, **2004**. http://onlinelibrary.wiley.com/book/10.1002/0471651141 [http://dx.doi.org/10.1002/0471651141]

[64] Araújo, A.S.; da Rocha, L.L.; Tomazela, D.M.; Sawaya, A.C.; Almeida, R.R.; Catharino, R.R.; Eberlin, M.N. Electrospray ionization mass spectrometry fingerprinting of beer. *Analyst (Lond.),* **2005**, *130*(6), 884-889. [http://dx.doi.org/10.1039/b415252b] [PMID: 15912237]

[65] Isidorov, V.A.; Bagan, R.; Bakier, S.; Swiecicka, I. Chemical composition and antimicrobial activity of Polish herbhoneys. *Food Chem.,* **2015**, *171*, 84-88. [http://dx.doi.org/10.1016/j.foodchem.2014.08.112] [PMID: 25308646]

[66] Aljadi, A.M.; Yusoff, K.M. Isolation and identification of phenolic acids in Malaysian honey with antibacterial properties. *Turk. J. Med. Sci.,* **2003**, *33*, 229-236.

[67] Gómez-Caravaca, A.M.; Gómez-Romero, M.; Arráez-Román, D.; Segura-Carretero, A.; Fernández-Gutiérrez, A. Advances in the analysis of phenolic compounds in products derived from bees. *J. Pharm. Biomed. Anal.,* **2006**, *41*(4), 1220-1234. [http://dx.doi.org/10.1016/j.jpba.2006.03.002] [PMID: 16621403]

[68] Popova, M.; Silici, S.; Kaftanoglu, O.; Bankova, V. Antibacterial activity of Turkish propolis and its qualitative and quantitative chemical composition. *Phytomedicine,* **2005**, *12*(3), 221-228. [http://dx.doi.org/10.1016/j.phymed.2003.09.007] [PMID: 15830845]

[69] Massaro, F.C.; Brooks, P.R.; Wallace, H.M.; Russell, F.D. Cerumen of Australian stingless bees (Tetragonula carbonaria): gas chromatography-mass spectrometry fingerprints and potential anti-inflammatory properties. *Naturwissenschaften,* **2011**, *98*(4), 329-337. [http://dx.doi.org/10.1007/s00114-011-0770-7] [PMID: 21347735]

[70] Waksmundzka-Hajnos, M.; Sherma, J.; Kowalska, T., Eds. *Thin layer chromatography in phytochemistry*; CRC Press: New York, **2008**, p. 896. [http://dx.doi.org/10.1201/9781420046786]

[71] Sajewicz, M.; Staszek, D. Waksmundzka-, M.; Kowalska, T., Comparison of TLC and HPLC fingerprints of phenolic acids and flavonoids fractions derived from selected sage (Salvia) species. *J. Liquid Chromatogr. Relat. Technol.,* **2012**, *35*(10), 1388-1403.

[72] Medić-Šarić, M.; Jasprica, I.; Mornar, A.; Smolčić-Bubalo, A.; Golja, P. Quantitative analysis of flavonoids and phenolic acids in propolis by two-dimensional thin layer chromatography. *JPC - J. Planar Chromatogr. -. Mod. TLC.,* **2004**, *17*(100), 459-463.

[73] Park, Y.K.; Koo, M.H.; Abreu, J.A.; Ikegaki, M.; Cury, J.A.; Rosalen, P.L. Antimicrobial activity of propolis on oral microorganisms. *Curr. Microbiol.,* **1998**, *36*(1), 24-28. [http://dx.doi.org/10.1007/s002849900274] [PMID: 9405742]

[74] Park, Y.K.; Alencar, S.M.; Aguiar, C.L. Botanical origin and chemical composition of Brazilian propolis. *J. Agric. Food Chem.,* **2002**, *50*(9), 2502-2506. [http://dx.doi.org/10.1021/jf011432b] [PMID: 11958612]

[75] Medić-Šarić, M.; Jasprica, I.; Smolčić-Bubalo, A.; Mornar, A. Optimization of chromatographic

conditions in thin layer chromatography of flavonoids and phenolic acids. *Croat. Chem. Acta,* **2004**, *77*(1-2), 361-366.

[76] Berezkin, V.G.; Kormishkina, E.V. Study of a new version of classical thin-layer chromatography with a closed adsorbent layer. *J. Planar Chromatogr. -. Mod. TLC.,* **2007**, *19*(February), 81-85.

[77] Tibor, C.; Szőgyi, M.; Cserháti, T.; Szőgyi, M. Eevaluation of chromatographic retention data by cluster analysis. New achievements. *Eur. Chem. Bull.,* **2012**, *1*(3-4), 69-74.

[78] Berezkin, V.G.; Chausov, A.V. The Simple chromatographic chamber and its application in circular and linear TLC. *J. Liquid Chromatogr. Relat. Technol.,* **2012**, *35*(2), 294-307. [http://dx.doi.org/10.1080/10826076.2011.597077]

[79] Morlock, G.E.; Ristivojevic, P.; Chernetsova, E.S. Combined multivariate data analysis of high-performance thin-layer chromatography fingerprints and direct analysis in real time mass spectra for profiling of natural products like propolis. *J. Chromatogr. A,* **2014**, *1328*, 104-112. [http://dx.doi.org/10.1016/j.chroma.2013.12.053] [PMID: 24440096]

[80] Kuś, P.M.; Congiu, F.; Teper, D.; Sroka, Z.; Jerković, I.; Tuberoso, C.I. Antioxidant activity, color characteristics, total phenol content and general HPLC fingerprints of six Polish unifloral honey types. *LWT -. Food Sci. Technol. (Campinas.),* **2014**, *55*(1), 124-130.

[81] Jasicka-Misiak, I.; Poliwoda, A.; Dereń, M.; Kafarski, P. Phenolic compounds and abscisic acid as potential markers for the floral origin of two Polish unifloral honeys. *Food Chem.,* **2012**, *131*(4), 1149-1156. [http://dx.doi.org/10.1016/j.foodchem.2011.09.083]

[82] Awale, S.; Li, F.; Onozuka, H.; Esumi, H.; Tezuka, Y.; Kadota, S. Constituents of Brazilian red propolis and their preferential cytotoxic activity against human pancreatic PANC-1 cancer cell line in nutrient-deprived condition. *Bioorg. Med. Chem.,* **2008**, *16*(1), 181-189. [http://dx.doi.org/10.1016/j.bmc.2007.10.004] [PMID: 17950610]

[83] Liao, S-G.; Zhang, L-J.; Li, C-B.; Lan, Y-Y.; Wang, A-M.; Huang, Y.; Zhen, L.; Fu, X-Z.; Zhou, W.; Qi, X-L.; Guan, Z-Z.; Wang, Y-L. Rapid screening and identification of caffeic acid and its esters in Erigeron breviscapus by ultra-performance liquid chromatography/tandem mass spectrometry. *Rapid Commun. Mass Spectrom.,* **2010**, *24*(17), 2533-2541. [http://dx.doi.org/10.1002/rcm.4662] [PMID: 20740527]

[84] Waksmundzka-Hajnos, M.; Sherma, J. *High performance liquid chromatography in phytochemical analysis*; CRC Press: USA, **2011**.

[85] Biesaga, M.; Pyrzynska, K. Liquid chromatography/tandem mass spectrometry studies of the phenolic compounds in honey. *J. Chromatogr. A,* **2009**, *1216*(38), 6620-6626. [http://dx.doi.org/10.1016/j.chroma.2009.07.066] [PMID: 19691961]

[86] Pasini, F.; Gardini, S.; Marcazzan, G.L.; Caboni, M.F. Buckwheat honeys: screening of composition and properties. *Food Chem.,* **2013**, *141*(3), 2802-2811. [http://dx.doi.org/10.1016/j.foodchem.2013.05.102] [PMID: 23871027]

[87] Falcão, S.I.; Vale, N.; Gomes, P.; Domingues, M.R.; Freire, C.; Cardoso, S.M.; Vilas-Boas, M. Phenolic profiling of Portuguese propolis by LC-MS spectrometry: uncommon propolis rich in flavonoid glycosides. *Phytochem. Anal.,* **2013**, *24*(4), 309-318.

[http://dx.doi.org/10.1002/pca.2412] [PMID: 23172843]

[88] Kečkeš, S.; Gašić, U.; Veličković, T.Ć.; Milojković-Opsenica, D.; Natić, M.; Tešić, Ž. The determination of phenolic profiles of Serbian unifloral honeys using ultra-high-performance liquid chromatography/high resolution accurate mass spectrometry. *Food Chem.,* **2013**, *138*(1), 32-40. [http://dx.doi.org/10.1016/j.foodchem.2012.10.025] [PMID: 23265452]

[89] Trautvetter, S.; Koelling-Speer, I.; Speer, K. Confirmation of phenolic acids and flavonoids in honeys by UPLC-MS. *Apidologie (Celle),* **2009**, *40*(2), 140-150. [http://dx.doi.org/10.1051/apido/2008072]

[90] Oelschlaegel, S.; Gruner, M.; Wang, P.N.; Boettcher, A.; Koelling-Speer, I.; Speer, K. Classification and characterization of manuka honeys based on phenolic compounds and methylglyoxal. *J. Agric. Food Chem.,* **2012**, *60*(29), 7229-7237. [http://dx.doi.org/10.1021/jf300888q] [PMID: 22676798]

[91] Li, W.; Sun, Y.; Liang, W.; Fitzloff, J.F.; van Breemen, R.B. Identification of caffeic acid derivatives in Actea racemosa (Cimicifuga racemosa, black cohosh) by liquid chromatography/tandem mass spectrometry. *Rapid Commun. Mass Spectrom.,* **2003**, *17*(9), 978-982. [http://dx.doi.org/10.1002/rcm.1008] [PMID: 12717772]

[92] Escuredo, O.; Seijo, M.C.; Salvador, J.; González-Martín, M.I. Near infrared spectroscopy for prediction of antioxidant compounds in the honey. *Food Chem.,* **2013**, *141*(4), 3409-3414. [http://dx.doi.org/10.1016/j.foodchem.2013.06.066] [PMID: 23993500]

[93] Khalil, M.I.; Alam, N.; Moniruzzaman, M.; Sulaiman, S.A.; Gan, S.H. Phenolic acid composition and antioxidant properties of Malaysian honeys. *J. Food Sci.,* **2011**, *76*(6), C921-C928. [http://dx.doi.org/10.1111/j.1750-3841.2011.02282.x] [PMID: 22417491]

[94] Liang, Y.; Cao, W.; Chen, W.; Xiao, X.; Zheng, J. Simultaneous determination of four phenolic components in citrus honey by high performance liquid chromatography using electrochemical detection. *Food Chem.,* **2009**, *114*(4), 1537-1541. [http://dx.doi.org/10.1016/j.foodchem.2008.11.024]

[95] Tomás-Barberán, F.A.; Martos, I.; Ferreres, F.; Radovic, B.S.; Anklam, E. HPLC flavonoid profiles as markers for the botanical origin of European unifloral honeys. *J. Sci. Food Agric.,* **2001**, *81*(5), 485-496. [http://dx.doi.org/10.1002/jsfa.836]

[96] Escriche, I.; Kadar, M.; Juan-Borrás, M.; Domenech, E. Suitability of antioxidant capacity, flavonoids and phenolic acids for floral authentication of honey. Impact of industrial thermal treatment. *Food Chem.,* **2014**, *142*, 135-143. [http://dx.doi.org/10.1016/j.foodchem.2013.07.033] [PMID: 24001823]

[97] Ramanauskiene, K.; Stelmakiene, A.; Briedis, V.; Ivanauskas, L.; Jakštas, V. The quantitative analysis of biologically active compounds in Lithuanian honey. *Food Chem.,* **2012**, *132*(3), 1544-1548. [http://dx.doi.org/10.1016/j.foodchem.2011.12.007]

[98] Zhang, X-H.; Wu, H-L.; Wang, J-Y.; Tu, D-Z.; Kang, C.; Zhao, J.; Chen, Y.; Miu, X-X.; Yu, R-Q. Fast HPLC-DAD quantification of nine polyphenols in honey by using second-order calibration method based on trilinear decomposition algorithm. *Food Chem.,* **2013**, *138*(1), 62-69. [http://dx.doi.org/10.1016/j.foodchem.2012.10.033] [PMID: 23265456]

[99] Chan, C.W.; Deadman, B.J.; Manley-Harris, M.; Wilkins, A.L.; Alber, D.G.; Harry, E. Analysis of the flavonoid component of bioactive New Zealand mānuka (Leptospermum scoparium) honey and the isolation, characterisation and synthesis of an unusual pyrrole. *Food Chem.,* **2013**, *141*(3), 1772-1781. [http://dx.doi.org/10.1016/j.foodchem.2013.04.092] [PMID: 23870890]

[100] Borsato, D.M.; Prudente, A.S.; Döll-Boscardin, P.M.; Borsato, A.V.; Luz, C.F.; Maia, B.H.; Cabrini, D.A.; Otuki, M.F.; Miguel, M.D.; Farago, P.V.; Miguel, O.G. Topical anti-inflammatory activity of a monofloral honey of Mimosa scabrella provided by Melipona marginata during winter in southern Brazil. *J. Med. Food,* **2014**, *17*(7), 817-825. [http://dx.doi.org/10.1089/jmf.2013.0024] [PMID: 24650139]

[101] Alvarez-Suarez, J.M.; González-Paramás, A.M.; Santos-Buelga, C.; Battino, M. Antioxidant characterization of native monofloral Cuban honeys. *J. Agric. Food Chem.,* **2010**, *58*(17), 9817-9824. [http://dx.doi.org/10.1021/jf1018164] [PMID: 20701246]

[102] Perna, A.; Intaglietta, I.; Simonetti, A.; Gambacorta, E. A comparative study on phenolic profile, vitamin C content and antioxidant activity of Italian honeys of different botanical origin. *Int. J. Food Sci. Technol.,* **2013**, *48*(9), 1899-1908. [http://dx.doi.org/10.1111/ijfs.12169]

[103] Gambacorta, E.; Simonetti, A.; Garrisi, N.; Intaglietta, I.; Perna, A. Antioxidant properties and phenolic content of sulla (*Hedysarum* spp.) honeys from Southern Italy. *Int. J. Food Sci. Technol.,* **2014**, *49*(10), 2260-2268. [http://dx.doi.org/10.1111/ijfs.12541]

[104] Hussein, S.Z.; Yusoff, K.M.; Makpol, S.; Yusof, Y.A. Antioxidant capacities and total phenolic contents increase with gamma irradiation in two types of Malaysian honey. *Molecules,* **2011**, *16*(8), 6378-6395. [http://dx.doi.org/10.3390/molecules16066378] [PMID: 21796076]

[105] Jaganathan, S.K.; Mandal, S.M.; Jana, S.K.; Das, S.; Mandal, M. Studies on the phenolic profiling, anti-oxidant and cytotoxic activity of Indian honey: *in vitro* evaluation. *Nat. Prod. Res.,* **2010**, *24*(14), 1295-1306. [http://dx.doi.org/10.1080/14786410903184408] [PMID: 20803373]

[106] Weston, R.J.; Mitchell, K.R.; Allen, K.L. Antibacterial phenolic components of New Zealand manuka honey. *Food Chem.,* **1999**, *64*(3), 295-301. [http://dx.doi.org/10.1016/S0308-8146(98)00100-9]

[107] Wang, J.; Xue, X.; Du, X.; Cheng, N.; Chen, L.; Zhao, J.; Zheng, J.; Cao, W. Identification of acacia honey adulteration with rape honey using liquid chromatography–electrochemical detection and chemometrics. *Food Anal. Methods.,* **2014**, *7*(10), 2003-2012. [http://dx.doi.org/10.1007/s12161-014-9833-7]

[108] Campone, L.; Piccinelli, A.L.; Pagano, I.; Carabetta, S.; Di Sanzo, R.; Russo, M.; Rastrelli, L. Determination of phenolic compounds in honey using dispersive liquid-liquid microextraction. *J. Chromatogr. A,* **2014**, *1334*, 9-15. [http://dx.doi.org/10.1016/j.chroma.2014.01.081] [PMID: 24565235]

[109] de Funari, C.S.; de Oliveira Ferro, V.; Mathor, M.B. Analysis of propolis from Baccharis dracunculifolia DC. (Compositae) and its effects on mouse fibroblasts. *J. Ethnopharmacol.,* **2007**,

111(2), 206-212.
[http://dx.doi.org/10.1016/j.jep.2006.11.032] [PMID: 17207952]

[110] Erdogan, S.; Ates, B.; Durmaz, G.; Yilmaz, I.; Seckin, T. Pressurized liquid extraction of phenolic compounds from Anatolia propolis and their radical scavenging capacities. *Food Chem. Toxicol.,* **2011**, *49*(7), 1592-1597.
[http://dx.doi.org/10.1016/j.fct.2011.04.006] [PMID: 21530603]

[111] Kumazawa, S.; Bonvehí, J.S.; Torres, C.; Mok-Ryeon, A.; Bermejo, F.J. Chemical and functional characterisation of propolis collected from East Andalusia (southern Spain). *Phytochem. Anal.,* **2013**, *24*(6), 608-615.
[http://dx.doi.org/10.1002/pca.2439] [PMID: 23670998]

[112] Miorin, P.L.; Levy Junior, N.C.; Custodio, A.R.; Bretz, W.A.; Marcucci, M.C.; Levy Junior, N.C.; Custodio, A.R.; Bretz, W.A.; Marcucci, M.C.; Levy, N.C.; Custodio, A.R.; Bretz, W.A.; Marcucci, M.C. Antibacterial activity of honey and propolis from Apis mellifera and Tetragonisca angustula against *Staphylococcus aureus. J. Appl. Microbiol.,* **2003**, *95*(5), 913-920.
[http://dx.doi.org/10.1046/j.1365-2672.2003.02050.x] [PMID: 14633019]

[113] de Sousa, J.P.; Bueno, P.C.; Gregório, L.E.; da Silva Filho, A.A.; Furtado, N.A.; de Sousa, M.L.; Bastos, J.K. A reliable quantitative method for the analysis of phenolic compounds in Brazilian propolis by reverse phase high performance liquid chromatography. *J. Sep. Sci.,* **2007**, *30*(16), 2656-2665.
[http://dx.doi.org/10.1002/jssc.200700228] [PMID: 17880032]

[114] de Sousa, J.P.; da Silva Filho, A.A.; Bueno, P.C.; Gregório, L.E.; Furtado, N.A.; Jorge, R.F.; Bastos, J.K. A validated reverse-phase HPLC analytical method for the quantification of phenolic compounds in Baccharis dracunculifolia. *Phytochem. Anal.,* **2009**, *20*(1), 24-32.
[http://dx.doi.org/10.1002/pca.1087] [PMID: 18752260]

[115] Aliyazıcıoglu, R.; Sahin, H.; Erturk, O.; Ulusoy, E.; Kolayli, S. Properties of phenolic composition and biological activity of propolis from Turkey. *Int. J. Food Prop.,* **2013**, *16*(2), 37-41.
[http://dx.doi.org/10.1080/10942912.2010.551312]

[116] Jasprica, I.; Smolčić-Bubalo, A.; Mornar, A.; Medić-Šarić, M. Investigation of the flavonoids in Croatian propolis by thin-layer chromatography. *JPC - J. Planar Chromatogr. -. Mod. TLC.,* **2004**, *17*(2), 95-101.

[117] Cao, W.; Liu, H-F.; Cheng, N.; Gao, H.; Wang, B-N.; Zheng, J-B. LC with electrochemical detection for analysis of caffeic acid and caffeic acid phenyl ester in propolis. *Chromatographia,* **2011**, *73*(3-4), 411-414.
[http://dx.doi.org/10.1007/s10337-010-1864-7]

[118] Pellati, F.; Orlandini, G.; Pinetti, D.; Benvenuti, S. HPLC-DAD and HPLC-ESI-MS/MS methods for metabolite profiling of propolis extracts. *J. Pharm. Biomed. Anal.,* **2011**, *55*(5), 934-948.
[http://dx.doi.org/10.1016/j.jpba.2011.03.024] [PMID: 21497475]

[119] Kumazawa, S.; Ahn, M-R.; Fujimoto, T.; Kato, M. Radical-scavenging activity and phenolic constituents of propolis from different regions of Argentina. *Nat. Prod. Res.,* **2010**, *24*(9), 804-812.
[http://dx.doi.org/10.1080/14786410802615270] [PMID: 20461626]

[120] Kao, Y.T.; Lu, M.J.; Chen, C. Preliminary analyses of phenolic compounds and antioxidant activities

in tea pollen extracts. *J. Food Drug Anal.,* **2011**, *19*(4), 470-477.

[121] Habib, H.M.; Al Meqbali, F.T.; Kamal, H.; Souka, U.D.; Ibrahim, W.H. Bioactive components, antioxidant and DNA damage inhibitory activities of honeys from arid regions. *Food Chem.,* **2014**, *153*, 28-34.
[http://dx.doi.org/10.1016/j.foodchem.2013.12.044] [PMID: 24491696]

[122] Dimitrova, B.; Gevrenova, R.; Anklam, E. Analysis of phenolic acids in honeys of different floral origin by solid-phase extraction and high-performance liquid chromatography. *Phytochem. Anal.,* **2007**, *18*(1), 24-32.
[http://dx.doi.org/10.1002/pca.948] [PMID: 17260695]

[123] Pichichero, E.; Canuti, L.; Canini, A. Characterisation of the phenolic and flavonoid fractions and antioxidant power of Italian honeys of different botanical origin. *J. Sci. Food Agric.,* **2009**, *89*(4), 609-616.
[http://dx.doi.org/10.1002/jsfa.3484]

[124] Escuredo, O.; Silva, L.R.; Valentão, P.; Seijo, M.C.; Andrade, P.B. Assessing Rubus honey value: Pollen and phenolic compounds content and antibacterial capacity. *Food Chem.,* **2012**, *130*(3), 671-678.
[http://dx.doi.org/10.1016/j.foodchem.2011.07.107]

[125] Kassim, M.; Achoui, M.; Mustafa, M.R.; Mohd, M.A.; Yusoff, K.M. Ellagic acid, phenolic acids, and flavonoids in Malaysian honey extracts demonstrate *in vitro* anti-inflammatory activity. *Nutr. Res.,* **2010**, *30*(9), 650-659.
[http://dx.doi.org/10.1016/j.nutres.2010.08.008] [PMID: 20934607]

[126] Righi, A.A.; Negri, G.; Salatino, A. Comparative chemistry of propolis from eight brazilian localities. *Evidence-Based Complement. Altern. Med.,* **2013**.

[127] Rocha, B.A.; Bueno, P.C.P.; Vaz, M.M.D.O.L.L.; Nascimento, A.P.; Ferreira, N.U.; Moreno, G.D.P.; Rodrigues, M.R.; Costa-Machado, A.R.D.M.; Barizon, E.A.; Campos, J.C.L.; De Oliveira, P.F.; Acésio, N.D.O.; Martins, S.D.P.L.; Tavares, D.C.; Berretta, A.A. Evaluation of a propolis water extract using a reliable RP-HPLC methodology and *in vitro* and *in vivo* efficacy and safety characterisation. *Evidence-Based Complement. Altern. Med.,* **2013**, *2013*

[128] Medić-Šarić, M.; Bojić, M.; Rastija, V.; Cvek, J. Polyphenolic Profiling of Croatian Propolis and Wine. *Food Technol. Biotechnol.,* **2013**, *51*(2), 159-170.

[129] Sarić, A.; Balog, T.; Sobocanec, S.; Kusić, B.; Sverko, V.; Rusak, G.; Likić, S.; Bubalo, D.; Pinto, B.; Reali, D.; Marotti, T. Antioxidant effects of flavonoid from Croatian *Cystus incanus* L. rich bee pollen. *Food Chem. Toxicol.,* **2009**, *47*(3), 547-554.
[http://dx.doi.org/10.1016/j.fct.2008.12.007] [PMID: 19124059]

[130] Fernandes-Silva, C.C.; Salatino, A.; Salatino, M.L.; Breyer, E.D.; Negri, G. Chemical profiling of six samples of brazilian propolis. *Quim. Nova,* **2013**, *36*(2), 237-240.
[http://dx.doi.org/10.1590/S0100-40422013000200006]

[131] Mishima, S.; Narita, Y.; Chikamatsu, S.; Inoh, Y.; Ohta, S.; Yoshida, C.; Araki, Y.; Akao, Y.; Suzuki, K.M.; Nozawa, Y. Effects of propolis on cell growth and gene expression in HL-60 cells. *J. Ethnopharmacol.,* **2005**, *99*(1), 5-11.
[http://dx.doi.org/10.1016/j.jep.2005.02.005] [PMID: 15848013]

[132] Salonen, A.; Saarnio, S.; Julkunen-Tiitto, R. Phenolic Compounds of Propolis from the Boreal Coniferous Zone. *J. Apic. Sci.,* **2012**, *56*(1), 13-22.

[133] Medana, C.; Carbone, F.; Aigotti, R.; Appendino, G.; Baiocchi, C. Selective analysis of phenolic compounds in propolis by HPLC-MS/MS. *Phytochem. Anal.,* **2008**, *19*(1), 32-39. [http://dx.doi.org/10.1002/pca.1010] [PMID: 17654520]

[134] Zhou, J.; Yao, L.; Li, Y.; Chen, L.; Wu, L.; Zhao, J. Floral classification of honey using liquid chromatography-diode array detection-tandem mass spectrometry and chemometric analysis. *Food Chem.,* **2014**, *145*, 941-949. [http://dx.doi.org/10.1016/j.foodchem.2013.08.117] [PMID: 24128567]

[135] Soler, C.; Gil, I. M.; García-Viguera, C.; Tomás-Barberán A., F.; Gil, M.; García-Viguera, C.; Tomás-Barberán, F.A., Flavonoid patterns of French honeys with different floral origin. *Apidologie (Celle),* **1995**, *26*(1), 53-60. [http://dx.doi.org/10.1051/apido:19950107]

[136] Daher, S.; Gülaçar, F.O. Analysis of phenolic and other aromatic compounds in honeys by solid-phase microextraction followed by gas chromatography-mass spectrometry. *J. Agric. Food Chem.,* **2008**, *56*(14), 5775-5780. [http://dx.doi.org/10.1021/jf8006745] [PMID: 18588314]

[137] Salgueiro, F.B.; Lira, A.F.; Rumjanek, V.M.; Castro, R.N. Phenolic composition and antioxidant properties of Brazilian honeys. *Quim. Nova,* **2014**, *37*(5), 821-826.

[138] Alissandrakis, E.; Kibaris, A.C.; Tarantilis, P.A.; Harizanis, P.C.; Polissiou, M. Flavour compounds of Greek cotton honey. *J. Sci. Food Agric.,* **2005**, *85*(9), 1444-1452. [http://dx.doi.org/10.1002/jsfa.2124]

[139] Montenegro, G.; Díaz-Forestier, J.; Fredes, C.; Rodríguez, S. Phenolic profiles of nectar and honey of Quillaja saponaria Mol. (Quillajaceae) as potential chemical markers. *Biol. Res.,* **2013**, *46*(2), 177-182. [http://dx.doi.org/10.4067/S0716-97602013000200009] [PMID: 23959016]

[140] Fanali, C.; Dugo, L.; Rocco, A. Nano-liquid chromatography in nutraceutical analysis: determination of polyphenols in bee pollen. *J. Chromatogr. A,* **2013**, *1313*, 270-274. [http://dx.doi.org/10.1016/j.chroma.2013.06.055] [PMID: 23880468]

[141] Gheldof, N.; Wang, X-H.; Engeseth, N.J. Identification and quantification of antioxidant components of honeys from various floral sources. *J. Agric. Food Chem.,* **2002**, *50*(21), 5870-5877. [http://dx.doi.org/10.1021/jf0256135] [PMID: 12358452]

[142] Baltrušaitytė, V.; Venskutonis, P.R.; Čeksterytė, V. Radical scavenging activity of different floral origin honey and beebread phenolic extracts. *Food Chem.,* **2007**, *101*(2), 502-514. [http://dx.doi.org/10.1016/j.foodchem.2006.02.007]

[143] Lianda, R. Sant' ana, L.; Echevarria, A.; Castro, R., Antioxidant Activity and Phenolic Composition of Brazilian Honey and their Extracts. *J. Braz. Chem. Soc.,* **2012**, *23*(4), 618-627.

[144] Asmussen, C.A.N.R.; Ang, X.I.A.O.O.N.G.W.; Eung, S.O.L.; Ightingale, M.A.N.; Chmidt, S.H.J.S.; Ngeseth, N.I.J.E. Selection and Use of Honey as an Antioxidant in a French Salad Dressing System. *J Agric Food Chem.,* **2008**, *56*(18), 8650-8657.

[http://dx.doi.org/10.1021/jf800635d]

[145] Silici, S.; Sarioglu, K.; Karaman, K. Determination of polyphenols of some turkish honeydew and nectar honeys using HPLC-DAD. *J. Liquid Chromatogr. Relat. Technol.,* **2013**, *36*(September), 2330-2341.

[146] Cheng, N.; Ren, N.; Gao, H.; Lei, X.; Zheng, J.; Cao, W. Antioxidant and hepatoprotective effects of Schisandra chinensis pollen extract on CCl4-induced acute liver damage in mice. *Food Chem. Toxicol.,* **2013**, *55*, 234-240.
[http://dx.doi.org/10.1016/j.fct.2012.11.022] [PMID: 23201450]

[147] Serra Bonvehí, J.; Soliva Torrentó, M.; Centelles Lorente, E. Evaluation of polyphenolic and flavonoid compounds in honeybee-collected pollen produced in Spain. *J. Agric. Food Chem.,* **2001**, *49*(4), 1848-1853.
[http://dx.doi.org/10.1021/jf0012300] [PMID: 11308335]

[148] da Silva, I.A.; da Silva, T.M.; Camara, C.A.; Queiroz, N.; Magnani, M.; de Novais, J.S.; Soledade, L.E.; Lima, Ede.O.; de Souza, A.L.; de Souza, A.G. Phenolic profile, antioxidant activity and palynological analysis of stingless bee honey from Amazonas, Northern Brazil. *Food Chem.,* **2013**, *141*(4), 3552-3558.
[http://dx.doi.org/10.1016/j.foodchem.2013.06.072] [PMID: 23993520]

[149] Marcucci, M.C.; Ferreres, F.; García-Viguera, C.; Bankova, V.S.; De Castro, S.L.; Dantas, A.P.; Valente, P.H.; Paulino, N. Phenolic compounds from Brazilian propolis with pharmacological activities. *J. Ethnopharmacol.,* **2001**, *74*(2), 105-112.
[http://dx.doi.org/10.1016/S0378-8741(00)00326-3] [PMID: 11167028]

[150] Kumazawa, S.; Yoneda, M.; Shibata, I.; Kanaeda, J.; Hamasaka, T.; Nakayama, T. Direct evidence for the plant origin of Brazilian propolis by the observation of honeybee behavior and phytochemical analysis. *Chem. Pharm. Bull. (Tokyo),* **2003**, *51*(6), 740-742.
[http://dx.doi.org/10.1248/cpb.51.740] [PMID: 12808259]

[151] Vit, P.; Soler, C.; Tomás-Barberán, F.A. Profiles of phenolic compounds of Apis mellifera and *Melipona* spp. honeys from Venezuela. *Zeitschrift Für Leb. Und -Forsch. A.,* **1997**, *204*(1), 43-47.

[152] Moura, S.A.L.; De, ; Negri, G.; Salatino, A.; Lima, L.D.D.C.; Dourado, L.P.A.; Mendes, J.B.; Ferreira, M.A.N.D.; Cara, D.C. Aqueous extract of Brazilian green propolis: Primary components, evaluation of inflammation and wound healing by using subcutaneous implanted sponges. *Evidence-Based Complement. Altern. Med.,* **2011**, *2011*, 8.

[153] De Maria, C.A.; Moreira, R.F. Analytical methods for chlorogenic acid. *Quim. Nova,* **2004**, *27*(4), 586-592.

[154] Bandoniene, D.; Murkovic, M.; Venskutonis, P.R. Determination of rosmarinic acid in sage and borage leaves by high-performance liquid chromatography with different detection methods. *J. Chromatogr. Sci.,* **2005**, *43*(7), 372-376.
[http://dx.doi.org/10.1093/chromsci/43.7.372] [PMID: 16176651]

[155] Messerli, S.M.; Ahn, M.R.; Kunimasa, K.; Yanagihara, M.; Tatefuji, T.; Hashimoto, K.; Mautner, V.; Uto, Y.; Hori, H.; Kumazawa, S.; Kaji, K.; Ohta, T.; Maruta, H.; Artepillin, C. Artepillin C (ARC) in Brazilian green propolis selectively blocks oncogenic PAK1 signaling and suppresses the growth of NF tumors in mice. *Phytother. Res.,* **2009**, *23*(3), 423-427.

[http://dx.doi.org/10.1002/ptr.2658] [PMID: 19003952]

[156] Holser, R. a. Principal Component Analysis of Phenolic Acid Spectra. *ISRN Spectrosc.,* **2012**, *2012*, 1-5.

[157] Pati, S.; Crupi, P.; Benucci, I.; Antonacci, D.; Di Luccia, A.; Esti, M. HPLC-DAD–MS/MS characterization of phenolic compounds in white wine stored without added sulfite. *Food Res. Int.,* **2014**, *66*, 207-215. [http://dx.doi.org/10.1016/j.foodres.2014.09.017]

[158] Matsuda, A.H.; de Almeida-Muradian, L.B. Validated method for the quantification of artepillin-C in Brazilian propolis. *Phytochem. Anal.,* **2008**, *19*(2), 179-183. [http://dx.doi.org/10.1002/pca.1043] [PMID: 18200605]

[159] Lee, I.; Ahn, S.; Kim, B.; Hwang, E.; Kim, Y.S. Analysis of Benzoic Acid in Quasi-Drug Drink Using Isotope Dilution Liquid Chromatography Mass Spectrometry. *Bull. Korean Chem. Soc.,* **2008**, *29*(11), 2125-2128. [http://dx.doi.org/10.5012/bkcs.2008.29.11.2125]

[160] Theerasin, S. Analysis and identification of phenolic compounds in Dioscorea hispida Dennst. *Asian J. Food Agro-Industry.,* **2009**, *2*(04), 547-560. http://www.ajofai.info/Abstract/Analysis and identification of phenolic compounds in dioscorea hispida dennst.pdf.

[161] Kuś, P.M.; Jerković, I.; Tuberoso, C.I.; Marijanović, Z.; Congiu, F. Cornflower (*Centaurea cyanus* L.) honey quality parameters: chromatographic fingerprints, chemical biomarkers, antioxidant capacity and others. *Food Chem.,* **2014**, *142*, 12-18. [http://dx.doi.org/10.1016/j.foodchem.2013.07.050] [PMID: 24001807]

[162] Tsiapara, A.V.; Jaakkola, M.; Chinou, I.; Graikou, K.; Tolonen, T.; Virtanen, V.; Moutsatsou, P. Bioactivity of Greek honey extracts on breast cancer (MCF-7), prostate cancer (PC-3) and endometrial cancer (Ishikawa) cells: Profile analysis of extracts. *Food Chem.,* **2009**, *116*(3), 702-708. [http://dx.doi.org/10.1016/j.foodchem.2009.03.024]

[163] Stephens, J.M.; Schlothauer, R.C.; Morris, B.D.; Yang, D.; Fearnley, L.; Greenwood, D.R.; Loomes, K.M. Phenolic compounds and methylglyoxal in some New Zealand manuka and kanuka honeys. *Food Chem.,* **2010**, *120*(1), 78-86. [http://dx.doi.org/10.1016/j.foodchem.2009.09.074]

[164] Karabagias, I.K.; Vavoura, M.V.; Nikolaou, C.; Badeka, A.V.; Kontakos, S.; Kontominas, M.G. Floral authentication of Greek unifloral honeys based on the combination of phenolic compounds, physicochemical parameters and chemometrics. *Food Res. Int.,* **2014**, *62*, 753-760. [http://dx.doi.org/10.1016/j.foodres.2014.04.015]

[165] Karabagias, I.; Vavoura, M.; Badeka, A.; Kontakos, S.; Kontominas, M.G. Differentiation of greek thyme honeys according to geographical origin based on the combination of phenolic compounds and conventional quality parameters. *Food Anal. Methods,* **2014**, *7*, 2113-2121. [http://dx.doi.org/10.1007/s12161-014-9851-5]

[166] Ververidis, F.; Trantas, E.; Douglas, C.; Vollmer, G.; Kretzschmar, G.; Panopoulos, N. Biotechnology of flavonoids and other phenylpropanoid-derived natural products. Part I: Chemical diversity, impacts on plant biology and human health. *Biotechnol. J.,* **2007**, *2*(10), 1214-1234. [http://dx.doi.org/10.1002/biot.200700084] [PMID: 17935117]

[167] Tanwar, B.; Modgil, R. Flavonoids: dietary occurrence and health benefits. *Spat. DD - Peer Rev. J. Complement. Med. Drug Discov.,* **2012**, *2*(1), 59.

[168] Bohm, B.A. *Introduction to flavonoids*; Harwood academic publishers: Amsterdam, **1998**.

[169] Mabry, T.; Markham, K.R.; Thomas, M.B. The Ultraviolet Spectra of Flavones and Flavonols In: *Syst. Identif. Flavonoids SE - 5*; Springer : Berlin Heidelberg, **1970**. [http://dx.doi.org/10.1007/978-3-642-88458-0_5]

[170] Santos-Buelga, C.; García-Viguera, C.; Tomás-Barberán, F.A. On-Line Identification of Flavonoids by HPLC Coupled to Diode Array Detection. In: *Methods Polyphen. Anal*; Santos-Buelga, C.; Williamson, G., Eds.; The Royal Society of Chemistry: Cambridge, United Kingdom, **2003**.

[171] Cuyckens, F.; Claeys, M. Mass spectrometry in the structural analysis of flavonoids. *J. Mass Spectrom.,* **2004**, *39*(1), 1-15. [http://dx.doi.org/10.1002/jms.585] [PMID: 14760608]

[172] Anklam, E. A review of the analytical methods to determine the geographical and botanical origin of honey. *Food Chem.,* **1998**, *63*(4), 549-562. [http://dx.doi.org/10.1016/S0308-8146(98)00057-0]

[173] Uthurry, C.; Hevia, D.; Gomez-Cordoves, C. Role of honey polyphenols in health. *J. ApiProduct ApiMedical Sci.,* **2011**, *3*(4), 141-159. [http://dx.doi.org/10.3896/IBRA.4.03.4.01]

[174] Justesen, U.; Knuthsen, P.; Leth, T. Quantitative analysis of flavonols, flavones, and flavanones in fruits, vegetables and beverages by high-performance liquid chromatography with photo-diode array and mass spectrometric detection. *J. Chromatogr. A,* **1998**, *799*(1-2), 101-110. [http://dx.doi.org/10.1016/S0021-9673(97)01061-3] [PMID: 9550103]

[175] Tao, W.; Yang, N.; Duan, J.A.; Wu, D.; Guo, J.; Tang, Y.; Qian, D.; Zhu, Z. Simultaneous determination of eleven major flavonoids in the pollen of Typha angustifolia by HPLC-PDA-MS. *Phytochem. Anal.,* **2011**, *22*(5), 455-461. [http://dx.doi.org/10.1002/pca.1302] [PMID: 22033915]

[176] Martos, I.; Ferreres, F.; Yao, L.; D'Arcy, B.; Caffin, N.; Tomás-Barberán, F.A. Flavonoids in monospecific eucalyptus honeys from Australia. *J. Agric. Food Chem.,* **2000**, *48*(10), 4744-4748. [http://dx.doi.org/10.1021/jf000277i] [PMID: 11052728]

[177] Campillo, N.; Viñas, P.; Férez-Melgarejo, G.; Hernández-Córdoba, M. Dispersive liquid-liquid microextraction for the determination of flavonoid aglycone compounds in honey using liquid chromatography with diode array detection and time-of-flight mass spectrometry. *Talanta,* **2015**, *131*, 185-191. [http://dx.doi.org/10.1016/j.talanta.2014.07.083] [PMID: 25281091]

[178] Silva, T.M.; dos Santos, F.P.; Evangelista-Rodrigues, A.; da Silva, E.M.; da Silva, G.S.; de Novais, J.S.; dos Santos, F.D.; Camara, C.A. Phenolic compounds, melissopalynological, physicochemical analysis and antioxidant activity of jandaíra (Melipona subnitida) honey. *J. Food Compos. Anal.,* **2013**, *29*(1), 10-18. [http://dx.doi.org/10.1016/j.jfca.2012.08.010]

[179] Daugsch, A.; Moraes, C.S.; Fort, P.; Park, Y.K. Brazilian red propolis--chemical composition and

botanical origin. *Evid. Based Complement. Alternat. Med.,* **2008**, *5*(4), 435-441. [http://dx.doi.org/10.1093/ecam/nem057] [PMID: 18955226]

[180] Freire, K.R.; Lins, A.C.; Dórea, M.C.; Santos, F.A.; Camara, C.A.; Silva, T.M. Palynological origin, phenolic content, and antioxidant properties of honeybee-collected pollen from Bahia, Brazil. *Molecules,* **2012**, *17*(2), 1652-1664. [http://dx.doi.org/10.3390/molecules17021652] [PMID: 22314384]

[181] Ferreres, F.; Tomáas-Barberáan, F.; Gil, M.I.; Tomáas-Lorente, F. An HPLC technique for flavonoid analysis in honey. *J. Sci. Food Agric.,* **1991**, *56*(1), 49-56. [http://dx.doi.org/10.1002/jsfa.2740560106]

[182] Piccinelli, A.L.; Lotti, C.; Campone, L.; Cuesta-Rubio, O.; Campo Fernandez, M.; Rastrelli, L. Cuban and Brazilian red propolis: botanical origin and comparative analysis by high-performance liquid chromatography-photodiode array detection/electrospray ionization tandem mass spectrometry. *J. Agric. Food Chem.,* **2011**, *59*(12), 6484-6491. [http://dx.doi.org/10.1021/jf201280z] [PMID: 21598949]

[183] Campos, M.; Markham, K.R.; Mitchell, K. An Approach to the Characterization of Bee Pollens *via* their Flavonoid / Phenolic Profiles. *Phytochemical Analysis,* **1997**, *8*(4), 184-185.

[184] Almaraz-Abarca, N.; da Graça Campos, M.; Ávila-Reyes, J.A.; Naranjo-Jiménez, N.; Herrera Corral, J.; González-Valdez, L.S. Antioxidant activity of polyphenolic extract of monofloral honeybee-collected pollen from mesquite (*Prosopis juliflora,* Leguminosae). *J. Food Compos. Anal.,* **2007**, *20*(2), 119-124. [http://dx.doi.org/10.1016/j.jfca.2006.08.001]

[185] Biesaga, M.; Pyrzyńska, K. Stability of bioactive polyphenols from honey during different extraction methods. *Food Chem.,* **2013**, *136*(1), 46-54. [http://dx.doi.org/10.1016/j.foodchem.2012.07.095] [PMID: 23017391]

[186] Truchado, P.; Ferreres, F.; Tomas-Barberan, F.A. Liquid chromatography-tandem mass spectrometry reveals the widespread occurrence of flavonoid glycosides in honey, and their potential as floral origin markers. *J. Chromatogr. A,* **2009**, *1216*(43), 7241-7248. [http://dx.doi.org/10.1016/j.chroma.2009.07.057] [PMID: 19683245]

[187] Arráez-Román, D.; Zurek, G.; Bässmann, C.; Almaraz-Abarca, N.; Quirantes, R.; Segura-Carretero, A.; Fernández-Gutiérrez, A. Identification of phenolic compounds from pollen extracts using capillary electrophoresis-electrospray time-of-flight mass spectrometry. *Anal. Bioanal. Chem.,* **2007**, *389*(6), 1909-1917. [http://dx.doi.org/10.1007/s00216-007-1611-6] [PMID: 17899027]

[188] Piccinelli, A.L.; Mencherini, T.; Celano, R.; Mouhoubi, Z.; Tamendjari, A.; Aquino, R.P.; Rastrelli, L. Chemical composition and antioxidant activity of Algerian propolis. *J. Agric. Food Chem.,* **2013**, *61*(21), 5080-5088. [http://dx.doi.org/10.1021/jf400779w] [PMID: 23650897]

[189] Sun, Y-M.; Wu, H-L.; Wang, J-Y.; Liu, Z.; Zhai, M.; Yu, R-Q. Simultaneous determination of eight flavonoids in propolis using chemometrics-assisted high performance liquid chromatography-diode array detection. *J. Chromatogr. B Analyt. Technol. Biomed. Life Sci.,* **2014**, *962*, 59-67. [http://dx.doi.org/10.1016/j.jchromb.2014.05.027] [PMID: 24907544]

[190] Truchado, P.; Erreres, F.E.; Ortolotti, L.A. *Nectar Flavonol Rhamnosides Are Floral Markers of Acacia (Robinia pseudacacia)*; Honey, **2008**, pp. 8815-8824.

[191] Ferreres, F.; Andrade, P.; Gil, M.I.; Tomás-Barberán, F.A. Floral nectar phenolics as biochemical markers for the botanical origin of heather honey. *Z. Lebensm. Unters. Forsch.,* **1996**, *202*(1), 40-44. [http://dx.doi.org/10.1007/BF01229682]

[192] Ferreres, F.; Andrade, P.; Tomás-Barberán, F.a. Flavonoids from Portuguese heather honey. *Z. Lebensm. Unters. Forsch.,* **1994**, *199*(1), 32-37. [http://dx.doi.org/10.1007/BF01192949]

[193] Martos, I.; Cossentini, M.; Ferreres, F.; Tomasbarberan, F. a, Flavonoid composition of Tunisian honeys and propolis. *J. Agric. Food Chem.,* **1997**, *45*(8), 2824-2829. [http://dx.doi.org/10.1021/jf9609284]

[194] Cavazza, A.; Corradini, C.; Musci, M.; Salvadeo, P. High-performance liquid chromatographic phenolic compound fingerprint for authenticity assessment of honey. *J. Sci. Food Agric.,* **2013**, *93*(5), 1169-1175. [http://dx.doi.org/10.1002/jsfa.5869] [PMID: 22968998]

[195] Usia, T.; Banskota, A.H.; Tezuka, Y.; Midorikawa, K.; Matsushige, K.; Kadota, S. Constituents of Chinese propolis and their antiproliferative activities. *J. Nat. Prod.,* **2002**, *65*(5), 673-676. [http://dx.doi.org/10.1021/np010486c] [PMID: 12027739]

[196] Ahn, M.; Kumazawa, S.; Usui, Y.; Nakamura, J.; Matsuka, M.; Zhu, F.; Nakayama, T. Antioxidant activity and constituents of propolis collected in various areas of China. *Food Chem.,* **2007**, *101*(4), 1383-1392. [http://dx.doi.org/10.1016/j.foodchem.2006.03.045]

[197] Fabre, N.; Rustan, I.; de Hoffmann, E.; Quetin-Leclercq, J. Determination of flavone, flavonol, and flavanone aglycones by negative ion liquid chromatography electrospray ion trap mass spectrometry. *J. Am. Soc. Mass Spectrom.,* **2001**, *12*(6), 707-715. [http://dx.doi.org/10.1016/S1044-0305(01)00226-4] [PMID: 11401161]

[198] Valencia, D.; Alday, E.; Robles-Zepeda, R.; Garibay-Escobar, A.; Galvez-Ruiz, J.C.; Salas-Reyes, M.; Jiménez-Estrada, M.; Velazquez-Contreras, E.; Hernandez, J.; Velazquez, C. Seasonal effect on chemical composition and biological activities of Sonoran propolis. *Food Chem.,* **2012**, *131*(2), 645-651. [http://dx.doi.org/10.1016/j.foodchem.2011.08.086]

[199] Vijaya Bhaskar Reddy, M.; Shen, Y-C.; Ohkoshi, E.; Bastow, K.F.; Qian, K.; Lee, K-H.; Wu, T-S. Bis-chalcone analogues as potent NO production inhibitors and as cytotoxic agents. *Eur. J. Med. Chem.,* **2012**, *47*(1), 97-103. [http://dx.doi.org/10.1016/j.ejmech.2011.10.026] [PMID: 22115618]

[200] Gardana, C.; Scaglianti, M.; Pietta, P.; Simonetti, P. Analysis of the polyphenolic fraction of propolis from different sources by liquid chromatography-tandem mass spectrometry. *J. Pharm. Biomed. Anal.,* **2007**, *45*(3), 390-399. [http://dx.doi.org/10.1016/j.jpba.2007.06.022] [PMID: 17935924]

[201] Kumazawa, S.; Hamasaka, T.; Nakayama, T. Antioxidant activity of propolis of various geographic

origins. *Food Chem.,* **2004**, *84*(3), 329-339.
[http://dx.doi.org/10.1016/S0308-8146(03)00216-4]

[202] Gregoris, E.; Stevanato, R. Correlations between polyphenolic composition and antioxidant activity of Venetian propolis. *Food Chem. Toxicol.,* **2010**, *48*(1), 76-82.
[http://dx.doi.org/10.1016/j.fct.2009.09.018] [PMID: 19766694]

[203] Coneac, G.; Gafiţanu, E.; Hădărugă, D.I.; Hădărugă, N.G.; Pînzaru, I.a.; Bandur, G.; Urşica, L.; Păunescu, V.; Gruia, A. Flavonoid contents of propolis from the West Side of Romania and correlation with the antioxidant activity. *Chem. Bull. "POLITEHNICA. Univ. (Timişoara),* **2008**, *53*(67), 56-60.

[204] Marghitas, L.A.; Dezmirean, D.S.; Pocol, C.B.; Ilea, M.; Bobis, O.; Gergen, I. FlThe development of a biochemical profile of acacia honey by identifying biochemical determinants of its quality. *Not. Bot. Horti Agrobot. Cluj-Napoca.,* **2010**, *38*(2), 84-90.

[205] Lima, B.; Tapia, A.; Luna, L.; Fabani, M.P.; Schmeda-Hirschmann, G.; Podio, N.S.; Wunderlin, D.A.; Feresin, G.E. Main flavonoids, DPPH activity, and metal content allow determination of the geographical origin of propolis from the Province of San Juan (Argentina). *J. Agric. Food Chem.,* **2009**, *57*(7), 2691-2698.
[http://dx.doi.org/10.1021/jf803866t] [PMID: 19334753]

[206] Zhang, T.; Omar, R.; Siheri, W.; Al Mutairi, S.; Clements, C.; Fearnley, J.; Edrada-Ebel, R.; Watson, D. Chromatographic analysis with different detectors in the chemical characterisation and dereplication of African propolis. *Talanta,* **2014**, *120*, 181-190.
[http://dx.doi.org/10.1016/j.talanta.2013.11.094] [PMID: 24468358]

[207] Siess, M.H. Le Bon, a M.; Canivenc-Lavier, M.C.; Amiot, M.J.; Sabatier, S.; Aubert, S.; Suschetet, M., Flavonoids of Honey and Propolis : Characterization and Effects on Hepatic Drug-Metabolizing Enzymes and Benzo [a] pyrene-DNA Binding in Rats. *J. Agric. Food Chem.,* **1996**, *44*, 2297-2301.
[http://dx.doi.org/10.1021/jf9504733]

[208] Ferreres, F.; García-Viguera, C.; Tomás-Lorente, F.; Tomás-Barberán, F.A. Hesperetin: A marker of the floral origin of citrus honey. *J. Sci. Food Agric.,* **1993**, *61*(1), 121-123.
[http://dx.doi.org/10.1002/jsfa.2740610119]

[209] Petrus, K.; Schwartz, H.; Sontag, G. Analysis of flavonoids in honey by HPLC coupled with coulometric electrode array detection and electrospray ionization mass spectrometry. *Anal. Bioanal. Chem.,* **2011**, *400*(8), 2555-2563.
[http://dx.doi.org/10.1007/s00216-010-4614-7] [PMID: 21229237]

[210] Falcão, S.I.; Vilas-Boas, M.; Estevinho, L.M.; Barros, C.; Domingues, M.R.; Cardoso, S.M. Phenolic characterization of Northeast Portuguese propolis: usual and unusual compounds. *Anal. Bioanal. Chem.,* **2010**, *396*(2), 887-897.
[http://dx.doi.org/10.1007/s00216-009-3232-8] [PMID: 19902191]

CHAPTER 5

Valuable Analytical Tools in Analysis of Honeybee Plant-Derived Compounds: Nuclear Magnetic Resonance Spectroscopy

Clementina M.M. Santos[1], Artur M.S. Silva[2,*]

[1] *School of Agriculture, Polytechnic Institute of Bragança, 5300-253 Bragança, Portugal*

[2] *Department of Chemistry & QOPNA, University of Aveiro, Campus Universitário de Santiago, 3810-193 Aveiro, Portugal*

Abstract: Over the past fifty years, Nuclear Magnetic Resonance (NMR) spectroscopy proved to be a powerful tool in the structural characterisation of bioactive compounds from natural sources. In this chapter we cover the basic theory behind each NMR technique used to determine the structure of several families of compounds (*e.g.* carbohydrates, phenolics and sesquiterpenoids) present in honey and propolis. We also provide basic information how 1D and 2D NMR techniques can help in the structure establishment of honeybee constituents. The ^{1}H and ^{13}C NMR data of several of these constituents are compiled and described, being some of them used as botanical and geographical markers. In the case of propolis, a list of compounds identified by NMR is presented. A basic overview in quantitative NMR determinations and in NMR coupled to chemometric methodologies highlights their use to detect honey adulteration and assign their authenticity.

Keywords: Adulteration, Authenticity, Botanical marker, ^{13}C NMR, COSY, DEPT, DOSY, Geographical marker, HMBC, HMQC, ^{1}H NMR, Honeybee, HSQC, Natural products, NMR spectroscopy, NOESY, Propolis, ROESY, Structure elucidation, TOCSY.

* **Correspondence to A.M.S. Silva:** Department of Chemistry, University of Aveiro, Campus Universitário de Santiago, 3810-193 Aveiro, Portugal; Tel: +351 234 379714; Fax: +351 234 370084; Email: artur.silva@ua.pt

Susana M. Cardoso (Ed)

1. INTRODUCTION

The most important tools used by chemists for the structural elucidation of natural and synthetic compounds are based on spectroscopic techniques, being NMR spectroscopy a useful resource on a routine basis. The data provided by 1D NMR techniques, such as ^{1}H and ^{13}C NMR experiments, is the first step to achieve this goal. The complexity of molecular structures implies the use of more advanced 1D and 2D NMR techniques for the complete assignment of proton, carbon or other nuclei resonances. It is a non-destructive technique and with the greatest advances over the past years, excellent results can be obtained from samples weighing less than a milligram.

Thus, a brief discussion on the basic concepts of the commonly used NMR experiments from the point of view of structural elucidation of organic compounds is required to highlights the usefulness of these spectroscopic techniques.

The ^{1}H nuclei is the most commonly observed nuclei in NMR spectroscopy and is the starting point for most structure determinations. Valuable information can be obtained from a ^{1}H NMR spectrum: the chemical shift (δ, ppm; using TMS as standard) that is correlated with its chemical environment; the coupling constant (J, Hz) that are determined by the interactions between individual nuclei and promoted by electrons in a chemical bond and finally, under suitable conditions, the area of a resonance that is related to the number of nuclei giving rise to the ^{1}H NMR signal [1, 2].

The ^{13}C NMR spectrum offers further characterization of a molecule and is directly related to the carbon skeleton. Typically, this spectrum is recorded with broadband decoupling of all protons, appearing each carbon resonance as a single line (δ, ppm; using TMS as standard) [1, 2]. The chemical shift of each resonance is once more indicative of their environment, being possible to identify certain functional groups that are not detected in a ^{1}H spectrum (*e.g.* carbonyls).

The ^{15}N NMR spectrum although having great importance for structural NMR analysis, since *N*-containing functional groups and N atoms are present in several molecular skeletons, the low natural abundance of ^{15}N (about 0.4%) and its extremely low relative sensitivity make difficult these measurements [1, 2]. The typical range of ^{15}N chemical shift is about 600 ppm and the recommend reference is nitromethane ($\delta = 0$ ppm), although these values can also be quoted with respect to saturated aqueous solution of ammonium chloride (δ = - 359.5 ppm) or ammonium nitrate (δ = - 3.9 ppm). Unfortunately, the analysis and comparison of the reported data is many times hampered due to the lack of information of the reference standard used.

The ^1H-^1H COSY (COrrelation SpectroscopY) spectrum provides a means of identifying mutually coupled protons (typically geminal and vicinal couplings), allowing the assignment of all proton resonances [1, 3]. In this technique, 1D spectrum is displayed along each axis with a contour projection of this spectrum along the diagonal axis. Off-diagonal peaks represent proton shift correlations (or proton couplings). A related COSY technique, DQF-COSY (Double Quantum Filtered-COrrelated SpectroscopY), is sometimes used to simplify the diagonal of the COSY spectrum where the peaks are greatly reduced in intensity with consequent clarification of this region. A further advantage of DQF-COSY is that in the phase sensitive mode, both diagonal and cross peaks can be adjusted to have a pure absorption line shape. A similar spectrum to ^1H-^1H COSY can be obtained by a TOCSY (TOtal Correlated SpectroscopY) experiment. The correlations observed between all the protons within a given spin system is irrespective of whether they are directly coupled or not. This technique is quite useful in severe resonance overlapping, in which ^1H-^1H COSY spectra can leave to ambiguous assignments.

NOE (Nuclear Overhauser Effect) difference experiment provides information about the spatial proximity of two protons within a molecule [1, 3]. It involves the application of a radio-frequency field to a single resonance in which the corresponding protons become saturated (it means that the difference of population between their high and low energy levels are forced to zero). Recording of the proton spectrum after the period of saturation may, therefore, show changes in signal intensities for the protons in the vicinity of the saturated proton. Thus, 1D NOE studies generally presents the differences between the original proton spectrum and the irradiated proton spectra. In the 2D NOESY (Nuclear Overhauser Effect SpectroscopY) experiments, only one spectrum is recorded and provides all the information from the NOE studies [1, 3].

2D ROESY (Rotating-frame Overhauser Effect SpectroscopY) is used to measure homonuclear ROE effects under spin-locked conditions, these cross-peaks arise from proton resonances that are spatially close even if they are not bonded [3]. It is a quite useful for molecules with large molecular sizes (NOEs are too weak to be detectable). Since in NOESY the cross-relaxation rate constant goes from positive (for small molecules) to negative (for large molecules) as the correlation time increases, or even null, in ROESY the cross-relaxation rate constant is positive for all rotational correlation times.

The first attempt to assign the signals of a complicated ^{13}C NMR spectrum is through DEPT (Distortionless Enhancement by Polarization Transfer) experiments which provide information about the number of protons attached to a carbon atom (C, CH, CH_2, CH_3) [1, 3]. This technique is more sensitive than the normal ^{13}C and also very useful to assign the type of carbons in molecules bearing aromatic, aliphatic and/or olefinic moieties. In a DEPT experiment, sequences of pulses with various delay times are used to create the DEPT spectra. This requires acquiring and processing three separate spectra termed as DEPT-45, DEPT-90 and DEPT-135 (the number indicates the flip angle of the editing proton pulse in the sequence). A DEPT-45 spectrum presents positive signals for all H-bearing carbon peaks; in a DEPT-90 spectrum only methine carbons are seen as positive signals and for DEPT-135 spectrum methyl and methine carbons peaks appear as positive signals and methylene carbons peaks give negative signals. Quaternary carbons are not detected in DEPT spectra because the technique relies on polarization transfer, which in this case is the transfer of proton magnetization onto the directly boundcarbon atoms.

Similar information is obtained from the 2D HSQC (Heteronuclear Single Quantum Correlation) and HMQC (Heteronuclear Multiple Quantum Coherence) experiments [3]. These two techniques provide single bond heteronuclear shift correlation, identifying one-bond H-C or H-N connectivities within a molecule. The resulting 2D spectrum displays in one axis ^1H chemical shifts and the other for the heteronuclei, most often ^{13}C or ^{15}N chemical shifts. The cross-peaks in the contour plot define to which carbon or nitrogen a particular proton (or group of protons) is directly attached.

HMBC (Heteronuclear Multiple Bond Correlation) is closely related to HMQC, although much less sensitive, that operates in essentially the same manner [1, 3]. In this experiment, however, the sequence timings are optimised for much smaller coupling constants and therefore establish long-range correlations. These ^{1}H-^{13}C couplings typically occur with significant intensity over 2 and 3 bonds ($^{2,3}J_{C\text{-}H}$ usually < 10 Hz), but may be apparent over 4 bonds in conjugated systems. Correlations can also be observed to quaternary centres (for example in the assignment of carbonyl resonances) and across heteroatoms other than carbon (for example hydroxyl protons).

Taking into consideration the information provided by the different techniques previously described, we will highlight below their application in the identification of several families of compounds present in honeybee plant-derived products. Commonly, the identification of the constituents of honeybee products by NMR can be done through three different approaches: i) isolation and complete structure elucidation of pure compounds by using 1D and 2D NMR techniques; ii) isolation and simple structure elucidation of pure compounds by using 1D NMR techniques, which data are compared with those of authentic samples; or iii) using NMR as a complementary tool of hyphenated chromatographic and/or chemometric techniques, identifying characteristic ^{1}H and/or ^{13}C NMR signals from complex mixtures of compounds. In the first case we will exhaustible describe the structure elucidation of such compounds and depict the corresponding structure. In the second case, the structure will be depicted and some important NMR features will be described. In the last case, we will only list the identified compounds.

2. HONEY

In the last few decades, alternative or complementary methodologies to mellissopalinology have been developed in order to identify and classify honeys origin focused on the presence of unique molecules or their mixtures. In this context, a series of compounds were isolated and characterized by NMR spectroscopy and used as markers of their botanical origins. In 1996, abscisic acid was detected in Portuguese heather honeys (*Erica* spp.) and identified for the first time in a monofloral honey [4]. The authors isolated two isomers of abscisic acid, (±)-2-*trans*-4-*trans*-abscisic-**1** and (±)-2-*cis*-4-*trans*-abscisic acid-**1** (Fig. **1**), being their NMR spectra very similar (Table **1**). The signal of H-4 of the *cis*, *trans*-**1** isomer is deshielded and that of C-4 are shielded compared with those of the *trans*,*trans*-**1** isomer. It is important to highlight that this work has a problem in the assignment of carbons C12-C15 of both isomers. The authors missed the assignment of C-15 and according to a deep analysis of the NMR data of *cis*, *trans*-**1** these assignments must be: i) C-12 and C-13 at 23.1 and 24.3 ppm, ii) C-14 at 19.1 ppm and iii) C-15 at 21.5 ppm [5].

Table 1. ^{1}H and ^{13}C NMR spectroscopic data (δ in CD_3OD, ppm; *J* in parenthesis, Hz) of (±)-2-*trans*-4-*trans*-abscisic acid **1** and (±)-2-*cis*-4-*trans*-abscisic acid **1**. (Adapted from [4]).

Position	(±)-2-*trans*-4-*trans*-1		(±)-2-*cis*-4-*trans*-1	
	^{1}H	^{13}C	^{1}H	^{13}C
1		171.51		170.60
2	5.87, s	119.94	5.78 s	118.10
3		161.95		163.00
4	6.47 d (15.9)	134.18	7.82 d (15.9)	128.50
5	6.18 d (15.9)	135.66	6.18 d (15.9)	136.90
6		79.69		79.90
7		153.04		151.50
8	5.95 s	127.31	5.98 s	127.10
9		197.70		198.30
10	2.49 d (17.1)	49.75	2.50 d (17.1)	49.70
	2.31 d (17.1)		2.30 d (17.1)	
11		41.63		41.70
12	1.11 s	14.38	1.12 s	19.10
13	1.02 s	23.08	1.04 s	23.10
14	1.90 s	24.36	1.93 s	21.40
15	2.29 s		2.05 s	

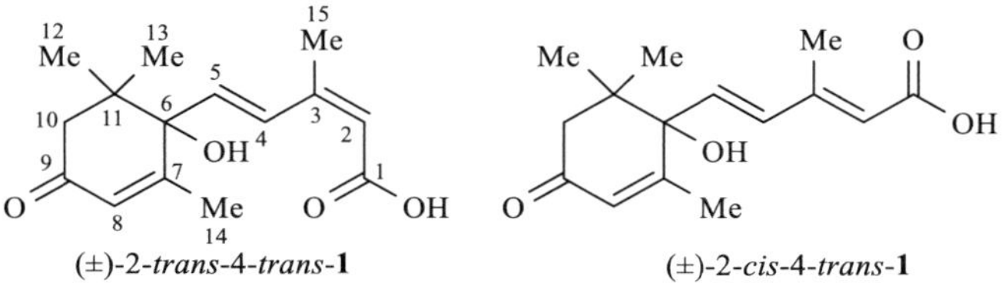

(±)-2-*trans*-4-*trans*-**1** (±)-2-*cis*-4-*trans*-**1**

Fig. (1). Structures of (±)-2-*trans*-4-*trans*-abscisic-**1** and (±)-2-*cis*-**4**-*trans*-abscisic acid-**1**.

In a strawberry-tree (*Arbutus unedo*) honey the phenolic acid homogentisic acid (2,5-dihydroxyphenylacetic acid), which structure was assigned based in the MS and NMR (^{1}H, ^{13}C and DEPT-135 spectra) data and confirmed by spectral comparison with an authentic sample, was used as marker of this monofloral honey [6]. Complementary markers of strawberry-tree honeys were (±)-2-*cis*-4-*tran*s-abscisic acid, (±)-2-*trans*-4-*tran*s-abscisic acid and unedone **2** (Fig. **2**) [7]. The structure of unedone **2** was identified by MS, HPLC-MS and NMR (^{1}H, ^{13}C, DEPT-90 and 135, COSY, HSQC and HMBC) (Table **2**).

Table 2. ^{1}H and ^{13}C NMR spectroscopic data (δ in $CDCl_3$, ppm; *J.* in parenthesis, Hz) of unedone **2**. (Adapted from [7]).

Position	δ_H	δ_C[a]	$^{2,3}J_{H\text{-}C}$ connectivities
1		37.1 s	C-1, C-3, C-4, C-6, C-11, C-12
2	2.38 dd (17.1, 1.2) (H-2α)	51.2 t	C-1, C-3, C-6, C-11, C-12
	2.50 br d (17.1) (H-2β)		
3		197.0 s	
4	6.03 br s	129.6 d	C-2, C-6, C-13
5		161.2 s	
6		68.7 s	
7	3.10 d (8.7)	66.0 d	C-5, C-6
7	4.01 dd (8.7, 3.9)	72.1 d	C-7, C-9, C-10
8	3.84 m	69.2 d	C-7, C-10
10	1.25 d (6.6)	19.7 q	C-8, C-9
11	1.16 s	25.7 q	C-1, C-2, C-6, C-12
12	1.01 s	26.9 q	C-1, C-2, C-3, C-6, C-11
13	1.78 d (1.2)	18.0 q	C-4, C-5, C-6

[a] The multiplicity was determined by DEPT experiments.

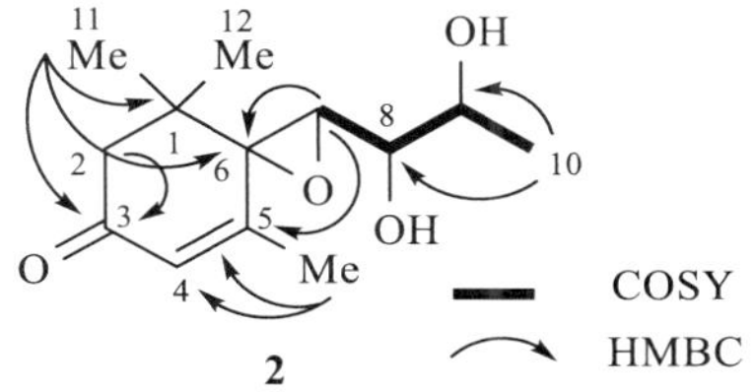

2

Fig. (2). Main COSY and HMBC correlations of unedone **2**.

Later on, the phenolic ester methyl syringate (methyl 4-hydroxy-3-5-dimethoxybenzoate) was identified as a chemical marker of the asphodel (*Asphodelus microcarpus* Salm. Et Viv.) monofloral honey [8]. The ^{1}H NMR spectrum of this compound was particularly important in the structure elucidation because of their four singlets (3.87 and 3.91 ppm of the methoxyl groups, 5.88 ppm of the hydroxyl group and 7.30 ppm corresponding to the symmetrical aromatic methines). 4-Hydroxyquinaldic acid (kynurenic acid) **3** and their tautomer 4-quinolone-2-carboxylic acid **4** (Fig. **3**) were reported as markers of the floral origin of chestnut honey [9]. Their structures were assigned based on the HPLC-DAD-MS-MS and NMR data (Table **3**). From the same type of honey, Beretta *et al.* isolated and elucidated the structure of two kynurenic acid related compounds 3-(2'-pyrrolidinyl)kynurenic acid **5** and its γ-lactam derivative **6** (Fig. **3**) [10]. A detailed NMR study was necessary to establish these structures 1D (^{1}H, ^{13}C and ^{15}N) and 2D (COSY, HMQC and HMBC) NMR techniques (Table **3**). In the ^{15}N NMR of compound **5**, the authors assigned the protonated aliphatic secondary amine N-1'' resonance at 435 ppm (through direct acquisition) and that of the quinolone heterocycle ring N-1 at 130 ppm (through a HMBC detection) [10]. However, they did not indicate the reference used in the ^{15}N NMR spectrum as well as the coupling constants used in the HMBC spectrum acquisition.

Table 3. ^{1}H, ^{13}C and ^{15}N NMR spectroscopic data (δ in DMSO-d_6, ppm) of compounds **3-6**.

Position	3		4		5			6		
	^{1}H	^{13}C	^{1}H	^{13}C	^{1}H	^{15}N	^{13}C	^{1}H	^{15}N	^{13}C
1						130			n.d.	

(Table 3) contd.....

Position	3		4		5			6		
	^{1}H	^{13}C	^{1}H	^{13}C	^{1}H	^{15}N	^{13}C	^{1}H	^{15}N	^{13}C
2		140.0		145.3			139.2			140.9
3	6.68	109.0	8.88	107.5			113.4			114.9
4		178.0		178.1			177.8			174.0
5	8.09	123.5	8.28	124.4	8.16		125.2	8.17		125.5
6	7.32	119.6	7.85	119.9	7.48		119.9	7.41		119.8
7	7.66	132.0	7.59	133.8	7.79		132.8	7.72		133.7
8	7.99	124.6	7.86	125.0	7.93		125.0	7.92		124.6
9		125.6		126.0			125.9			126.8
10		142.8		139.8			141.9			141.1
2’		163.8		166.5			164.1			166.7
1’’					8.88, 9.95	435			n.d.	
2’’					5.11		57.6	4.66		62.3
3’’					2.10, 2.20		24.9	2.30-2.40		29.2
4’’					2.20		29.6	1.14		29.4
5’’					3.32		46.1	3.35, 3.60		42.0

n.d. = not detected

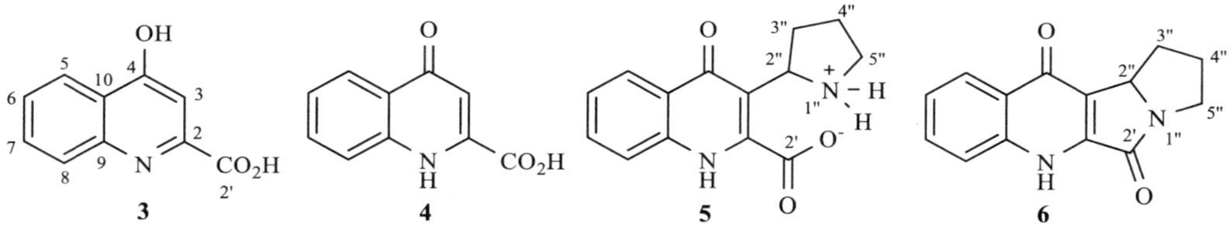

Fig. (3). Structures of compounds **3-6**.

Later, the same group of researchers used DOSY NMR analysis to demonstrate that among various honeys of different botanical origins chestnut honey showed the highest content of quinolone alkaloids (compounds **4-6** and 4-quinolone) [11]. This technique uses the molecular diffusion rate in solution to separate resonances generated by different compounds in a mixture, with no need of separation.

A rapid and easy differentiation of oak honeydew honey from other honey types can be based on the methylene ^{1}H (two multiplets at 1.98 and 1.80 ppm) and ^{13}C (37.87 ppm) NMR resonance of quercitol (a glucosidase inhibitor that blocks the absorption and metabolism of carbohydrates) [12]. These resonances only appear in the NMR spectra of the oak honeydew honey being quercitol a good taxonomic marker of this honey and of the genus *Quercus*. The semi-quantitative evaluation of quercitol, using the unambiguously distinguished signal at 1.80 ppm (H-6 axial), can be determined by the ratio intensity to the anomeric proton of the α- and β-glucopyranose, always present in high amounts in honey [12].

Low Field ^{1}H Nuclear Magnetic Resonance (LF ^{1}H NMR) was used to differentiate the botanical origin of eight (eucalyptus, "assa-lipto", oranges, Barbados cherry, cashew tree, "assa-peixe", "cipó-uva", and polyfloral) Brazilian honeys [13]. For instance, comparison of the continuous distributed NMR (T_2) relaxation curves revealed differences in the water mobility among the different botanical sources. Cashew honey that has higher water activity exhibited broader T_2 distribution than did the other with lower water activity. The bi-exponential fitting of the transversal relaxation (T_2) data revealed two water populations T_{21} and T_{22} in all the samples, corresponding to relaxation times of 0.6-1.8 ms and 2.3-5.4 ms, respectively (Table **4**). Good linear correlations were observed between the T_2 and T_{21} parameters and the physical and chemical data, including water contents, water activity, pH and colour [13].

Table 4. Range of the LF ^{1}H NMR parameters obtained in honeys according to the different botanical origins. (Adapted from [13]).

Types of honeys	T_{21} [ms]	T_{22} [ms]
Cashew tree	1.88 ± 0.10^a (1.65-2.06)	5.49 ± 0.12^a (5.15-5.68)
Polyfloral	1.58 ± 0.07^b (1.44-1.71)	4.91 ± 0.08^b (4.71-5.05)
Barbados cherry	1.36 ± 0.12^c (1.11-1.63)	3.53 ± 0.10^c (3.33-3.72)
"Assa-Peixe"	1.31 ± 0.07^c (1.16-1.42)	4.07 ± 0.10^c (3.89-4.24)
"Assa-Lipto"	1.07 ± 0.07^d (0.96-1.22)	3.43 ± 0.09^c (3.32-3.59)
Eucalyptus	1.09 ± 0.12^d (0.89-1.32)	2.95 ± 0.09^d (2.75-3.10)
Oranges	0.84 ± 0.05^e (0.73-0.94)	3.30 ± 0.12^c (3.07-3.50)
"Cipó-Uva"	0.64 ± 0.04^f (0.57e0.74)	2.43 ± 0.06^e (2.29-2.54)

$^{a, b, c, ...}$ Different letters in a column indicate significant differences ($p < 0.01$) within each treatment (ANOVA).

The LF ^{1}H NMR was also used to detect honey adulteration of high fructose corn syrup [14].

Markers of geographical origin could also be characterized by a combination of NMR data and chemometric analysis. Unsupervised principal component analysis (PCA) is a powerful tool to distinguish ^{1}H NMR spectra of polyfloral and of acacia honey samples and of geographical differentiation for the later ones [15].

^{13}C NMR spectra of the acacia honey allow to identify sugar isoforms. The β-fructopyranose (βFP) and β-fructofuranose (βFF) ratio and the αFF, βFP, βFF and α-glucopyranose (αGP) (Fig. **4**) chemical shifts suggested possible geographical markers of Argentinian and Hungarian honey samples [15]. ^{1}H NMR spectroscopy and multivariate analysis were used to classify the geographical differentiation of Corsican and non-Corsican honeys [16]. In this case, 2D TOCSY spectra were used to determine the molecular structural characteristics of the main components of the honey samples.Comparative analysis of ^{13}C NMR spectra of basic and depolymerized galactomannan polymers (Table **5**), obtained from a hot water extract of the seeds of Chinese honey locust (*Gleditsia sinensis* Lam.), with those of the corresponding monosaccharides revealed that the galactomannan macromolecule possesses residues of 1,4-β-D-mannopyranose, substituted with residues of α-D-galactopyranose at C-6 [17].

In the carbohydrate region of the ^{1}H NMR spectra of several Brazilian honeys, the resonances of the main monosaccharides were identified [18]. Those signals are almost equal to all analyzed honeys with only small intensity variations.

Table 5. ^{13}C NMR spectroscopic data (δ in DMSO-d_6, ppm) of depolymerized galactomannan from the seeds of *Gleditsia sinensis*. (Adapted from [17]).

	C-1	C-2	C-3	C-4	C-5	C-6
α-D-galactopyranosyl	99.7	69.5	70.0	70.6	72.2	62.1
α-D-galactopyranose	93.5	69.6	70.4	70.6	71.7	62.4
4-*O*-β-D-mannopyranosyl	101.2	71.0	72.3	77.3	76.1	61.4
				77.6		
4,6-D-di-*O*-β-D-mannopyranosyl	101.0	71.0	72.4	77.6	74.4	67.4
				78.0		
α-mannopyranose	94.9	72.5	74.3	67.9	77.4	62.3

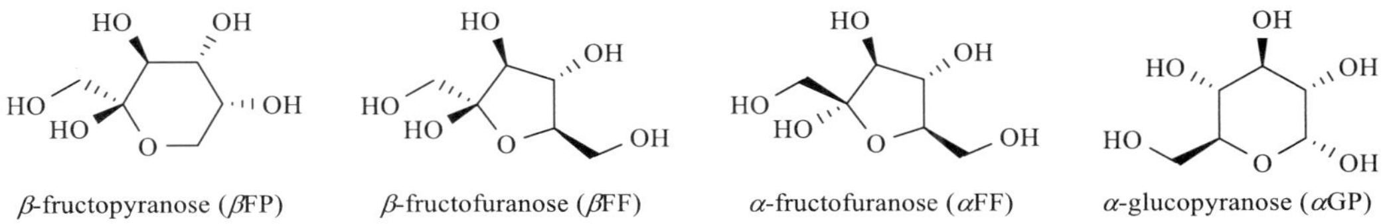

Fig. (4). Structures of β-fructopyranose (βFP), β-fructofuranose (βFF),α-fructo-furanose (αFF) andα-glucopyranose (αGP).

The assignments of the major ^{1}H and ^{13}C NMR signals of the major constituents were done by 2D NMR experiments (*g*COSY, TOCSY, *g*HSQC, *g*HMBC) (Table **6**). The presence of typical resonances (chemical shift and coupling constant) allowed the assignment of some minor constituents (Table **7**). Chemometric analysis PCA and Hierarchical Cluster Analysis (HCA) of these ^{1}H NMR spectra discriminate its botanic origin, namely eucalyptus (higher amount of lactic acid), citrus (higher amount of sucrose) and wildflower (higher amount of phenylalanine and tyrosine) [18]. A similar study conducted by Beretta *et al.* of solid-phase extracted samples allowed the identification of the botanical origin of Italian honeys [19]. Honeydew honey showed typical resonances of an aliphatic component (sets of signals at 0.85-1.60, 2.15, 4.60-5.20 ppm, one doublet at 5.70 ppm and a double of triplets at 6.75 ppm). Chestnut honey presented typical resonances of kynurenic acid **3** and a structurally related metabolite while linden honey showed signals assigned to cyclohexane-1,3-diene-1-carboxylic acid and its 1-*O*-β-gentiobiosyl ester [19].

Table 6. ^{1}H and ^{13}C NMR spectroscopic data (δ in D_2O, ppm; *J* in parenthesis, Hz) of the main saccharides present in Brazilian honeys: α-glucopyranose (αGP), β-glucopyranose (βGP), β-fructopyranose (βFP), β-fructofuranose (βFF) and α-fructo-furanose (αFF). (Adapted from [18]).

Position	αGP		βGP		βFP	
	^{1}H	^{13}C	^{1}H	^{13}C	^{1}H	^{13}C
1	5.22 d (3.70)	94.6	4.63 d (8.0)	98.4	3.52-3.57 m	66.5
1'					3.66-3.72 m	66.5
2	3.49-3.54 m	74.0	3.23 dd (9.2, 8.0)	76.7		110.6
3	3.66-3.73 m	75.4	3.30-3.50 m	78.3	3.74-3.79 m	70.2
4	3.35-3.42 m	72.1 or 72.2	3.35-3.46 m	72.1 or 72.2	3.85-3.90 m	72.3

(Table 6) contd.....

Position	*α*GP		*β*GP		*β*FP	
	^{1}H	^{13}C	^{1}H	^{13}C	^{1}H	^{13}C
5	3.78-3.84 m	73.9	3.40-3.50 m	78.4	3.97-3.99 m	71.8
6	3.67-3.73 m	63.3	3.72-3.77 m	63.4	3.65-3.72 m	65.9
6’	3.79-3.85 m	63.3	3.85-3.90 m	63.4	3.97-4.03 m	65.9
Position	*β*FF		*α*FF			
	^{1}H	^{13}C	^{1}H	^{13}C		
1	3.53-3.58 m	65.4	3.62-3.65 m	65.6		
1’						
2		104.1		107.0		
3	4.07-4.10 m	78.0	4.07-4.10 m	84.5		
4	4.07-4.10 m	77.0	3.95-4.00 m	78.6		
5	3.77-3.85 m	83.2	4.02-4.07 m	83.8		
6	3.75-3.82 m	65.0	3.64-3.68 m	63.7		
6’	3.62-3.68 m	65.0	3.77-3.80 m	63.7		

Table 7. ^{1}H and ^{13}C NMR spectroscopic data (δ in D_2O, ppm; *J* in parenthesis, Hz) of the minor compounds present in Brazilian honeys. (Adapted from [18]).

Position	^{1}H	^{13}C	**Position**	^{1}H	^{13}C
Acetic acid			**Alanine**		
CO_2H		179.8	CO_2H		178.3
CH_3	2.00 s	23.6	CH	3.75-3.85 m	53.1
Citric acid			CH_3	1.46 d (7.3)	18.9
1,5-CO_2H		176.5	**5-Hydroxymethylfurfural**		
2-CH	2.79 d (15.5)	46.1	2-C		154.3
2’-CH	2.94 d (15.5)	46.1	3-CH	7.54 d (3.7)	129.7
3-C		76.1	4-CH	6.68 d (3.7)	113.7
4-CH	2.79 d (15.5)	46.1	5-C		164.1
4’-CH	2.94 d (15.5)	46.1	CH_2	4.69 s	58.8
6-CO_2H		180.2	CHO	9.45 s	183.2
Formic acid			**Phenylalanine**		
HCO_2H	8.45 s	173.5	Aromatic 2,6-CH	7.31 d (7.2)	132.1
Lactic acid			Aromatic 3,5-CH	7.41 d (7.2)	131.8

(Table 7) contd.....

Position	^{1}H	^{13}C	Position	^{1}H	^{13}C
CO_2H		183.7	Aromatic 4-CH	7.35-7.39 m	130.4
CH	4.30-4.40 m	70.6	**Tyrosine**		
CH_3	1.35 d (6.9)	22.5	Aromatic 1-C		129.4
Ethanol			Aromatic 2,6-CH	7.18 d (8.4)	133.5
CH_2	3.56-3.66 m	60.1	Aromatic 3,5-CH	6.88 d (8.4)	118.6
CH_3	1.15 t (7.1)	19.6	Aromatic 4-COH		157.5

Honey adulteration increased exponentially both in terms of geographic and botanical origin, leading to a huge demand for novel analytical methods to effectively control the quality of honey. Israeli citrus honeys authenticity were determined by the D/H(CH_3) isotope ratio of the ethanols, produced by alcoholic fermentation (dry yeast *Saccharomyces bayanos*), measured by deuterium NMR (Table **8**) [20]. Ethanols obtained from fermentation of citrus honeys have D/H(CH_3) values similar to those obtained from citrus juice and 5 ppm higher than the values obtained for other honeys.

Table 8. Isotope ratio parameters in various samples of honey. (Adapted from [20]).

Source	D/H(CH_3)[a]	δ^{13}C%[b]
citrus	105.4	-23.7
citrus	105.2	-24.1
citrus	105.2	
citrus	104.9	-24.0
citrus	104.8	-24.4
citrus	104.7	-23.3
citrus	104.6	-23.9
citrus	104.1	-24.3
field flowers[c]	100.0	-25.7
field flowers[c]	100.0	-24.1
field flowers[c]	99.6	-23.6
field flowers[c]	99.2	-24.4
field flowers[c]	98.9	-25.0
field flowers[c]	97.8	
onion	100.3	-24.1

(Table 8) contd.....

Source	D/H(CH_3)[a]	$\delta^{13}C$%[b]
cotton	99.1	-25.1
eucalyptus	98.9	-24.1
zaatar[d]	98.2	-24.1
thorns[e]	97.8	-24.1
beet sugar[f]	95.3	

[a] ppm D/H of methyl group of ethanols obtained by alcoholic fermentation.
[b] $\delta^{13}C$) [[($^{13}C/^{12}C$)sample/($^{13}C/^{12}C$)standard] - 1] x 1000.
[c] Mainly Crucifeara, Compositae, Labiatae, and Leguminosae. [d]*Majorana syriaca*. [e]*Centaurea verutum L.*[f] The bees were fed with beet sugar solutions.

The application of multivariate analysis *e.g.* PCA, Cluster Analysis (CLA), Correspondence Analysis (CA), Factor Analysis (FA), and General Discriminant Analysis (GDA) to the chromatographic and spectroscopic (*e.g.* 1D and 2D NMR) data proved to be extremely useful to group and detect honeys of different origins [21 - 23].

PCA and Partial Least Squares Discriminant Analysis (PLS-DA) of ^1H and ^{13}C NMR based sugar profiles of herbhoneys and Polish honeys did not identify any distinct cluster between them but some tendency appeared regarding the *β* GP and *β* FP content [24]. The traditional honeys seem to be a little bit rich in fructose than the herbhoneys. In both PCA and PLS-DA methods there is a cluster related to sucrose content allowing the fast detection of adulteration. The authors collected 1D and 2D NMR spectra of mono and disaccharides from artificial honey to allow a reliable assignment of these compounds in real honey samples (Table **9** only shows the NMR data of disaccharides which were not yet described in this chapter) [24].

Table 9. ^1H and ^{13}C NMR spectroscopic data (δ in DMSO-d_6, ppm) of disaccharides present in an artificial honey sample. (Adapted from [27]).

Position	Sucrose			*α*-Maltose			*β*-Maltose		
	^1H	^1H (OH)	^{13}C	^1H	^1H (OH)	^{13}C	^1H	^1H (OH)	^{13}C
1	4.86		92.27	4.88	6.35	92.52	4.29	6.71	97.11
2	3.17	5.11	71.95	3.16	4.72	72.23	2.92	5.06	74.76
3	3.50	n.d.	73.20	3.44	5.34	73.39	3.36	5.47	76.85
4	3.10	4.88	70.18	3.27		80.59	3.24		80.06

(Table 9) contd.....

Position	Sucrose			*α*-Maltose			*β*-Maltose		
	^{1}H	^{1}H (OH)	^{13}C	^{1}H	^{1}H (OH)	^{13}C	^{1}H	^{1}H (OH)	^{13}C
5	3.21		72.95	3.17		75.40	3.77		75.82
6	3.41, 3.46	n.d.	60.92	3.55	4.52	61.05	3.48, 3.62	4.59	61.19
1'	3.36, 3.51	n.d.	62.51	4.94		101.20	4.98		101.13
2'			104.36	3.44	5.43	71.03	3.21	5.45	72.78
3'	3.84	4.65	77.44	3.21	4.68	72.82	3.33	4.66	73.82
4'	3.74	5.32	74.66	3.52	4.87	66.98	3.03	4.88	70.26
5'	3.54		82.78	3.63		73.33	3.43		73.79
6'	3.47, 3.70	4.49	62.56	3.26, 3.36	4.50	63.69	3.41, 3.57	4.60	61.31

n.d. = not detected

^{1}H and 2D TOCSY NMR spectroscopy allowed structure elucidation (by comparing with chemical shifts and *J* coupling constants of authentic samples) of several biomarkers of Corsican honeys, namely, kynurenic acid **3** (chestnut honey), α-isophorone and 2,5-dihydroxyphenylacetic acid (strawberry-tree honey) [25]. A complete NMR study (^{1}H, ^{13}C, TOCSY, COSY, NOESY, HMQC and HMBC spectra) of semi-purified compounds allowed to identify some known and new biomarkers of different types of honeys [26]. The same authors have characterized a complete set of biomarkers and other compounds **7-27** (Fig. **5**) extracted from monofloral honeys [27]. The complete assignment of proton and carbon resonances were carried out with the aid of 2D NMR techniques (DQF-COSY, TOCSY, NOESY, HMQC and HMBC spectra) (Tables **10-12**).

Table 10. ^{1}H and ^{13}C NMR spectroscopic data (δ in $CDCl_3$, ppm; *J* in parenthesis, Hz) of compounds **7-12**.

Position	7		8^a		9^a	
	^{1}H	^{13}C	^{1}H	^{13}C	^{1}H	^{13}C
1		154.6		154.3		142.1
2	7.18 dt (5.8, 1.7)	136.8	7.18 d (5.8)	116.0	7.24 d (5.8)	120.0
3	6.18 dt (5.8, 1.7)	115.8	6.18 d (5.8)	136.2	6.19 d (5.8)	136.5
4		125.3		125.3		126.2
5	2.47 t (9.6)	21.6	2.48 m	23.8	2.52 m	23.9
6	2.32 t (9.6)	23.6	2.32 m	21.7	2.52 m	21.5
7		170.9		172.0		172.3

(Table 10) contd.....

Position	7		8[a]		9[a]	
	^{1}H	^{13}C	^{1}H	^{13}C	^{1}H	^{13}C
8		73.2		73.0		144.3
9	1.42 s	28.1	1.40 s	28.6	5.31, 5.16 s	115.5
10	1.42 s	28.1	1.40 s	28.6	1.97 s	20.3
11	3.76 s	63.4				
Position	**10**		**11**		**12**	
	^{1}H	^{13}C	^{1}H	^{13}C	^{1}H	^{13}C
1		125.0	1.33	29.3		
2	8.08 d (8.1)	130.1		71.2		159.8
3	7.62 d (8.1)	124.4	5.73 dt (15.7, 1.1)	143.4		
			5.64 ddd (15.7, 7.9, 6.6)	121.9		
4		155.3	2.25 dd (7.9, 1.1)	44.9		n.d.
5	7.62 d (8.6)	130.1	2.30 dd (6.6, 1.1)	44.9	8.3 dd (8.0, 1.1)	126.2
6	8.08 d (8.6)	124.4		73.0	7.66 d (8.1)	127.0
7	n.d.	170.1	5.93 dd (17.3, 10.8)	144.5	7.46 t (7.5)	125.9
8		73.0	5.07 dd (10.8, 1.1)	112.2	7.74 td (7.0, 1.4)	134.2
			5.22 dd (17.3, 1.1)	112.2		
9	1.61 s	31.0				149.2
10	1,61 s	31.0				120.9
11						
1’					3.2 t (7.9)	32.9
2’			1.33	29.3	2.3 t (7.6)	19.8
3’					4.2 t (7.2)	46.4
6’			1.29	27.4		

[a] These data were taken from reference [28].

Table 11. ^{1}H and ^{13}C NMR spectroscopic data (δ in $CDCl_3$, ppm; *J* in parenthesis, Hz) of compounds **13-19**.

Position	13		14		15[a]		16[a]	
	^{1}H	^{13}C	^{1}H	^{13}C	^{1}H	^{13}C	^{1}H	^{13}C
1	n.d.		n.d.					
2		162.5	7.95 d (6.6)	139.5	5.44	78.8		163.1
3	6.68 d (9.5)	121.8	6.3 d (6.6)	109	3.10, 2.84	43.8, 43.3	6.68	103.0
4	7.78 d (9.5)	141.0		180.1		197.3		181.8

(Table 11) contd.....

Position	13		14		15[a]		16[a]	
	^{1}H	^{13}C	^{1}H	^{13}C	^{1}H	^{13}C	^{1}H	^{13}C
5	7.58 dd (8.1, 1.2)	127.9	8.07 dd (7.8, 1.3)	130.7	12.05	165.4	12.81	161.4
6	7.24 td (8.1, 1.2)	122.4	7.59 td	121.6	6.01	95.6	6.31	98.1
7	7.52 td (8.4, 1.2)	130.5	7.59 td	132.4	12.05	168.3	12.81	164.3
8	7.17 d (8.4)	115.1	7.89 dd	115.8	6.01	97.2	6.48	94.1
9		137.8		144.8		164.6		157.0
10		120.2		129.9		103.3		105.1
1'								130.6
2'					7.44	127.3	7.89	126.3
3',5'					7.44	129.6	7.54	129.1
4'					7.44	129.7	7.54	132.0
6'					7.44	127.3	7.89	126.3
Position	**17**		**18[a]**		**19[a]**			
	^{1}H	^{13}C	^{1}H	^{13}C	^{1}H	^{13}C		
1					4.00	67.9		
2	5.04 d (11.9)	83.1		151.8		136.8		
3	4.57 d (11.9)	72.4			5.43	125.7		
4		198.0		148.0	2.08	23.2		
5	11.25 s	162.8		107.8	1.61	42.4		
6	6.08 d (2.2)	94.4		155.1		73.1		
7		163.7			5.93	145.8		
8	6.11 d (2.2)	95.4	7.55	141.2	5.09, 5.24	111.2		
9		169.0						
10		101.0	3.42	27.7				
11	3.81 s	55.6	3.61	29.8				
12			4.01	33.2				
1'		138.9						
2'	7.44 m	128.7			1.68	13.3		
3',5'	7.46 m	125.8						
4'	7.42 m	128.2						
6'	7.44 m	128.7			1.30	13.3		

[a] These data were taken from reference [29].

Table 12. ^{1}H and ^{13}C NMR spectroscopic data (δ in $CDCl_3$, ppm; *J* in parenthesis, Hz) of compounds **20-27**.

Position	20		21		22		23		24		25		26	
	^{1}H	^{13}C	^{1}H	^{13}C	^{1}H	^{13}C	^{1}H	^{13}C	^{1}H	^{13}C	^{1}H	^{13}C	^{1}H	^{13}C
1		42.5		n.d.		170.4		170.4		178.0		178.0		121.0
2	2.51 d (17.2)	49.8	2.51 d (17.2)	n.d.	5.85 dt (15.6)	120.1	5.85 dt (15.6)	120.1	2.37 m	33.2	2.37 m	33.2	7.34 s	106.5
	2.34 d (17.2)	49.8	2.34 d (17.2)											
3		196.8		n.d.	7.07 dt (15.6)	151.7	7.07 dt (15.6)	151.7	1.66 m	24.4	1.66 m	24.4		147.5
4	5.97 s	128.2	6.01 s	126.1	2.26 m	31.8	2.26 m	31.8	1.38 m	29.1	1.38 m	29.1		139.9
5		160.8		n.d.	1.51 m	27.4	1.51 m	27.4	1.51 m	27.4	1.51 m	27.4		147.5
6		79.3	2.74 d (9.6)	55.0	1.38 m	29.1	1.38 m	29.1	1.51 m	27.4	1.38 m	29.1	7.34 s	106.5
7	6.47 d (15.9)	130.3	6.68 dd (15.7, 9.6)	143.5	1.37 m	29.3	1.66 m	24.4	1.38 m	29.1	1.66 m	24.4	3.96 s	56.3
8	6.84 d (15.9)	145.0	6.2 d (15.7)	133.2	1.66 m	24.4	2.37 m	33.2	1.66 m	24.4	2.37 m	33.2	3.96 s	56.3
9		198.0		n.d.	2.37 m	33.2		178.0	2.37 m	33.2		178.0	3.91 s	51.8
10	2.31 s	28.4	2.31 s	34.0		178.0				178.0				167.3
1'	1.03 s	23.8	1.03 s	n.d.										
1''	1.11 s	24.7	1.11 s	n.d.										
5'	1.89 s	18.8	1.92 s	23.0										

Position	27		Position	27	
	^{1}H	^{13}C		^{1}H	^{13}C
1	3.56 dd (5.4, 2.8)	68.2	2a (R3, R2)	2.33 dt (15.1, 7.4)	34.5
2	5.21 m	69.5	2b (R3, R2)	2.30 dt (15.1, 7.4)	34.5
3	4.18 dd (11.9, 6.4)	61.8	3 (R2, R3)	1.64 m	25.2
3'	4.35 dd (11.9, 6.4)	61.8	4-7, 12-17 (R2, R3)	1.31 m	29.1
1 (R1)	3.45 m	71.4	9, 10 (R2, R3)	5.36 m	129.9

(Table 12) contd.....

Position	27		Position	27	
	^{1}H	^{13}C		^{1}H	^{13}C
2(R2)	1.56 m	29.8	8, 11 (R2, R3)	2.04 b	27.8
3 (R3)	1.3 m	22.5	18 (R1, R2, R3)	0.91 t (7.2)	14.0

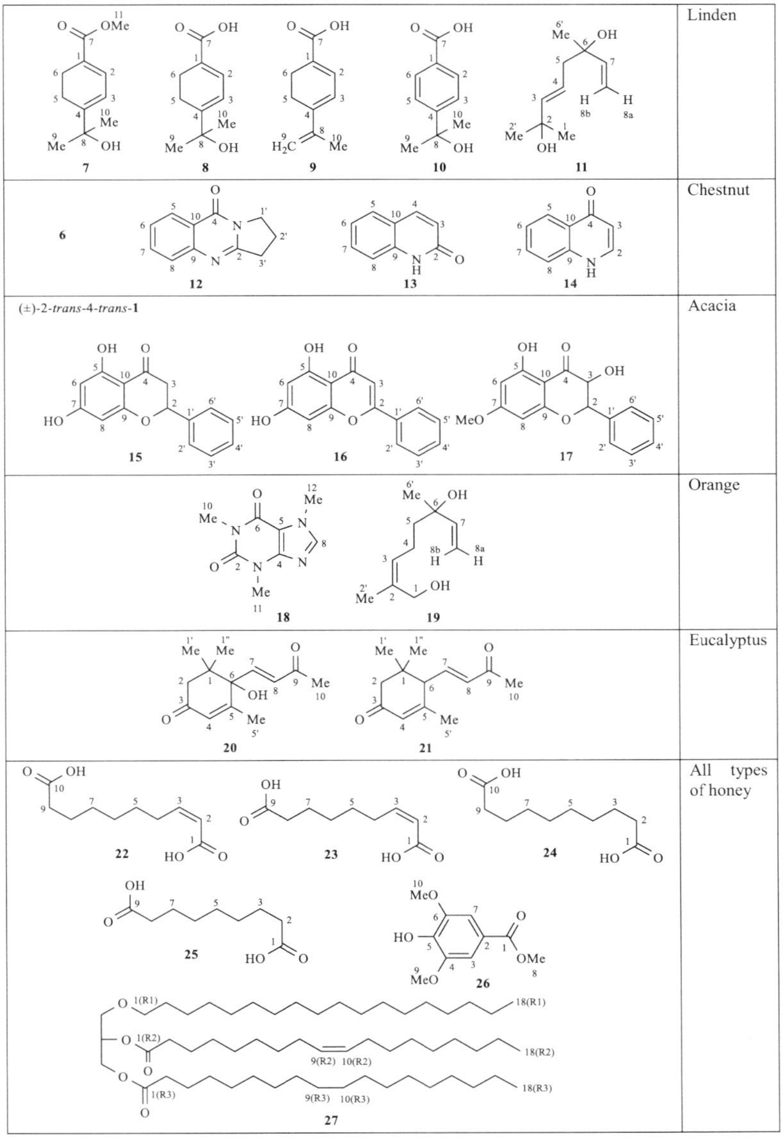

Fig. (5). Structures of compounds **7-27**.

Lianda and Castro, isolated for the first time from a Brazilian *Citrus* honey, the flavonol morin (2',3,4',5,7-pentahydroxyflavone) [30]. The characterization of this compound was based on its ^{1}H and ^{13}C NMR spectra and further comparison with the literature.

The combination of NMR spectroscopy (^{1}H saccharide content) and chemometrics allowed a very good differentiation for multifloral honey samples from different countries and also for rhododendron and "high mountain multifloral" honeys from very closely related regions. NMR appears to be one of the elective techniques to identify the authenticity of honey, offering the possibility to assign the metabolite contents of honeys (^{1}H and ^{13}C anomeric resonances of several identified saccharides) with a single experiment and avoiding derivatization (Table **13**) [31]. The saccharide contents can be used to characterize honey samples and to construct an identity card of saccharides for each floral source, being the adulteration by carbohydrate addition identified by analysing the different ratios of the present saccharides [32].

Table 13. ^{1}H and ^{13}C NMR spectroscopic data (δ in D_2O, ppm) of all the identified saccharides anomeric signals. (Adapted from [31]).

Saccharide	Abbreviation	^{1}H	^{13}C
Erlose	Erl	5.3505	102.53 94.71
Fructose	Frc	4.0524	77.64
Gentiobiose	Gnt	5.1732 4.5860 4.4455	94.71 98.63 105.46
Glucose	α/βGl	5.1732 4.5860	94.71 98.63
Isomaltose	Imt	5.1874 4.9055 4.6194	94.71 101.07 98.63
Isomaltotriose	Im3	5.1874 4.8988 4.6194	94.71 101.12 98.63
Kojibiose	Kjb	5.3841 5.3304 5.0404 4.7506	92.27 100.57 99.11 99.11

(Table 13) contd.....

Saccharide	Abbreviation	^{1}H	^{13}C
Kestose	Kst	5.3729 4.2284	95.19 79.45
Maltose	Mlt	5.3505 5.1732 4.5860	102.53 94.71 98.63
Maltotetraose	Ml4	5.3567 5.1732 4.5860	102.53 94.71 98.63
Maltotriose	Ml3	5.3383 5.1732 4.5860	102.53 94.71 98.63
Maltulose	Mtl	5.1874 5.1536 5.1034	103.23 101.06 100.59
Melezitose	Mlz	5.3924 5.1316	94.71 103.01
Melibiose	Mlb	5.1732 4.9251 4.6150	101.28 94.71 98.63
Nigerose	Ngr	5.3159 5.3003 5.1732 4.6115	102.04 102.04 94.71 98.63
Palatinose	Plt	4.9183	101.06
Raffinose	Rff	5.3729 4.9397	95.19 101.06
Sucrose	Scr	5.3505	94.71
Turanose	Trn	5.2444 5.2444 5.1536 5.1034	101.07 103.49 99.59 100.57

The origin of Polish monofloral and multifloral honeys has been done with the application of chemometric studies using NMR spectroscopy [33]. The markers of dark monofloral honeys were: i) heather honey presents two dominant chemical markers, dehydrovomifoliol **28** (formed from abscisic acid **1**) and phenyl acetic acid **29** (formed from phenylpyruvic acid); and ii) in the buckwheat honey formic acid **30** and tyrosine **31** were the identified new chemical markers (Fig. **6**) since the expected ones were not observed. For light monofloral honeys we can conclude: i) in the case of rape honey it was not possible to identify any markers although there are two typical ^{1}H (8.704 and 8.328 ppm) and ^{13}C (178.2 and 113.1 ppm) NMR resonances; and ii) in the case of lime honey there are high levels of acetic acid and 4-(1-hydroxy-1-methylethyl)cyclohexane-1,3-diene-1-carboxylic acid **32** (Fig. **6**). In this case there are also some sets of ^{1}H and ^{13}C characteristic chemical shifts (1.369/29.98 and 6.16/118.56 ppm). The chemical composition of multifloral honeys is similar to that of light honeys. Even so, the authors have constructed a heatmap from the characteristic NMR peaks (Table **14**) to differentiate the different types of multifloral honeys [33].

Table 14. ^{1}H and ^{13}C NMR spectroscopic data (δ in D_2O, ppm) of compounds **28-32**.

Position	28		29		30		31		32	
	^{1}H	^{13}C	^{1}H	^{13}C	^{1}H	^{13}C	^{1}H	^{13}C	^{1}H	^{13}C
1									2.3	24
2			7.30	129.1	8.06	166.8	6.74	118.9	2.3	24
3	2.53	50.4	7.29	131			7.04	133.8		
4			7.37	131					6.1	123
5	5.94	128.8	7.29	131			7.04	133.8	6.9	123
6			7.30	129.1			6.74	118.9		
7			3.37	50.7						
8	6.27	130.8					2.62	46.1		
9	6.91	145								
12	1.03	24.2							1.2	28
13	1.11	22.8							1.2	28
15	1.89	18.6								
16	2.32	28.3								

The identification of individual carbohydrates of multicomponent artificial mixtures using a computer-aided analysis of their ^{13}C NMR spectra, without previous purification, is based in the assignment of each component by comparison the signals of the mixture spectrum with those of the pure reference spectra compiled in a library. Each compound is identified by taking into consideration three parameters: i) the number of observed carbons with respect with the number of expected signals; ii) the number of overlapped signals of carbons which possess the same chemical shift; and iii) the difference in the chemical shift of each signal in the mixture and in the reference [34, 35]. The quantitation of mono-, di- and trisaccharides from the referred artificial mixtures was made using 1,4-dioxane as internal standard, without the need to use inverse gated decoupling technique (technique used to acquire quantitative ^{13}C NMR spectra) or a relaxation reagent (a paramagnetic species allowing the shortening of the relaxation time) [36].

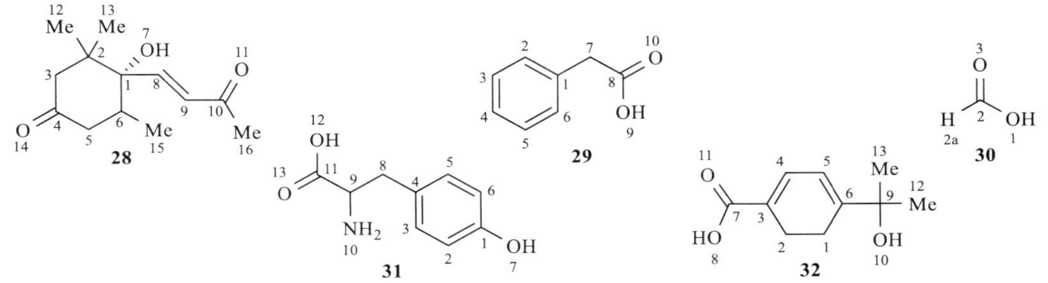

Fig. (6). Structures of compounds **28-32**.

The referred two procedures were used to the direct identification and quantitation of carbohydrates present in six Corsican authentic honeys of different floral types (mixed floral, *Castanea*, *Robinia*, *Asphodelus*, *Anthyllus* and *Clementina*) (Table **15**) [36]. Together with the mono- and disaccharides, several oligosaccharides (identification of some of these is not easy by chromatographic methods) were observed in levels ranging from 0.4 to 3.3% [36].

Table 15. ^{13}C NMR quantitative analysis of carbohydrates (%) in six honeys[a]. (Adapted from [36]).

Components	Honey floral type					
	Mixed floral	*Castanea*	*Robinia*	*Asphodelus*[b]	*Anthyllis*	*Clementina*
Fructose	35.6	37.1	32.4	32.7	31.5	33.4
Glucose	26.8	23.0	25.9	22.7	28.7	28.4
Turanose	2.2[c]	1.7[c]	0.8[c]	1.8[c]	1.7[c]	1.8[c]
Maltulose	0.8[c]	1.0[c]	0.6[c]	0.7[c]	0.6[c]	1.8[c]
Maltose			0.7	0.9		
Isomaltose	1.4				1.7	

(Table 15) contd.....

Components	Honey floral type					
	Mixed floral	*Castanea*	*Robinia*	*Asphodelus*[b]	*Anthyllis*	*Clementina*
Nigerose					2.6	2.3
Isomaltotriose	2.5	1.7		1.7		3.3
Melezitose				1.1		
Erlose		0.4		1.7		
Carbohydrates	69.3	64.9	60.4	63.3	66.8	71.0
Oligosaccharides	6.9	4.8	2.1	7.9	6.6	9.2
F+G	62.4	60.1	58.3	55.4	60.2	61.8
F/G	1.33	1.6	1.25	1.44	1.1	1.17

[a] All percentages are expressed with respect to fresh matter (raw honey).
[b] Repeatability of the measurement given at 95% for four analyses of Asphodelus honey: fructose 32.2±1.7%; glucose, 24.6±1.8%; turanose, 1.9±0.2%; maltulose, 0.9±0.2%; maltose, 0.7±0.2%; isomaltotriose, 1.8±0.2%; melezitose, 1.0±0.2%; erlose, 1.4+0.3%. [c] Correction factors applied since the minor anomeric forms (α-FF) of maltulose and of turanose are not detected and estimated at 8.9% and 17.0%, respectively, in the solution of maltulose and turanose.

Independent Component Analysis (ICA) has been used to solve the overlap of ^{1}H NMR spectra of foodstuffs, including honey. The ICA strategy has been proven to effectively extract spectra of pure components and concentrations from the NMR of mixtures [37].

Selective 1D TOCSY experiments were used to isolate the features of a single spin system from a crowded forest of peaks (*e.g.* amino acids from honey), thus allowing for more certain identification of a species, and cleaner and more accurate quantification (Table **16**) [38]. PCA of data sets (integrals of minor honey components, namely amino acids) derived from selective TOCSY spectra of commercial honey samples constitute a more sensitive way to distinguish the origin of honey than the classical metabonomics approach [38].

Table 16. Amino acids observed in honey by 1D TOCSY. (Adapted from [38]).

Amino acid	Excitation (ppm)		TOCSY peaks (ppm)	
Pro	β1	(2.33)	α δ1 δ2 β2 γ	(4.12) (3.40) (3.32) (2.06) (1.98)

(Table 16) contd.....

Amino acid	Excitation (ppm)		TOCSY peaks (ppm)	
Pro	γ	(1.98)	α δ1 δ2 β1	(4.12)[a] (3.40) (3.32) (2.33)
Ala	β	(1.46)	α	(3.80)[a]
Thr	CH_3 γ	(1.31)	α β	(3.60)[a] (4.25)
Tyr	CH	(6.90)	CH	(7.19)[a]
Tyr	CH	(7.19)	CH	(6.90)
ethanol	CH_3	(1.17)	CH_2	(3.65)[a]
Ile	CH_2 γ1	(1.46)	β CH_2 γ2 CH_3 γ CH_3 δ	(1.98) (1.25) (0.99) (0.91)
Ile	CH_2 γ2	(1.25)	β CH_2 γ1 CH_3 γ CH_3 δ	(1.98) (1.46) (0.99) (0.91)
Phe[a,b]	na[c]		na[c]	

[a] Used in PCA analysis. [b] Phe quantitation based on peak height at 7.42 ppm. [c] Not applicable.

DOSY (Diffusion Ordered SpectroscopY) was directly applied to the monofloral manuka honey for simultaneously identify and in part quantify their components and also to correlate the antibacterial activity with the components identified. It was shown that the use of an internal standard is essential for the calibration and quantisation of diffusion, and the relationship between diffusion and molecular mass needs to be established for each component [39].

The concentration of methylglyoxal of manuka honey has been determined for the first time by a quantitative NMR method (qNMR), without the need of chromatographic separation or a derivatisation procedure [40]. Methylglyoxal **33** is a naturally occurring dicarbonyl compound that undergoes a "spontaneous" reaction with water, leading to the two major compounds methylglyoxal monohydrate **34** and methylglyoxal dihydrate **35**, remaining less than 1% unreacted (Fig. **7**). To determine the concentration of methylglyoxal **33** in a sample, the methyl peaks of methylglyoxal monohydrate **34** (2.297-2.314 ppm, 3H, s), methylglyoxal dihydrate **35** (1.369-1.386 ppm, 3H, s) and TSP (3-

trimethylsilyl[2,2,3,3-D_4]propionic acid) normally used as reference in aqueous solution) (-0.01 to 0.01 ppm, 9H, s) was integrated [40]. The concentration of methylglyoxal **33** in the NMR sample was calculated using equation 1.

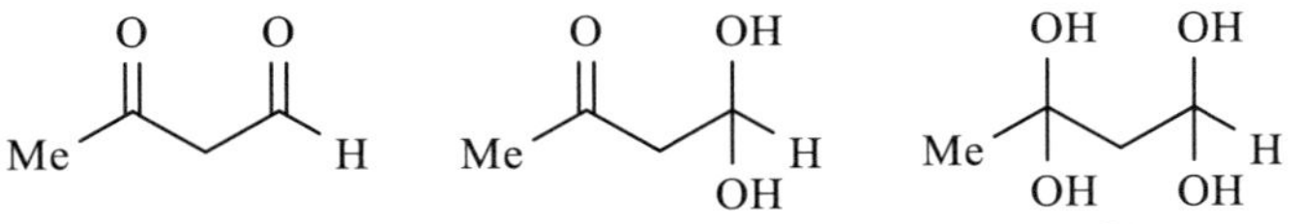

Fig. (7). Structures of methylglyoxal **33**, methylglyoxal monohydrate **34** and methylglyoxal dihydrate **35**.

$$3\left(\frac{A_m+A_d}{A_{TSP}}\right)\times[TSP]=[MGO_{NMR}] \quad \textbf{(1)}$$

where A_m corresponds to the area of the methyl resonance of methylglyoxal monohydrate **34**; A_d to the area of methylglyoxal dihydrate **35**; A_{TSP} to the area of the TSP internal standard; TSP to the concentration of TSP (M); and MGO_{NMR} the concentration of methylglyoxal **33** (M). This value was converted to a concentration of methylglyoxal **33** in undiluted honey using equation 2. This approach has shown that previously applied methodology may overestimated the methylglyoxal concentration in manuka honey [40].

$$72.06\times\left(\frac{[MGO_{NMR}]}{Honey}\right)\times 1000=\left[MGO_{Honey}\right] \quad \textbf{(2)}$$

3. PROPOLIS

A wide variety of compounds can be found in propolis samples and their chemical composition varies according to its variety and environmental factors. Generally, propolis from Europe and China contains flavonoid aglycones (flavones and flavanones), phenolic acids and their esters. In Japanese propolis many prenylflavonoids has been identified. Propolis from Solomon Islands seems to contain prenylflavonoids similar to those found in Japanese samples. In a study conducted in thirty three propolis samples, the application of PCA to ^1H NMR data allowed to identify different geographical origins and divided the samples into three groups, African, Asian and European propolis [41].

Papotti *et al.* [42] stated that the harvesting method used for gathering propolis can also lead to variations on their chemical composition. For that, they analyzed sixty propolis samples and used NMR and multivariate statistical methods to develop an efficient and appropriate model for classifying them. NMR spectroscopy coupled to an appropriate spectral analytical tool (for spectra pre-treatment and analysis) was used for the recognition and detection of several phenolic compounds in propolis extracts [43]. This work demonstrated that ten (caffeic acid, *p*-coumaric acid, naringenin, quercetin, apigenin, pinostrobin, pinocembrin, kaempferol, chrysin and galangin) out of twelve typical phenolic propolis compounds were identified as statistically significant in most of the sixty five samples tested.

Since it is possible to identify in propolis samples several families of compounds and would be a hard work the description of their structural characterization according to their origin, we will summarize below the compounds identified in American (Table **17**), Asian (Table **18**), African and Oceana (Table **19**) propolis, in which NMR techniques were used to elucidate their structures.

Table 17. Compounds identified in American propolis, which structures were determined using NMR spectroscopy.

Propolis type	Compounds	Ref.
Cuban (methanol extract)	isoprenylated benzophenone	[44]
Red Cuban (methanol extract)	gallic acid isoliquiritigenin (–)-liquiritigenin formononetin biochanin A (3*S*)-vestitol (3*S*)-7-*O*-methylvestitol (3*S*)-7,4'-dihydroxy-2'-methoxyisoflavan (6a*S*,11a*S*)-medicarpin (6a*S*,11a*S*)-homopterocarpin (6a*R*,11a*R*)-vesticarpan (6a*R*,11a*R*)-3,8-dihydroxy-9-methoxypterocarpan (6a*R*,11a*R*)-3-hydroxy-8,9-dimethoxypterocarpan (6a*R*,11a*R*)-3,4-dihydroxy-9-methoxypterocarpan	[45]

(Table 17) contd.....

Propolis type	Compounds	Ref.
Cuban (methanol extract)	Brown-type – isoprenylated benzophenones (nemorosone, scrobiculatones A and B) Red-type – isoflavonoids (isoflavans, isoflavanones and pterocarpans) Yellow-type – aliphatic compounds (terpenoids and sterols)	[46]
Yellow Cuban (methanol extract)	Type A – 3-hydroxylated triterpenoids Type B – 3-acetylated triterpenoids	[47]
Red Cuban (ethanol extract)	retusapurpurin A and retusapurpurin B	[48]
Red Mexican (methanol extract)	1-(3',4'-dihydroxy-2'-methoxyphenyl)-3-phenylpropane (*Z*)-1-(2'-methoxy-4',5'-dihydroxyphenyl)-3-phenylprop-2-ene 3-hydroxy-5,6-dimethoxyflavan (–)-7-hydroxyflavanone (+)-pinocembrin (–)-mucronulatol (–)-arizonicanol (+)-vestitol (–)-melilotocarpan A (–)-melilotocarpan D	[49]
Hondoriun(ethanol extract)	(*E*)-cinnamyl (*E*)-cinnamate (*E*)-cinnamyl (*Z*)-cinnamate hydrocinnamyl (*E*)-cinnamate benzyl (*E*)-cinnamate (*E*)-cinnamyl (*E*)-*p*-coumarate (*E*)-cinnamyl (*Z*)-*p*-coumarate (*E*)-cinnamyl benzoate 6*β*-hydroxy-3-oxolup-20(29)-en-28-oic acid 3-oxo-oleanoic acid sakuranetin liquiritigenin kukulkanin B (*E*)-*p*-coumaric acid (*E*)-cinnamic acid *p*-hydroxybenzoic acid	[50]
Brazilian (methanol extract)	3-hydroxy-2,2-dimethyl-8-prenylchroman-6-propenoic acid	[51]
Brazilian (methanol extract)	(*Z*)-2,2-dimethyl-8-(3-methyl-2-butenyl)benzopyran-6-propenoic acid (*E*)-2,2-dimethyl-8-(3-methyl-2-butenyl)benzopyran-6-propenoic acid	[52]
Brazilian (methanol extract)	3-prenyl-4-hydroxycinnamic acid 2,2-dimethyl-6-carboxyethenyl-2*H*-1-benzopyran 3,5-diprenyl-4-hydroxycinnamic acid 2,2-dimethyl-6-carboxyethenyl-8-prenyl-2*H*-1-benzopyran	[53]

(Table 17) contd.....

Propolis type	Compounds	Ref.
Brazilian (São Paulo) (methanol extract)	isomangiferolic acid 24-methylenecycloartane-3β,26-diol mangiferolic acid mangiferonic acid ambonic acid ambolic acid	[54]
Brazilian (Minas Gerais) (water extract)	chlorogenic acid 4-caffeoylquinic acid 5-caffeoylquinic acid 3,4-di-*O*-caffeoylquinic acid 3,5-di-*O*-caffeoylquinic acid 4,5-di-*O*-caffeoylquinic acid	[55]
Brazilian (Minas Gerais) (water extract)	3,4-di-*O*-caffeoylquinic acid 3,5-di-*O*-caffeoylquinic acid 4,5-di-*O*-caffeoylquinic acid	[56]
Brazilian (Minas Gerais) (methanol and chloroform extract)	bauer-7-en-3β-yl acetate	[57]
Brazilian (Ceará) (ethanol extract)	canaric acid lupeol lupenone germanicone quercetin kaempferol acacetin	[58]
Brazilian type 6 (Bahia) (ethanol 80% extract)	hyperibone A	[59]
Green Brazilian (Minas Gerais) (ethanol extract)	dihydrokaempferide kaempferide isosakuranetin betuletol	[60]
Green Brazilian (Minas Gerais) (ethanol extract)	mixture of α- and β-amyrin lupeol ramnocitrin eupalitin acacetin 3-prenyl-4-hydroxycinnamic acid 3,5-diprenyl-4-hydroxycinnamic acid (*E*)-3-[4-(3-phenylpropanoyloxy)]-3,5-diprenylcinnamic acid	[61]
Red Brazilian (Alagoas) (ethanol 70% extract)	2,3-epoxy-2-(3-methyl-2-butenyl)naphthalene-1,4-dione	[62]
Red Brazilian (Alagoas) (ethanol 80% extract)	vestitol neovestitol isoliquiritigenin	[63]

(Table 17) contd.....

Propolis type	Compounds	Ref.
Brazilian (Manaus-Amazon) (ethanol 80% extract)	xanthochymol gambogenone 7-*epi*-nemorosone 7-*epi*-clusianone	[64]
Red Brazilian (ethanol extract)	(6a*S*,11a*S*)-6a-ethoxymedicarpan 2-(2',4'-dihydroxyphenyl)-3-methyl-6-methoxybenzofuran 2,6-dihydroxy-2-[(4-hydroxyphenyl)methyl]-3-benzofuranone	[65]
Red Brazilian (ethanol extract)	guttiferone E/xanthochymol, oblongifolin A, 7,5'-dihydroxy-3'-methoxyisoflavone, retusapurpurin A and retusapurpurin B	[37]
Geopropolis Brazilian (ethanol extract)	6-*O*-*p*-coumaroyl-*D*-galactopyranose 6-*O*-cinnamoyl-1-*O*-*p*-coumaroyl-β-*D*-glucopyranose 7-*O*-methylnaringenin 7-*O*-methylaromadendrin 7,4'-di-*O*-methylaromadendrin 4'-*O*-methylkaempferol 3-*O*-methylquercetin 5-*O*-methylaromadendrin 5-*O*-methylkaempferol	[66]
Geopropolis Brazilian (methanol extract)	5,7,4'-trihydroxyflavanone 3,5,6,7,4'-pentahydroxyflavonol naringenin-4'-*O*-β-D-glucopyranoside myricetin-3-*O*-β-D-glucopyranoside	[67]

Trying to relate the secondary metabolites of propolis with the vegetation surrounded of the bee hive, Edrada-Ebel *et al.* isolated two new prenylated stilbene derivatives (Table **20**) from the Ghanian propolis and a new phloroglucinone from a Cameroon propolis [88]. The *trans* olefinic configuration of protons H-α and H-β of (*E*)-5-{2-[8-hydroxy-2-methyl-2-(4-methy-pent-3-en-1-yl)-2*H*-chromen-6-yl]vinyl}-2-(3-methylbut-2-en-1-yl)benzene-1,3-diol **36** and 5-[(*E*)-3,5-dihydroxystyryl]-3-[(*E*)-3,7-dimethylocta-2,6-dien-1-yl]-benzene-1,2-diol **37** was established based on their ^{1}H-^{1}H coupling constant of ~ 16 Hz.

These structures were unequivocally assigned by using 2D NMR, mainly based on the HMBC correlations. The key HMBC correlations are those of H-β with C-1', C-2',6' of the resorcinol ring and of the H-α with C-1 and C-5 of chromenol moiety (Fig. **8**).

Table 18. Compounds identified in Asian propolis, which structures were determined using NMR spectroscopy.

Propolis type	Compounds	Ref.
Taiwanese (ethanol extract)	propolins A, B	[68]
Taiwanese (ethanol extract)	propolin C	[69]
Taiwanese (ethanol extract)	propolins A, B, C, D, E, F	[70]
Taiwanese (ethanol extract)	propolin G	[71]
Japanese (ethanol extract)	propolin C (or nymphaeol-A) propolin D (or nymphaeol-B) propolin F (or isonymphaeol-B) propolin G (or nymphaeol-C)	[72]
Japanese (ethanol extract)	propolins A, B, C, D, E, F, H (or 3'-geranylnaringenin) prokinawan	[73]
Thai (methanol extract)	(7''*S*)-8-[1-(4'-hydroxy-3'-methoxyphenyl)prop-2-en-1-yl]-(2*S*)-pinocembrin (*E*)-cinnamyl (*E*)-cinnamylidenate (*E*)-cinnamyl (*E*)-*p*-methoxycinnamate (*E*)-cinnamyl (*E*)-cinnamate (*E*)-cinnamyl (*E*)-ferulate (*E*)-cinnamyl (*E*)-*p*-coumarate	[74]
Thai (methanol extract) (*cont.*)	phenethyl caffeate phenethyl ferulate benzyl (*E*)-ferulate benzyl (*E*)-p-coumarate benzyl (*E*)-isoferulate (2*S*)-pinostrobin (2*S*)-pinocembrin (2*R*,3*R*)-alpinone-3-acetate (2*R*,3*R*)-pinobanksin 3-acetate (2*R*,3*R*)-pinobanksin 3-isobutyrate (2*R*,3*R*)-pinobanksin 3-(2-methyl)butyrate (2*R*,3*R*)-pinobanksin 3-propanoate tectochrysin chrysin izalpinin	
Thai (methanol 80% extract)	cardanol	[75]
Jeju Island, Korean (ethanol extract)	4'-methoxybavachromanol laserpitin isolaserpitin	[76]

(Table 18) contd.....

Propolis type	Compounds	Ref.
Korean (ethanol extract)	phenethyl caffeate benzyl caffeate ethyl caffeate benzyl ferulate 3',3'-dimethylallyl ferulate cinnamyl 3,4-dimethoxycaffeate cinnamyl coumarate benzyl coumarate phenethyl cinnamate cinnamyl cinnamate	[77]
Wuhan, Chinese (methanol extract)	2-acetyl-1-coumaroyl-3-cinnamoylglycerol (+)-2-acetyl-1-feruloyl-3-cinnamoylglycerol (–)-2-acetyl-1-feruloyl-3-cinnamoylglycerol 2-acetyl-1,3-dicinnamoylglycerol (–)-2-acetyl-1-(*E*)-feruloyl-3-[3''(ζ),16'']-dihydroxypalmitoylglycerol	[78]
Al-Baha, Saudi Arabian (ethylacetate extract)	(12*E*)-communic acid (12*Z*)-communic acid sandaracopimaric acid (+)-ferruginol (+)-totarol 3β-acetoxy-19(29)-taraxasten-20*a*-ol cycloartenol 24-methylene-cycloartenol β-amyrin-3β-*O*-acetate α-amyrin-3β-*O*-acetate taraxasterol-3β-*O*-acetate pseudotaraxasterol-3β-*O*-acetate lupeol-3β-*O*-acetate	[79]

Table 19. Compounds identified in African and Oceana propolis, which structures were determined using NMR spectroscopy.

Propolis type	Compounds	Ref.
Tunisian (methanol extract)	5,3'-dihydroxy-3,7,4',5'-tetramethoxyflavone 5,4'-dihydroxy-3,7,3'-trimethoxyflavone (or pachypodol)	[80]

(Table 19) contd.....

Propolis type	Compounds	Ref.
Jordanian (methanol extract)	cinnamyl cinnamate tetracosanoic acid tectochrysin pinocembrin pinobanksin-3-*O*-acetate 3-methylethergalangin chrysin galangin genkwanin alpinone naringenin apigenin cryptomeridiol agathadiol 24-(*Z*)-3-oxolanosta-1,7,24-trien-26-oic acid 24(*Z*)-1β,3β-dihydroxyeupha-7,24-dien-26-oic acid	[81]
Algerian (methanol extract)	3-methyl-3-butenyl (*E*)-caffeate 2-methyl-2-butenyl (*E*)-caffeate phenethyl (*E*)-caffeate chrysin apigenin kaempferol galangin pinobanksin pinobanksin 3-acetate pinobanksin 3-(*E*)-caffeate cupressic acid isocupressic acid imbricatoloic acid torulosal isoagathotal torulosol agathadiol cistadiol 18-hydroxy-*cis*-clerodan-3-ene-15-oic acid myricetin 3,7,4',5'-tetramethyl ether	[82]
Cameroonian (70% ethanol extract)	13 alk(en)ylphenols α-amyrin β-amyrin lupeol cycloartenol mangiferonic acid mangiferolic acid ambonic acid ambolic acid isomagiferolic acid 9 alk(en)ylresorcinols	[83]

(Table 19) contd.....

Propolis type	Compounds	Ref.
Solomon Islands (95% ethanol extract)	propolins C, D, G, H	[84]
Solomon Islands (ethanol extract)	solophenol A bonannione A sophoraflavanone A (2*S*)-5,7-dihydroxy-4'-methoxy-8-prenylflavanone	[85]
Solomon Islands (ethanol extract)	solophenol B, C, D solomonin	[86]
Australian [hexane:methanol(1:1)]	(2*S*)-cryptostrobin (2*S*)-stroboponin (2*S*)-cryptostrobin 7-methyl ether (2*S*)-desmethoxymatteucinol (2*S*)-pinostrobin (2*S*)-pinocembrin	[87]

Table 20. ^{1}H and ^{1}C NMR spectroscopic data (δ, ppm; *J* in parenthesis, Hz) of compounds **36-38**.

Position	36 (in $CDCl_3$)		37 [in $(CD_3)_2CO$]		Position	38 (in $CDCl_3$)	
	^{1}H	^{13}C	^{1}H	^{13}C		^{1}H	^{13}C
1	6.68 d (1.7)	116.9	6.82 d (1.8)	120.0	1		66.5
2		121.3		119.6	2A	3.52 dd (15.0, 11.7)	31.0
3		139.9		143.5	2B	1.58 m	31.0
3-OH			8.62 s		3	4.94 dd (11.7, 2.9)	85.7
4		145.0		140.0	6		88.6
4-OH	5.42 s		7.39 s		7	2.78 dd (10.7, 8.2)	42.6
5	6.94 d (1.7)	113.1	6.95 d (1.8)	110.5	8A	2.39 m	31.5
6		130.5		128.6	8B	1.88 m	31.5
α	6.86 d (16.2)	128.7	6.91 d (16.3)	129.3	9	2.10 m	44.5
β	6.76 d (16.2)	126.5	6.76 d (16.3)	126.3	10		50.4
1'		137.9		125.6	11		81.9
2'	6.53 br s	106.7	6.49 br d (2.0)	104.9	12		208.4
3'		155.7		159.3	13		205.2
3'-OH	5.35 s		8.37 s		14		68.5
4'		113.2	6.24 br t (2.0)	101.1	15A	2.57 dd (16.3, 6.6)	41.7
5'		155.7		159.3	15B	1.85 d (16.3)	41.7
5'-OH	5.35 s		8.37 s		16		204.4
6'	6.53 br s	106.7	6.49 br d (2.0)	104.9	17		84.3

(Table 20) contd.....

Position	36 (in $CDCl_3$)		37 [in $(CD_3)_2CO$]		Position	38 (in $CDCl_3$)	
	^1H	^{13}C	^1H	^{13}C		^1H	^{13}C
1''	6.34 d (10.0)	123.1	3.34 d (7.2)	28.7	18	1.20 s	21.5
2''	5.59 d (10.0)	130.3	5.36 t (7.2)	122.5	19	1.16 s	20.6
3''		80.8		135.4	20	1.29 s	28.2
4''	1.74 m	41.6	2.02 m	40.1	21	1.09 s	17.8
5''	2.06 m	31.0	2.05 m	29.7	22	1.48 s	24.9
6''	5.07 br t (6.3)	124.2	5.08 m	124.7	23	1.35 s	22.7
7''		132.6		130.6	24		192.2
8''	1.65 s	26.1	1.74 s	15.9	25/33		135.3
9''	1.57 s	18.1	1.57 s	17.5	26/30	7.16 d (8.0)	128.9
10''	1.41 s	27.0	1.63 s	25.6	27/29	7.28 t (8.0)	128.2
1'''	3.41 br d (7.0)	22.9			28	7.40 t (7.2)	132.4
2'''	5.26 br t (7.0)	122.0			30		128.9
3'''		135.2			31A	2.59 d (6.5)	29.8
4'''	1.75 s	26.2			31B	1.25 m	29.8
5'''	1.81 s	18.3			32	5.09 t (6.9)	118.9
					33		135.3
					34	1.69 s	26.1
					35	1.68 s	18.2

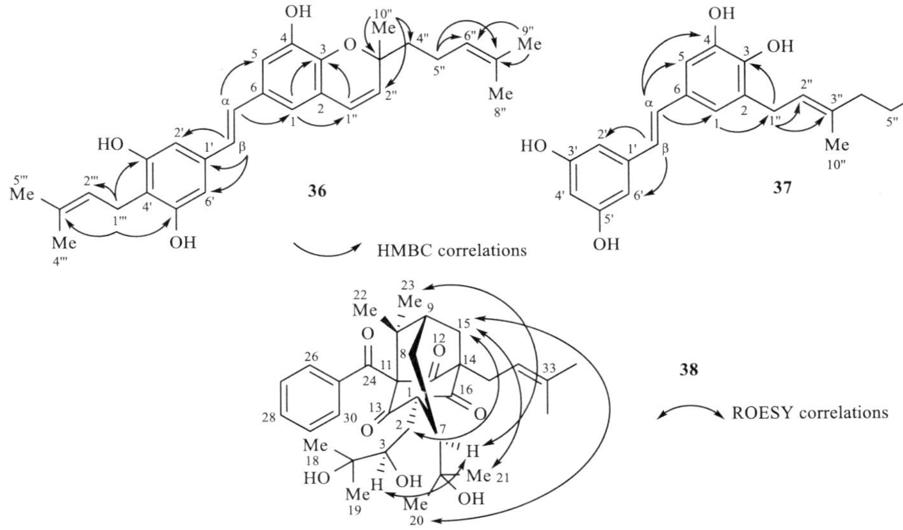

Fig. (8). HMBC correlations of compounds **36** and **37** and ROESY correlations of compound **38**.

The NMR data of the prenylated phloroglucinone **38** (Table **20**) was assigned based on those of plukenetione C isolated before from *Clusia plukenetii* [89]. The relative stereochemistry of compound **38** was determined by its ROESY correlations (Fig. **8**).

In conclusion, we have shown that NMR spectroscopy is a powerful analytical tool to: identify and quantify the constituents of honey and propolis and to elucidate their structures; detect their adulteration and differentiate their botanical and geographical origin.

CONFLICT OF INTEREST

The authors confirm that there is no conflict of interest to declare for this publication.

ACKNOWLEDGEMENTS

Thanks are due to the University of Aveiro and to FCT/MEC for the financial support of the QOPNA research Unit (FCT UID/QUI/00062/2013) through national founds and, where applicable, co-financed by the FEDER, within the PT2020 Partnership Agreement and also to Polytechnic Institute of Bragança.

ABBREVIATIONS

δ	= Chemical shift
CA	= Correspondence analysis
CLA	= Cluster analysis
^{13}C NMR	= Carbon-13 nuclear magnetic resonance spectroscopy
cont.	= Continuation
COSY	= Correlation spectroscopy
d	= Doublet
dd	= Double doublet
ddd	= Double doublet of doublets
dt	= Double triplet
DEPT	= Distortionless enhancement by polarization transfer
DQF	= Double quantum filtered
DOSY	= Diffusion ordered spectroscopy

FA	=	Factor analysis
GDA	=	General discriminant analysis
HCA	=	Hierarchical cluster analysis
HMBC	=	Heteronuclear multiple bond correlation
HMQC	=	Heteronuclear multiple quantum coherence
^{1}H NMR	=	Proton nuclear magnetic resonance spectroscopy
HPLC-DAD-MS	=	High pressure liquid chromatography-diode array detector-mass spectrometry
HPLC-MS	=	High pressure liquid chromatography-mass spectrometry
HSQC	=	Heteronuclear single quantum correlation
Hz	=	Hertz
ICA	=	Independent component analysis
J	=	Coupling constant
m	=	Multiplet
n.d.	=	Not detected
NOESY	=	Nuclear Overhauser effect spectroscopy
PCA	=	Principal component analysis
PLS-DA	=	Partial least squares discriminant analysis
ppm	=	Parts-per-million
ROESY	=	Rotating-frame Overhauser effect spectroscopy
s	=	Singlet
t	=	Triplet
td	=	Triplet of doublets
TMS	=	Tetramethylsilane
TOCSY	=	Total correlated spectroscopy

REFERENCES

[1] Balci, M. *Basic ^1H-^{13}C-NMR spectroscopy*; Elsevier: Amsterdam, **2005**, p. 430. [PMID: 25532]

[2] Hess, M.; Meier, H.; Zeeh, B. *Spectroscopic methods in organic chemistry. Chapter 3*; Murray, M., Ed.; Thieme: Leipzig, **1997**, pp. 71-218.

[3] Simpson, J.H. *Organic structure determination using 2D NMR spectroscopy*; Elsevier, Academic Press: Canada, **2008**.

[4] Ferreres, F.; Andrade, P.; Tomás-Barberán, F.A. Natural occurrence of abscisic acid in heather honey and floral nectar. *J. Agric. Food Chem.*, **1996**, *44*, 2053-2056. [http://dx.doi.org/10.1021/jf9507553]

[5] Milborrow, B.V. The conformation of abscisic acid by n.m.r. and a revision of the proposed mechanism for cyclization during its biosynthesis. *Biochem. J.*, **1984**, *220*(1), 325-332. [http://dx.doi.org/10.1042/bj2200325] [PMID: 6234887]

[6] Cabras, P.; Angioni, A.; Tuberoso, C.; Floris, I.; Reniero, F.; Guillou, C.; Ghelli, S. Homogentisic acid: a phenolic acid as a marker of strawberry-tree (*Arbutus unedo*) honey. *J. Agric. Food Chem.*, **1999**, *47*(10), 4064-4067. [http://dx.doi.org/10.1021/jf990141o] [PMID: 10552766]

[7] Tuberoso, C.I.; Bifulco, E.; Caboni, P.; Cottiglia, F.; Cabras, P.; Floris, I. Floral markers of strawberry tree (*Arbutus unedo* L.) honey. *J. Agric. Food Chem.*, **2010**, *58*(1), 384-389. [http://dx.doi.org/10.1021/jf9024147] [PMID: 19919097]

[8] Tuberoso, C.I.; Bifulco, E.; Jerković, I.; Caboni, P.; Cabras, P.; Floris, I. Methyl syringate: a chemical marker of asphodel (*Asphodelus microcarpus* Salzm. et Viv.) monofloral honey. *J. Agric. Food Chem.*, **2009**, *57*(9), 3895-3900. [http://dx.doi.org/10.1021/jf803991j] [PMID: 19309074]

[9] Truchado, P.; Martos, I.; Bortolotti, L.; Sabatini, A.G.; Ferreres, F.; Tomás-Barberan, F.A. Use of quinoline alkaloids as markers of the floral origin of chestnut honey. *J. Agric. Food Chem.*, **2009**, *57*(13), 5680-5686. [http://dx.doi.org/10.1021/jf900766v] [PMID: 19530652]

[10] Beretta, G.; Vistoli, G.; Caneva, E.; Anselmi, C.; Maffei Facino, R. Structure elucidation and NMR assignments of two new pyrrolidinyl quinoline alkaloids from chestnut honey. *Magn. Reson. Chem.*, **2009**, *47*(5), 456-459. [http://dx.doi.org/10.1002/mrc.2407] [PMID: 19253309]

[11] Beretta, G.; Artali, R.; Caneva, E.; Orlandini, S.; Centini, M.; Facino, R.M. Quinoline alkaloids in honey: further analytical (HPLC-DAD-ESI-MS, multidimensional diffusion-ordered NMR spectroscopy), theoretical and chemometric studies. *J. Pharm. Biomed. Anal.*, **2009**, *50*(3), 432-439. [http://dx.doi.org/10.1016/j.jpba.2009.05.029] [PMID: 19560302]

[12] Simova, S.; Atanassov, A.; Shishiniova, M.; Bankova, V. A rapid differentiation between oak honeydew honey and nectar and other honeydew honeys by NMR spectroscopy. *Food Chem.*, **2012**, *134*(3), 1706-1710. [http://dx.doi.org/10.1016/j.foodchem.2012.03.071] [PMID: 25006002]

[13] Ribeiro, R.O.; Mársico, E.T.; Carneiro, C.S.; Monteiro, M.L.; Conte Júnior, C.A.; Mano, S.; Jesus, E.F. Classification of Brazilian honeys by physical and chemical analytical methods and low field nuclear magnetic resonance (LF ^{1}H NMR). *LWT - Food Sci. Technol. (Campinas.)*, **2014**, *55*, 90-95.

[14] Ribeiro, R.O.; Mársico, E.T.; Carneiro, C.S.; Monteiro, M.L.; Conte Júnior, C.A.; Mano, S.; Jesus, E.F. Detection of honey adulteration of high fructose corn syrup by Low Field Nuclear Magnetic Resonance (LF ^{1}H NMR). *J. Food Eng.*, **2014**, *135*, 39-43. [http://dx.doi.org/10.1016/j.jfoodeng.2014.03.009]

[15] Consonni, R.; Cagliani, L.R. Geographical characterization of polyfloral and acacia honeys by nuclear magnetic resonance and chemometrics. *J. Agric. Food Chem.*, **2008**, *56*(16), 6873-6880. [http://dx.doi.org/10.1021/jf801332r] [PMID: 18627162]

[16] Donarski, J.A.; Jones, S.A.; Charlton, A.J. Application of cryoprobe ^{1}H nuclear magnetic resonance spectroscopy and multivariate analysis for the verification of corsican honey. *J. Agric. Food Chem.,* **2008**, *56*(14), 5451-5456.
[http://dx.doi.org/10.1021/jf072402x] [PMID: 18564849]

[17] Oleynikov, D.N.; Rohin, A.V. Galactomannan from seeds of Chinese honey locust (*Gleditsia sinensis* Lam.). *Appl. Biochem. Microbiol.,* **2010**, *46*, 103-107.
[http://dx.doi.org/10.1134/S0003683810010187]

[18] Boffo, E.F.; Tavares, L.A.; Tobias, A.C.; Ferreira, M.M.; Ferreira, A.G. Identification of components of Brazilian honey by ^{1}H NMR and classification of its botanical origin by chemometric methods. *LWT- Food Sci. Technol. (Campinas.),* **2012**, *49*, 55-63.

[19] Beretta, G.; Caneva, E.; Regazzoni, L.; Bakhtyari, N.G.; Maffei Facino, R. A solid-phase extraction procedure coupled to ^{1}H NMR, with chemometric analysis, to seek reliable markers of the botanical origin of honey. *Anal. Chim. Acta,* **2008**, *620*(1-2), 176-182.
[http://dx.doi.org/10.1016/j.aca.2008.05.025] [PMID: 18558139]

[20] Lindner, P.; Bermann, E.; Gamarnik, B. Characterization of citrus honey by deuterium NMR. *J. Agric. Food Chem.,* **1996**, *44*, 139-140.
[http://dx.doi.org/10.1021/jf950359k]

[21] Arvanitoyannis, I.S.; Chalhoub, C.; Gotsiou, P.; Lydakis-Simantiris, N.; Kefalas, P. Novel quality control methods in conjunction with chemometrics (multivariate analysis) for detecting honey authenticity. *Crit. Rev. Food Sci. Nutr.,* **2005**, *45*(3), 193-203.
[http://dx.doi.org/10.1080/10408690590956369] [PMID: 16048148]

[22] Lolli, M.; Bertelli, D.; Plessi, M.; Sabatini, A.G.; Restani, C. Classification of Italian honeys by 2D HR-NMR. *J. Agric. Food Chem.,* **2008**, *56*(4), 1298-1304.
[http://dx.doi.org/10.1021/jf072763c] [PMID: 18205311]

[23] Bertelli, D.; Lolli, M.; Papotti, G.; Bortolotti, L.; Serra, G.; Plessi, M. Detection of honey adulteration by sugar syrups using one-dimensional and two-dimensional high-resolution nuclear magnetic resonance. *J. Agric. Food Chem.,* **2010**, *58*(15), 8495-8501.
[http://dx.doi.org/10.1021/jf101460t] [PMID: 20681637]

[24] Jamróz, M.K.; Paradowska, K.; Zawada, K.; Makarova, K.; Kaźmierski, S.; Wawer, I. ^{1}H and ^{13}C NMR-based sugar profiling with chemometric analysis and antioxidant activity of herbhoneys and honeys. *J. Sci. Food Agric.,* **2014**, *94*(2), 246-255.
[http://dx.doi.org/10.1002/jsfa.6241] [PMID: 23712445]

[25] Donarski, J.A.; Jones, S.A.; Harrison, M.; Driffield, M.; Charlton, A.J. Identification of botanical biomarkers found in Corsican honey. *Food Chem.,* **2010**, *118*, 987-994.
[http://dx.doi.org/10.1016/j.foodchem.2008.10.033]

[26] Schievano, E.; Peggion, E.; Mammi, S. ^{1}H nuclear magnetic resonance spectra of chloroform extracts of honey for chemometric determination of its botanical origin. *J. Agric. Food Chem.,* **2010**, *58*(1), 57-65.
[http://dx.doi.org/10.1021/jf9022977] [PMID: 19938861]

[27] Schievano, E.; Morelato, E.; Facchin, C.; Mammi, S. Characterization of markers of botanical origin

and other compounds extracted from unifloral honeys. *J. Agric. Food Chem.,* **2013**, *61*(8), 1747-1755. [http://dx.doi.org/10.1021/jf302798d] [PMID: 23360363]

[28] Frérot, E.; Velluz, A.; Decorzant, E.; Naef, R. From linden flower to linden honey. Part 2: Glycosidic precursors of cyclohexa-1,3-diene-1-carboxylic acids. *Chem. Biodivers.,* **2006**, *3*(1), 94-100. [http://dx.doi.org/10.1002/cbdv.200690012] [PMID: 17193221]

[29] Schievano, E.; Stocchero, M.; Morelato, E.; Facchin, C.; Mammi, S. An NMR-based metabolomics approach to identify the botanical origin of honey. *Metabolomics,* **2012**, *8*, 679-690. [http://dx.doi.org/10.1007/s11306-011-0362-8]

[30] Lianda, R.L.; Castro, R.N. Isolamento e identificação da morina em mel brasileiro de *Apis mellifera. Quim. Nova,* **2008**, *31*, 1472-1474. [http://dx.doi.org/10.1590/S0100-40422008000600036]

[31] Consonni, R.; Cagliani, L.R.; Cogliati, C. Geographical discrimination of honeys by saccharides analysis. *Food Contr.,* **2013**, *32*, 543-548. [http://dx.doi.org/10.1016/j.foodcont.2013.01.038]

[32] Consonni, R.; Cagliani, L.R.; Cogliati, C. NMR characterization of saccharides in Italian honeys of different floral sources. *J. Agric. Food Chem.,* **2012**, *60*(18), 4526-4534. [http://dx.doi.org/10.1021/jf3008713] [PMID: 22509771]

[33] Zieliński, L.; Deja, S.; Jasicka-Misiak, I.; Kafarski, P. Chemometrics as a tool of origin determination of Polish monofloral and multifloral honeys. *J. Agric. Food Chem.,* **2014**, *62*(13), 2973-2981. [http://dx.doi.org/10.1021/jf4056715] [PMID: 24641200]

[34] Tomi, F.; Bradesi, P.; Bighelli, A.; Casanova, J. Computer-aided identification of individual components of essential oil using carbon-13 NMR spectroscopy. *J. Magn. Reson. A,* **1995**, *1*, 25-34.

[35] Bighelli, A.; Tomi, F.; Casanova, J. Computer-aided carbon-13 NMR study of phenols contained in liquids produced by pyrolysis of biomass. *Biomass Bioenergy,* **1994**, *6*, 461-464. [http://dx.doi.org/10.1016/0961-9534(94)00083-6]

[36] Mazzoni, V.; Bradesi, P.; Tomi, F. Casanova, Joseph. Direct qualitative and quantitative analysis of carbohydrate mixtures using ^{13}C NMR spectroscopy: application to honey. *Magn. Reson. Chem.,* **1997**, *35*, S81-S90. [http://dx.doi.org/10.1002/(SICI)1097-458X(199712)35:13<S81::AID-OMR202>3.0.CO;2-E]

[37] Monakhova, Y.B.; Tsikin, A.M.; Kuballa, T.; Lachenmeier, D.W.; Mushtakova, S.P. Independent component analysis (ICA) algorithms for improved spectral deconvolution of overlapped signals in ^{1}H NMR analysis: application to foods and related products. *Magn. Reson. Chem.,* **2014**, *52*(5), 231-240. [http://dx.doi.org/10.1002/mrc.4059] [PMID: 24604756]

[38] Sandusky, P.; Raftery, D. Use of selective TOCSY NMR experiments for quantifying minor components in complex mixtures: application to the metabonomics of amino acids in honey. *Anal. Chem.,* **2005**, *77*(8), 2455-2463. [http://dx.doi.org/10.1021/ac0484979] [PMID: 15828781]

[39] Gresley, A.L.; Kenny, J.; Cassar, C.; Kelly, A.; Sinclair, A.; Fielder, M.D. The application of high resolution diffusion NMR to the analysis of manuka honey. *Food Chem.,* **2012**, *135*(4), 2879-2886. [http://dx.doi.org/10.1016/j.foodchem.2012.07.072] [PMID: 22980885]

[40] Donarski, J.A.; Roberts, D.P.; Charlton, A.J. Quantitative NMR spectroscopy for the rapid measurement of methylglyoxal in manuka honey. *Anal. Methods,* **2010**, *2*, 1479-1483. [http://dx.doi.org/10.1039/c0ay00125b]

[41] Watson, D.G.; Peyfoon, E.; Zheng, L.; Lu, D.; Seidel, V.; Johnston, B.; Parkinson, J.A.; Fearnley, J. Application of principal components analysis to ^{1}H-NMR data obtained from propolis samples of different geographical origin. *Phytochem. Anal.,* **2006**, *17*(5), 323-331. [http://dx.doi.org/10.1002/pca.921] [PMID: 17019933]

[42] Papotti, G.; Bertelli, D.; Plessi, M.; Rossi, M.C. Use of HR-NMR to classify propolis obtained using different harvesting methods. *Int. J. Food Sci. Technol.,* **2010**, *45*, 1610-1618. [http://dx.doi.org/10.1111/j.1365-2621.2010.02310.x]

[43] Bertelli, D.; Papotti, G.; Bortolotti, L.; Marcazzan, G.L.; Plessi, M. ^{1}H-NMR simultaneous identification of health-relevant compounds in propolis extracts. *Phytochem. Anal.,* **2012**, *23*(3), 260-266. [http://dx.doi.org/10.1002/pca.1352] [PMID: 21853496]

[44] Rubio, O.C.; Cuellar Cuellar, A.; Rojas, N.; Castro, H.V.; Rastrelli, L.; Aquino, R. A polyisoprenylated benzophenone from Cuban propolis. *J. Nat. Prod.,* **1999**, *62*(7), 1013-1015. [http://dx.doi.org/10.1021/np980339n] [PMID: 10425129]

[45] Piccinelli, A.L.; Campo Fernandez, M.; Cuesta-Rubio, O.; Márquez Hernández, I.; De Simone, F.; Rastrelli, L. Isoflavonoids isolated from Cuban propolis. *J. Agric. Food Chem.,* **2005**, *53*(23), 9010-9016. [http://dx.doi.org/10.1021/jf0518756] [PMID: 16277396]

[46] Cuesta-Rubio, O.; Piccinelli, A.L.; Fernandez, M.C.; Hernández, I.M.; Rosado, A.; Rastrelli, L. Chemical characterization of Cuban propolis by HPLC-PDA, HPLC-MS, and NMR: the brown, red, and yellow Cuban varieties of propolis. *J. Agric. Food Chem.,* **2007**, *55*(18), 7502-7509. [http://dx.doi.org/10.1021/jf071296w] [PMID: 17691806]

[47] Márquez Hernández, I.; Cuesta-Rubio, O.; Campo Fernández, M.; Rosado Pérez, A.; Montes de Oca Porto, R.; Piccinelli, A.L.; Rastrelli, L. Studies on the constituents of yellow Cuban propolis: GC-MS determination of triterpenoids and flavonoids. *J. Agric. Food Chem.,* **2010**, *58*(8), 4725-4730. [http://dx.doi.org/10.1021/jf904527n] [PMID: 20230059]

[48] Piccinelli, A.L.; Lotti, C.; Campone, L.; Cuesta-Rubio, O.; Campo Fernandez, M.; Rastrelli, L. Cuban and Brazilian red propolis: botanical origin and comparative analysis by high-performance liquid chromatography-photodiode array detection/electrospray ionization tandem mass spectrometry. *J. Agric. Food Chem.,* **2011**, *59*(12), 6484-6491. [http://dx.doi.org/10.1021/jf201280z] [PMID: 21598949]

[49] Lotti, C.; Campo Fernandez, M.; Piccinelli, A.L.; Cuesta-Rubio, O.; Márquez Hernández, I.; Rastrelli, L. Chemical constituents of red Mexican propolis. *J. Agric. Food Chem.,* **2010**, *58*(4), 2209-2213. [http://dx.doi.org/10.1021/jf100070w] [PMID: 20121106]

[50] Lotti, C.; Piccinelli, A.L.; Arevalo, C.; Ruiz, I.; Migliani De Castro, G.M.; Figueira Reis De Sá, L.; Tessis, A.C.; Ferreira-Pereira, A.; Rastrelli, L. Constituents of Hondurian propolis with inhibitory effects on *Saccharomyces cerevisiae* multidrug resistance protein Pdr5p. *J. Agric. Food Chem.,* **2012**, *60*(42), 10540-10545.

[http://dx.doi.org/10.1021/jf302578r] [PMID: 23004023]

[51] Banskota, A.H.; Tezuka, Y.; Prasain, J.K.; Matsushige, K.; Saiki, I.; Kadota, S. Chemical constituents of Brazilian propolis and their cytotoxic activities. *J. Nat. Prod.,* **1998**, *61*(7), 896-900. [http://dx.doi.org/10.1021/np980028c] [PMID: 9677271]

[52] Hirota, M.; Matsuno, T.; Fujiwara, T.; Sugiyama, H.; Mineshita, S. Enhanced cytotoxicity in a *Z*-photoisomer of a benzopyran derivative of propolis. *J. Nat. Prod.,* **2000**, *63*(3), 366-370. [http://dx.doi.org/10.1021/np990463m] [PMID: 10757720]

[53] Marcucci, M.C.; Ferreres, F.; García-Viguera, C.; Bankova, V.S.; De Castro, S.L.; Dantas, A.P.; Valente, P.H.; Paulino, N. Phenolic compounds from Brazilian propolis with pharmacological activities. *J. Ethnopharmacol.,* **2001**, *74*(2), 105-112. [http://dx.doi.org/10.1016/S0378-8741(00)00326-3] [PMID: 11167028]

[54] da Silva, M.S.; Citó, A.M.; Chaves, M.H.; Lopes, J.A. Triterpenoides tipo cicloartano de própolis de Teresina – PI. *Quim. Nova,* **2005**, *28*, 801-804. [http://dx.doi.org/10.1590/S0100-40422005000500013]

[55] Tatefuji, T.; Izumi, N.; Ohta, T.; Arai, S.; Ikeda, M.; Kurimoto, M. Isolation and identification of compounds from Brazilian propolis which enhance macrophage spreading and mobility. *Biol. Pharm. Bull.,* **1996**, *19*(7), 966-970. [http://dx.doi.org/10.1248/bpb.19.966] [PMID: 8839971]

[56] Mishima, S.; Inoh, Y.; Narita, Y.; Ohta, S.; Sakamoto, T.; Araki, Y.; Suzuki, K-M.; Akao, Y.; Nozawa, Y. Identification of caffeoylquinic acid derivatives from Brazilian propolis as constituents involved in induction of granulocytic differentiation of HL-60 cells. *Bioorg. Med. Chem.,* **2005**, *13*(20), 5814-5818. [http://dx.doi.org/10.1016/j.bmc.2005.05.044] [PMID: 15993085]

[57] Teixeira, É.W.; Message, D.; Negri, G.; Salatino, A. Bauer-7-en-3β-yl acetate: a major constituent of unusual samples of Brazilian propolis. *Quim. Nova,* **2006**, *29*, 245-246. [http://dx.doi.org/10.1590/S0100-40422006000200013]

[58] Albuquerque, I.L.; Alves, L.A.; Lemos, T.L.; Monte, F.J. Ácido canárico (3,4-seco derivado do lupano) em própolis do Ceará. *Quim. Nova,* **2007**, *30*, 828-831. [http://dx.doi.org/10.1590/S0100-40422007000400013]

[59] Castro, M.L.; do Nascimento, A.M.; Ikegaki, M.; Costa-Neto, C.M.; Alencar, S.M.; Rosalen, P.L. Identification of a bioactive compound isolated from Brazilian propolis type 6. *Bioorg. Med. Chem.,* **2009**, *17*(14), 5332-5335. [http://dx.doi.org/10.1016/j.bmc.2009.04.066] [PMID: 19497755]

[60] Maruyama, H.; Sumitou, Y.; Sakamoto, T.; Araki, Y.; Hara, H. Antihypertensive effects of flavonoids isolated from Brazilian green propolis in spontaneously hypertensive rats. *Biol. Pharm. Bull.,* **2009**, *32*(7), 1244-1250. [http://dx.doi.org/10.1248/bpb.32.1244] [PMID: 19571393]

[61] Tavares, L.C.; Lemos, T.L.; Arriaga, A.M.; Santiago, G.M.; Braz-Filho, R. Estudo químicos de uma amostra de própolis verde de Passa Quatro, Minas Gerais, Brasil. *Quim. Nova,* **2010**, *33*, 2051-2054. [http://dx.doi.org/10.1590/S0100-40422010001000009]

[62] Trusheva, B.; Popova, M.; Bankova, V.; Simova, S.; Marcucci, M.C.; Miorin, P.L.; Pasin, F.R.; Tsvetkova, I. Bioactive constituents of Brazilian red propolis. *eCAM,* **2006**, *3*, 249-254.

[63] Oldoni, T.L.; Cabral, I.S.; d'Arce, M.A.; Rosalen, P.L.; Ikegaki, M.; Nascimento, A.M.; Alencar, S.M. Isolation and analysis of bioactive isoflavonoids and chalcone from a new type of Brazilian propolis. *Sep. Pur. Technol.,* **2011**, *77*, 208-214. [http://dx.doi.org/10.1016/j.seppur.2010.12.007]

[64] Ishida, V.F.; Negri, G.; Salatino, A.; Bandeira, M.F. A new type of Brazilian propolis: prenylated benzophenones in propolis from Amazon and effects against cariogenic bacteria. *Food Chem.,* **2011**, *125*, 966-972. [http://dx.doi.org/10.1016/j.foodchem.2010.09.089]

[65] Awale, S.; Li, F.; Onozuka, H.; Esumi, H.; Tezuka, Y.; Kadota, S. Constituents of Brazilian red propolis and their preferential cytotoxic activity against human pancreatic PANC-1 cancer cell line in nutrient-deprived condition. *Bioorg. Med. Chem.,* **2008**, *16*(1), 181-189. [http://dx.doi.org/10.1016/j.bmc.2007.10.004] [PMID: 17950610]

[66] de Souza, S.A.; Camara, C.A.; Silva, E.M.; Silva, T.M. Composition and antioxidant activity of geopropolis collected by Melipona subnitida (Jandaíra) bees. *Evid.-Based Compl. Altern. Med.,* **2013**, *8011383*, 5.

[67] da Silva, E.C.; Muniz, M.P.; Nunomura, R.C.; Nunomura, S.M.; Zilse, G.A. Constituintes fenólicos e actividade antioxidante da geoprópolis de duas espécies de abelhas sem ferrão amazónicas. *Quim. Nova,* **2013**, *36*, 628-633. [http://dx.doi.org/10.1590/S0100-40422013000500003]

[68] Chen, C-N.; Wu, C-L.; Shy, H-S.; Lin, J-K. Cytotoxic prenylflavanones from Taiwanese propolis. *J. Nat. Prod.,* **2003**, *66*(4), 503-506. [http://dx.doi.org/10.1021/np0203180] [PMID: 12713401]

[69] Chen, C-N.; Wu, C-L.; Lin, J-K. Propolin C from propolis induces apoptosis through activating caspases, Bid and cytochrome *c* release in human melanoma cells. *Biochem. Pharmacol.,* **2004**, *67*(1), 53-66. [http://dx.doi.org/10.1016/j.bcp.2003.07.020] [PMID: 14667928]

[70] Chen, C.-N.; Weng, M.-S.; Wu, C.-L.; Lin, J.-K. Comparison of radical scavenging activity, cytotoxic effects and apoptosis induction in human melanoma cells by Taiwanese propolis from different sources. *eCAM,* **2004**, *1*, 175-185.

[71] Huang, W-J.; Huang, C.H.; Wu, C-L.; Lin, J-K.; Chen, Y-W.; Lin, C-L.; Chuang, S-E.; Huang, C-Y.; Chen, C-N. Propolin G, a prenylflavanone, isolated from Taiwanese propolis, induces caspase-dependent apoptosis in brain cancer cells. *J. Agric. Food Chem.,* **2007**, *55*(18), 7366-7376. [http://dx.doi.org/10.1021/jf0710579] [PMID: 17685631]

[72] Kumazawa, S.; Goto, H.; Hamasaka, T.; Fukumoto, S.; Fujimoto, T.; Nakayama, T. A new prenylated flavonoid from propolis collected in Okinawa, Japan. *Biosci. Biotechnol. Biochem.,* **2004**, *68*(1), 260-262. [http://dx.doi.org/10.1271/bbb.68.260] [PMID: 14745198]

[73] Kumazawa, S.; Ueda, R.; Hamasaka, T.; Fukumoto, S.; Fujimoto, T.; Nakayama, T. Antioxidant

prenylated flavonoids from propolis collected in Okinawa, Japan. *J. Agric. Food Chem.,* **2007**, *55*(19), 7722-7725.
[http://dx.doi.org/10.1021/jf071187h] [PMID: 17711339]

[74] Athikomkulchai, S.; Awale, S.; Ruangrungsi, N.; Ruchirawat, S.; Kadota, S. Chemical constituents of Thai propolis. *Fitoterapia,* **2013**, *88*, 96-100.
[http://dx.doi.org/10.1016/j.fitote.2013.04.008] [PMID: 23660244]

[75] Boonsai, P.; Phuwapraisirisan, P.; Chanchao, C. Antibacterial activity of a cardanol from Thai *Apis mellifera* propolis. *Int. J. Med. Sci.,* **2014**, *11*(4), 327-336.
[http://dx.doi.org/10.7150/ijms.7373] [PMID: 24578609]

[76] Kumazawa, S.; Suzuki, S. A. M.-R.; Kamihira, M.; Udagawa, Y.; Bang, K.-S.; Nakayama, T. A new chalcone from propolis collected on Jeju Island, Korea. *Food Sci. Technol. Res.,* **2006**, *12*, 67-69.
[http://dx.doi.org/10.3136/fstr.12.67]

[77] Lee, I-K.; Han, M-S.; Kim, D-W.; Yun, B-S. Phenylpropanoid acid esters from Korean propolis and their antioxidant activities. *Bioorg. Med. Chem. Lett.,* **2014**, *24*(15), 3503-3505.
[http://dx.doi.org/10.1016/j.bmcl.2014.05.065] [PMID: 24928402]

[78] Shi, H.; Yang, H.; Zhang, X.; Sheng, Y.; Huang, H.; Yu, L. Isolation and characterization of five glycerol esters from Wuhan propolis and their potential anti-inflammatory properties. *J. Agric. Food Chem.,* **2012**, *60*(40), 10041-10047.
[http://dx.doi.org/10.1021/jf302601m] [PMID: 22978445]

[79] Jerz, G.; Elnakady, Y.A.; Braun, A.; Jäckel, K.; Sasse, F.; Al Ghamdi, A.A.; Omar, M.O.; Winterhalter, P. Preparative mass-spectrometry profiling of bioactive metabolites in Saudi-Arabian propolis fractionated by high-speed countercurrent chromatography and off-line atmospheric pressure chemical ionization mass-spectrometry injection. *J. Chromatogr. A,* **2014**, *1347*, 17-29.
[http://dx.doi.org/10.1016/j.chroma.2014.04.068] [PMID: 24831423]

[80] Martos, I.; Cossentini, M.; Ferreres, F.; Tomás-Barberán, F.A. Flavonoid composition of Tunisian honeys and propolis. *J. Agric. Food Chem.,* **1997**, *45*, 2824-2829.
[http://dx.doi.org/10.1021/jf9609284]

[81] Shaheen, S.A.; Zarga, M.H.; Nazer, I.K.; Darwish, R.M.; Al-Jaber, H.I. Chemical constituents of Jordanian propolis. *Nat. Prod. Res.,* **2011**, *25*(14), 1312-1318.
[http://dx.doi.org/10.1080/14786419.2010.509060] [PMID: 21732907]

[82] Piccinelli, A.L.; Mencherini, T.; Celano, R.; Mouhoubi, Z.; Tamendjari, A.; Aquino, R.P.; Rastrelli, L. Chemical composition and antioxidant activity of Algerian propolis. *J. Agric. Food Chem.,* **2013**, *61*(21), 5080-5088.
[http://dx.doi.org/10.1021/jf400779w] [PMID: 23650897]

[83] Kardar, M.N.; Zhang, T.; Coxon, G.D.; Watson, D.G.; Fearnley, J.; Seidel, V. Characterisation of triterpenes and new phenolic lipids in Cameroonian propolis. *Phytochemistry,* **2014**, *106*, 156-163.
[http://dx.doi.org/10.1016/j.phytochem.2014.07.016] [PMID: 25104230]

[84] Raghukumar, R.; Vali, L.; Watson, D.; Fearnley, J.; Seidel, V. Antimethicillin-resistant *Staphylococcus aureus* (MRSA) activity of 'pacific propolis' and isolated prenylflavanones. *Phytother. Res.,* **2010**, *24*(8), 1181-1187.
[PMID: 20077439]

[85] Inui, S.; Shimamura, Y.; Masuda, S.; Shirafuji, K.; Moli, R.T.; Kumazawa, S. A new prenylflavonoid isolated from propolis collected in the Solomon Islands. *Biosci. Biotechnol. Biochem.,* **2012**, *76*(5), 1038-1040.
[http://dx.doi.org/10.1271/bbb.120021] [PMID: 22738984]

[86] Inui, S.; Hosoya, T.; Shimamura, Y.; Masuda, S.; Ogawa, T.; Kobayashi, H.; Shirafuji, K.; Moli, R.T.; Kozone, I.; Shin-ya, K.; Kumazawa, S. Solophenols B-D and solomonin: new prenylated polyphenols isolated from propolis collected from the Solomon Islands and their antibacterial activity. *J. Agric. Food Chem.,* **2012**, *60*(47), 11765-11770.
[http://dx.doi.org/10.1021/jf303516w] [PMID: 23067056]

[87] Massaro, C.F.; Katouli, M.; Grkovic, T.; Vu, H.; Quinn, R.J.; Heard, T.A.; Carvalho, C.; Manley-Harris, M.; Wallace, H.M.; Brooks, P. Anti-staphylococcal activity of *C*-methyl flavanones from propolis of Australian stingless bees (*Tetragonula carbonaria*) and fruit resins of *Corymbia torelliana* (Myrtaceae). *Fitoterapia,* **2014**, *95*, 247-257.
[http://dx.doi.org/10.1016/j.fitote.2014.03.024] [PMID: 24704551]

[88] Almutairi, S.; Eapen, B.; Chundi, S.M.; Akhalil, A.; Siheri, W.; Clements, C.; Fearnley, J.; Watson, D.G.; Edrada-Ebel, R. New anti-trypanosomal active prenylated compounds from African propolis. *Phytochem. Lett.,* **2014**, *10*, 35-39.
[http://dx.doi.org/10.1016/j.phytol.2014.07.002]

[89] Henry, G.E.; Jacobs, H.; Carrington, C.M.; McLean, S.; Reynolds, W.F. Prenylated benzophenone derivatives from Caribbean *Clusia* species (Guttiferae). Plukenetiones B-G and Xerophenone A. *Tetrahedron,* **1999**, *55*, 1581-1596.
[http://dx.doi.org/10.1016/S0040-4020(98)01203-4]

CHAPTER 6

Electrochemical Sensors for Assessing Antioxidant Capacity of Bee Products

António M. Peres[1], Mara E.B. Sousa[2], Ana C.A. Veloso[3,4], Letícia Estevinho[5], Luís G. Dias[5,6,*]

[1] *LSRE- Laboratory of Separation and Reaction Engineering - Associate Laboratory LSRE-LCM, School of Agriculture, Polytechnic Institute of Bragança, Campus de St[a] Apolónia, Apartado 1172, 5301-855 Bragança, Portugal*

[2] *CIMO - Mountain Research Centre, School of Agriculture, Polytechnic Institute of Bragança, Campus de St[a] Apolónia, Apartado 1172, 5301-855 Bragança, Portugal*

[3] *Polytechnic Institute of Coimbra, ISEC, DEQB, Rua Pedro Nunes, Quinta da Nora, 3030-199 Coimbra, Portugal*

[4] *CEB - Centre of Biological Engineering, University of Minho, Campus de Gualtar, 4710-057 Braga, Portugal*

[5] *School of Agriculture, Polytechnic Institute of Bragança, Campus de St[a] Apolónia, Apartado 1172, 5301-855 Bragança, Portugal*

[6] *CQ-VR, Center of Chemistry – Vila Real, University of Trás-os-Montes e Alto Douro, Apartado 1013, 5001-801 Vila Real, Portugal*

Abstract: This chapter is focused on the application of electrochemical techniques (*e.g.*, sensors and biosensors), as the predominant methodology, to the quantification of individual or total phenolic compounds, either in standard solutions or in real matrices (*e.g.*, plants, fruits and beverages) and their capability for assessing antioxidant activity/capacity. Specially, the potential application to evaluate antioxidant capacity of bee-hives products (*e.g.*, propolis, honey) is addressed. Finally, the voltammetric behavior of Portuguese monofloral honeys is discussed for the first time, taking into account the expected effects of honey color and floral origin.

[*] **Address correspondence to Luis G. Dias:** Escola Superior Agrária, Instituto Politécnico de Bragança, Campus Santa Apolónia, 5301-855 Bragança, Portugal; Tel: +351273303220; Fax: +351273325405; Email: ldias@ipb.pt

Susana M. Cardoso (Ed)

Also, a possible relation with the expected antioxidant capacity of honeys is discussed, considering their floral origin. Works describing the use of electrochemical detection imbibed on liquid chromatographic or capillary electrophoretic configurations among other analytical methods will not be focused in this review, although their undoubtedly potentials and proved applications.

Keywords: Antioxidant activity, Bee products, Cyclic voltammetry, Differential pulse voltammetry, Electrochemical techniques, Honey, Phenolic compounds, Pollen, Propolis, Square wave voltammetry.

1. INTRODUCTION

Honey, propolis and pollen are plant-derived products and a source of polyphenolic compounds. Honey has a high content of sugars and small amounts of minerals, proteins and other constituents namely flavonoids, phenolic acids, enzymes, amino acids and vitamins [1, 2]. In the case of honey, the total phenolic content appears to be strongly correlated with the antioxidant activity, being the highest values found in darker honeys [1, 3 - 6]. Propolis is a resinous material, which pharmaceutical properties are well known and attributed to the high contents in polyphenols [7] (flavonoids, phenolic acids and their esters), as well as terpenoids, steroids and amino acids [8]. Bee pollen also contains considerable amounts of phytochemicals and nutrients, being rich in carotenoids, flavonoids and phytosterols [9]. The floral and geographical origins affect greatly the quantity and composition of polyphenol compounds in these bee's products, which reflects on the antioxidant capacity of each product [1, 8].

Electroanalytical methods are well-known tools used to study chemical and biological systems, allowing evaluating the antioxidant capacity of compounds that act as reducing agents, like phenolic compounds, which are easily oxidized on the surface of electrodes. These methods have several advantages when compared to other more traditional techniques (for example, spectrophotometric methods) since, they are simple methodologies, have low detection limits, good selectivity, reduced time of analysis, low consumption of reagents, having an overall lower environmental impact. Since, the electrochemical signals are due to the presence of analytes with electrical properties (antioxidants), it is not necessary to generate or use oxidized species. Among the electroanalytical techniques, three voltammetric methods have been particularly reported as fast screening tools for assessing the composition quality and bioactivity of samples: cyclic voltammetry (CV), differential pulse voltammetry (DPV) and square-wave voltammetry (SWV). These techniques use electrolysis conditions to study the phenomena occurring between the electrode surface and the thin solution layer in contact, differing in the way that potential is applied [10 - 13]. They allow to obtain quantitative and qualitative information of chemical species during the electrolysis using the current-potential curve generated (*i.e.*, the voltammograms).

In the literature, some works report the direct application of electrochemical techniques for bioactivity assessment and/or quality evaluation of bee-hive products, being mainly focused in honey and propolis analysis. Indeed, to the best of the authors' knowledge there is no study concerning pollen evaluation. Despite not being widespread used [14], these analytical techniques are an attractive approach to characterize compounds that act as reducing agents in natural products, being envisaged an increase of their application in a near future for bee products analysis.

2. VOLTAMMETRIC TECHNIQUES: GENERAL CONCEPTS

The voltammetric techniques that use electrochemical cells are based on two or three electrodes, immersed in a solution, which allows the movement of ions by charge transfer (electrolytes). The electrodes (metals or semiconductors, solid or liquid) allow the charge transfer through the electrons movement.

2.1. Electrochemical Cells

The electrochemical cells can be galvanic or electrolytic cells. In the galvanic cells, the reactions occur spontaneously in the electrodes converting the energy generated in a chemical reaction into electrical energy. Applying a potential to the cell beyond the potential of the reversible reaction, the reaction direction is changed, being possible the conversion of electrical energy into chemical energy. In these conditions, it is an electrolytic cell. The electrolytic cells allow studying the reduction (electron capture) and oxidation (release of electrons) phenomena, in general, under the action of an external controlled potential, enabling to control the reaction's direction and their intensity. So, accurate information about the oxidation-reduction reactions (for example, on the concentration, thermodynamics, kinetics and reaction mechanisms) can be obtained [10 - 13].

In the electrochemical cell, the electrolyte solution (supporting electrolyte) reduces the medium resistance, eliminating the migration current and, depending on the analysis, keeps constant the solution pH (buffer solution). A three-electrode cell (Fig. **1**) should be used when the product between the current and the resistance component ($i \times R$) is high. Usually, there is a reference electrode (RE), a counter electrode (CE) and a working electrode (WE). The WE corresponds to an inert electrode and can be a metal (mercury, platinum or gold), a glassy carbon or a carbon paste. The WE's size, structure and material depend on the stability and selectivity of the WE surface towards the compound to be analyzed. The CE can be of any type (usually platinum), should be a good conductor and must not interfere with the reactions occurring in the solution. The RE may be a saturated calomel (Hg/Hg_2Cl_2) electrode or a silver/silver chloride (Ag/AgCl) electrode.

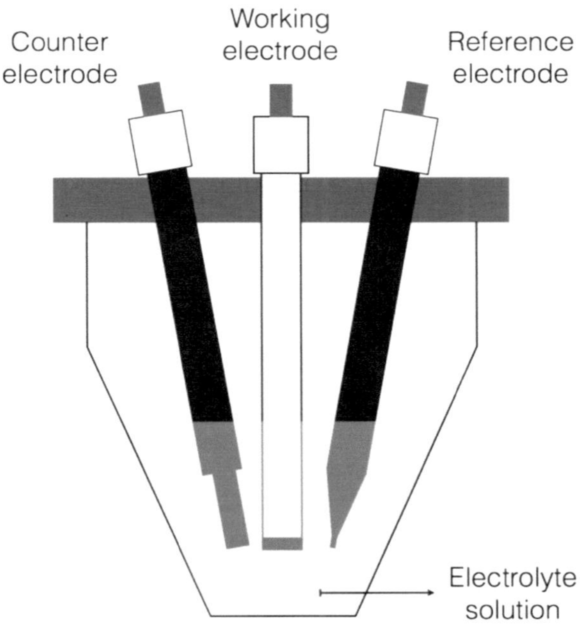

Fig. (1). Three electrodes typical electrochemical cell for voltammetry assays: working electrode (WE); counter electrode (CE); reference electrode (RE).

In this system, the current flows between the WE (usually, it has a small surface to quickly achieve the potential imposed on it) and the CE (the driving current from the WE). The potential is measured between the WE and the RE, which should be located as close as possible to the WE to reduce the resistance between them. Potentiostats are used in these methods, allowing the measurement of the applied potential (using high input impedance) and of the resulting current. This allows the passage of current through the RE, being the potential kept at a constant value. The potential and current are measured simultaneously being usually the recorded data visualized as a voltammogram (*i.e.*, the plot of current *vs*. potential scan).

As mentioned earlier, CV is one of the three most common electrochemical methods used for assessing antioxidant capacity of single- or multi-components of natural products, as well as for assessment of the total phenolic contents of natural samples.

2.2. Cyclic Voltammetry

The CV is a technique for acquiring quality information on the redox states, the oxidation state stability and electron transfer kinetics. In this technique, the potential is applied in two directions, namely in the form of a triangular wave, while the current is monitored. Fig. (**2A**) shows the potential parameters set over time, for a CV analysis and Fig. (**2B**) shows a typical cyclic voltammogram for a reversible redox process, being indicated the parameters that can be drawn from the chart: the cathodic peak potential (E_{cp}); the anodic peak potential (E_{ap}); the cathodic peak current (i_{cp}); and the anodic peak current (i_{ap}).

The analysis generates voltammograms where both the oxidation (anodic) and the reduction (cathodic) waves are plotted. In the absence of electroactive species, the electrochemical cell functions as a capacitor, so cations and anions migrate to the cathode and anode respectively, without an effective charge transfer, resulting in a voltage-current curve (voltammogram) without peaks. The presence of electroactive compounds in the interface between the solution thin layer and the electrode surface, leads to the appearance of current peaks, generated at their reduction and/or oxidation potentials, for the analyzed compound.

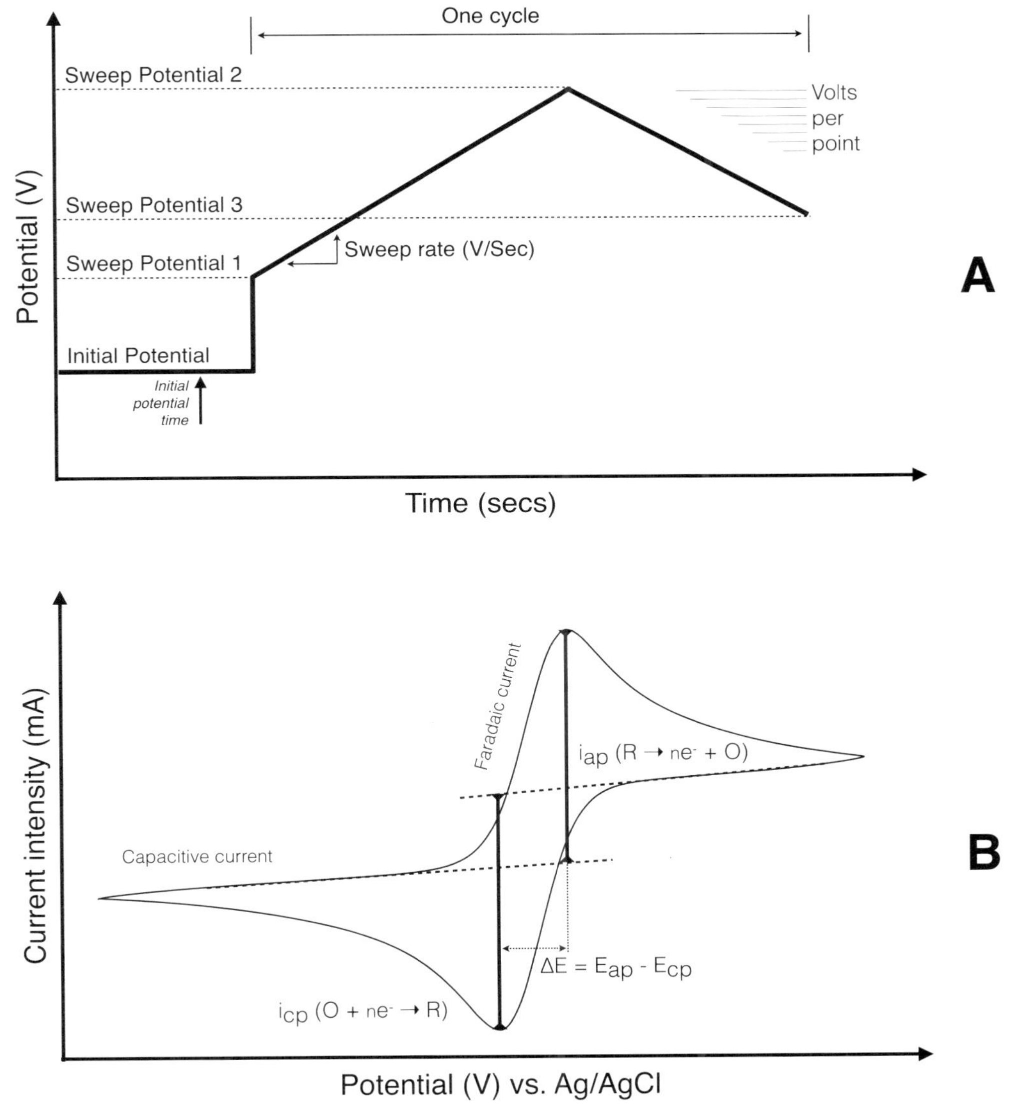

Fig. (2). CV analysis: scheme of the potential parameters to be set (**A**) and typical voltammogram (**B**).

For reversible redox processes, when the potential is reversed the newly oxidized species are reduced at the electrode interface, achieving the original state if the electrode reaches the polarization level at a sufficiently negative potential. In reversible processes, the difference between the anodic and cathodic potentials is zero ($\Delta E = E_{cp} - E_{ap} = 0$) and the ratio between the reduction and oxidation current peaks is equal to 1 ($i_{ap} / i_{cp} = 1$). In some cases, the cyclic voltammograms show a quasi-reversible process, when differences from those values are observed. If an oxidized or reduced compound does not regenerate completely, by reversing the potential applied to the electrode, the redox process is irreversible [11]. Besides this qualitative information, if the peak current is proportional to the analyte's concentration, a calibration curve can be established enabling quantifying the redox active compounds present in the sample.

Overall, CV is commonly used to characterize the redox system [15 - 18], being a non-destructive electrochemical analytical technique with good sensitivity. In general, when CV is applied for analyzing natural products extracts, a voltage scanning is applied at the WE and the current observed due to the oxidation of an antioxidant compound (reducing agents that are able to donate an electron) is measured. For single compound analysis, the maximum current response at the anodic peak is proportional to its concentration but, when analyzing an extract containing a mixture of compounds, the area under the curve shows a better correlation with the total antioxidant capacity [16]. In antioxidant properties studies, a standard compound (for instance, caffeic acid), with a similar chemical structure as the target compounds, is used to establish a calibration curve, allowing quantifying the total phenolic content and the related antioxidant activity of the biological samples [19]. Also, it allows a comparison with the results from antioxidant spectrophotometric assays (for instance, DPPH (2,2-diphenyl-1-picrylhydrazyl) and ferrous chelation assays). These general procedures are also used in DPV and SWV.

2.3. Differential Pulse Voltammetry

The DPV is a pulse voltammetry technique that has the capacity to discriminate charging (capacitance) current resulting in a more selective and sensitive technique towards oxidation or reduction currents (faradaic currents) than conventional voltammetry. This technique measures the difference of the current before (i(1)) and after a small potential pulse (i(2)) is applied, with amplitudes between 10 and 100 mV, for several milliseconds, superimposed on an applied linear potential sweep (potential is changing linearly with time) [11, 12]. Fig. (**3A**) shows the potential parameters usually used to carry out a typical DPV analysis, being an example of voltammogram shown in Fig. (**3B**).

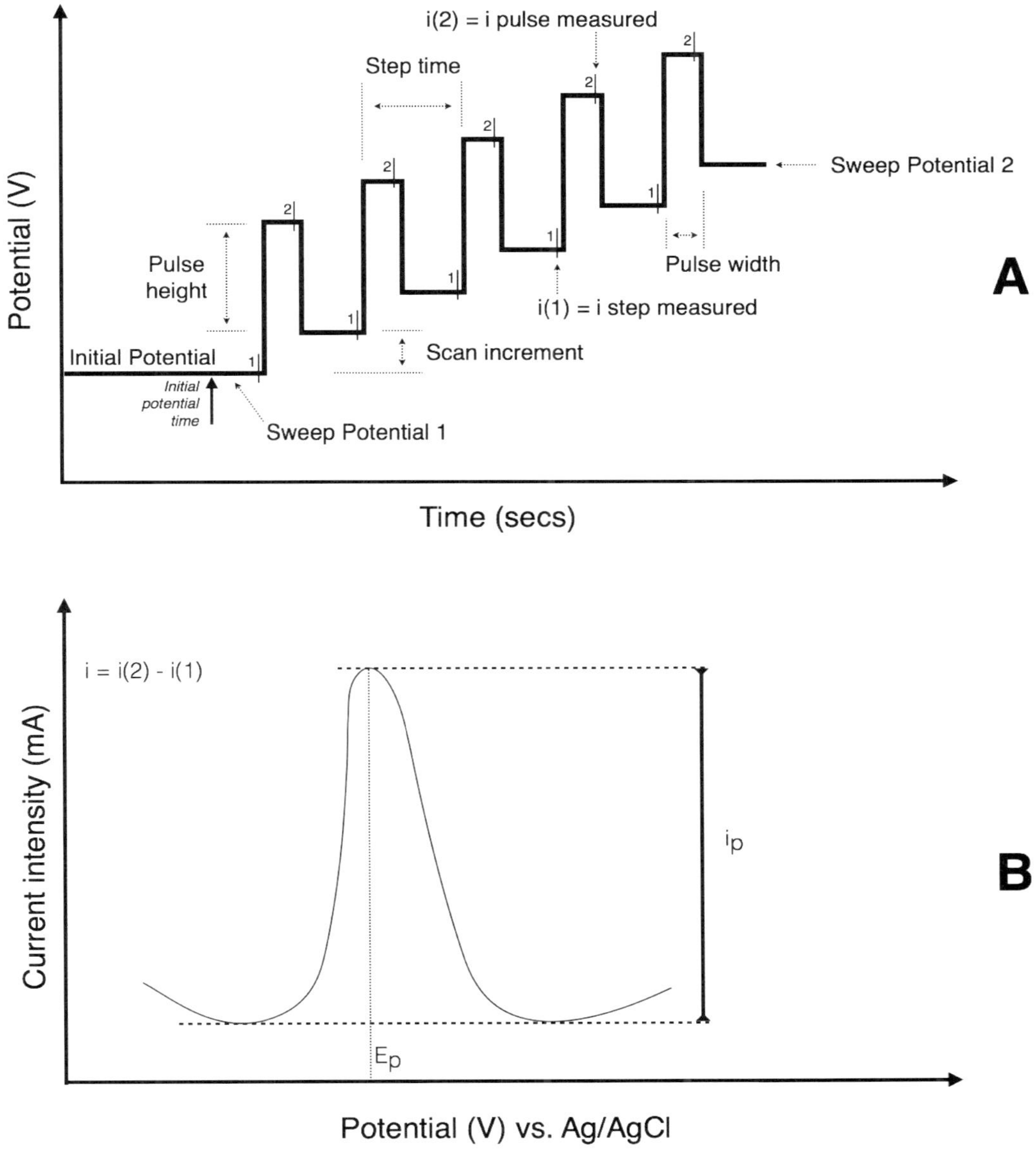

Fig. (3). DPV analysis: scheme of the potential parameters to be set (**A**) and typical voltammogram (**B**).

The voltammogram represents the difference between the two currents obtained on two applied potentials. In each pulse it goes through a maximum, showing a peak shape, being the peak's position on the potential (E_p) axis dependent of the type of analyte under study, and its height (i_p) on its concentration [20].

2.4. Square Wave Voltammetry

The SWV is another pulse voltammetry technique, which uses a potential waveform enabling to obtain more defined peaks, a good discrimination against background currents, larger dynamic concentration range and more sensitive detection of analytes than CV [11, 12]. Moreover, SWV is a faster (due to the use of frequencies between 1 and 100 Hz) technique compared with DPV [11, 12, 21], which reduces the consumption of electroactive compounds, leading to lower adsorption of species at the electrode surface (lower blocking of the electrode surface), turning out a more sensitive technique. The great advantage of pulse techniques in relation to CV is the greater capacity to discriminate the influence of capacitive current, resulting in a wider dynamic range and higher sensitivity. This technique is usually used in combination with CV because the latter provides information about the reversibility of the anodic waves.

With SWV technique, the current is measured in the WE, while between the WE and the RE, the current is swept by a symmetrical square wave, with ΔEp of amplitude, superimposed on a voltage ramp with a staircase-shaped. The current is sampled at the forward pulse (i_{fwd}) and at the end of the reverse pulse (i_{rev}) in each square-wave cycle. Measuring the current in both directions (positive towards oxidation and negative to reduction generating a peak for each of the processes) allows obtaining information concerning the oxidation or reduction of the electroactive species at the electrode surface. This dual measuring minimizes the capacitive current contribution on the total current reading. The difference between these two currents (response wave $\Delta i = i_{fwd} - i_{rev}$) is plotted *versus* the sweep potential obtaining a peak-shaped voltammogram display [11, 12]. Fig. (**4**) shows the potential parameters that are defined for the SWV analysis and the typical voltammograms obtained from the analysis: i_{fwd} curve, i_{rev} curve and Δi.

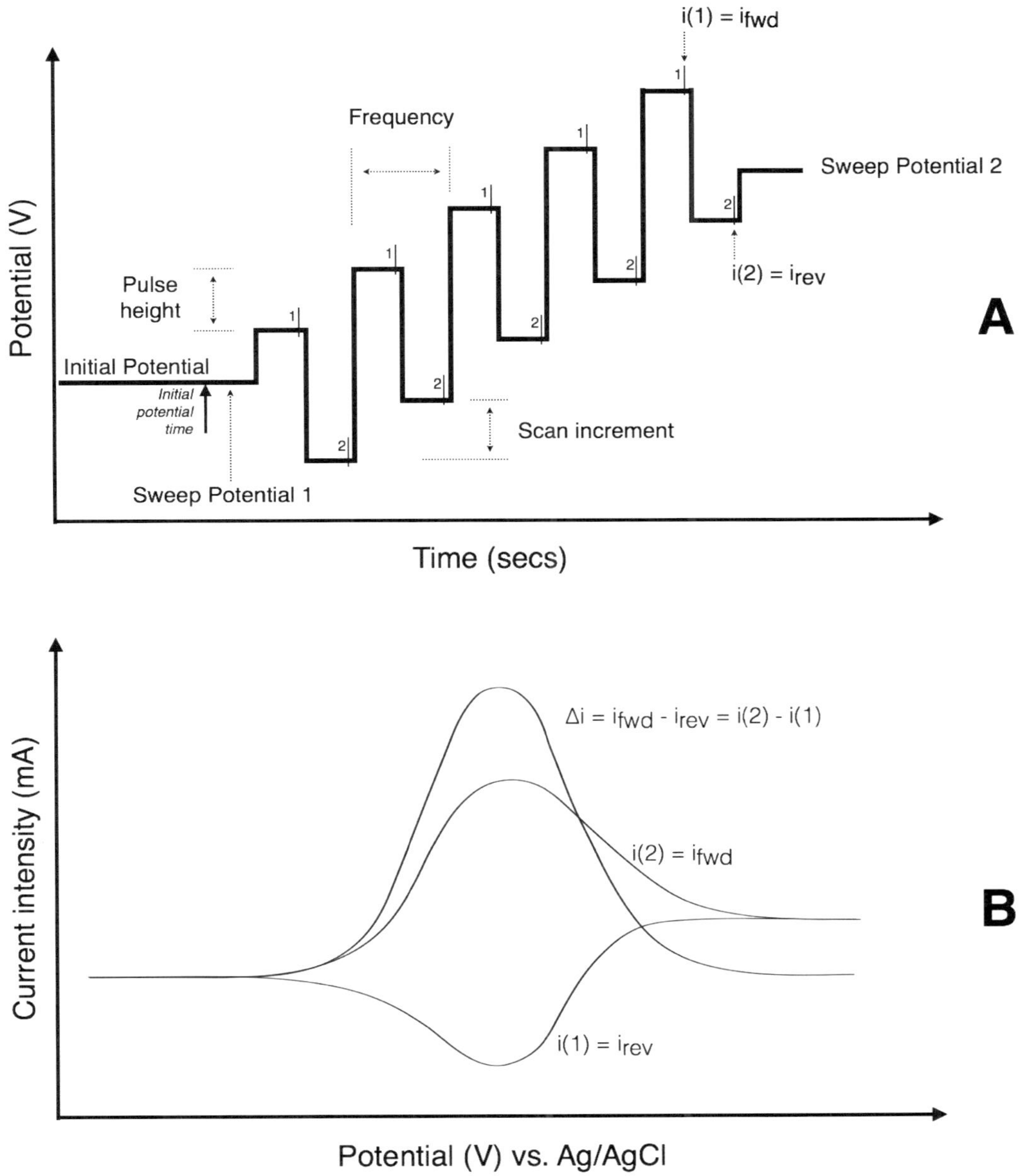

Fig. (4). SWV analysis: scheme of the potential parameters to be set (**A**) and typical voltammogram (**B**).

As DPV technique, the SWV produces faradaic response current peaks (the Δi follows Faraday's law), being the peak height directly proportional to the concentrations of electroactive compounds [10, 11]. Also, these two pulse techniques can be referred as polarography when using a dropping mercury electrode or a static mercury drop electrode as the WE [10, 11].

3. ANTIOXIDANT CAPACITY ASSESSEMENT USING VOLTAMMETRIC DEVICES: HONEY AND PROPOLIS ANALYZES

The availability of analytical tools for antioxidant screening and total phenolic compounds quantification is of major importance. The antioxidant activity of polyphenols can be evaluated according to their reactivity as a hydrogen- or electron-donating agent; capability to stabilize and delocalize the unpaired electron of radicals; reactivity with other antioxidants; and, transition metal-chelating potential [22]. So, different methods can be used for antioxidant evaluation namely, chromatography, spectrophotometry and electroanalytical techniques.

As previously mentioned, this work is focused in the use of electrochemical methodologies, not coupled to any other analytical method, for evaluating the antioxidant capacity or the quality of bee-hive products. In the first case, the studies reported in the literature only deal with antioxidant capacity assessment of propolis and honey, which could be due to the fact that, accurate and quantitative methods for antioxidant capacity measurement using electrochemical methodologies are still an object of study because of the diversity of samples and related overall mixture properties [23 - 25]. Regarding the direct use of electrochemical approaches for bee-hive products quality evaluation, only one work could be found for honey [26], where it was referred the application of differential pulse polarography (voltammetry using a mercury drop electrode) to quantify hydroxymethylfurfural and fructose levels. This is unexpected since these techniques do not require a time-consuming sample treatment and are cost-effective methods enabling fast, simple and sensitive analysis of bioactive compounds associated with radicals scavenging and so, to the antioxidant capacity [20, 27].

In the literature [28 - 34], the use of electrochemical devices for bee-hive products analysis is mainly focused in antioxidant capacity assessment (Table **1**). For other matrices, like plants, fruits, beverages or algae, voltammetric methods and electrochemical biosensors have been described for quantifying single or total phenolic compounds as well as to assess the antioxidant capacity in standard solutions or in extracts [15, 35 - 44]. Indeed, voltammetric methods (*e.g.*, CV, DPV or SWV) have been applied as a fast screening tool for identifying novel antioxidants, being an attractive approach although they have not yet found a widespread use [14]. The capability of electrochemical devices to assess antioxidant capacity relies in the fact that phenolic compounds are prone to redox inter-conversion. Indeed, most of the phenolic compounds are electrochemically active at moderate oxidation potentials and so, the use of electrochemical methods may be preferable compared to spectrophotometric, chromatographic, capillary electrophoretic or chemiluminescent techniques, since they are less prone to interferences from non-electroactive substances [40]. Also, due to the different mechanisms of antioxidant action of some phenolic compounds (*e.g.*, flavonoids), electroanalytical methods are viewed as the most suitable techniques for antioxidant capacity evaluation and for electroactive species characterization [45].

Table 1. Electrochemical studies in bee-hive products and their isolated compounds.

Electrochemical technique	Sample	Type and conditions of analysis	Instrumental conditions	Ref.
Direct current polarography	Honey of different floral sources and main constituents (flavonoids, phenolic acids, amino and organic acids, and carbohydrates).	Hydrogen peroxide scavenging (HPS) activity of honey and isolated compounds. Analysis with Clark-Lubs borate buffer (pH 9.8) using H_2O_2 concentrations higher than $1x10^{-3}$ M.	Three-electrode system with a dropping mercury WE, a saturated calomel RE and Pt-foil as AE. Mercury dropping time of 1 s and the current-potential curves recorded at room temperature using starting potential at 0.10 V and scan rate of 10 mV/s. Calibration curve with decrease of the anodic current peak of H_2O_2 in presence of honey samples (signal) *vs.* mass of added samples. The slope used was the measure of HPS activity.	[28]

(Table 1) contd.....

Electrochemical technique	Sample	Type and conditions of analysis	Instrumental conditions	Ref.
Direct current polarography	Propolis	Hydrogen peroxide scavenging (HPS) activity of propolis using Clark-Lubs Borate buffer with pH 9.8 and H_2O_2 concentrations higher than $1x10^{-3}$ M.	Three-electrode cell with a dropping mercury WE, a saturated calomel RE and Pt foil as AE. Mercury dropping time of 1 s and the current-potential curves recorded at room temperature using starting potential at 0.10 V and scan rate of 10 mV/s. Calibration curve with decrease of the anodic current peak of H_2O_2 in presence of propolis samples (signal) *vs*. mass of added samples. The slope used was the measure of HPS activity.	[29]
Cyclic voltammetry	Flavonoids and caffeic acid esters isolated from propolis	Redox properties of isolated compounds from propolis using acetonitrile under Argon atmosphere	Electrochemical cell of three electrodes containing a working and auxiliary Pt and reference saturated calomel electrode. Voltammograms were recorded between -3 and +3 V.	[30]
Cyclic voltammetry	Chrysin isolated from propolis	Redox properties of an isolated compound extracted from propolis and dissolved in ethanol was analyzed in a series of Britton-Robinson buffer solutions from pH 2.0 to 9.0.	A conventional three-electrode system with a saturated calomel RE, and Pt wire CE and static mercury drop WE. Analysis were carried out in the potential interval of -1.8 to -1.2 V.	[31]
Cyclic voltammetry	Propolis	Antioxidant capacity of methanolic extracts of propolis in pH 7 phosphate buffer solution. Ascorbic and gallic acids used as standards in the calculation of antioxidant capacity.	Electrochemical cell with glassy carbon WE, a Pt wire CE, and a saturated calomel RE. The potential was swept in inverse scanning mode starting from -0.2 to +0.8 V with a scanning rate of 100 mV/s. Antioxidant capacity calibration curve using the area below the anodic curve of the voltammogram, as signal, *vs* standard's concentrations.	[32]

(Table 1) contd.....

Electrochemical technique	Sample	Type and conditions of analysis	Instrumental conditions	Ref.
Cyclic voltammetry	Propolis and pinocembrin and galangin isolated from propolis	Antioxidant activity of propolis extracts and isolated compounds. Analysis using hydro-methanolic solutions with pH=6.6 phosphate buffer. Ascorbic acid used as standard compound for antioxidant capacity comparison.	A three-electrode system applied with an Ag/AgCl RE, a Pt WE and a Pt wire CE. Voltammograms were recorded from -0.1 to +1.3 V with a scan rate of 100 mV/s. Total charge (peak area) below the anodic wave curve of the voltammogram was used *vs* standard's concentrations for curve fitting method.	[33]
Cyclic voltammetry	Propolis	Antioxidant capacity of methanolic propolis extract in pH=7 phosphate buffer aqueous solution. Ascorbic and gallic acids used as standards for evaluating antioxidant capacity.	A three-electrode cell was used with a glassy carbon WE, a Pt-wire CE and a saturated calomel RE. The potential sweep was in the interval -0.2 to 0.8 V with a scanning rate of 0.1 mV/s. Calculations using anodic area *vs*standard's concentrations.	[34]

However, the presence of two or more species in a sample, with similar redox properties, may be a major problem, namely when CV is applied, which can be partially overcome by applying high resolution pulse voltammetric methods like DPV and SWV that are able to enhance the discrimination between the target and the interferent molecules [46]. Nevertheless, in some situations, it is reported that, contrary to what should be expectable, CV may have an apparent higher sensitivity compared to DPV technique, providing information on the reversibility of the anodic waves [41]. So, the simultaneous use of different voltammetric methodologies can be a clear advantage [14]. Even so, the potential user of these electroanalytical methods must be aware of some possible limitations, namely the occurrence of overlapping signals for multi-component mixtures analysis, especially if one compound is present in excess and the electrode sensitivity is reduced due to the sluggish electron transfer kinetics or fouling of the electrode, through contamination or passivation of the electrode surface [14, 42, 46]. The latter leads to the formation of insulating films at the electrode surface due to adsorption effects, which may cause to a non detectable anodic wave potential signal [14, 42, 46].

Finally, it should be kept in mind that these electroanalytical methods, although being inexpensive, simple, rapid, sensitivity and portable, should be used mostly for basic analyses since the selective determination of a particular compound is, usually, not feasible, being in general not possible to establish which compound is responsible for each peak observed, and so only some reasonable assumptions might be tentatively highlighted when analyzing real samples [42, 44]. Nevertheless, voltammetric methods, used individually or in combination, are attractive screening techniques for raw samples enabling the identification of potential antioxidants with a greater confidence compared to well-known chemical assays (*e.g.*, DPPH or the ferrous chelation methods) [14]. In addition, electrochemical techniques (*e.g.*, CV and SWV) may even be used to guide the isolation of antioxidant natural products from complex crude samples (*e.g.*, marine algae extracts) contributing to the isolation of antioxidant molecules [47].

Recently, the CV technique was used to determine the antioxidant capacity of flavonoids-metal ions complexes as well to discuss flavonoids/metal ions interactions [48]. Also, electrochemical ultra-micro sensors were developed and satisfactorily applied in the determination of synthetic and natural antioxidants in oils, based on SWV data [49]. So, electrochemical approaches (*e.g.*, CV and SWV) may provide new insights of the process and kinetic related to the electrochemical oxidation of phenolic compounds, namely flavonoids, contributing to a deep knowledge of physical and chemical properties of antioxidants as well as for understanding the mechanisms of their oxidation or reduction processes [50]. Also, several works using DNA- or purine-based electrochemical biosensors have been applied to evaluate antioxidant capacities of different matrices (*e.g.*, plant extracts, beverages) [42, 51 - 56]. Nanomaterials, such as metal nanoparticles (MNPs) and quantum dots (QDs), applications for *in vitro* antioxidant capacity assessment in complex samples were recently reviewed and discussed [57]. So, electrochemical techniques, alone or combined to traditional analytical methods (*e.g.*, chromatography), can be a practical tool for evaluating the antioxidant of different matrices.

However, the direct application of electrochemical methods for bee-hives products antioxidant capacity assessment has been rarely reported in the literature. The studies used voltammetric or polarographic assays, and were focused mainly in the analysis of propolis [29, 30, 32 - 34, 58, 59] and less frequently of honey

[28]. Phenolic compounds isolated from propolis were also electrochemically characterized by CV, and in particular the antioxidant activity of chrysin, a widely distributed flavonoid in propolis, was evaluated by linear sweep voltammetry (LSV) on a static mercury electrode [31]. Unfortunately, no work has been found by the authors regarding the direct application of an electrochemical device for bioactivity capacity assessment of pollen.

In the next section, the application of electrochemical devices to assess antioxidant capacity of bee-hive products is reviewed, namely for honey and propolis. Concerning honey, CV assays of our research group are also presented and discussed.

3.1. Honey's Antioxidant Capacity Evaluation using Voltammetric Sensors

In the literature, it was only possible to find one work that describes the application of a polarographic method for assessing antioxidant activity of honeys. Gorjanović *et al.* [28] proposed a direct current polarographic assay to evaluate HPS activity of honey from different floral sources and its main constituents (*e.g.*, flavonoids, phenolic acids, amino acids, organic acids and carbohydrates). As for propolis extracts analysis [29], the assay was based on the decrease of anodic current of hydrogen peroxide complex, formed in alkaline solution, at the potential of mercury dissolution. Antioxidant activity of honey reflected an integrated action of different constituents, both phenolics and non-phenolics [28]. Also, the potential of polarography for antioxidant capacity assessment, already demonstrated for propolis extracts, was confirmed for honey analysis [28, 29].

To further investigate the potential applicability of a voltammetric approach to evaluate redox activity of honey, CV preliminary assays were conducted by our research team in Portuguese honeys. The pollinic profiles (based on the melissopalynology analysis), color classification (according to the quantitative mm Pfund scale based on spectrophotometry assays) and the potentiometric behavior (recorded using an electronic tongue) were previously reported by the research group [60]. Honeys were classified as monofloral, according to their floral origin (*e.g.*, *Castanea* sp., *Echium* sp., *Erica* sp., *Lavandula* sp., *Prunus* sp.

and *Rubus* sp.) and their color range included white (17 < mm Pfund ≤ 34), extra light amber (34 < mm Pfund ≤ 50), light amber (50 < mm Pfund ≤ 85), amber (85 < mm Pfund ≤ 114) and dark amber (> 114mm Pfund). Honey samples were dissolved in water (5 g of honey with 50 g of water) before being electrochemically analyzed. The voltammetric equipment consisted of a potentiostat-Galvanostat device (PG580, Uniscan) with a typical 3-electrodes system, using the American current polarity convention. A silver electrode (M295Ag, Radiometer) was used as the WE, a Pt electrode (M241Pt, Radiometer) used as the CE and an Ag/AgCl electrode (M90-02, Orion) as the RE (Fig. **5**).

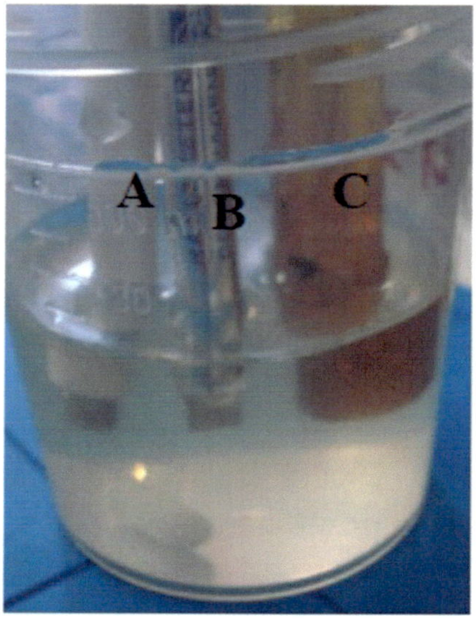

Fig. (5). The 3-electrodes system used: **A**) silver (M295Ag, Radiometer) working electrode; **B**) Pt (M241Pt, Radiometer) counter electrode; **C**) Ag/AgCl (M90-02, Orion) reference electrode.

An example of the cyclic voltammograms recorded during the aqueous honey solution analysis is shown in Fig. (**6**). As can be seen, only one anodic wave curve was obtained, showing a typical irreversible oxidation electrochemical process with one anodic peak at a negative potential (≈ -250 mV). Similar electrochemical profiles (data not shown) were observed for all monofloral honeys analyzed, although with slight different oxidation potentials, varying from -250 up to -100 mV, depending on the color and floral origin of the honey sample. The wide anodic curve may be attributed to the combined response of several electro-active chemicals present in the honey samples (*e.g.* flavonoids, phenolic acids and water soluble vitamins) that had different oxidation potentials [33].

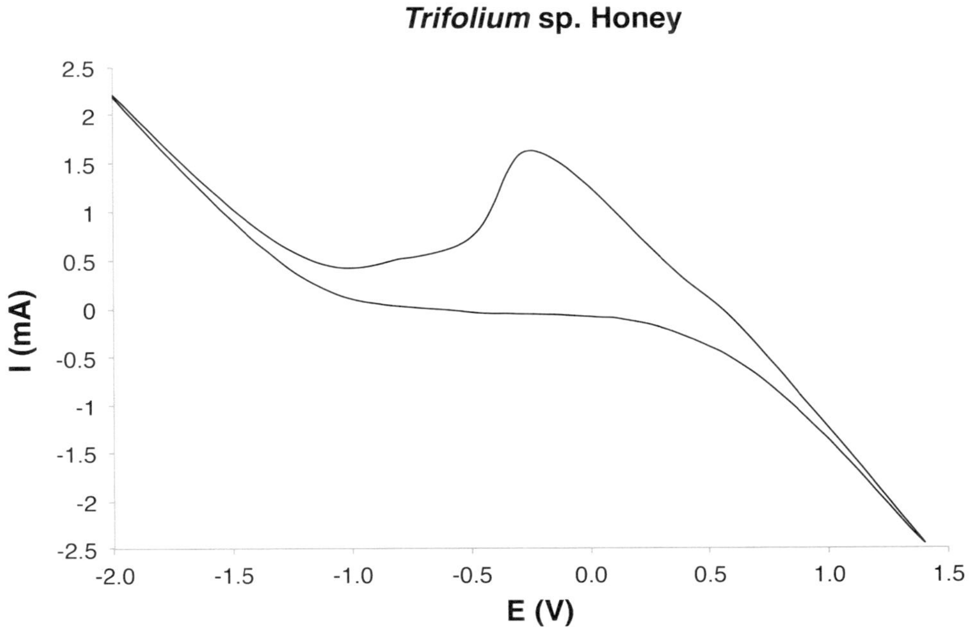

Fig. (6). A typical cyclic voltammograms recorded of an aqueous honey solution (*e.g. Trifolium* sp. monofloral honey).

Fig. (7) exemplifies the cyclic voltammograms recorded for monofloral honeys of *Lavandula* sp. with different colors (varying from white to dark amber). The similarity observed in the oxidation potentials and global voltammetric profiles may indicate that Portuguese *Lavandula* sp. monofloral honeys have analogous chemical composition regarding electro-active species, regardless honey color. However, the anodic peak current and the anodic curve area increased with the increasing darkness of the *Lavandula* sp. honey (from white to dark amber, Fig. (7), which could be related to the known higher content of phenolic compounds (and by consequence of the related antioxidant capacity) of dark colored honeys compared to light colored honeys [61 - 65]. Indeed, for the honey samples shown in Fig. (7), a positive linear correlation could be established between the mm Pfund values of the *Lavandula* sp. honeys and the respective anodic peak current intensities (R-Pearson = 0.968), confirming the previous conclusion (*i.e.*, > peak current → > mm Pfund).

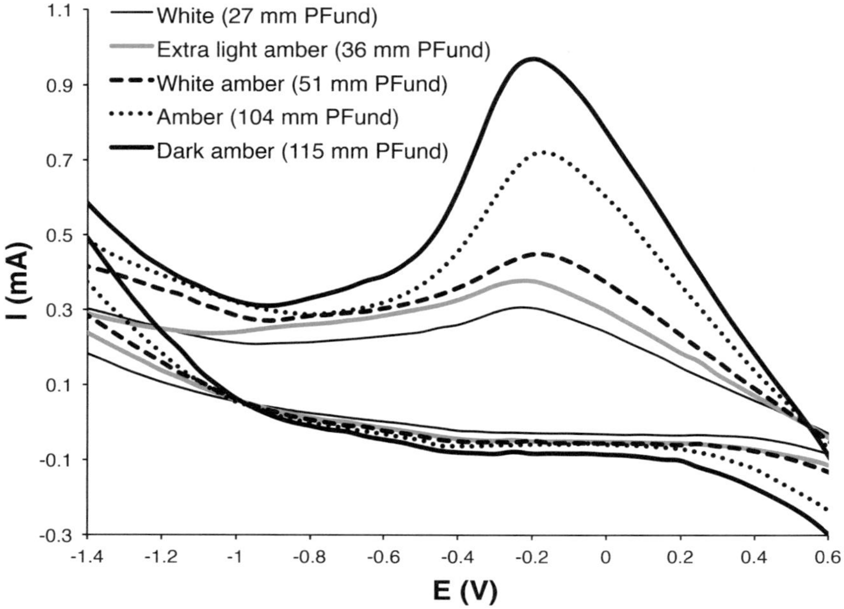

Fig. (7). Cyclic voltammograms recorded for monofloral honeys of *Lavandula* sp. with different colors (varying from white to dark amber).

Similar behaviors could be found for the others Portuguese monofloral honeys (data not shown). Moreover, similar voltammetric behaviors were observed for honey samples with different floral origins but with the same color classification (based on the mm Pfund scale), although with anodic peak current at slight different potentials and with different anodic curves areas. This fact may indicate that the Portuguese monofloral honeys with the same color may contain similar electro-active species although in different levels, and so, with different expected antioxidant capacities. As an example, Fig. (**8**) depicts the voltammetric profiles of dark amber honeys with different floral origins. From the figure it may be inferred that both the anodic peak current and the anodic curve area increase in the order *Castanea* sp. < *Echium* sp. < *Rubus* sp. < *Lavandula* sp. < *Prunus* sp. < *Erica* sp. < *Trifolium* sp. honeys, which may indicate that both the phenolic content and the antioxidant capacity of Portuguese monofloral honeys within the same color range may vary. Indeed, it has been reported that extracts with lower oxidation potential values may exhibit higher antioxidant capacity [66].

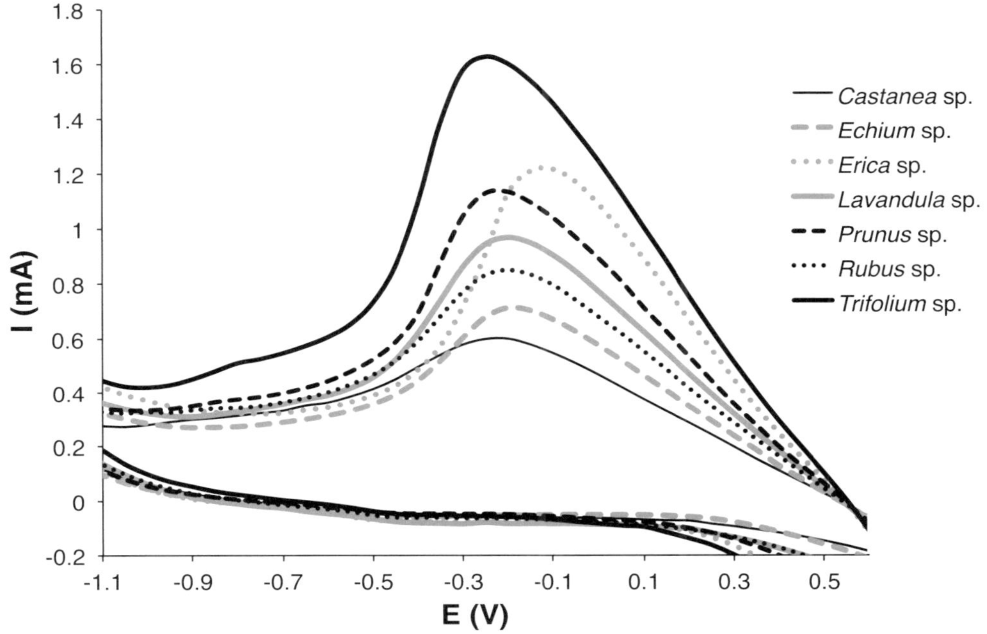

Fig. (8). Voltammetric profiles of dark amber honeys with different floral origins.

3.2. Propolis' Antioxidant Capacity Evaluation using Voltammetric Sensors

Several works report the capability of direct voltammetric methods (including CV, DPV and SWV) and polarographic approaches to assess the antioxidant activity of propolis extracts, demonstrating the possibility of applying these electrochemical techniques on resinous substances. Also, the quality of the results obtained with these electroanalytical tools demonstrate that these methodologies could be recommended as appropriate for determination of antioxidant capacity of propolis extracts being a possible, fast and cost-effective alternative to widely accepted assays.

One of the first works reported in the literature dates back to 1995 [30] and describes the evaluation of the redox properties of flavonoids isolated from propolis, by CV in acetonitrile extracts. The aprotic solvent reduced the radical intermediates reactivity, enabling the identification of the redox steps and intermediates compounds. The work of Rapta *et al.* [30] demonstrated the existence of a negative correlation between lipid antioxidant properties of flavonoids and caffeic acid esters isolated from propolis and their oxidation potential. After, Zheng *et al.*[31] studied the electrochemical behavior of chrysin, a flavonoid isolated from propolis, by LSV and CV. The assays were conducted using an electrochemistry workstation and a conventional three-electrode system (a standard saturated calomel as the RE, a platinum (Pt) wire as the CE and a static mercury drop as the WE). The approach allowed to propose an electrochemical reduction mechanism of chrysin [31]. Also, the ability of chrysin for scavenging active oxygen radicals yielded by the autoxidation of pyrogallol was evaluated showing the good antioxidant capacity of this flavonoid. More recently, Laskar *et al.* [33] studied by CV the *in vitro* antioxidant capacity of aqueous and ethanol extracts of propolis from India. A potentiostat–galvanostat apparatus was used coupled to a three-electrode system, including an Ag/AgCl as the RE, a platinum (Pt) electrode as a WE and a Pt wire as a CE. Each extract was diluted using the same volume of 0.2 M phosphate buffer (pH 6.6). Voltammograms were recorded from -100 to +1300 mV with a scan rate of 100 mV s^{-1}. Both type of extracts showed an irreversible electrochemical behavior with one anodic peak at oxidation potentials less positive than that recorded for ascorbic acid standard solutions. Furthermore, Falcão *et al.* [58] characterized Portuguese propolis based on its electrochemical behavior. The voltammetric evaluation was performed by CV, DPV and SWV, in ethanolic extracts (-0.5 to +200 mV). A potentiostat with a typical three-electrode cell (Ag/AgCl as the RE, a Pt wire as CE and a glassy carbon disk as the WE) was used. The redox profiles of propolis phenolic extracts were studied by CV, while DPV and SWV enabling the quantification of electroactive species present in the different extracts of propolis. Irreversible oxidation processes were observed at different potentials depending on the geographical origin of the samples allowing the discrimination of propolis samples by geographical origin. Rebiai *et al.* [34] evaluated the antioxidant capacity of methanolic Algerian propolis extracts by CV (-200 to +800 mV, at a scan rate of 100 mV/s). A potentiostat device connected to a typical 3-electrodes system was used (a glassy carbon electrode as the WE, a Pt wire as the CE and a Hg/Hg_2Cl_2 as the RE). The CV results showed that propolis extracts had electrochemical behaviors slight different from those recorded for standard phenolic solutions, suggesting a different electroactive chemical composition, although possessing antioxidant capacity under *in vitro* conditions. Lourenço *et al.* [59] also applied CV to assess the antioxidant capacity of propolis ethanol extracts from Azores (Portugal). The preliminary study showed that the extracts possessing higher antioxidant activities also had higher antibacterial activities. More recently, Rebiai *et al.* [32] tested polyphenols extracted from Algerian propolis, using CV in aqueous media. Also, antioxidant capacities of propolis methanol extracts were evaluated by CV. This electrochemical technique provided a qualitative composition of each extract as well as an estimative of the total polyphenols content in each extract. The propolis methanol extracts presented typical irreversible oxidation processes similarly to that recorded for standard solutions, although with oxidation potentials more positive than ascorbic acid and lower than gallic acid. Under the electrochemical conditions used, the CV data did not indicate that propolis extracts had an antioxidant capacity lower than gallic acid and greater than ascorbic acid-Indeed, propolis extracts showed a higher antioxidant capacity compared with that of gallic acid standard solution, contrary to the expected behavior.

One work reported the application of direct current (DC) polarography for assessing the antioxidant capacity of commercial propolis extracts available in Serbia [29]. In this study, the antioxidant activity was evaluated by plotting of the polarographic anodic current decrease of an initial alkaline solution of H_2O_2 due to the gradually addition of pre-established volumes of propolis ethanol extracts. A Polarographic Analyzer PAR was used for the electrochemical measurements, with a conventional three-electrode cell (a dropping mercury electrode was used as the WE, a saturated calomel electrode as the RE and the Pt foil as the CE).

CONCLUSION

The examples presented on electrochemical techniques application in the evaluation of antioxidant capacity or determination of polyphenolic compounds in food, beverages and related plant extracts showed the added value of these techniques. They provide supplementary insight about redox-processes. These techniques have been considered as complementary or alternative in the evaluation of antioxidant power, mainly due to the speed of analysis with a reduced sample treatment. The number of studies shows that the application of these methodologies in the study of antioxidant properties in honey and propolis samples is in an early stage. Considering the economic, nutritional and human health importance of these products, it is expected that the electrochemical studies become an essential analytical tool for the characterization of these samples.

CONFLICT OF INTEREST

The authors declare no conflict of interest regarding this publication.

ACKNOWLEDGEMENTS

This work was supported by Fundação para a Ciência e a Tecnologia (FCT) and the European Community fund FEDER, under the Program PT2020 (Project UID/EQU/50020/2013) and by the strategic funding of UID/BIO/04469/2013 unit.

REFERENCES

[1] Bertoncelj, J.; Dobersek, U.; Jamnik, M.; Golob, T. Evaluation of the phenolic content, antioxidant activity and colour of Slovenian honey. *Food Chem.,* **2007**, *105*, 822-828. [http://dx.doi.org/10.1016/j.foodchem.2007.01.060]

[2] Escuredo, O.; Míguez, M.; Fernández-González, M.; Carmen Seijo, M. Nutritional value and antioxidant activity of honeys produced in a European Atlantic area. *Food Chem.,* **2013**, *138*(2-3), 851-856. [http://dx.doi.org/10.1016/j.foodchem.2012.11.015] [PMID: 23411187]

[3] Aljadi, A.M.; Kamaruddin, M.Y. Evaluation of the phenolic contents and antioxidant capacities of two Malaysian floral honeys. *Food Chem.,* **2004**, *85*, 513-518. [http://dx.doi.org/10.1016/S0308-8146(02)00596-4]

[4] Al-Mamary, M.; Al-Meeri, A.; Al-Habori, M. Antioxidant activities and total phenolics of different types of honey. *Nutr. Res.,* **2002**, *22*, 1041-1047. [http://dx.doi.org/10.1016/S0271-5317(02)00406-2]

[5] Blasa, M.; Candiracci, M.; Accorsi, A.; Piacentini, M.P.; Albertini, M.C.; Piatti, E. Raw Millefiori honey is packed full of antioxidants. *Food Chem.,* **2006**, *97*, 217-222. [http://dx.doi.org/10.1016/j.foodchem.2005.03.039]

[6] Gheldof, N.; Wang, X-H.; Engeseth, N.J. Identification and quantification of antioxidant components of honeys from various floral sources. *J. Agric. Food Chem.,* **2002**, *50*(21), 5870-5877. [http://dx.doi.org/10.1021/jf0256135] [PMID: 12358452]

[7] Basim, E.; Basim, H.; Ozcan, M. Antibacterial activities of Turkish pollen and propolis extracts against plant bacterial pathogens. *J. Food Eng.,* **2006**, *77*, 992-996. [http://dx.doi.org/10.1016/j.jfoodeng.2005.08.027]

[8] Kumazawa, S.; Hamasaka, T.; Nakayama, T. Antioxidant activity of propolis of various geographic origins. *Food Chem.,* **2004**, *84*, 329-339. [http://dx.doi.org/10.1016/S0308-8146(03)00216-4]

[9] Broadhurts, C.L. Bee products: medicine from the hive. *Nutr. Sci. News,* **1999**, *4*, 366-368.

[10] Zoski, C.G. *Handbook of electrochemistry*; Elsevier: Netherlands, **2007**.

[11] Thomas, F.G.; Henze, G. *Introduction to Voltammetric Analysis: Theory and Practice*; CSIRO Publishing: Australia, **2001**.

[12] Compton, R.G.; Laborda, E.; Ward, K.R. *Understanding Voltammetry: Simulation of Electrode Processes,* 2nd; Imperial College Press: London, **2013**, pp. 1-260.

[13] Scholz, F. *Electroanalytical Methods - Guide to Experiments and Applications*; Springer: Germany, **2010**.

[14] Ragubeer, N.; Beukes, D.R.; Limson, J.L. Critical assessment of voltammetry for rapid screening of antioxidants in marine algae. *Food Chem.,* **2010**, *121*, 227-232. [http://dx.doi.org/10.1016/j.foodchem.2009.11.076]

[15] Barros, L.; Falcão, S.; Baptista, P.; Freire, C.; Vilas-Boas, M.; Ferreira, I.C. Antioxidant activity of *Agaricus* sp. mushrooms by chemical, biochemical and electrochemical assays. *Food Chem.,* **2008**, *111*, 61-66. [http://dx.doi.org/10.1016/j.foodchem.2008.03.033]

[16] Chevion, S.; Roberts, M.A.; Chevion, M. The use of cyclic voltammetry for the evaluation of antioxidant capacity. *Free Radic. Biol. Med.,* **2000**, *28*(6), 860-870. [http://dx.doi.org/10.1016/S0891-5849(00)00178-7] [PMID: 10802216]

[17] Blasco, A.J.; Rogerio, M.C.; González, M.C.; Escarpa, A. "Electrochemical Index" as a screening method to determine "total polyphenolics" in foods: A proposal. *Anal. Chim. Acta,* **2005**, *539*, 237-244. [http://dx.doi.org/10.1016/j.aca.2005.02.056]

[18] Cosio, M.S.; Buratti, S.; Mannino, S.; Benedetti, S. Use of an electrochemical method to evaluate the antioxidant activity of herb extracts from the Labiatae family. *Food Chem.,* **2006**, *97*, 725-731. [http://dx.doi.org/10.1016/j.foodchem.2005.05.043]

[19] Photinon, K.; Chalermchart, Y.; Khanongnuch, C.; Wang, S.H.; Liu, C.C. A thick-film sensor as a novel device for determination of polyphenols and their antioxidant capacity in white wine. *Sensors (Basel),* **2010**, *10*(3), 1670-1678. [http://dx.doi.org/10.3390/s100301670] [PMID: 22294893]

[20] Sochor, J.; Dobes, J.; Krystofova, O.; Ruttkay-Nedecky, B.; Babula, P.; Pohanka, M.; Jurikova, T.; Zitka, O.; Adam, V.; Klejdus, B.; Kizek, R. Electrochemistry as a tool for studying antioxidant properties. *Int. J. Electrochem. Sci.,* **2013**, *8*, 8464-8489.

[21] Blasco, A.J.; Crevillen, A.G.; Gonzalez, M.C.; Escarpa, A. Direct electrochemical sensing and

detection of natural antioxidants and antioxidant capacity *in vitro* systems. *Electroanalysis,* **2007**, *19*, 2275-2286.
[http://dx.doi.org/10.1002/elan.200704004]

[22] Rice-Evans, C.A.; Miller, N.J.; Paganga, G. Antioxidant properties of phenolic compounds. *Trends Plant Sci.: rev,* **1997**, *2*, 152-159.

[23] Sochor, J.; Salas, P.; Zehnalek, J.; Krska, B.; Adam, V.; Havel, L.; Kizek, R. An assay for spectrometric determination of antioxidant activity of a biological extract. *Lis. Cukrov. Repar.,* **2010**, *126*, 416-417.

[24] Frankel, E.N.; Finley, J.W. How to standardize the multiplicity of methods to evaluate natural antioxidants. *J. Agric. Food Chem.,* **2008**, *56*(13), 4901-4908.
[http://dx.doi.org/10.1021/jf800336p] [PMID: 18553885]

[25] Prior, R.L.; Wu, X.; Schaich, K. Standardized methods for the determination of antioxidant capacity and phenolics in foods and dietary supplements. *J. Agric. Food Chem.,* **2005**, *53*(10), 4290-4302.
[http://dx.doi.org/10.1021/jf0502698] [PMID: 15884874]

[26] Reyes-Salas, E.O.; Gazcón-Orta, N.E.; Manzanilla-Cano, J.A.; Reyes-Salas, A.M.; Camou, A.; Reyes-González, A.; Caballero-Puente, H.D. Electrochemical evaluation of quality characteristics in honey from *Meliponini* and *Apis Mellifera* bees. *Annals Food Sci. Tech.,* **2014**, *15*, 35-40.

[27] Prieto-Simon, B.; Cortina, M.; Campas, M.; Calas-Blanchard, C. Electrochemical biosensors as a tool for antioxidant capacity assessment. *Sens. Actuators B Chem.,* **2008**, *129*, 459-466.
[http://dx.doi.org/10.1016/j.snb.2007.08.004]

[28] Gorjanović, S.Ž.; Alvarez-Suarez, J.M.; Novaković, M.M.; Pastor, F.T.; Pezo, L.; Battino, M.; Sužnjevic, D.Ž. Comparative analysis of antioxidant activity of honey of different floral sources using recently developed polarographic and various spectrophotometric assays. *J. Food Compos. Anal.,* **2013**, *30*, 13-18.
[http://dx.doi.org/10.1016/j.jfca.2012.12.004]

[29] Potkonjak, N.I.; Veselinović, D.S.; Novaković, M.M.; Gorjanović, S.Ž.; Pezo, L.L.; Sužnjević, D.Ž. Antioxidant activity of propolis extracts from Serbia: a polarographic approach. *Food Chem. Toxicol.,* **2012**, *50*(10), 3614-3618.
[http://dx.doi.org/10.1016/j.fct.2012.07.029] [PMID: 22842121]

[30] Rapta, P.; Misík, V.; Staško, A.; Vrábel, I. Redox intermediates of flavonoids and caffeic acid esters from propolis: an EPR spectroscopy and cyclic voltammetry study. *Free Radic. Biol. Med.,* **1995**, *18*(5), 901-908.
[http://dx.doi.org/10.1016/0891-5849(94)00232-9] [PMID: 7797098]

[31] Zheng, J-B.; Zhang, H-F.; Gao, H. Investigation on electrochemical behavior and scavenging superoxide anion ability of chrysin at mercury electrode. *Chin. J. Chem.,* **2005**, *23*, 1042-1046.
[http://dx.doi.org/10.1002/cjoc.200591042]

[32] Rebiai, A.; Lanez, T.; Belfar, M.L. Total polyphenol contents, radical scavenging and cyclic voltammetry of algerian propolis. *Int. J. Pharm. Pharm. Sci.,* **2014**, *6*, 395-400.

[33] Laskar, R.A.; Sk, I.; Roy, N.; Begum, N.A. Antioxidant activity of Indian propolis and its chemical constituents. *Food Chem.,* **2010**, *122*, 233-237.

[http://dx.doi.org/10.1016/j.foodchem.2010.02.068]

[34] Rebiai, A.; Lanez, T.; Belfar, M.L. *In vitro* evaluation of anitoxidant capacity of algerian propolis by spectrophotometrical and electrochemical assays. *Int. J. Pharmacol.,* **2011**, *7*, 113-118. [http://dx.doi.org/10.3923/ijp.2011.113.118]

[35] Campanella, L.; Martini, E.; Tomassetti, M. Antioxidant capacity of the algae using a biosensor method. *Talanta,* **2005**, *66*(4), 902-911. [http://dx.doi.org/10.1016/j.talanta.2004.12.052] [PMID: 18970070]

[36] Piljac-Žegarac, J.; Valek, L.; Stipčević, T.; Martinez, S. Electrochemical determination of antioxidant capacity of fruit tea infusions. *Food Chem.,* **2010**, *121*, 820-825. [http://dx.doi.org/10.1016/j.foodchem.2009.12.090]

[37] Barros, L.; Cabrita, L.; Vilas Boas, M.; Carvalho, A.M.; Ferreira, I.C. Chemical, biochemical and electrochemical assays to evaluate phytochemicals and antioxidant activity of wild plants. *Food Chem.,* **2011**, *127*, 1600-1608. [http://dx.doi.org/10.1016/j.foodchem.2011.02.024]

[38] Tyurin, V.Yu.; Meleshonkova, N.N.; Dolganov, A.V.; Glukhova, A.P.; Milaeva, E.R. Electrochemical method in determination of antioxidative activity using ferrocene derivatives as examples. *Russ. Chem. B+. Int. Ed.,* **2011**, *60*, 647-655.

[39] Barroso, M.F.; Delerue-Matos, C.; Oliveira, M.B. Electrochemical evaluation of total antioxidant capacity of beverages using a purine-biosensor. *Food Chem.,* **2012**, *132*, 1055-1062. [http://dx.doi.org/10.1016/j.foodchem.2011.10.072]

[40] Magarelli, G.; da Silva, J.G.; Filho, I.A.; Lopes, I.S. SouzaDe, J.R.; Hoffmann, L.V.; de Castro, C.S.P. Development and validation of a voltammetric method for determination of total phenolic acids in cotton cultivars. *Microchem. J.,* **2013**, *109*, 23-28. [http://dx.doi.org/10.1016/j.microc.2012.05.014]

[41] Rebelo, M.J.; Rego, R.; Ferreira, M.; Oliveira, M.C. Comparative study of the antioxidant capacity and polyphenol content of Douro wines by chemical and electrochemical methods. *Food Chem.,* **2013**, *141*(1), 566-573. [http://dx.doi.org/10.1016/j.foodchem.2013.02.120] [PMID: 23768395]

[42] de Sá, L.Z.; Castro, P.F.; Lino, F.M.; Bernardes, M.J.; Viegas, J.C.; Dinis, T.C.; Santana, M.J.; Romao, W.; Vaz, B.G.; Lião, L.M.; Ghedini, P.C.; Rocha, M.L.; Gil, E.S. Antioxidant potential and vasodilatory activity of fermented beverages of jabuticaba berry (*Myrciaria jaboticaba*). *J. Funct. Foods,* **2014**, *8*, 169-179. [http://dx.doi.org/10.1016/j.jff.2014.03.009]

[43] Pisoschi, A.M.; Pop, A.; Serban, A.I.; Fafaneata, C. Electrochemical methods for ascorbic acid determination. *Electrochim. Acta,* **2014**, *121*, 443-460. [http://dx.doi.org/10.1016/j.electacta.2013.12.127]

[44] Jeszka-Skowron, M.; Zgoła-Grześkowiak, A.; Grześkowiak, T. Analytical methods applied for the characterization and the determination of bioactive compounds in coffee. *Eur. Food Res. Technol.,* **2015**, *240*, 19-31. [http://dx.doi.org/10.1007/s00217-014-2356-z]

[45] Gil, E.S.; Couto, R.O. Flavonoid electrochemistry: a review on the electroanalytical applications. *Braz. J. Pharmacog.,* **2013**, *23*, 542-558.

[46] Lawrence, N.S.; Beckett, E.L.; Davis, J.; Compton, R.G. Advances in the voltammetric analysis of small biologically relevant compounds. *Anal. Biochem.,* **2002**, *303*(1), 1-16. [http://dx.doi.org/10.1006/abio.2002.5584] [PMID: 11906145]

[47] Ragubeer, N.; Limson, J.L.; Beukes, D.R. Electrochemistry-guided isolation of antioxidant metabolites from *Sargassum elegans*. *Food Chem.,* **2012**, *131*, 286-290. [http://dx.doi.org/10.1016/j.foodchem.2011.08.037]

[48] Porfírio, D.A.; Ferreira, R.Q.; Malagutti, A.R.; Valle, E.M. Electrochemical study of the increased antioxidant capacity of lavonoids through complexation with iron(II) ions. *Electrochim. Acta,* **2014**, *141*, 33-38. [http://dx.doi.org/10.1016/j.electacta.2014.07.046]

[49] Robledo, S.N.; Tesio, A.Y.; Ceballosa, C.D.; Zon, M.A.; Fernández, H. Electrochemical ultra-micro sensors for the determination of synthetic and natural antioxidants in edible vegetable oils. *Sens. Actuators B Chem.,* **2014**, *192*, 467-473. [http://dx.doi.org/10.1016/j.snb.2013.11.023]

[50] Masek, A.; Chrzescijanska, E.; Zaborski, M. Electrooxidation of morin hydrate at a Pt electrode studied by cyclic voltammetry. *Food Chem.,* **2014**, *148*, 18-23. [http://dx.doi.org/10.1016/j.foodchem.2013.10.003] [PMID: 24262520]

[51] Mello, L.D.; Hernandez, S.; Marrazza, G.; Mascini, M.; Kubota, L.T. Investigations of the antioxidant properties of plant extracts using a DNA-electrochemical biosensor. *Biosens. Bioelectron.,* **2006**, *21*(7), 1374-1382. [http://dx.doi.org/10.1016/j.bios.2005.05.012] [PMID: 16002275]

[52] Kamel, A.H.; Moreira, F.T.; Delerue-Matos, C.; Sales, M.G. Electrochemical determination of antioxidant capacities in flavored waters by guanine and adenine biosensors. *Biosens. Bioelectron.,* **2008**, *24*(4), 591-599. [http://dx.doi.org/10.1016/j.bios.2008.06.007] [PMID: 18640022]

[53] Barroso, M.F.; Delerue-Matos, C.; Oliveira, M.B. Electrochemical DNA-sensor for evaluation of total antioxidant capacity of flavours and flavoured waters using superoxide radical damage. *Biosens. Bioelectron.,* **2011**, *26*(9), 3748-3754. a [http://dx.doi.org/10.1016/j.bios.2011.02.015] [PMID: 21474298]

[54] Barroso, M.F.; de-los-Santos-Álvarez, N.; Lobo-Castañón, M.J.; Miranda-Ordieres, A.J.; Delerue-Matos, C.; Oliveira, M.B.; Tuñón-Blanco, P. DNA-based biosensor for the electrocatalytic determination of antioxidant capacity in beverages. *Biosens. Bioelectron.,* **2011**, *26*(5), 2396-2401. b [http://dx.doi.org/10.1016/j.bios.2010.10.019] [PMID: 21067909]

[55] Barroso, M.F.; de-los-Santos-Álvarez, N.; Lobo-Castañón, M.J.; Miranda-Ordieres, A.J.; Delerue-Matos, C.; Oliveira, M.B. Electrocatalytic evaluation of DNA damage by superoxide Radical for antioxidant capacity assessment. *J. Electroanal. Chem.,* **2011**, *659*, 43-49. c [http://dx.doi.org/10.1016/j.jelechem.2011.04.022]

[56] Barroso, M.F.; Delerue-Matos, C.; Oliveira, M.B. Evaluation of the total antioxidant capacity of

flavored water and electrochemical purine damage by sulfate radicals using a purine-based sensor. *Electrochim. Acta,* **2011**, *56*, 8954-8961. d
[http://dx.doi.org/10.1016/j.electacta.2011.07.135]

[57] Vilela, D.; González, M.C.; Escarpa, A. Nanoparticles as analytical tools for in-vitro antioxidant-capacity assessment and beyond. *Trends Analyt. Chem.,* **2015**, *64*, 1-16.
[http://dx.doi.org/10.1016/j.trac.2014.07.017]

[58] Falcão, S.I.; Vilas-Boas, M.; Estevinho, L.M.; Barros, C.; Domingues, M.R.; Cardoso, S.M. Phenolic characterization of Northeast Portuguese propolis: usual and unusual compounds. *Anal. Bioanal. Chem.,* **2010**, *396*(2), 887-897.
[http://dx.doi.org/10.1007/s00216-009-3232-8] [PMID: 19902191]

[59] Lourenço, T.; Oliveira, T.; Ferreira, A.M.; Oliveira, R.; Bento, F.; Geraldo, D.; Aguiar, C.A.; Cunha, A. Antimicrobial and antioxidant properties of propolis ethanol extracts from Terceira Island (Azores, Portugal). *Planta Medica,* **2014**, *80*, P1L17.

[60] Sousa, M.E.; Dias, L.G.; Veloso, A.C.; Estevinho, L.; Peres, A.M.; Machado, A.A. Practical procedure for discriminating monofloral honey with a broad pollen profile variability using an electronic tongue. *Talanta,* **2014**, *128*, 284-292.
[http://dx.doi.org/10.1016/j.talanta.2014.05.004] [PMID: 25059162]

[61] Ferreira, I.C.; Aires, E.; Barreira, J.C.; Estevinho, L.M. Antioxidant activity of Portuguese honey samples: Different contributions of the entire honey and phenolic extract. *Food Chem.,* **2009**, *114*, 1438-1443.
[http://dx.doi.org/10.1016/j.foodchem.2008.11.028]

[62] Isla, M.I.; Craig, A.; Ordoñez, R.; Zampini, C.; Sayago, J.; Bedascarrasbure, E.; Alvarez, A.; Salomón, V.; Maldonado, L. Physico chemical and bioactive properties of honeys from Northwestern Argentina. *LWT-. Food Sci. Technol.,* **2011**, *44*, 1922-1930.

[63] Özcan, M.M.; Ölmez, C. Some qualitative properties of different monofloral honeys. *Food Chem.,* **2014**, *163*, 212-218.
[http://dx.doi.org/10.1016/j.foodchem.2014.04.072] [PMID: 24912718]

[64] Jerković, I.; Kuś, P.M.; Tuberoso, C.I.; Šarolić, M. Phytochemical and physical-chemical analysis of Polish willow (*Salix* spp.) honey: identification of the marker compounds. *Food Chem.,* **2014**, *145*, 8-14.
[http://dx.doi.org/10.1016/j.foodchem.2013.08.004] [PMID: 24128442]

[65] Kuś, P.M.; Congiu, F.; Teper, D.; Sroka, Z.; Jerković, I.; Tuberoso, C.I. Antioxidant activity, color characteristics, total phenol content and general HPLC fingerprints of six Polish unifloral honey types. *LWT-. Food Sci. Technol.,* **2014**, *55*, 124-130.

[66] Kilmartin, P.A.; Hsu, C.F. Characterisation of polyphenols in green, oolong, and black teas, and in coffee, using cyclic voltammetry. *Food Chem.,* **2003**, *82*, 501-512.
[http://dx.doi.org/10.1016/S0308-8146(03)00066-9]

CHAPTER 7

Infrared Spectroscopy as a Valuable Tool for the Analysis of Honey Bee Plant-Derived Products

Daniel Cozzolino*

School of Medical and Applied Sciences, Central Queensland University, Rockhampton, 4701, QLD, Australia

Abstract: The importance of honey has been recently promoted due to its nutritional, pharmaceutical and therapeutical characteristics. In recent years, the combination of novel and rapid instrumental techniques based on infrared spectroscopy [mid infrared (MIR), near infrared (NIR)] combined with multivariate data analysis has resulted in the development of both qualitative and quantitative methods for the analysis of honey and bee products. The most important applications of these technologies in honey have been associated with authenticity, discrimination or traceability issues. However, few reports can be found on the use of both NIR and MIR to quantitatively analyse honey composition and less information is available for the analysis of other bee products. This chapter aims to describe and discuss different applications on the use of NIR and MIR spectroscopies to analyse honey, pollen and bee derived products. A brief description of some qualitative applications will be also discussed.

Keywords: Adulteration, Artificial neural networks, Authenticity, Chemometrics, Composition, Fourier, Glucose, Honey, Infrared, Mid infrared, Multivariate data, Near infrared, PLS, Pollen, Principal component, Propolis.

1. INTRODUCTION

The importance of honey has been recently promoted due to its nutritional, pharmaceutical and therapeutical characteristics [1]. In order to maintain high quality standards in the honey industry analytical methods are needed.

* **Address correspondence to D. Cozzolino:** School of Medical and Applied Sciences, Central Queensland University, Rockhampton, 4701, QLD, Australia; Email: d.cozzolino@cqu.edu.au

Susana M. Cardoso (Ed)

Therefore, effective quality control methods like high performance liquid chromatography (HPLC), gas chromatography (GC), mass spectrometry (MS), thin layer chromatography (TLC), enzymatic tests and physical tests, nuclear magnetic resonance (NMR), mid infrared (MIR), and near infrared (NIR) have enriched the diversity of analytical tools available to monitor and analyse honey and derived products [3 - 6].

In recent years, the combination of novel rapid instrumental techniques with multivariate data methods (MVA) has resulted in the development of rapid and inexpensive methods of analysis. Techniques such as near infrared and mid infrared have been the most widely used due to their intrinsic characteristics (*e.g.* low cost, non-destructive) in order to qualitatively and quantitatively analyse several food matrices, including honey [4 - 8]. This chapter aims to describe and discuss different applications on the use of NIR and MIR spectroscopy to analyse honey, pollen and bee derived products. A brief description of some qualitative applications will be also discussed.

2. INFRARED SPECTROSCOPY AND MULTIVARIATE ANALYSIS

Infrared radiation (IR) lays between the visible (VIS) and the microwave wavelengths regions of the electromagnetic spectrum. The nominal range of wavelengths for NIR range is between 750 and 2,500 nm (13,400 to 4,000 cm^{-1}), while for the MIR, the spectral range is from 2,500 to 25,000 nm (4,000 to 400 cm^{-1}) [4 - 8]. In this wavelength range, solid, liquid or gaseous samples can absorb some of the incoming infrared radiation at specific wavelengths or frequencies resulting in a 'fingerprint' or spectrum of the sample [5 - 8]. Infrared spectroscopy is the absorption measurement of different IR frequencies by a sample positioned in the path of an IR beam (*i.e.* NIR and MIR beams); when the frequency of a specific vibration is equal to the frequency of the IR radiation directed at the molecule, this molecule absorbs the radiation [5 - 8].

Mid-infrared spectroscopy allows structural elucidation and compound identification; functional groups absorb photons at characteristic frequencies of MIR radiation and include mainly bands that come from stretching and bending fundamental vibrations. Stretching vibrations are those where the distance between atoms decreases or increases while atoms remain in the same bond axis [5 - 8]. However, in order to evaluate the composition of a complex food sample like honey and honey products, spectral interpretations should not be limited to one or two bands where the whole spectrum needs to be taken into consideration. In NIR the overlapping of many different overtone and combination vibrations results in broad bands with low structural selectivity in NIR spectra compared with MIR spectra where fundamentals are more resolved, allowing the structure of a sample to be better elucidated [4 - 8]. NIR spectroscopy is widely used to determine organic matter constituents and it is based on the absorption of electromagnetic radiation by a sample at wavelengths in the 800-2500 nm range. NIR spectra are composed of broad bands arising from overlapping absorption corresponding mainly to overtones and combinations of vibrational mode C-H, N-H, and O-H chemical bonds [4 - 8]. Overtones correspond to energy transitions that are higher than those for fundamentals and the frequencies of first and second overtones correspond to about two or three times that of the fundamentals [4 - 8]. Combination bands result from transitions involving two or more different vibrational modes of one functional group occurring simultaneously; the frequency of a combination band is the sum or the multiples of the relevant frequencies [4 - 8]. In addition the existence of combination bands (*e.g.* C-O stretch and N-H bend in protein), gives rise to a crowded NIR spectrum with strongly overlapping bands. A major disadvantage of this characteristic overlap and complexity feature of the NIR spectra has been the difficulty of quantification and interpretation of the spectra without the use of MVA methods. On the other hand, the broad overlapping bands can diminish the need for using a large number of wavelengths in calibration and analysis routines [4 - 8].

The combination of MVA with analytical instruments such as NIR or MIR spectroscopy provide with the ability to determine more than one component at a time [1 - 8]. The use of MVA also provides with the ability to detect patterns in a given data set as well as with different algorithms that can be used to develop mathematical models to predict or monitor composition or other quality characteristics (*e.g.* origin, traceability) [4 - 8]. Multivariate data or chemometrics covers quite a broad range of methods such as exploratory data analysis, pattern recognition (PR), and statistical experimental design (DoE) [1 - 6]. The most commonly used multivariate data analysis techniques applied are principal component analysis (PCA), partial least squares (PLS), principal component regression (PCR), discriminant analysis (DA) and artificial neural networks (ANN) [1 - 6]. Chemometrics, unlike classical statistics, considers multiple variables simultaneously and takes collinearity (the variation in one variable, or a group of variables, in terms of co-variation with other variables) into account [1 - 6]. Calibration development can mathematically describe the co-variation (degree of association) between variables, or find a mathematical function (regression model), by which the values of the dependent variables are calculated from values of the measured (independent) variables [1 - 6].

Principal component analysis (PCA) is used as a tool for screening, extracting and compressing multivariate data [1 - 6]. It employs a mathematical procedure that transforms a set of possibly correlated response variables into a new set of non-correlated variables, called principal components. This analysis can be performed on either a data matrix or a correlation matrix depending on the type of variables being measured [1 - 6]. It produces linear combinations of variables that are useful descriptors or even predictors of some particular structure in the data matrix.

Discriminant analysis (DA) and partial least squares discriminant analysis (PLS-DA) can be considered as a qualitative calibration method [1 - 6]. Instead of calibrating for a continuous variable, one calibrates for group membership (categories) [1 - 6]. The resulting models are evaluated in terms of their predictive ability to predict the new and unknown samples (standard error of prediction, SEP) [1 - 6]. Discrimination models are usually developed using PLS regression technique as described elsewhere [1 - 6]. Other discriminant technique extensively used is linear discriminant analysis (LDA). This is a supervised classification technique [1 - 6]. The criterion of LDA for selection of latent variables is maximum differentiation between the categories and minimal variance within categories [1 - 6]. This method produces a number of orthogonal linear discriminant functions, equal to the number of categories minus one, that allow the samples to be classified in one or another category [1 - 6].

Application of ANN is a more recent technique for data and knowledge processing that is characterised by its analogy with a biological neuron [1 - 6].

Unlike linear regression (*e.g.* PCR and PLS), ANN can deal with nonlinear relationships between variables and both PLS and PCA are the two most powerful tools for data analysis which require to extract the information about quality attributes hidden in the data obtained from instrumental methods ('model calibration') [1 - 6].

3. APPLICATIONS OF INFRARED IN HONEY AND PRODUCTS

3.1. Honey Chemical Composition

The feasibility of using NIR spectroscopy to determine chemical composition of commercial honey was evaluated by several authors [9 - 7]. The effect of various sample presentation methods as well as the regression model used to develop NIR calibrations for the measurement of chemical parameters in honey was examined by Qiu *et al.* [9]. Modified partial least squares (mPLS) was selected as regression algorithm for the calibration of all honey parameters with the exception of moisture, for which the optimal calibration was developed using PLS regression [9]. Validation of the calibration models was achieved using an independent set of honey samples. The results obtained by these authors showed that NIR spectroscopy could accurately (coefficient of determination) determine moisture (R^2 = 1), fructose (R^2 = 0.97), glucose (R^2 = 0.91), sucrose (R^2 = 0.86), and maltose (R^2 = 0.93) content in honey (Table **1**). However, the accuracy of the prediction models for the determination of other parameters such as free acidity, lactone, and hydroxylmethylfurfural (HMF) contents was considered poor or unreliable accordingly to the authors [9]. Still, this study demonstrated that NIR spectroscopy can be used for rapid determination of major components in commercial honey.

The use of NIR transflectance spectroscopy was also evaluated for the measurement of fructose, glucose and moisture [10]. A total of 161 honey samples were collected during three consecutive harvests and analyzed using instrumental, enzymatic (fructose and glucose), and refractometric (moisture) methods [10]. Robust calibration (n = 121) and validation (n = 40) models were developed (see Table **1**).

Table 1. Summary of the calibration and prediction statistics reported by different authors on the use of infrared spectroscopy to predict chemical composition in honey and bee products.

Sample/Matrix	Parameters	Infrared technique	SECV/RMSECV	RPD	Reference
Honey	Moisture (%)	VIS - NIR	0.08		[9]
	HMF (mg/Kg)		60		[9]
	Fructose (%)		0.57		[9]
	Glucose (%)		0.52		[9]
	Sucrose (%)		0.28		[9]
	Maltose (%)		0.31		[9]
	Free acidity (mequiv/kg)		3.51		[9]
	Lactose (mequiv/kg)		0.44		[9]
Honey (n=118)	Fructose	NIR	0.48		[10]
	Glucose		0.071		[10]
	Moisture		0.14		[10]
Honey (n=138)	Direct polarization (degrees)	NIR	0.049		[11]
	Sucrose (% w/w)		0.38		[11]
Honey (n=42)	Sugars (sucrose, maltose, glucose, fructose) (%)	MIR	1.2 and 1.5		[12]
Honey	Water (%)	NIR	1.0		[13]
	Fructose (%)		1.7		[13]
	Glucose (%)		1.5		[13]
	Sucrose (%)		1.1		[13]
	Melizitose		0.8		[13]
	Free acidity		7		[13]
	Proline		192		[13]
	Electric conductivity		0.29		[13]
	pH value		0.4		[13]
Honey (n=63)	Glucose	ATR-MIR	0.97	2.88	[14]
	Fructose		0.90	2.52	[14]
	Melizitose		0.39	3.03	[14]
	Turanose		0.21	2.41	[14]
	Maltose		0.42	1.52	[14]

(Table 1) contd.....

Sample/Matrix	Parameters	Infrared technique	SECV/RMSECV	RPD	Reference
Honey (n=60)	Phenol (mg/100 g)	NIR	33.74		[17]
	Flavonoid (mg/100 g)		1.07		[17]
	Vitamin C (mg/100 g)		0.99		[17]
	Oxidation index		3.68		[17]
	Cu (mg/100 g)		0.06		[17]
Honey (n=42)	K (mg/100 g)	NIR	28	5.2	[18]
	Ca (mg/100 g)		2.8	4.7	[18]
	Mg (mg/100 g)		2.3	4.7	[18]
	P (mg/100 g)		1.3	4.2	[18]
Honey (n=48)	Humidity (%)	ATR-MIR*	0.32-0.61		[22]
	Sugars (%)		1.03-1.31		[22]
	Acidity (%)		2.20-3.35		[22]
Pollen (n=52)	Protein (%)	NIR	0.77		[23]
	Moisture (%)		0.69		[23]
	Ash (%)		0.11		[23]
	Reducing sugars		3.54		[23]
	pH		0.18		[23]

ATR: attenuated total reflectance; HMF: hydroxylmethylfurfural; NIR: near infrared; MIR: mid infrared; SECV: standard error in cross validation; RMSECV: root mean square error of cross validation; RPD: residual predictive deviation (SD/SECV); VIS: visible; * different algorithms were used.

In another study, the same authors reported the use of NIR transflectance spectroscopy to determine polarimetric parameters (direct polarization, polarization after inversion, specific rotation in dry matter, and polarization due to non-monosaccharides) and sucrose [11]. Calibration models were developed using mPLS regression and scatter correction using standard normal variation (SNV) and detrend. For direct polarization, polarization after inversion, specific rotation in dry matter, and polarization due to non-monosaccharides, good statistics were reported by the authors. In opposition, reported statistics for the prediction of sucrose were not as good as those obtained for the other chemical parameters. Hence, according to these authors, NIR spectroscopy was not an effective method for quantitative analysis of sucrose in the set of honey samples analysed due to the range in composition observed. However, the authors suggested that NIR spectroscopy may be an acceptable method for a semi-quantitative evaluation of sucrose in honey.

Fourier transform infrared (FTIR) spectroscopy combined with micro attenuated total reflectance (mATR) as sampling accessory was used for the simultaneous determination of saccharides such as fructose, glucose, sucrose, and maltose [12]. Two calibration models were developed using either PLS or principal component regression (PCR). The first model used a data set of 42 standard mixtures of fructose, glucose, sucrose and maltose, prepared over the range of concentrations normally present in honey, whereas the second model used a set of 45 honey samples composed of various floral sources and regions. The calibration models developed were validated using a different data sets and verified using HPLC as the reference method. The R^2 values reported for the different sugars were between 0.97 and 0.99, demonstrating the good predictive ability and accuracy of the method proposed [12].

Fourier transform NIR spectroscopy (FT-NIR) was also evaluated to quantitatively determine 24 different compositional parameters in honey samples from different botanical origins [13]. Calibration models for the chemical parameters were developed using PLS regression. These calibrations were then validated using an independent set of samples which yield satisfactory accuracies for the determination of parameters such as water [standard error of prediction (SEP) = 0.3 g/100 g], glucose (SEP = 1.3 g/100 g), fructose (SEP = 1.6 g/100 g), sucrose (SEP = 0.4 g/100 g), total monosaccharide content (SEP = 2.6 g/100 g) as well as fructose/glucose (SEP = 0.09) and glucose/water ratios (SEP = 0.12) (Table **1**). According to these authors, the prediction accuracy for hydroxymethylfurfural (HMF), proline, pH-value, electrical conductivity (EC), free acidity as well as other minor sugars such as maltose, turanose, nigerose, erlose, trehalose, isomaltose, kojibiose, melezitose, raffinose, gentiobiose, melibiose, maltotriose was poor and unreliable [13]. Hence, overall these results demonstrated that NIR spectroscopy is a valuable, rapid and non-destructive tool for the quantitative analysis of only some chemical properties related to the main components in honey [13].

The use of FTIR-ATR spectroscopy combined with PLS regression was reported as a method for the prediction of several sugars in standard sugar mixtures and honey samples [14]. In their study, the authors used standards of trehalose, glucose, fructose, sucrose, melezitose, turanose and maltose in order to identify and quantify these sugar components in 63 honey samples using HPLC as the reference method. The first derivative was used and the MIR range from 1500 to 750 cm^{-1} selected to develop the PLS models. The best calibration models reported by the authors were for fructose ($R^2 = 0.86$, RPD = 2.6) and for glucose ($R^2 = 0.86$, RPD = 2.55), respectively and thus, FTIR-ATR was suggested by the authors to be a good methodology to quantify the main sugar content in honey and easily adapted to routine analysis [14].

The use of FTIR-ATR spectroscopy was also reported for the determination of 14 different chemical components in honey samples sourced from Brazil [15]. Nine different honey samples (six monofloral and three polyfloral) were analysed using palynological, color, and sensorial analysis to obtain preliminary results for these types of honey. The results reported by these authors showed that FTIR-ATR can be used as a screening method for the routine analysis of Brazilian honey samples. The authors also compared the efficiency of FTIR-ATR method with the recommended methodologies for physicochemical parameters of eighteen samples of *Melipona subnitida* honey. Although significant differences were found between the values obtained using those techniques for HMF, ash and EC, the results for the other parameters did not differ significantly, suggesting that this rapid and non-destructive methodology may predict parameters usually used to assess honeys' quality.

The application of FTIR spectroscopy on this topic is evident. In another study de-Almeida-Muradian and collaborators studied the effects of different treatments such as room temperature, fridge and freezer storage on the quality parameters of honey measured using FTIR spectroscopy [16]. These authors showed that darkening of the honey was observed, particularly in samples stored in the fridge or freezer. However, the changes observed in the honey samples kept on the fridge were not statistically different from those observed in the honey samples kept on the freezer, except for free acidity [16]. Overall, this work suggested that storage at room temperature might be the best way to storage [16].

NIR spectroscopy has also been used to predict chemical parameters associated with the antioxidant activity of honeys. In the work of Escuredo *et al.* [17], a total of 60 honey samples were sourced in order to develop NIR calibration models using mPLS regression and validated using 15 samples. Calibration models were developed for the determination of phenols, flavonoids, vitamin C, antioxidant capacity (DPPH assay), oxidation index and copper. These models were optimised using cross-validation, and the best model was evaluated using the R^2, the standard error of cross-validation (SECV), RPD and the root mean standard error (RMSE) in the prediction set. The results reported by the authors suggested that the NIR calibrations developed could be used for the rapid determination of chemical compounds associated with antioxidant capacity in honey. The same authors also reported the use of NIR spectroscopy for the rapid prediction of mineral content in honey samples (see Table **1**) [18].

The analysis of soluble solids content (SSC) and moisture was examined using NIR spectroscopy [19]. Calibration models for SSC and moisture were developed using PLS regression by combining different mathematical pre-treatments as well as different spectral ranges. The R^2, root mean square error of cross validation (RMSECV), and root mean square error of validation sets (RMSEP) reported were for SSC 0.99, 0.190 and 0.127 while for moisture the reported statistics were 0.998, 0.187 and 0.125 respectively.

Honey samples sourced from China were analysed using a FT-NIR spectrometer in order to determinate fructose and glucose content. Two different modes namely transflectance (800 - 2500 nm, 2 mm optical path length) and transmittance (800 - 1370 nm, 20 mm optical path length) were compared by the authors [20]. These authors reported significant differences in the prediction accuracies obtained for the prediction of fructose and glucose when both modes were compared [20]. Support vector machine (SVM) (non-linear), and genetic algorithm (GA) were used to develop different calibration models. The results reported by these authors showed differences in the calibration models for fructose and glucose content [20]. In fact, these authors found that for the determination of glucose, short wavelength and long optical path length should be used, while the whole NIR wavelength range and short optical path length should be recommended for the measurement of fructose [20]. The same authors also evaluated the potential use of FT-NIR spectroscopy as a non-destructive method for determining the main chemical components in samples sources from unifloral and multifloral honey samples [21]. In this particular case, calibration models were developed for water fructose and glucose contents using PLS regression and the best calibration models yield a R^2 of 0.97 (RMSEP = 0.41%) for water, R^2 of 0.93 (RMSEP = 0.1.91%) for fructose and R^2 of 0.89 (RMSEP = 2.15%) for glucose [21].

The use of FTIR-ATR combined with MVA was also reported by Pataca *et al.* [22] for determination of reducing sugars, humidity and acidity in honey bee samples. For that, multivariate calibration models were developed using PLS regression and were refined through variable selection per interval (iPLS) and GA. According to the authors the calibration models showed satisfactory results for all parameters with average relative errors of 6% for acidity, 1% for reducing sugars and 2% for humidity.

3.2. Bee Pollen and Propolis

Although much less studied, the use of infrared spectroscopy has also been applied for pollen and propolis samples. In particular, NIR spectroscopy together with a remote reflectance fibre-optic probe was evaluated as a method for the determination of the major components in bee pollen [23]. The method allows immediate control of the bee pollen without prior sample treatment or destruction through direct application of the fibre-optic probe to the sample using mPLS as regression method. The calibration results obtained using 45 samples of bee pollen allowed the measurement of protein, moisture, ash, reducing sugars, and pH, yield a R^2 and prediction corrected standard errors (SEPC) of 0.91, 0.56% for protein, of 0.78 and 0.49% for moisture; 0.92 and 0.049% for ash; 0.81 and 1.32 g of glucose/100 g of bee pollen; 0.84 and 0.15 for pH, respectively [23].

Colombia exhibits an extraordinarily high species diversity of the subfamily *Apinae* (honeybees, bumblebees, stingless bees, orchid bees) [24]. This fact makes it worthwhile to look for beeswax as biological material produced by these insects and to prove possible applications in technique and human life [24]. These authors examined for the first time waxes from pollen and honey pots, brood cells, as well as different species of the tribes *Bombini*, *Meliponini,* and *Apini* native to South America [24]. A FTIR method was used to obtain infrared spectra of the waxes

[24]. The authors reported that the IR spectra of waxes are more similar within the taxonomic groups than between them and are related to the altitude where bees live [24]. According to these authors this work contributed to the achievement of information that will serve to establish energy mechanisms used by these insects and to set up conservation strategies to protect these species [24].

In addition, bees wax content derived from raw propolis samples were analysed using NIR spectroscopy [25]. Confirmation of the identity of beeswax isolated from different propolis samples sourced from 27 Dutch apiaries from various locations were analyzed using the propose method [25]. The beeswax content varied between 1.0 and 42.5% with an average of 11.1% [25].

3.3. Honey Adulteration

Several reports can be found in the literature on the use of vibrational spectroscopy (*e.g.* NIR, MIR) to authenticate or discriminate honey from different botanical or geographical origins as well as a tool to target adulteration issues [1, 26 - 40]. Numerous classification tools have been used to target these issues such as Fisher discriminat analysis (FDA), partial least squares discriminat analysis (PLS-DA), PCA, linear discriminat analysis (LDA), least squares support vector machine (LS-SVM), K-nearest neighbours (KNN) and soft independent modelling of class analogy (SIMCA). Most of these tools can be used alone or combined with various mathematical pre-treatments, wavelength or wave number ranges [1, 26 - 40]. The combination of NIR spectroscopy and chemometrics has been evaluated to detect adulteration in honey samples by Zhu *et al.* [40]. Several algorithms and pre-processing methods were used by these authors such as wavelet transformation (WT), PCA, LS-SVM, SVM, back propagation artificial neural network (BP-ANN), LDA, and KNN [40]. These authors reported that WT was the most effective method for variables selection and the best classification models were achieved using LS-SVM (accuracy of 95.1%) [40]. The implementation of VIS and NIR spectroscopy for the detection of glucose concentration in a mixture of Saudi and imported honey samples adulterated by glucose syrup using five concentrations: 0, 5, 12, 19, and 33 g/100 g was reported [42]. The results reported by these authors suggest that NIR spectroscopy is a powerful technique for the quantification of glucose adulteration in Saudi honey

[42]. Information contained in the NIR spectra of honey samples sourced from protected geographical indication (PGI) ''Mel de Galicia'' was processed by means of different chemometric techniques to develop an authentication system [43]. According to these authors classification based on SIMCA achieved the best PGI-model with 93.3% of sensitivity and 100% of specificity [43]. The main conclusion derived from this study was that the combination of NIR information data with SIMCA allowed the development of a single and fast method to differentiate between genuine PGI-Galician honey samples and other commercial honey samples from other origins .

Botanical origin of the nectar predominantly affects the chemical composition of honey [47]. The discrimination of Anatolian honey samples from different botanical origins were attempted using hierarchical clustering and PCA using MIR spectra [47]. According to the authors discrimination of sample groups was achieved successfully with hierarchical clustering over the spectral range of 1800–750 cm^{-1} [47]. Table **2** presents some other examples on the application of NIR or MIR to authenticate, trace and monitor adulteration in honey as reported by different authors.

Table 2. Examples of applications of near and mid infrared spectroscopy to authenticate or trace honey samples.

Sample matrix	Infrared region	Algorithm used	Application	Reference
Honey	MIR	PCA, SVM	Identification of botanical origin in unifloral honey	[41]
Honey	VIS-NIR	PLS	Adulteration with glucose	[42]
Honey	NIR	SIMCA, PLS-DA	Geographical origin	[43]
Honey	MIR	PCA	Adulteration with sugars	[44]
Honey	VIS-NIR	LDA, PCA, SVM	Discrimination	[45]
Honey	MIR	Direct observation spectra	Glucose adulteration	[46]

LDA: linear discriminant analysis, NIR: near infrared, MIR: mid infrared, VIS: visible, PCA: principal component analysis, PLS: partial least squares, SVM: support vector machines, PLS-DA: partial least squares discriminant analysis.

CONCLUSIONS

Infrared and other vibrational spectroscopy (*e.g.* UV, Raman) technologies have been introduced as fingerprinting techniques to analyse agricultural products and foods, and in recent years have been rediscover by several research groups to be used in honey applications. The main advantages of these techniques over the traditional chemical and chromatographic methods are the rapidity and the ease of use in routine operations. In addition, infrared spectroscopy based methods are well known to be non-destructive techniques requiring minimal or zero sample preparation. Overall, from the reports and information published, it is clear that the breadth of this type of applications for the analysis of honey and other bee products, either in routine use or under development, is showing no sign of diminishing.

CONFLICT OF INTEREST

The author confirms that author has no conflict of interest to declare for this publication.

ACKNOWLEDGEMENTS

Declared none.

REFERENCES

[1] Arvanitoyannis, I.S.; Chalhoub, C.; Gotsiou, P.; Lydakis-Simantiris, N.; Kefalas, P. Novel quality control methods in conjunction with chemometrics (multivariate analysis) for detecting honey authenticity. *Crit. Rev. Food Sci. Nutr.,* **2005**, *45*(3), 193-203. [http://dx.doi.org/10.1080/10408690590956369] [PMID: 16048148]

[2] Cordella, C.; Moussa, I.; Martel, A-C.; Sbirrazzuoli, N.; Lizzani-Cuvelier, L. Recent developments in food characterization and adulteration detection: technique-oriented perspectives. *J. Agric. Food Chem.,* **2002**, *50*(7), 1751-1764. [http://dx.doi.org/10.1021/jf011096z] [PMID: 11902909]

[3] Cozzolino, D. Recent trends on the use of infrared spectroscopy to trace and authenticate natural and agricultural food products. *Appl. Spectrosc. Rev.,* **2012**, *47*, 518-530. [http://dx.doi.org/10.1080/05704928.2012.667858]

[4] Cozzolino, D.; Corbella, E.; Smyth, H.E. Quality control of honey using infrared spectroscopy: a review. *Appl. Spectrosc. Rev.,* **2011**, *46*, 523-538. [http://dx.doi.org/10.1080/05704928.2011.587857]

[5] Hashimoto, A.; Kameoka, T. Applications of infrared spectroscopy to biochemical, food, and agricultural processes. *Appl. Spectrosc. Rev.,* **2008**, *43*, 416-451. [http://dx.doi.org/10.1080/05704920802108131]

[6] Karoui, R.; Downey, G.; Blecker, C. Mid-infrared spectroscopy coupled with chemometrics: a tool for the analysis of intact food systems and the exploration of their molecular structure-quality relationships - a review. *Chem. Rev.,* **2010**, *110*(10), 6144-6168. [http://dx.doi.org/10.1021/cr100090k] [PMID: 20804166]

[7] Li-Chan, E.C. *Applications of Vibrational Spectroscopy in Food Science, Li-Chan. E.C.Y*; Griffiths, P.R.; Chalmers, J.M., Eds.; Wiley and Sons: UK, **2010**, Vol. 14, pp. 3-46.

[8] Subramanian, A.; Rodrigez-Saona, L. *Infrared spectroscopy for food quality analysis and control*; Sun, D.W., Ed.; Academic Press: UK, **2009**, pp. 146-174.

[9] Qiu, P.Y.; Ding, H.B.; Tang, Y.K.; Xu, R.J. Determination of chemical composition of commercial honey by near-infrared spectroscopy. *J. Agric. Food Chem.,* **1999**, *47*(7), 2760-2765. [http://dx.doi.org/10.1021/jf9811368] [PMID: 10552561]

[10] García-Alvarez, M.; Huidobro, J.F.; Hermida, M.; Rodríguez-Otero, J.L. Major components of honey analysis by near-infrared transflectance spectroscopy. *J. Agric. Food Chem.,* **2000**, *48*(11), 5154-5158. [http://dx.doi.org/10.1021/jf000170v] [PMID: 11087451]

[11] García-Alvarez, M.; Ceresuela, S.; Huidobro, J.F.; Hermida, M.; Rodríguez-Otero, J.L. Determination of polarimetric parameters of honey by near-infrared transflectance spectroscopy. *J. Agric. Food Chem.,* **2002**, *50*(3), 419-425. [http://dx.doi.org/10.1021/jf0105438] [PMID: 11804506]

[12] Tewari, J.; Irudayaraj, J. Quantification of saccharides in multiple floral honeys using fourier transform infrared microattenuated total reflectance spectroscopy. *J. Agric. Food Chem.,* **2004**, *52*(11), 3237-3243. [http://dx.doi.org/10.1021/jf035176+] [PMID: 15161176]

[13] Ruoff, K.; Luginbuehl, W.; Bogdanov, S.; Bosset, J-O.; Estermann, B.; Ziolko, T.; Kheradmandan, S.; Amado, R. Quantitative determination of physical and chemical measurands in honey by near-infrared spectrometry. *Eur. Food Res. Technol.,* **2007**, *225*, 415-423. [http://dx.doi.org/10.1007/s00217-006-0432-8]

[14] Anjos, O.; Campos, M-G.; Ruiz, P.C.; Antunes, P. Application of FTIR-ATR spectroscopy to the quantification of sugar in honey. *Food Chem.,* **2015**, *169*, 218-223. [http://dx.doi.org/10.1016/j.foodchem.2014.07.138] [PMID: 25236219]

[15] de Almeida-Muradian, L.B.; Sousa, R.J.; Barth, O.M.; Gallmann, P. Preliminary data on Brazilian monofloral honey from the northeast region using FT-IR ATR spectroscopic, palynological, and color analysis. *Quim. Nova,* **2014**, *37*, 716-719. [http://dx.doi.org/10.5935/0100-4042.20140115]

[16] de Almeida-Muradian, L.B.; Stramm, K.M.; Estevinho, L.M. Efficiency of the FT-IR ATR spectrometry for the prediction of the physicochemical characteristics of *Melipona subnitida* honey and study of the temperature's effect on those properties. *Int. J. Food Sci. Technol.,* **2014**, *49*, 188-

195.
[http://dx.doi.org/10.1111/ijfs.12297]

[17] Escuredo, O.; Seijo, M.C.; Salvador, J.; González-Martín, M.I. Near infrared spectroscopy for prediction of antioxidant compounds in the honey. *Food Chem.,* **2013**, *141*(4), 3409-3414.
[http://dx.doi.org/10.1016/j.foodchem.2013.06.066] [PMID: 23993500]

[18] Escuredo, O.; González-Martín, M.I.; Rodríguez-Flores, M.S.; Seijo, M.C. Near infrared spectroscopy applied to the rapid prediction of the floral origin and mineral content of honeys. *Food Chem.,* **2015**, *170*, 47-54.
[http://dx.doi.org/10.1016/j.foodchem.2014.08.061] [PMID: 25306316]

[19] Li, S.F.; Zhang, X.; Shan, Y.; Li, Z.H. [Prediction analysis of soluble solids content and moisture in honey by near infrared spectroscopy]. *Guangpuxue Yu Guangpu Fenxi,* **2010**, *30*(9), 2377-2380.
[PMID: 21105399]

[20] Tu, Z-H.; Zhu, D-Z.; Ji, B-P.; Meng, C-Y.; Wang, L-G.; Qing, Z-S. Difference analysis and optimization study for determination of fructose and glucose by near infrared spectroscopy. *Chin. J. Anal. Chem.,* **2010**, *38*, 45-50.

[21] Tu, Z.H.; Ji, B.P.; Meng, C.Y.; Zhu, D.Z.; Wang, L.G.; Qing, Z.S. [Possibilities of near-infrared spectroscopy for the assessment of principle components in honey]. *Guangpuxue Yu Guangpu Fenxi,* **2009**, *29*(12), 3291-3294.
[PMID: 20210153]

[22] Pataca, L.C.; Borges Neto, W.; Marcucci, M.C.; Poppi, R.J. Determination of apparent reducing sugars, moisture and acidity in honey by attenuated total reflectance-Fourier transform infrared spectrometry. *Talanta,* **2007**, *71*(5), 1926-1931.
[http://dx.doi.org/10.1016/j.talanta.2006.08.028] [PMID: 19071543]

[23] González-Martín, I.; Hernández-Hierro, J.M.; Barros-Ferreiro, N.; Cordón Marcos, C.; García-Villanova, R.J. Use of NIRS technology with a remote reflectance fibre-optic probe for predicting major components in bee pollen. *Talanta,* **2007**, *72*(3), 998-1003.
[http://dx.doi.org/10.1016/j.talanta.2006.12.039] [PMID: 19071715]

[24] Pinzon, F.; Torres, A.; Hoffmann, W.; Lamprecht, I. Thermo-analytical and infrared spectroscopic investigations on wax samples of native Colombian bees living in different altitudes. *Eng. Life Sci.,* **2013**, *13*, 520-527.
[http://dx.doi.org/10.1002/elsc.201200103]

[25] Hogendoorn, E.A.; Sommeijer, M.J.; Vredenbregt, M.J. Alternative method for measuring beeswax content in propolis from the Netherlands. *J. Apic. Sci.,* **2013**, *57*, 81-90.

[26] Hennessy, S.; Downey, G.; O'Donnell, C.P. Attempted confirmation of the provenance of Corsican PDO honey using FT-IR spectroscopy and multivariate data analysis. *J. Agric. Food Chem.,* **2010**, *58*(17), 9401-9406.
[http://dx.doi.org/10.1021/jf101500n] [PMID: 20695639]

[27] Hennessy, S.; Downey, G.; O'Donnell, C. Multivariate analysis of attenuated total reflection-Fourier transform infrared spectroscopic data to confirm the origin of honeys. *Appl. Spectrosc.,* **2008**, *62*(10), 1115-1123.
[http://dx.doi.org/10.1366/000370208786049033] [PMID: 18926021]

[28] Gallardo-Velazquez, T.; Osorio-Revilla, G.; Zuñiga-de Loa, M.; Rivera-Espinoza, Y. Application of FTIR-HATR spectroscopy and multivariate analysis to the quantification of adulterants in Mexican honeys. *Food Res. Int.,* **2009**, *42*, 313-318. [http://dx.doi.org/10.1016/j.foodres.2008.11.010]

[29] Etzold, E.; Lichtenberg-Kraag, B. Determination of the botanical origin of honey by Fourier-transformed infrared spectroscopy: An approach for routine analysis. *Eur. Food Res. Technol.,* **2008**, *227*, 579-586. [http://dx.doi.org/10.1007/s00217-007-0759-9]

[30] Irudayaraj, J.; Xu, F.; Tewari, J. Rapid determination of invert cane sugar adulteration in honey using FTIR spectroscopy and multivariate analysis. *J. Food Sci.,* **2003**, *68*, 2040-2045. [http://dx.doi.org/10.1111/j.1365-2621.2003.tb07015.x]

[31] Irudayaraj, J.; Sivakesava, S. Detection of adulteration in honey by discriminant analysis using FTIR spectroscopy. *Trans. Am. Soc. Agric. Eng,* **2001**, *44*, 643-650. [http://dx.doi.org/10.13031/2013.6092]

[32] Kelly, J.F.; Downey, G.; Fouratier, V. Initial study of honey adulteration by sugar solutions using midinfrared (MIR) spectroscopy and chemometrics. *J. Agric. Food Chem.,* **2004**, *52*(1), 33-39. [http://dx.doi.org/10.1021/jf034985q] [PMID: 14709010]

[33] Kelly, J.D.; Petisco, C.; Downey, G. Application of Fourier transform midinfrared spectroscopy to the discrimination between Irish artisanal honey and such honey adulterated with various sugar syrups. *J. Agric. Food Chem.,* **2006**, *54*(17), 6166-6171. [http://dx.doi.org/10.1021/jf0613785] [PMID: 16910703]

[34] Mishra, S.; Kamboj, U.; Kaur, H.; Kapur, P. Detection of jaggery syrup in honey using near-infrared spectroscopy. *Int. J. Food Sci. Nutr.,* **2010**, *61*(3), 306-315. [http://dx.doi.org/10.3109/09637480903476415] [PMID: 20109130]

[35] Ruoff, K.; Luginbühl, W.; Bogdanov, S.; Bosset, J.O.; Estermann, B.; Ziolko, T.; Amado, R. Authentication of the botanical origin of honey by near-infrared spectroscopy. *J. Agric. Food Chem.,* **2006**, *54*(18), 6867-6872. [http://dx.doi.org/10.1021/jf060770f] [PMID: 16939351]

[36] Ruoff, K.; Luginbühl, W.; Künzli, R.; Iglesias, M.T.; Bogdanov, S.; Bosset, J.O.; von der Ohe, K.; von der Ohe, W.; Amado, R. Authentication of the botanical and geographical origin of honey by mid-infrared spectroscopy. *J. Agric. Food Chem.,* **2006**, *54*(18), 6873-6880. [http://dx.doi.org/10.1021/jf060838r] [PMID: 16939352]

[37] Sivakesava, S.; Irudayaraj, J. Detection of inverted beet sugar adulteration of honeyby FTIR spectroscopy. *J. Sci. Food Agric.,* **2001**, *81*, 683-690. [http://dx.doi.org/10.1002/jsfa.858]

[38] Woodcock, T.; Downey, G.; O'Donnell, C.P. Near infrared spectral fingerprinting for confirmation of claimed PDO provenance of honey. *Food Chem.,* **2009**, *114*, 742-746. [http://dx.doi.org/10.1016/j.foodchem.2008.10.034]

[39] Woodcock, T.; Downey, G.; Kelly, J.D.; O'Donnell, C. Geographical classification of honey samples by near-infrared spectroscopy: a feasibility study. *J. Agric. Food Chem.,* **2007**, *55*(22), 9128-9134.

[http://dx.doi.org/10.1021/jf072010q] [PMID: 17927137]

[40] Zhu, X.; Li, S.; Shan, Y.; Zhang, Z.; Li, G.; Su, D.; Liu, F. Detection of adulterants such as sweeteners materials in honey using near-infrared spectroscopy and chemometrics. *J. Food Eng.,* **2010**, *101*, 92-97.
[http://dx.doi.org/10.1016/j.jfoodeng.2010.06.014]

[41] Lenhardt, L.; Zekovic, I.; Dramicanin, T. Authentication of the botanical origin of unifloral honey by infrared spectroscopy coupled with support vector machine algorithm. *Phys. Scr.,* **2014**, *162*, 014042.
[http://dx.doi.org/10.1088/0031-8949/2014/T162/014042]

[42] Mouazen, A.M.; Al-Walaan, N. Glucose adulteration in Saudi honey with visible and near infrared spectroscopy. *Int. J. Food Prop,* **2014**, *17*, 2263-2274.
[http://dx.doi.org/10.1080/10942912.2013.791837]

[43] Herrero Latorre, C.; Peña Crecente, R.M.; García Martín, S.; Barciela García, J. A fast chemometric procedure based on NIR data for authentication of honey with protected geographical indication. *Food Chem.,* **2013**, *141*(4), 3559-3565.
[http://dx.doi.org/10.1016/j.foodchem.2013.06.022] [PMID: 23993521]

[44] Rios-Corripio, M.A.; Rojas-Lopez, M.; Delgado-Macuil, R. Analysis of adulteration in honey with standard sugar solutions and syrups using attenuated total reflectance-Fourier transform infrared spectroscopy and multivariate methods. *CYTA-J. Food,* **2012**, *10*, 119-122.

[45] Li, Yun; Yang, Haiqing Honey Discrimination Using Visible and Near-Infrared Spectroscopy. *ISRN Spectroscopy,* **2012**, *2012*, 1-4.
[http://dx.doi.org/10.5402/2012/487040]

[46] Nikolova, K.; Panchev, I.; Sainov, S.; Gentscheva, G.; Ivanova, E. Selected physical properties of lime bee honey in order to discriminate between pure honey and honey adulterated with glucose. *Int. J. Food Prop.,* **2012**, *15*, 1358-1368.
[http://dx.doi.org/10.1080/10942912.2010.521965]

[47] Gok, S.; Severcan, M.; Goormaghtigh, E.; Kandemir, I.; Severcan, F. Differentiation of Anatolian honey samples from different botanical origins by ATR-FTIR spectroscopy using multivariate analysis. *Food Chem.,* **2015**, *170*, 234-240.
[http://dx.doi.org/10.1016/j.foodchem.2014.08.040] [PMID: 25306340]

CHAPTER 8(I)

Antioxidant Properties of Bee Products of Plant-Origin. Part 1. Honey

Marta Quicazán*, Carlos Zuluaga

Instituto de Ciencia y Tecnología de Alimentos – ICTA, Universidad Nacional de Colombia, Ciudad Universitaria, Av. Kr. 30 # 45-03 Ed. 500C, 111321. Bogotá, Colombia

Abstract: The study of antioxidant-antiradical activity of food products has received a rising interest since last decades, parallel to the boom of functional foods and healthy consumption trends, and to the increasing number of scientific evidence linking this physicochemical property to prevention of several degenerative diseases, in particular of cancer. Honey belongs to the category of natural foods showing high antioxidant activity, which depends largely on its botanical/geographical origin. Different studies have been conducted to describe the antioxidant activity of honey by both *in vitro* and *in vivo* techniques, which are reported in a vast extension of articles. However, the lack of a standard protocol for measuring the antioxidant activity has been one of the main drawbacks found. Techniques such as 2,2-diphenyl-1-picrylhydrazyl (DPPH), Trolox Equivalent Antioxidant Capacity (TEAC), Ferric Reducing Antioxidant Power (FRAP) and Oxygen Radical Absorbance Capacity (ORAC) are the most common *in vitro* methods. It has been suggested to use at least two techniques for measuring antioxidant activity, since these are only an approximation to what occurs in the body. It is known that biologically active ingredients which may contribute to the antioxidant effect of honey include vitamins, minerals, organic acids, flavonoids, phenolic compounds and even products derived from Maillard reaction.

Keywords: Antiradical, Bioactive compounds, Biological activity, DPPH, Flavonoids, FRAP, Free radical, Honey enzymes, Honeybee products, Maillard reaction, Minerals, Natural foods, ORAC, Origin, Oxidative stress, Phenolic compounds, Scavenging capacity, TEAC, Vitamins.

* **Address correspondence to M. Quicazán:** Instituto de Ciencia y Tecnología de Alimentos – ICTA, Universidad Nacional de Colombia. Ciudad Universitaria, Av. Kr. 30 # 45-03 Ed. 500C, 111321. Bogotá, Colombia; Tel: +571 3165000 ext. 19211; Fax: +571 3165300; Email: mcquicazand@unal.edu.co

Susana M. Cardoso (Ed)

1. INTRODUCTION

Since recent decades, there has been an increase in degenerative or chronic diseases such as diabetes mellitus, hypertension, cancer, Alzheimer's disease, atherosclerosis and heart disease as a consequence of oxidative stress [1 - 3]. Sies and Jones [4] stated that oxidative stress could be defined as "an imbalance between oxidants and anti-oxidants in favor of the oxidants, leading to a disruption of redox signaling and control and/or molecular damage". It is caused by a concentration of reactive oxygen species (ROS) or reactive nitrogen species (RNS) higher than antioxidants. ROS include superoxide ($O2^{\bullet -}$), hydroxyl ($^{\bullet}OH$) and hydrogen peroxide (H_2O_2), while RNS comprise nitric oxide ($NO^{\bullet}$), nitrogen dioxide ($NO_2^{\bullet -}$) and peroxynitrite ($OONO^{-}$) [4, 5]. Several authors [5 - 7] commented the ability of cells to scavenge excess reactive species is largely dependent on the efficiency of the overall antioxidant defense system.

Halliwell and Gutteridge defined an antioxidant as "any substance that delays, prevents or removes oxidative damage to a target molecule" [5]. The overall antioxidant defense network consists of endogenous and exogenous antioxidants. The endogenous antioxidants comprise the enzymatic antioxidants such as superoxide dismutase (SOD), catalase (CAT) and glutathione peroxidase (GPx) and non-enzymatic antioxidants including glutathione (GSH), vitamins C and E, and small molecules [4, 5]. The exogenous antioxidants comprise the micronutrients present in foodstuff [7].

Although there are a wide variety of natural antioxidant substances, the best sources of antioxidant compounds seems to be those from plant origin [8]. Some studies [9, 10] demonstrated that medicinal aromatic herbs, fruits and leaves of some berry plants, can biosynthesize phytochemicals possessing antioxidant activity and they may be used as a natural source of free radical scavenging compounds. Since the majority of these plants are used by the bees to collect honey nectar, bioactive components can be also found in bee products of plant-origin. For this reason, these products are recognized as a rich and natural source of bioactive compounds with potential antioxidant activity, similar to some fruits and vegetables [11].

Several methods for antioxidant activity evaluation have been carried out in order to assess potential biological effects of honey, propolis or pollen extracts. Often, as a first approach, the antioxidant potential of extracts is directly evaluated using chemical models, which in the case of hive products (as in general) include 2,2'-azinobis(3-ethylbenzothiazoline-6-sulphonic acid) (ABTS) assay, the 2,2-diphenyl-1-picrylhydrazyl (DPPH) assay, ferric reducing antioxidant power (FRAP) assay, β-carotene bleaching assay, inhibition of lipid peroxidation using thiobarbituric acid reactive substances (TBARS) [12], allowing to cover the different mechanisms of action of antioxidants, either by transfer of hydrogen atoms or electron transfer. However, as many of these methods are based on radicals not found in biological systems (that hinders their transposition to *in vivo* systems), many studies also focused on absorption and on the impacts at biological level, in particular oxidative stress through biological markers, by using cell lines and *in vivo* models. In these tests, the antioxidant activity of extracts is measured through assessing the activity of antioxidant enzymes such as GPx, SOD or CAT at the cellular level, often accompanied by monitoring lipid oxidation (*e.g.* malonaldehyde, isoprostanes, *etc*), protein oxidation (protein carbonyls) and DNA damages (8-oxo-2-deoxyguanosine) [13].

Nevertheless, the lack of a validated assay to measure the antioxidant capacity difficults the comparison of antioxidant activity data when bee products of plant-origin are analyzed, and even when investigators use the same method, different modifications to the technique are often introduced. Thus the results among samples from different origins, even if they are of the same botanical species, are hard to compare [14, 15]. Different reviews have been published discussing the chemistry of antioxidant assays or the advantages/disadvantages of each method [16 - 18]. Dezmirean *et al.* mentioned that "choosing the correct method depends on the food matrices due to the nature of biological antioxidants present in the sample to be analyzed (enzymatic, nonenzymatic, biological or just dietary)" [19].

2. ANTIOXIDANT PROPERTIES OF HONEY

Honey, as a source of antioxidants, has been proven to be effective against deteriorative oxidation reactions in food, such as enzymatic browning of fruit and vegetables [20, 21], lipid oxidation in meat [22, 23], and to inhibit the growth of foodborne pathogens and food spoilage organisms [24]. Besides, some reports have shown the health-promoting activity of honey, especially for reducing heart diseases [25] or inflammatory processes [26], and even for improve of the immune system [27] and for the treatment of gastric ulcers and gastritis [28].

This chapter summarizes existing information regarding the antioxidant activity of honeys worldwide. A brief description of the compounds with antioxidant activity found in honey is also performed. Due to the lack of space and the vast amount of information published around the world regarding the composition of honey, it is impossible to report all available articles and hence, this chapter focuses on the research performed mainly in last fifteen years (2001–2015).

2.1. Compounds with Antioxidant Activity in Honey

The antioxidants present in honey include both enzymatic (catalase, glucose oxidase and peroxidase) [29] and non-enzymatic (ascorbic acid, α-tocopherol, carotenoids, amino acids, proteins, organic acids, Maillard reaction products and polyphenolic compounds) substances [14, 30 - 35]. Damintoti *et al.* [36] suggested that "the organic acids present in honey, such as gluconic, malic, and citric acids, contribute to antioxidant activity through metal chelation and increase the effect of flavonoids by synergy enzymes also present in honey", so that one can protect other against oxidative destruction. Viuda Martos *et al.* [37] mention that "glucose oxidase and catalase also show antioxidant activity through their ability to eliminate oxygen from foods".

In more detail, polyphenols are plant secondary metabolites highly recognized as food antioxidants [38]. Polyphenols can range from simple molecules to highly polymerized compounds, being flavonoids the most common subgroup [39]. The main phenolic compounds in honey include ellagic acid, gallic acid, syringic acid, benzoic acid, cinnamic acid, ferulic acids, myricetin, chlorogenic acid, caffeic acid, hesperetin, coumaric acid, isorhamnetin, chrysin, quercetin, galangin, luteolin and kaempferol [40 - 43]. Some of these bioactive compounds, in particular galangin, kaempferol, quercetin, isorhamnetin and luteolin are found in most honey samples, meanwhile hesperetin and naringenin are found in few varieties [43, 44].

The chemical properties of phenolic acids or flavonoids, in terms of their availability as hydrogen donating radical scavengers, predict their antioxidant properties according to their chemical structure [45]. More recently, other studies confirmed that the structure of flavonoids and phenolic acids had an antioxidant power relationship due to the rule of (-OH) groups attached to the C2=C3 double bond and adjacent to the 4-carbonyl in C ring, and also demostrated the importance of the B-ring hydroxylation on the antioxidant activity of flavonoids [46, 47].

The quantity of phenolic compounds varies according to the floral and geographical origin of honey. In addition, processing, handling and storage of honey may influence its composition [14, 48, 49]. In general, higher antioxidant capacity is found for darker honey samples, as well as in honey with higher content of water [14, 30, 35, 50]. Honey color depends on the potential alkalinity and ash content, as well as on the antioxidant active pigments, such as carotenoids and flavonoids [51].

On the other hand, recent results published by Alvarez-Suarez *et al.* [18] indicated "a strong correlation between melanoidins and the total radical scavenging activity of unheated honeys. Melanoidins are formed in the final stage of the Maillard reaction, where their formation is initiated by the reactions between amino groups of an amino acid/proteins and the carbonyl group of reducing sugars". Moreover, products of the Maillard reaction have shown several biological activities such as antioxidant [52, 53], antimicrobial [54, 55], antihypertensive [56] and prebiotic properties [57, 58].

The relationship between antioxidant activity and Maillard reaction products has been investigated in model systems [54, 59, 60]. It is reported also that non-enzymatic browning is observed during prolong storage of honey, specifically for light-coloured honeys, which may indicate a spontaneous polymerization of Maillard reaction products to form melanoidin polymers [18]. Heating is also been referred to increase the brown pigment formation and this is associated to an increase in the antioxidant activity of honey [34, 48]. Brudzynski and Miotto [61] have found correlations between the Maillard reaction-like products, phenolic content, honey colour and antioxidant activity of unheated raw honeys.

In fact, some compounds with antioxidant capacity can arise from Maillard reactions or the pyrolysis of amino acids when honey is heated (*e.g.* pyrroles) [62]. 1*H*-Pyrrole has been found in *Robinia pseudoacacia* L., *Castanea sativa* L. and *Salvia officinalis* L. honeys [62, 63], while 1*H*-pyrrole-2-carboxylic acid has been found in *Paliurus spinachristi* honeys [64] and 1*H*-pyrrole-3,4-diacetic acid has been found in pine honey (*Pinus brutia* Ten) [42].

In addition, Kowalski [65] found that conventional heating at 90°C contributes to increase the antioxidant properties of buckwheat and lime honey even when these were subjected to a microwave field at a constant power level of 1.26 W/g. Moreover, Brudzynski [33] found that the heat-treatment accelerated the formation of melanoidins and increased the antioxidant activity in light-coloured clover honey and medium-coloured manuka honey. On the contrary, heat-treatment caused a decrease of melanoidins and this was followed by a reduction in the antioxidant activity in dark buckwheat honey.

Moreover, Turkmen *et al.* [66] observed that despite the negative impact on acceptance by consumers, the heat treatment of honey resulted in a positive effect on human health, as a consequence of these Maillard-originated products. Concerning this issue, it is important to highlight the work of Wang *et al.* [49] whom mention that the global antioxidant capacity varies depending on the type of honey and on processing. In this case, buckwheat honey was more affected by processing than clover honey in terms of reduction in antioxidant capacity. These authors also showed that the impact of heat processing on the phenolic profile was complex and variable; for instance, compounds like quercetin and galangin only increased significantly in clover honey.

2.2. Assessment of Antioxidant activities of Honey in Chemical Models

It has been recommended that at least two test systems must be employed for evaluating the antioxidant activities of food and food extracts, in order to establish certain validity [18]. For this reason, and considering that there is no universal method, numerous tests have been developed for measuring the antioxidant activity of honeys and honey-based products. Various tests are in use, each one based on different principles and experimental conditions; the most common are the FRAP assay Ferric Reducing Antioxidant Power (FRAP) assay, the 1,1-diphenyl-2-picrylhydrazyl (DPPH) method, the Oxygen Radical Absorbance Capacity (ORAC), the evaluation of superoxide radical-scavenging activity and the measurement of Trolox Equivalent Antioxidant Activity (TEAC) [14]. In addition, Buratti *et al.* [45] developed the technique of amperometric flow injection analysis (FIA) in order to determine antioxidant activity of honey bee products, by using an electrochemical procedure based on oxidation current obtained at the fixed potential [67]. Also, Gorjanovic *et al.* [68] have applied DC polarography for determination of hydrogen peroxide scavenging (HPS) activity of honey. Beretta *et al.* [69] have mentioned that "only through a combination of antioxidant testings, comparative analyses and statistical evaluation, the antioxidant behavior of honey can be understood".

The aforementioned methods are based on assessing how an oxidizing agent, which induces oxidative stress and damage to an oxidizable substrate, can be inhibited or reduced in the presence of an antioxidant, in a proportional manner to its antioxidant activity [70, 71]. *In vitro* methodologies provide useful information for comparing the antioxidant activity of different samples, however they are limited from a biological point of view because they do not reproduce the physiological situation. Even, due to the lack of uniformity in procedures and the absence of consensus about the methodologies which could be included in regulation of honeys, results of these parameters are usually reported in a wide variety of units; for instance, DPPH is usually reported as EC_{50}, IC_{50} or $SC_{50.}$ These indices tends to express the same result, defined as the effective concentration of honey that is required to scavenge 50% of radical. In some other cases, ascorbic acid equivalent per 100 g or μmol Trolox equivalent per g (or mg/mL) are also used. These discrepancies make almost impossible to compare results between honeys from different regions.

2.2.1. Trolox Equivalent Antioxidant Capacity (TEAC)

TEAC has been described by Van den Berg [72] and by Re *et al.* [73] and it has been used widely in honeys [68, 74 - 77]. TEAC assesses the ability of a compound to scavenge the stable ABTS radical cation ($ABTS^{\bullet+}$), which is produced by reacting ABTS with potassium persulphate ($K_2S_2O_8$) [18]. One drawback of this method is the fact that $ABTS^{\bullet+}$ is not found nor is similar to radicals found in biological systems and thus represents a "non-physiological" radical source. In spite of this, the TEAC assay is considered an easy and accurate method for determining the radical scavenging ability of honey samples, as by the hydrogen-donation reaction mechanism. TEAC can be evaluated over a wide pH range,and it also could be employed to a wide range of foodstuff, since the the $ABTS^{\bullet+}$ radical is soluble in water and organic solvents; however, it must be taken into account that the results provided by this assay are dependent on time of analysis [18]. In Table **1** some TEAC values of worldwide selected honeys are presented, according to the units employed to express the results.

Table 1. TEAC antioxidant activity reports for honeys worldwide.

TEAC	Honey	Reference
mg AAE/kg		
489.44 ±47.49	Floral, Czech	[76]
596.87 ±14.75	Lime, Czech	
658.73 ±4.77	Raspberry, Czech	
543.97 ±32.56	Rape, Czech	
814.77 ±64.12	Mixture, Czech	
982.93 ±32.18	Honeydew, Czech	
SC_{50} (mg/mL)		
43.25 ±0.68	Manuka, New Zealand	[75]
44.37 ±0.79	Acacia, Germany	
202.30 ±1.03	Wild carrot, Algeria	
µmol TE/100 g		
96.9 ±14.9	Linen vine, Cuba	[74]
53.6 ±7.8	Morning glory, Cuba	
72.4 ±11.9	Singing bean, Cuba	
39.5 ±9.3	Black mangrove, Cuba	
27.0 ±6.0	Christmas vine, Cuba	

(Table 1) contd.....

TEAC	Honey	Reference
582.0 ±4.0	Forest, Serbia	[68]
506.0 ±2.0	Pine, Serbia	
370.0 ±4.0	Urtica, Serbia	
352.0 ±8.0	Meadow, Serbia	
204.0 ±6.0	Tiglio, Serbia	
495.9 ±122.7	Eucalyptus, Brazil	[77]
575.5 ±80.9	Cambera, Brazil	

AAE: Ascorbic acid equivalents
TE: Trolox equivalent

2.2.2. 2,2-Diphenyl-1-picrylhydrazyl (DPPH) Assay

The DPPH analysis is a quick and simple test widely recognized for antioxidant screening. However, Álvarez-Suárez *et al.* mention that complications are found when the tested compounds have spectra that overlap DPPH at 515 nm, *e.g.* carotenoids [18].

In the DPPH assay, antioxidants reduce the free radical 2,2-diphenyl-1-picrylhydrazyl. As a consequence, the purple color of $DPPH^{\bullet}$ fades and the change of absorbance can be followed spectrophotometrically at 515 nm [18].

The results for evaluating the scavenging activity of honey for $DPPH^{\bullet}$ are reported as IC_{50} [14, 45, 68, 69, 75, 77 - 80]. Quercetin and ascorbic acid can be used as positive controls. Other authors express the scavenging capacity of honey samples as a % equivalent of ascorbic acid [76, 81, 82]. Table **2** resumes DPPH gathered data for worldwide selected honeys, according to the units employed to report results.

Despite the disadvantages of DPPH (for instance the fact that many antioxidants could be almost or completely inert to DPPH), the radical $DPPH^{\bullet}$ is stable, commercially available and does not have to be generated before the assay, like $ABTS^{\bullet +}$. Due to this, it could be explained why this method is widely employed for measuring the antioxidant capacity in honey [18].

Table 2. DPPH antioxidant activity reports for honeys worldwide.

DPPH	Honey	Reference
IC_{50}, mg/mL		
1.63 ±0.17	Strawberry tree, N.R.	[69]
4.00 ±0.44	Buckwheat, N.R.	
8.48 ±0.24	Honeydew, N.R.	
5.13 ±0.13	Chestnut, N.R.	
5.32 ±0.03	Multifloral, N.R.	
5.81 ±0.04	Chicory, N.R.	
16.90 ±0.11	Sulla, N.R.	
45.45 ±0.04	Acacia, N.R.	
25.00 ±0.01	Clover, N.R.	
5.80 ±0.12	Tualang, Malaysia	[78]
6.68 ±0.28	Gelam, Malaysia	
10.32 ±0.17	Indian forest, Malaysia	
10.86 ±0.38	Pineapple, Malaysia	
12.20 ±3.50	Multifloral, Croatia	[80]
53.80 ±8.50	Acacia, Slovenia	[14]
28.80 ±5.40	Lime, Slovenia	
10.00 ±1.80	Chestnut, Slovenia	
10.70 ±2.20	Multifloral, Slovenia	
13.46 ±0.07	Manuka, New Zealand	[75]
13.62 ±0.05	Acacia, Germany	
53.31 ±0.08	Wild carrot, Algeria	
15.25 ±4.27	Eucalyptus, Brazil	[77]
23.71 ±6.01	Cambara, Brazil	
mg AAE/kg		
9.65 ±0.57	Tualang, Malaysia	[82]
50.17 ±5.54	Gelam, Malaysia	
29.98 ±6.04	Acacia, Malaysia	

(Table 2) contd.....

DPPH	Honey	Reference
14.15 ±3.03	Floral, Czech	[76]
20.61 ±0.63	Raspberry, Czech	
16.68 ±1.80	Rape, Czech	
28.47 ±4.28	Mixture, Czech	
40.71 ±2.26	Honeydew, Czech	
EC_{50}, µmol TE/g		
1.21 ±0.02	Forest, Serbia	[68]
1.18 ±0.03	Pine, Serbia	
0.49 ±0.01	Urtica, Serbia	
0.22 ±0.01	Meadow, Serbia	
0.25 ±0.01	Tiglio, Serbia	
30.00 ±0.00	Black locust, Poland	[79]
20.00 ±10.00	Goldenrod, Poland	
40.00 ±10.00	Rapeseed, Poland	
40.00 ±10.00	Lime, Poland	
60.00 ±10.00	Heather, Poland	
120.00 ±20.00	Buckwheat, Poland	

N.R. Origin not reported. AAE: Ascorbic Acid Equivalents. TE: Trolox Equivalents

2.2.3. Ferric Reducing Ability of Plasma (FRAP)

The principle of FRAP method is based on the reduction of a ferric 2,4,6-tripyridyl-*s*-triazine complex (Fe^{3+}-TPTZ) to its ferrous colored form (Fe^{2+}-TPTZ) in the presence of antioxidants [18]. The FRAP assay was described by Benzie and Strain [83]. Table **3** shows some reports of antioxidant capacity of honeys, measured by FRAP technique. It is not established a standard unit for expressing the FRAP result, some authors report FRAP as the concentration of antioxidants having a ferric reducing ability equivalent to that of 1 mM $FeSO_4$ used as the standard solution [18]. Also, some other authors use Trolox and $Fe(NH_4)_2(SO_4)_2.6H_2O$ (ammonium ferrous sulfate) for expressing the results as micromoles of Trolox Equivalent (TE) or ammonium ferrous sulfate per 100 g of honey [18].

Table 3. FRAP antioxidant activity reports for honeys worldwide.

FRAP	Honey	Reference
ABS_{700}		
1.21 ±0.01	Manuka	[75]
1.37 ±0.06	Acacia	
0.64 ±0.05	Wild carrot	
µmol Fe(II)/ 100 g		
196.7 ±20.0	Linen vine, Cuba	[84]
117.1 ±17.8	Morning glory, Cuba	
153.9 ±18.7	Singing bean, Cuba	
72.6 ±1.9	Black mangrove, Cuba	
54.6 ±14.1	Christmas vine, Cuba	
1501.4 ±60.2	Strawberry tree, N.R.	[69]
800.7 ±23.8	Buckwheat, N.R.	
772.0 ±21.5	Honeydew, N.R.	
388.6 ±8.2	Chestnut, N.R.	
361.9 ±10.8	Multifloral, N.R.	
209.5 ±2.8	Chicory, N.R.	
155.2 ±6.6	Sulla, N.R.	
79.5 ±3.7	Acacia, N.R.	
72.8 ±3.0	Clover, N.R.	
113.5 ±2.9	Jerusalem thorn, Croatia	[85]
113.8 ±9.7	Sunflower, Croatia	
121.3 ±4.7	Sage, Croatia	
118.6 ±4.5	Velebit winter, Croatia	
99.7 ±4.0	Winter savory, Croatia	
23.0 ±2.8	Amopha, Croatia	
84.6 ±2.6	Chestnut, Croatia	
73.8 ±6.8	Linden, Croatia	
12.1 ±2.0	Acasia, Croatia	
52.2 ±6.1	Oilseed rape, Croatia	
92.9 ±1.7	Goldenrod, Croatia	

(Table 3) contd.....

FRAP	Honey	Reference
121.9 ±3.9	Tualang, Malaysia	[78]
115.6 ±3.9	Gelam, Malaysia	
73.4 ±4.0	Indian Forest, Malaysia	
47.9 ±1.8	Pineapple, Malaysia	
32.2 ±1.0	Tualang, Malaysia	[86]
272.3 ±57.1	Eucalyptus, Brazil	[77]
241.5 ±39.3	Cambara, Brazil	
295.40 ±77.21	Multifloral, Croatia	[80]
71.0 ±10.2	Acacia, Slovenia	[14]
118.8 ±20.3	Lime, Slovenia	
360.1 ±66.5	Chestnut, Slovenia	
224.8 ±24.7	Multifloral, Slovenia	
60.0 ±10.0	Black locust, Poland	[79]
100.0 ±10.0	Goldenrod, Poland	
130.0 ±30.0	Rapeseed, Poland	
140.0 ±40.0	Lime, Poland	
173.0 ±157.3	Multifloral, Brazil	[87]
305.9 ±158.8	Orange blossom, Brazil	
µmol TE/100 g		
52.39 ±5.19	Tualang, Malaysia	[82]
82.53 ±5.03	Gelam, Malaysia	
82.39 ±5.93	Acacia, Malaysia	
498.00 ±3.00	Forest, Serbia	[68]
405.00 ±3.00	Pine, Serbia	
203.00 ±3.00	Urtica, Serbia	
67.00 ±2.00	Meadow, Serbia	
61.00 ±2.00	Tiglio, Serbia	
96.90 ±14.90	Linen vine, Cuba	[84]
53.60 ±7.80	Morning glory, Cuba	
72.40 ±11.90	Singing bean, Cuba	
39.50 ±9.30	Black mangrove, Cuba	
27.00 ±6.00	Christmas vine, Cuba	
mg AAE/kg		

(Table 3) contd.....

FRAP	Honey	Reference
295.4 ±48.7	Floral, Czech	[76]
415.6 ±31.2	Lime, Czech	
443.4 ±6.0	Raspberry, Czech	
370.3 ±27.1	Rape, Czech	
565.5 ±63.5	Mixture, Czech	
776.1 ±68.2	Honeydew, Czech	

ABS_{700}: The absorbance of the obtained solution measured at 700 nm. AAE: Ascorbic Acid Equivalents. TE: Trolox Equivalents. N.R. Origin not reported

Some disadvantages have been reported for FRAP technique. False results could be obtained if it is employed any substance with redox potential lower than that of the redox pair Fe(III)/Fe(II), so Fe(III) may be reduced to Fe(II). Also, the time of reaction is not the same for all antioxidants, and eventually some antioxidants could not be measured since they do not reduce Fe(III) within the observation time (typically 4 min), *e.g.* many polyphenols which may require more than 30 min to react [18]. In addition, it is important to mention that due to the chemical characteristics of this technique, carotenoids cannot be determined [88]. Finally, some interferences can be found with some substances that absorb at the same wavelength employed for FRAP method, which induce an overestimation of the result [18].

2.2.4. Oxygen Radical Absorbance Capacity (ORAC)

The ORAC assay has been used widely for assessing the antioxidant capacity of a variety of foodstuff, among them honey [18]. ORAC assay in honey can be carried out by using a modified procedure of Cao *et al.* [89] and Ou *et al.* [90, 91]. In ORAC assay, the free radicals are produced by 2,2'-azobis(--amidinopropane)dihydrochloride (AAPH) and the fluorescent indicator protein phycoerthrin (β-PE) is subsequently oxidized. This loss of fluorescence can be inhibited by antioxidants and is commonly monitored using a microplate fluorescence reader [18]. Trolox is usually used as the only standard, being a great advantage when comparing the activity of distinct honeys samples. Some reports for ORAC antioxidant activity in honey are shown in Table **4**.

Table 4. ORAC antioxidant activity reports for honeys worldwide.

Honey	ORAC, µmol TE/g	Reference
Strawberry tree, N.R.	21.07 ±0.34	[69]
Buckwheat, N.R.	11.60 ±0.03	
Honeydew, N.R.	6.30 ±0.22	
Chestnut, N.R.	8.90 ±0.45	
Multifloral, N.R.	8.22 ±0.42	
Chicory, N.R.	6.72 ±0.33	
Sulla, N.R.	5.66 ±0.13	
Acacia, N.R.	2.12 ±0.01	
Clover, N.R.	2.15 ±0.02	
Forest, Serbia	12.90 ±0.10	[68]
Pine, Serbia	11.60 ±0.20	
Urtica, Serbia	10.20 ±0.30	
Meadow, Serbia	10.00 ±0.30	
Tiglio, Serbia	9.50 ±0.10	
Buckwheat - Illinois, U.S.A.	16.95 ±0.76	[92]
Buckwheat - New York, U.S.A.	9.75 ±0.48	
Soy, U.S.A.	9.49 ±0.29	
Hawaiian Christmas Berry, U.S.A.	8.87 ±0.33	
Clover, U.S.A.	6.53 ±0.70	
Tupelo, U.S.A.	6.48 ±0.37	
Fireweed, U.S.A.	3.09 ±0.27	
Acacia, U.S.A.	3.00 ±0.16	

TE: Trolox equivalent. N.R. Origin not reported

The advantages of this method is the use of a protein (phycoerthrin) as substrate, because it prevents the own substrate to generate free radicals due to its oxidation and its capacity to react with hydrophilic and hydrophobic samples. As disadvantages, one can mentioned that the protein PE photobleached under plate-reader interacts with polyphenols due to non-specific protein binding and therefore losses fluorescence even without the added radical generator [18].

2.3. Assessment of Antioxidant Activities of Honey in Cellular Models

Cellular models are becoming important models to assess antioxidant activity in honeys. Different cell lines of human and/or animal origin have been employed to evaluate the effect of honey on the treatment of several oxidative-stress related diseases, such as cardiovascular diseases, hypertension, cancer, diabetes and microbial infections [93]. Table **5** resumes relevant studies focusing the antioxidant activity of honey in cellular models.

Table 5. Assessment of antioxidant properties of selected honeys, as demonstrated in cellular models.

Geographic Location	Model	Treatment Conditions	Effects	Ref.
Malaysia (*Melaleuca* sp.)	3Y1 rat fibroblast cell lines	Antioxidant activity assay 5 mL of phosphate buffered saline (PBS), followed by the addition of 1 mL of 25% trypsin	The production of hydrogen peroxide by the honey-glucose oxidase system stimulates cell proliferation. The presence of phenolics renders protection to the cells against the toxic effect of hydrogen peroxide.	[95]
Malaysia (Tualang)	HCEP cells	Oxidative stress assay 50 μM H2O2 for 24 h	Tualang honey contains active phytocompounds that enhance HCEP cell migration and resistance to oxidative stress.	[96]
Italy	Human endothelial cells (EA.hy926)	Study in removing and reducing ROS Honey 1% in PBS	Honey components spare or regenerate endothelial GSH, which can prevent the atherogenic action of oxidized LDL	[97]

HCEP : human corneal epithelial progenitor. ROS: Reactive Oxygen Species

The assessment of honey and honey extracts of distinct botanical and geographical origins in decreasing the content of cellular reactive oxygen species (ROS) have been observed for distinct cellular models, including corneal, fibroblast and endothelial (see Table **5**). In addition, bioactive compounds of honey have been extracted and employed, *e.g.* a flavonoid extract has been used to reduce inflammatory processes [94].

2.4. Assessment of Antioxidant Activities of Honey in *in vivo* Models

Studies focusing on the antioxidant ability of honey in *in vivo* models are still scarce, but results obtained after consumption of honey alone or with other antioxidant beverages gave promising results regarding the antioxidant capacity of human serum [98, 99]. Schramm *et al.* [98], evaluated maize syrup or buckwheat honeys with a different antioxidant capacity in a dose of 1.5 g/kg body weight. Results showed how honey caused an increase of both the antioxidant and the reducing serum capacity in comparison to the sugar control. In addition, Al-Waili *et al.* [100] elaborated an human diet supplemented with a daily honey serving of 1.2 g/kg body weight and their results indicated that honey could improve distinct body antioxidant agents, namely those of blood vitamin C concentration (by 47%), β-carotene concentration (by 3%), uric acid (12%) and of glutathione reductase (by 7%).

Further reported biological tests *in vivo* indicated that buckwheat honey intake increases serum antioxidant potential [101]. In fact, honeys from different floral sources (Buckwheat, Clover, and Sage) have been shown to inhibit formation of mutagens during frying [102, 103]. Some other studies are reported in Table **6**, where honey has been proven to have favorable effects against oxidative damage and degenerative diseases.

Table 6. Assessment of antioxidant properties of selected honeys, as demonstrated in *in vivo* models.

Geographic Location	Model	Treatment Conditions	Effects	Ref.
Bangladesh (Multifloral)	Adult male Wistar rats (180-210 g)	Group 1, normal saline for 4 weeks followed by a single dose of 0.5% gum tragacanth. Group 2, 5 g/kg honey for 4 weeks followed by a single dose of 0.5% gum tragacanth. Group 3, 5 g/kg honey for 4 weeks followed by a single dose of APAP. Group 4, normal saline for 4 weeks followed by a single dose of APAP.	Favorable hepatonephroprotective action of Sundarban honey against APAP-induced oxidative damage.	[104]
Malaysia (Gelam honey)	Sprague-Dawley rats	Control (n = 6) force-fed with water 2.5 mL/kg body weight Gelam honey (n = 6) group supplemented with 2.5mg/kg body weight of gelam for 8 months	Gelam honey reduced the oxidative damage through the modulation of antioxidant enzyme activity which was more prominent in young group compared to aged group.	[105]

(Table 6) contd.....

Geographic Location	Model	Treatment Conditions	Effects	Ref.
Turkey *(Rhododendron)*	Sprague-Dawley female rats (6-8 months old; 250-300g)	Group 1 (control): 1 ml of 0.9% NaCl solution Group 2 (GTX): 0.015 mg/kg/bw of Gray anotoxin-III Group 3 (RH1): 0.015 mg/kg/bw GTX + 0.1 g/kg/bw of RH Group 4 (RH2): 0.015 mg/kg/bw GTX +0.5 g/kg/bw of RH Group 5 (RH3): 0.015 mg/kg/bw GTX +2.5 g/kg/bw of RH	Both GTX and high dose Rhododendron honey treatments showed an oxidant effect on plasma and numerous tissues investigated. Lower dosages (RH3 and RH4) showed amelioration effects.	[106]
N.R.	Male Wistar rats diabetic (8-10 weeks old; 200-250 g)	Group 1: no treatment Group 2: treated with insulin Group 3: treated with natural honey Group 4: treated simultaneously with insulin and natural honey	Use of honey as an antioxidant can effectively inhibit apoptosis or neuronal cell death in the hippocampus of STZ-induced diabetic rats.	[107]
Iran	Male Wistar diabetic rats	Group 1: Diabetes-induced. Group 2. Honey-received diabetic. Group 3. Metformin-administrated diabetic. Group 4. Honey and metformin co-treated diabetic group. Group 5. Non-diabetic-hon-y-administrated group.	Honey could inhibit the diabetes-induced damages in testicular tissue. Honey and metformin co-administration showed better results *versus* other forms of application	[108]

APAP: *N*-acetyl-*p*-aminophenol; GTX: Gray anotoxin-III; RH: Rhododendron honey; N.R.: Not reported.

2.5. Correlation between Antioxidant Activity and Phenolic Compounds

Several studies have shown that antioxidant activity of honey is strongly correlated with the content of total phenolics and flavonoids. Some reports of correlations are shown in Table **7**. Besides this, some reports indicate a strong correlation between antioxidant activity and the colour of honey, as it was mentioned before. Many researchers found that dark honeys have a higher total phenolic content and consequently a higher antioxidant capacity [14, 69].

Table 7. Reports of correlations between antioxidant activity assays, total phenolics and flavonoid content.

Correlation	Type of honey	r	Reference
DPPH and Total Phenolics	Serbian honeys: (Dead nettle, Linden, Acacia)	0.988	[14, 41, 68, 77 - 79, 109, 110]
	Malaysian honeys: (Tualang, Gelam)	0.975	
	Polish honeys: (Rapeseed, Heather, Lime, Black locust, Buckwheat, Goldenrod)	0.943	
	Slovenian honeys: (Acacia, Lime, Chestnut, Fir, Spruce, Multifloral, Forest)	0.932	
	Malaysian honeys: (Gelam, Nenas)	0.855	
	Malaysian honey: (Acacia)	0.785	
	Algerian honeys	0.615	
	Brazilian honeys: (Morrao de Candeia, Eucalyptus, Cambara)	-0.428	
FRAP and Total Phenolics	Malaysian honeys: (Tualang, Gelam)	0.991	[14, 41, 68, 76 - 79, 109 - 111]
	Malaysian honeys: (Gelam, Nenas)	0.990	
	Serbian honeys: (Dead nettle, Linden, Acacia)	0.978	
	Polish honeys: (Rapeseed, Heather, Lime, Black locust, Buckwheat, Goldenrod)	0.975	
	Slovenian honeys (Acacia, Lime, Chestnut, Fir, Spruce, Multifloral, Forest)	0.966	
	Italian honeys: (Millefiori, Acacia)	0.938	
	Czech honeys: (Floral, Lime, Raspberry, Rape, Mixture, Honeydew)	0.852	
	Malaysian honey: (Acacia)	0.780	
	Algerian honeys	0.668	
	Brazilian honeys: (Morrao de Candeia, Eucalyptus, Cambara)	0.577	
ORAC and Total Phenolics	Canadian honeys: (Buckwheat, Clover)	0.95	[34, 68]
	Serbian honeys: (Dead nettle, Linden, Acacia)	0.935	
TEAC and Total Phenolics	Serbian honeys: (Dead nettle, Linden, Acacia)	0.977	[68, 77]
	Brazilian honeys: (Morrao de Candeia, Eucalyptus, Cambara)	0.393	
DPPH and flavonoids	Malaysian honeys: (Tualang, Gelam, Acacia)	0.928	[77, 82, 109]
	Algerian honeys	0.888	
	Brazilian honeys: (Morrao de Candeia, Eucalyptus, Cambara)	-0.771	

(Table 7) contd.....

Correlation	Type of honey	r	Reference
FRAP and flavonoids	Malaysian honeys: (Tualang, Gelam, Acacia)	0.991	[77, 82, 109, 111]
	Italian honeys: (Millefiori, Acacia)	0.983	
	Algerian honeys	0.893	
	Brazilian honeys: (Morrao de Candeia, Eucalyptus, Cambara)	0.817	
TEAC and flavonoids	Brazilian honeys: (Morrao de Candeia, Eucalyptus, Cambara)	0.752	[77]
DPPH and FRAP	Serbian honeys: (Dead nettle, Linden, Acacia)	0.987	[68, 69, 76, 77, 112, 113]
	European and African honeys: (Strawberry tree, Buckwheat, Chestnut, Sulla, Clover, Dandelion, Chicory, Acacia, Multi-flora, Honeydew, Tropical)	0.889	
	Malaysian honey: (Acacia)	0.850	
	Commercial Indian honeys	0.840	
	Czech honeys: (Floral, Lime, Raspberry, Rape, Mixture, Honeydew)	0.821	
	Brazilian honeys: (Morrao de Candeia, Eucalyptus, Cambara)	-0.841	
DPPH and ORAC	Serbian honeys: (Dead nettle, Linden, Acacia)	0.888	[68, 69]
	European and African honeys: (Strawberry tree, Buckwheat, Chestnut, Sulla, Clover, Dandelion, Chicory, Acacia, Multi-flora, Honeydew, Tropical)	0.861	
DPPH and TEAC	Serbian honeys: (Dead nettle, Linden, Acacia)	0.935	[15, 50, 6, 76, 77]
	Czech honeys: (Floral, Lime, Raspberry, Rape, Mixture, Honeydew)	0.865	
	Polish honeys	0.740	
	Polish honeys: (Acacia, Goldenrods, Rape, Lime, Nectar, Honeydew, Multifloral, Buckwheat, Honeydew, Phacelia, Heather)	0.715	
	Brazilian honeys: (Morrao de Candeia, Eucalyptus, Cambara)	-0.668	
FRAP and ORAC	Serbian honeys: (Dead nettle, Linden, Acacia)	0.859	[68, 69]
	European and African honeys: (Strawberry tree, Buckwheat, Chestnut, Sulla, Clover, Dandelion, Chicory, Acacia, Multi-flora, Honeydew, Tropical)	0.716	
FRAP and TEAC	Monofloral Cuban	0.96	[68, 76, 77, 84]
	Czech honeys: (Floral, Lime, Raspberry, Rape, Mixture, Honeydew)	0.937	
	Serbian honeys: (Dead nettle, Linden, Acacia)	0.925	
	Brazilian honeys: (Morrao de Candeia, Eucalyptus, Cambara)	0.761	
ORAC and TEAC	Serbian honeys: (Dead nettle, Linden, Acacia)	0.961	[68]

CONCLUSIONS

The determination of antioxidant activity has become a tool for generating added-value in honey extremely useful. Purchasing trends of consumers are directed towards the consumption of functional foods, and honey stands out for its content of bioactive compounds with antioxidant capacity. The scientific reports showed a great amount of compounds present in honey with antioxidant capacity, particularly enzymes, carotenoids, phenolic compounds, flavonoids, vitamins, minerals and even compounds derived from the Maillard reaction. Different *in vitro* methods have been employed to measure antioxidant activity in honey, the most used are DPPH, FRAP, TEAC and ORAC, which have shown a positive correlation with the content of phenolic compounds and flavonoids, for which are attributed largely the scavenging activity of honey. It is reported that the greatest influence on honey's antioxidant activity is due to the botanical origin, and in contrast, processing, handling and storage have a reduced effect in antioxidant activity. On the other hand, due to the lack of a standard analytical technique and the nature of both honey and reagents used in the measurements of antioxidant capacity for each technique, it is common that the results among one method and other are not at all similar. In any case, the values shown are for reference only, and necessarily they must be confirmed by *in vivo* techniques. For antioxidant activity is recommended, since the discrepancies previously mentioned, to employ for a same sample at least two techniques in order to have a reliable value.

CONFLICT OF INTEREST

The author confirms that author has no conflict of interest to declare for this publication.

ACKNOWLEDGEMENTS

The authors wish to thank the Administrative Department of Science, Technology and Innovation, COLCIENCIAS, the Ministry of Agriculture and Rural Development of Colombia, the Research Direction (DIB) of Universidad Nacional de Colombia, and to the following Colombian beekeepers associations: Asociación de Apicultores de Boyacá (ASOAPIBOY), the Asociación de Apicultores de la Región del Sumapaz (ASOAPIS), the Asociación de Apicultores

Conservacionistas de la Sierra Nevada de Santa Marta (APISIERRA), the Asociación Apícola Comunera (ASOAPICOM), and Apicultura Los Cerezos.

REFERENCES

[1] Albright, A. Biological and social exposures in youth set the stage for premature chronic diseases. *J. Am. Diet. Assoc.,* **2008**, *108*(11), 1843-1845.
[http://dx.doi.org/10.1016/j.jada.2008.09.017] [PMID: 18954573]

[2] Shibata, N.; Kobayashi, M. The role for oxidative stress in neurodegenerative diseases. *Brain Nerve,* **2008**, *60*(2), 157-170.
[PMID: 18306664]

[3] Kadenbach, B.; Ramzan, R.; Vogt, S. Degenerative diseases, oxidative stress and cytochrome c oxidase function. *Trends Mol. Med.,* **2009**, *15*(4), 139-147.
[http://dx.doi.org/10.1016/j.molmed.2009.02.004] [PMID: 19303362]

[4] Sies, H.; Jones, D. *Encyclopedia of stress*; Fink, , Ed.; Elsevier: Amsterdam, **2007**, 3, pp. 45-48.

[5] Halliwell, B.; Gutteridge, J. *Free Radicals in Biology and Medicine*; Clarendon Press: Oxford, **2007**.

[6] Halliwell, B. Free radicals and antioxidants - quo vadis? *Trends Pharmacol. Sci.,* **2011**, *32*(3), 125-130.
[http://dx.doi.org/10.1016/j.tips.2010.12.002] [PMID: 21216018]

[7] Erejuwa, O.O.; Sulaiman, S.A.; Ab Wahab, M.S. Honey: a novel antioxidant. *Molecules,* **2012**, *17*(4), 4400-4423.
[http://dx.doi.org/10.3390/molecules17044400] [PMID: 22499188]

[8] Hur, S.J.; Lee, S.Y.; Kim, Y.C.; Choi, I.; Kim, G.B. Effect of fermentation on the antioxidant activity in plant-based foods. *Food Chem.,* **2014**, *160*, 346-356.
[http://dx.doi.org/10.1016/j.foodchem.2014.03.112] [PMID: 24799248]

[9] Lin, H.H.; Charles, A.L.; Hsieh, C.W.; Lee, Y.C.; Ciou, J.Y. Antioxidant effects of 14 Chinese traditional medicinal herbs against human low-density lipoprotein oxidation. *J. Tradit. Complement. Med.,* **2015**, *5*(1), 51-55.
[http://dx.doi.org/10.1016/j.jtcme.2014.10.001] [PMID: 26151009]

[10] Wu, P.; Ma, G.; Li, N.; Deng, Q.; Yin, Y.; Huang, R. Investigation of *in vitro* and *in vivo* antioxidant activities of flavonoids rich extract from the berries of Rhodomyrtus tomentosa(Ait.) Hassk. *Food Chem.,* **2015**, *173*, 194-202.
[http://dx.doi.org/10.1016/j.foodchem.2014.10.023] [PMID: 25466012]

[11] Dezmirean, G.; Al, L.; Dezmirean, D. Honey Like Component of Functional Food. *Anim. Sci. Biotechnol,* **2011**, *44*, 406-411.

[12] Alvarez-Suarez, J.M.; Giampieri, F.; Damiani, E.; Astolfi, P.; Fattorini, D.; Regoli, F.; Quiles, J.L.; Battino, M. Radical-scavenging activity, protective effect against lipid peroxidation and mineral contents of monofloral Cuban honeys. *Plant Foods Hum. Nutr.,* **2012**, *67*(1), 31-38.
[http://dx.doi.org/10.1007/s11130-011-0268-7] [PMID: 22212954]

[13] Bartosz, G. Oxidative stress in plants. *Acta Physiol. Plant.,* **1997**, *19*, 47-64.

[http://dx.doi.org/10.1007/s11738-997-0022-9]

[14] Bertoncelj, J.; Dobersek, U.; Jamnik, M.; Golob, T. Evaluation of the phenolic content, antioxidant activity and colour of Slovenian honey. *Food Chem.,* **2007**, *105*, 822-828. [http://dx.doi.org/10.1016/j.foodchem.2007.01.060]

[15] Wilczyńska, A. Phenolic Content and Antioxidant Activity of Different Types of Polish Honey - A short report. *Pol. J. Food Nutr. Sci,* **2010**, *60*, 309-313.

[16] Huang, D.; Ou, B.; Prior, R.L. The chemistry behind antioxidant capacity assays. *J. Agric. Food Chem.,* **2005**, *53*(6), 1841-1856. [http://dx.doi.org/10.1021/jf030723c] [PMID: 15769103]

[17] Prior, R.L.; Wu, X.; Schaich, K. Standardized methods for the determination of antioxidant capacity and phenolics in foods and dietary supplements. *J. Agric. Food Chem.,* **2005**, *53*(10), 4290-4302. [http://dx.doi.org/10.1021/jf0502698] [PMID: 15884874]

[18] Alvarez-Suarez, J.; Tulipani, S.; Romandini, S.; Vidal, A.; Battino, M. Methodological Aspects about Determination of Phenolic Compounds and *In vitro* Evaluation of Antioxidant Capacity in the Honey: A Review. *Curr. Anal. Chem.,* **2009**, *5*, 293-302. [http://dx.doi.org/10.2174/157341109789077768]

[19] Dezmirean, G.I.; Mărghitaş, L.A.; Bobiş, O.; Dezmirean, D.S.; Bonta, V.; Erler, S. Botanical origin causes changes in nutritional profile and antioxidant activity of fermented products obtained from honey. *J. Agric. Food Chem.,* **2012**, *60*(32), 8028-8035. [http://dx.doi.org/10.1021/jf3022282] [PMID: 22835207]

[20] Nyawali, B.; Chungu, D.; Chisha-Kasumu, E.; Vinya, R.; Chileshe, F.; Ng'andwe, P. Enzymatic browning reduction in white cabbage (Brassica oleracea) using honey: Does honey color matter. *LWT -. Food Sci. Technol. (Campinas.),* **2015**, *61*, 543-549.

[21] de la Rosa, L.; Alvarez-Parrilla, E.; Moyers-Montoya, E.; Villegas-Ochoa, M.; Ayala-Zavala, J.; Hernandez, J.; Ruiz-Cruz, S.; Gonzalez-Aguilar, G. Mechanism for the inhibition of apple juice enzymatic browning by Palo Fierro (desert ironweed) honey extract and other natural compounds. *LWT -. Food Sci. Technol. (Campinas.),* **2011**, *44*, 269-276.

[22] Johnston, J.E.; Sepe, H.A.; Miano, C.L.; Brannan, R.G.; Alderton, A.L. Honey inhibits lipid oxidation in ready-to-eat ground beef patties. *Meat Sci.,* **2005**, *70*(4), 627-631. [http://dx.doi.org/10.1016/j.meatsci.2005.02.011] [PMID: 22063889]

[23] Nagai, T.; Inoue, R.; Kanamori, N.; Suzuki, N.; Nagashima, T. Characterization of honey from different floral sources. Its functional properties and effects of honey species on storage of meat. *Food Chem.,* **2006**, *97*, 256-262. [http://dx.doi.org/10.1016/j.foodchem.2005.03.045]

[24] Mundo, M.A.; Padilla-Zakour, O.I.; Worobo, R.W. Growth inhibition of foodborne pathogens and food spoilage organisms by select raw honeys. *Int. J. Food Microbiol.,* **2004**, *97*(1), 1-8. [http://dx.doi.org/10.1016/j.ijfoodmicro.2004.03.025] [PMID: 15527912]

[25] Tuberoso, C.I.; Boban, M.; Bifulco, E.; Budimir, D.; Pirisi, F.M. Antioxidant capacity and vasodilatory properties of Mediterranean food: the case of Cannonau wine, myrtle berries liqueur and strawberry-tree honey. *Food Chem.,* **2013**, *140*(4), 686-691.

[http://dx.doi.org/10.1016/j.foodchem.2012.09.071] [PMID: 23692754]

[26] Khiati, B.; Bacha, S.; Aissat, S.; Ahmed, M. The use of Algerian honey on cutaneous wound healing: a case report and review of the literature. *Asian Pacific J. Trop. Dis,* **2014**, *4*, S867-S869. [http://dx.doi.org/10.1016/S2222-1808(14)60748-9]

[27] Khalil, M.I.; Sulaiman, S.A. The potential role of honey and its polyphenols in preventing heart diseases: a review. *Afr. J. Tradit. Complement. Altern. Medicines,* **2010**, *7*(4), 315-321. [http://dx.doi.org/10.4314/ajtcam.v7i4.56693] [PMID: 21731163]

[28] Matongo, F.; Nwodo, U.U. *In vitro* assessment of Helicobacter pylori ureases inhibition by honey fractions. *Arch. Med. Res.,* **2014**, *45*(7), 540-546. [http://dx.doi.org/10.1016/j.arcmed.2014.09.001] [PMID: 25240315]

[29] Aliyu, M.; Odunola, O.; Farooq, A.; Mesaik, A.; Choudhary, M.; Erukainure, O.; Perveen, S.; Jabeen, A. Fractionation of acacia honey affects its antioxidant potential *in vitro. J. Acute Dis,* **2012**, *1*, 115-119. [http://dx.doi.org/10.1016/S2221-6189(13)60027-X]

[30] Aljadi, A.; Kamaruddin, M. Evaluation of the phenolic contents and antioxidant capacities of two Malaysian floral honeys. *Food Chem.,* **2004**, *85*, 513-518. [http://dx.doi.org/10.1016/S0308-8146(02)00596-4]

[31] Ferreira, I.; Aires, E.; Barreira, J.; Estevinho, L. Antioxidant activity of Portuguese honey samples: Different contributions of the entire honey and phenolic extract. *Food Chem.,* **2009**, *114*, 1438-1443. [http://dx.doi.org/10.1016/j.foodchem.2008.11.028]

[32] da Silva, V.; Cerqueira, M.; Lowinsohn, D.; Matos, M.; Matos, R. Amperometric detection of ascorbic acid in honey using ascorbate oxidase immobilised on amberlite IRA-743. *Food Chem.,* **2012**, *133*, 1050-1054. [http://dx.doi.org/10.1016/j.foodchem.2012.01.066]

[33] Brudzynski, K.; Miotto, D. The recognition of high molecular weight melanoidins as the main components responsible for radical-scavenging capacity of unheated and heat-treated Canadian honeys. *Food Chem.,* **2011**, *125*, 570-575. [http://dx.doi.org/10.1016/j.foodchem.2010.09.049]

[34] Brudzynski, K.; Miotto, D. Honey melanoidins: Analysis of the compositions of the high molecular weight melanoidins exhibiting radical-scavenging activity. *Food Chem.,* **2011**, *127*(3), 1023-1030. [http://dx.doi.org/10.1016/j.foodchem.2011.01.075] [PMID: 25214092]

[35] Baltrusaityte, V.; Venskutonis, P.; Ceksteryte, V. Radical scavenging activity of different floral origin honey and beebread phenolic extracts. *Food Chem.,* **2007**, *101*, 502-514. [http://dx.doi.org/10.1016/j.foodchem.2006.02.007]

[36] Damintoti, K.; Mamoudou, H.; Simpore, J.; Traore, A. Antioxidant and antibacterial activities of polyphenols from ethnomedicinal plants of Burkina Faso. *Afr. J. Biotechnol.,* **2005**, *4*, 823-828.

[37] Viuda-Martos, M.; Ruiz-Navajas, Y.; Fernández-López, J.; Pérez-Alvarez, J.A. Functional properties of honey, propolis, and royal jelly. *J. Food Sci.,* **2008**, *73*(9), R117-R124. [http://dx.doi.org/10.1111/j.1750-3841.2008.00966.x] [PMID: 19021816]

[38] Agati, G.; Azzarello, E.; Pollastri, S.; Tattini, M. Flavonoids as antioxidants in plants: location and

functional significance. *Plant Sci.,* **2012**, *196*, 67-76.
[http://dx.doi.org/10.1016/j.plantsci.2012.07.014] [PMID: 23017900]

[39] Blasa, M.; Candiracci, M.; Accorsi, A.; Piacentini, M.; Piatti, E. Honey flavonoids as protection agents against oxidative damage to human red blood cells. *Food Chem.,* **2007**, *104*, 1635-1640.
[http://dx.doi.org/10.1016/j.foodchem.2007.03.014]

[40] Kassim, M.; Achoui, M.; Mustafa, M.R.; Mohd, M.A.; Yusoff, K.M. Ellagic acid, phenolic acids, and flavonoids in Malaysian honey extracts demonstrate *in vitro* anti-inflammatory activity. *Nutr. Res.,* **2010**, *30*(9), 650-659.
[http://dx.doi.org/10.1016/j.nutres.2010.08.008] [PMID: 20934607]

[41] Hussein, S.Z.; Yusoff, K.M.; Makpol, S.; Yusof, Y.A. Antioxidant capacities and total phenolic contents increase with gamma irradiation in two types of Malaysian honey. *Molecules,* **2011**, *16*(8), 6378-6395.
[http://dx.doi.org/10.3390/molecules16066378] [PMID: 21796076]

[42] Eraslan, G.; Kanbur, M.; Silici, S.; Karabacak, M. Beneficial effect of pine honey on trichlorfon induced some biochemical alterations in mice. *Ecotoxicol. Environ. Saf.,* **2010**, *73*(5), 1084-1091.
[http://dx.doi.org/10.1016/j.ecoenv.2010.02.017] [PMID: 20303175]

[43] Petrus, K.; Schwartz, H.; Sontag, G. Analysis of flavonoids in honey by HPLC coupled with coulometric electrode array detection and electrospray ionization mass spectrometry. *Anal. Bioanal. Chem.,* **2011**, *400*(8), 2555-2563.
[http://dx.doi.org/10.1007/s00216-010-4614-7] [PMID: 21229237]

[44] Khalil, M.I.; Alam, N.; Moniruzzaman, M.; Sulaiman, S.A.; Gan, S.H. Phenolic acid composition and antioxidant properties of Malaysian honeys. *J. Food Sci.,* **2011**, *76*(6), C921-C928.
[http://dx.doi.org/10.1111/j.1750-3841.2011.02282.x] [PMID: 22417491]

[45] Buratti, S.; Benedetti, S.; Cosio, M.S. Evaluation of the antioxidant power of honey, propolis and royal jelly by amperometric flow injection analysis. *Talanta,* **2007**, *71*(3), 1387-1392.
[http://dx.doi.org/10.1016/j.talanta.2006.07.006] [PMID: 19071462]

[46] Heim, K.E.; Tagliaferro, A.R.; Bobilya, D.J. Flavonoid antioxidants: chemistry, metabolism and structure-activity relationships. *J. Nutr. Biochem.,* **2002**, *13*(10), 572-584.
[http://dx.doi.org/10.1016/S0955-2863(02)00208-5] [PMID: 12550068]

[47] Tsimogiannis, D.; Oreopoulou, V. The contribution of flavonoid C-ring on the DPPH free radical scavenging efficiency. A kinetic approach for the 3′,4′-hydroxy substituted members. *Innov. Food Sci. Emerg. Technol.,* **2006**, *7*, 140-146.
[http://dx.doi.org/10.1016/j.ifset.2005.09.001]

[48] Turkmen, N.; Sari, F.; Poyrazoglu, E.; Velioglu, Y. Effects of prolonged heating on antioxidant activity and colour of honey. *Food Chem.,* **2005**, *95*, 653-657.
[http://dx.doi.org/10.1016/j.foodchem.2005.02.004]

[49] Wang, X.; Ghedolf, N.; Engeseth, N. Effect of processing and storage on antioxidant capacity of honey. *J. Food Sci.,* **2004**, *69*, 96-101.

[50] Wilczyńska, A. Effect of filtration on colour, antioxidant activity and total phenolics of honey. *LWT -. Food Sci. Technol. (Campinas.),* **2014**, *57*, 767-774.

[51] Frankel, S.; Robinson, G.; Berenbaum, M. Antioxidant capacity and correlated characteristics of 14 unifloral honeys. *Apic. Res,* **1998**, *37*, 27-31.

[52] Delgado-Andrade, C.; Rufián-Henares, J.A.; Morales, F.J. Assessing the antioxidant activity of melanoidins from coffee brews by different antioxidant methods. *J. Agric. Food Chem.,* **2005**, *53*(20), 7832-7836.
[http://dx.doi.org/10.1021/jf0512353] [PMID: 16190638]

[53] Gniechwitz, D.; Reichardt, N.; Ralph, J.; Blaut, M.; Steinhart, H.; Bunzel, M. Isolation and characterization of a coffee melanoidin fraction. *J. Sci. Food Agric.,* **2008**, *88*, 2153-2160.
[http://dx.doi.org/10.1002/jsfa.3327]

[54] Wu, S.; Hu, J.; Wei, L.; Du, Y.; Shi, X.; Zhang, L. Antioxidant and antimicrobial activity of Maillard reaction products from xylan with chitosan/chitooligomer/glucosamine hydrochloride/taurine model systems. *Food Chem.,* **2014**, *148*, 196-203.
[http://dx.doi.org/10.1016/j.foodchem.2013.10.044] [PMID: 24262546]

[55] Hauser, C.; Müller, U.; Sauer, T.; Augner, K.; Pischetsrieder, M. Maillard reaction products as antimicrobial components for packaging films. *Food Chem.,* **2014**, *145*, 608-613.
[http://dx.doi.org/10.1016/j.foodchem.2013.08.083] [PMID: 24128521]

[56] Rufian-Henares, J.; Morales, F. Functional properties of melanoidins: *In vitro* antioxidant, antimicrobial and antihypertensive activities. *Food Res. Int.,* **2007**, *40*, 995-1002.
[http://dx.doi.org/10.1016/j.foodres.2007.05.002]

[57] Borrelli, R.C.; Fogliano, V. Bread crust melanoidins as potential prebiotic ingredients. *Mol. Nutr. Food Res.,* **2005**, *49*(7), 673-678.
[http://dx.doi.org/10.1002/mnfr.200500011] [PMID: 15986388]

[58] Corzo-Martinez, M.; Hernandez-Hernandez, O.; Villamiel, M.; Rastall, R.; Moreno, J. *In vitro* bifidogenic effect of Maillard-type milk protein–galactose conjugates on the human intestinal microbiota. *Int. Dairy J.,* **2013**, *31*, 127-131.
[http://dx.doi.org/10.1016/j.idairyj.2013.01.004]

[59] Limsuwanmanee, J.; Chaijan, M.; Manurakchinakorn, S.; Panpipat, W.; Klomklao, S.; Benjakul, S. Antioxidant activity of Maillard reaction products derived from stingray (Himantura signifier) non-protein nitrogenous fraction and sugar model systems. *LWT -. Food Sci. Technol. (Campinas.),* **2014**, *57*, 718-724.

[60] Vhangani, L.N.; Van Wyk, J. Antioxidant activity of Maillard reaction products (MRPs) derived from fructose-lysine and ribose-lysine model systems. *Food Chem.,* **2013**, *137*(1-4), 92-98.
[http://dx.doi.org/10.1016/j.foodchem.2012.09.030] [PMID: 23199995]

[61] Brudzynski, K.; Miotto, D. The relationship between the content of Maillard reaction-like products and bioactivity of Canadian honeys. *Food Chem.,* **2010**, *124*, 869-874.
[http://dx.doi.org/10.1016/j.foodchem.2010.07.009]

[62] Jerković, I.; Mastelić, J.; Marijanović, Z.; Klein, Z.; Jelić, M. Comparison of hydrodistillation and ultrasonic solvent extraction for the isolation of volatile compounds from two unifloral honeys of Robinia pseudoacacia L. and Castanea sativa L. *Ultrason. Sonochem.,* **2007**, *14*(6), 750-756.
[http://dx.doi.org/10.1016/j.ultsonch.2006.12.014] [PMID: 17321190]

[63] Jerković, I.; Mastelić, J.; Marijanović, Z. A variety of volatile compounds as markers in unifloral honey from dalmatian sage (Salvia officinalis L.). *Chem. Biodivers.,* **2006**, *3*(12), 1307-1316. [http://dx.doi.org/10.1002/cbdv.200690134] [PMID: 17193245]

[64] Jerkovic, I.; Tuberoso, C.; Marijanovic, Z.; Jelic, M.; Kasum, A. Headspace, volatile and semi-volatile patterns of Paliurus spina-christi unifloral honey as markers of botanical origin. *Food Chem.,* **2009**, *112*, 239-245. [http://dx.doi.org/10.1016/j.foodchem.2008.05.080]

[65] Kowalski, S. Changes of antioxidant activity and formation of 5-hydroxymethylfurfural in honey during thermal and microwave processing. *Food Chem.,* **2013**, *141*(2), 1378-1382. [http://dx.doi.org/10.1016/j.foodchem.2013.04.025] [PMID: 23790927]

[66] Turkmen, N.; Sari, F.; Poyrazoglu, E.; Velioglu, Y. Effects of prolonged heating on antioxidant activity and colour of honey. *Food Chem.,* **2006**, *95*, 653-657. [http://dx.doi.org/10.1016/j.foodchem.2005.02.004]

[67] Ávila, M.; Crevillen, A.; González, M.; Escarpa, A. Electroanalytical approach to evaluate antioxidant capacity in honeys: Proposal of an antioxidant index. *Electroanalysis,* **2006**, *18*, 1821-1826. [http://dx.doi.org/10.1002/elan.200603599]

[68] Gorjanović, S.; Alvarez-Suarez, J.; Novaković, M.; Pastor, F.; Pezo, L.; Battino, M.; Sužnjević, D. Comparative analysis of antioxidant activity of honey of different floral sources using recently developed polarographic and various spectrophotometric assays. *J. Food Compos. Anal.,* **2013**, *30*, 13-18. [http://dx.doi.org/10.1016/j.jfca.2012.12.004]

[69] Beretta, G.; Granata, P.; Ferrero, M.; Orioli, M.; Maffei, R. Standardization of antioxidant properties of honey by a combination of spectrophotometric/fluorometric assays and chemometrics. *Anal. Chim. Acta,* **2005**, *533*, 185-191. [http://dx.doi.org/10.1016/j.aca.2004.11.010]

[70] Arnao, M.; Cano, A.; Acosta, M. The hydrophilic and lipophilic contribution to total antioxidant activity. *Food Chem.,* **2001**, *73*, 239-244. [http://dx.doi.org/10.1016/S0308-8146(00)00324-1]

[71] Schwarz, K.; Bertelsen, G.; Nissen, L.; Gardner, P.; Heinonen, M.; Hopia, A.; Huynh-Ba, T.; Lambelet, P.; McPhail, D.; Skibsted, L.; Tijburg, L. Investigation of plant extracts for the protection of processed foods against lipid oxidation. Comparison of antioxidant assays based on radical scavenging, lipid oxidation and analysis of the principal antioxidant compounds. *Eur. Food Res. Technol.,* **2001**, *212*, 319-328. [http://dx.doi.org/10.1007/s002170000256]

[72] Van den Berg, R.; Haenen, G.; Van den Berg, H.; Bast, A. Applicability of an improved Trolox equivalent antioxidant capacity (TEAC) assay for evaluation of antioxidant capacity measurements of mixtures. *Food Chem.,* **1999**, *66*, 511-517. [http://dx.doi.org/10.1016/S0308-8146(99)00089-8]

[73] Re, R.; Pellegrini, N.; Proteggente, A.; Pannala, A.; Yang, M.; Rice-Evans, C. Antioxidant activity applying an improved ABTS radical cation decolorization assay. *Free Radic. Biol. Med.,* **1999**, *26*(9-10), 1231-1237.

[http://dx.doi.org/10.1016/S0891-5849(98)00315-3] [PMID: 10381194]

[74] Alvarez-Suarez, J.; Tulipani, S.; Romandini, S.; Bertoli, E.; Battino, M. Contribution of honey in nutrition and human health: a review. *Med. J. Nutrition Metab.,* **2010**, *3*, 15-23. [http://dx.doi.org/10.1007/s12349-009-0051-6]

[75] Alzahrani, H.A.; Boukraa, L.; Bellik, Y.; Abdellah, F.; Bakhotmah, B.A.; Kolayli, S.; Sahin, H. Evaluation of the antioxidant activity of three varieties of honey from different botanical and geographical origins. *Glob. J. Health Sci.,* **2012**, *4*(6), 191-196. [http://dx.doi.org/10.5539/gjhs.v4n6p191] [PMID: 23121756]

[76] Lachman, J.; Orsák, M.; Hejtmánková, A.; Kovářová, E. Evaluation of antioxidant activity and total phenolics of selected Czech honeys. *LWT -. Food Sci. Technol. (Campinas.),* **2010**, *43*, 52-58.

[77] Sant'Ana, L.D.; Sousa, J.P.; Salgueiro, F.B.; Lorenzon, M.C.; Castro, R.N. Characterization of monofloral honeys with multivariate analysis of their chemical profile and antioxidant activity. *J. Food Sci.,* **2012**, *77*(1), C135-C140. [http://dx.doi.org/10.1111/j.1750-3841.2011.02490.x] [PMID: 22133147]

[78] Kishore, R.K.; Halim, A.S.; Syazana, M.S.; Sirajudeen, K.N. Tualang honey has higher phenolic content and greater radical scavenging activity compared with other honey sources. *Nutr. Res.,* **2011**, *31*(4), 322-325. [http://dx.doi.org/10.1016/j.nutres.2011.03.001] [PMID: 21530807]

[79] Kuś, P.; Congiu, F.; Teper, D.; Sroka, Z.; Jerković, I.; Tuberoso, C. Antioxidant activity, color characteristics, total phenol content and general HPLC fingerprints of six Polish unifloral honey types. *LWT -. Food Sci. Technol. (Campinas.),* **2014**, *55*, 124-130.

[80] Marković, K.; Škaričić, Z.; Šarić, G.; Krpan, M.; Major, N.; Hruškar, M.; Vahčić, N. *7th International Congress of Food Technologists, Biotechnologists and Nutritionists,* **2011**, , pp. 167-172.

[81] Vela, L.; Lorenzo, C.; Perez, R. Antioxidant capacity of Spanish honeys and its correlation with polyphenol content and other physicochemical properties. *J. Sci. Food Agric.,* **2007**, *87*, 1069-1075. [http://dx.doi.org/10.1002/jsfa.2813]

[82] Chua, L.S.; Rahaman, N.L.; Adnan, N.A.; Eddie Tan, T.T. Antioxidant activity of three honey samples in relation with their biochemical components. *J. Anal. Methods Chem.,* **2013**, *2013*, 313798. [http://dx.doi.org/10.1155/2013/313798] [PMID: 24027653]

[83] Benzie, I.F.; Strain, J.J. The ferric reducing ability of plasma (FRAP) as a measure of "antioxidant power": the FRAP assay. *Anal. Biochem.,* **1996**, *239*(1), 70-76. [http://dx.doi.org/10.1006/abio.1996.0292] [PMID: 8660627]

[84] Alvarez-Suarez, J.M.; Tulipani, S.; Díaz, D.; Estevez, Y.; Romandini, S.; Giampieri, F.; Damiani, E.; Astolfi, P.; Bompadre, S.; Battino, M. Antioxidant and antimicrobial capacity of several monofloral Cuban honeys and their correlation with color, polyphenol content and other chemical compounds. *Food Chem. Toxicol.,* **2010**, *48*(8-9), 2490-2499. [http://dx.doi.org/10.1016/j.fct.2010.06.021] [PMID: 20558231]

[85] Piljac-Zegarac, J.; Stipcevic, T.; Belscak, A. Antioxidant properties and phenolic content of different floral origin honeys. *J. ApiProduct ApiMedical Sci,* **2009**, *1*, 43-50. [http://dx.doi.org/10.3896/IBRA.4.01.2.04]

[86] Mohamed, M.; Sirajudeen, K.; Swamy, M.; Yaacob, N.S.; Sulaiman, S.A. Studies on the antioxidant properties of Tualang honey of Malaysia. *Afr. J. Tradit. Complement. Altern. Medicines,* **2010**, *7*(1), 59-63.
[http://dx.doi.org/10.4314/ajtcam.v7i1.57256] [PMID: 21304614]

[87] Lianda, R.; Sant'Ana, L.; Echevarria, L.; Castro, R. Antioxidant Activity and Phenolic Composition of Brazilian Honeys and their Extracts. *J. Braz. Chem. Soc.,* **2012**, *23*, 618-627.

[88] Pulido, R.; Bravo, L.; Saura-Calixto, F. Antioxidant activity of dietary polyphenols as determined by a modified ferric reducing/antioxidant power assay. *J. Agric. Food Chem.,* **2000**, *48*(8), 3396-3402.
[http://dx.doi.org/10.1021/jf9913458] [PMID: 10956123]

[89] Cao, G.; Prior, R.L. Comparison of different analytical methods for assessing total antioxidant capacity of human serum. *Clin. Chem.,* **1998**, *44*(6 Pt 1), 1309-1315.
[PMID: 9625058]

[90] Ou, B.; Hampsch-Woodill, M.; Prior, R.L. Development and validation of an improved oxygen radical absorbance capacity assay using fluorescein as the fluorescent probe. *J. Agric. Food Chem.,* **2001**, *49*(10), 4619-4626.
[http://dx.doi.org/10.1021/jf010586o] [PMID: 11599998]

[91] Ou, B.; Hampsch-Woodill, M.; Flanagan, J.; Deemer, E.K.; Prior, R.L.; Huang, D. Novel fluorometric assay for hydroxyl radical prevention capacity using fluorescein as the probe. *J. Agric. Food Chem.,* **2002**, *50*(10), 2772-2777.
[http://dx.doi.org/10.1021/jf011480w] [PMID: 11982397]

[92] Gheldof, N.; Engeseth, N.J. Antioxidant capacity of honeys from various floral sources based on the determination of oxygen radical absorbance capacity and inhibition of *in vitro* lipoprotein oxidation in human serum samples. *J. Agric. Food Chem.,* **2002**, *50*(10), 3050-3055.
[http://dx.doi.org/10.1021/jf0114637] [PMID: 11982440]

[93] Alvarez-Suarez, J.M.; Giampieri, F.; Battino, M. Honey as a source of dietary antioxidants: structures, bioavailability and evidence of protective effects against human chronic diseases. *Curr. Med. Chem.,* **2013**, *20*(5), 621-638.
[http://dx.doi.org/10.2174/092986713804999358] [PMID: 23298140]

[94] Candiracci, M.; Piatti, E.; Dominguez-Barragán, M.; García-Antrás, D.; Morgado, B.; Ruano, D.; Gutiérrez, J.F.; Parrado, J.; Castaño, A. Anti-inflammatory activity of a honey flavonoid extract on lipopolysaccharide-activated N13 microglial cells. *J. Agric. Food Chem.,* **2012**, *60*(50), 12304-12311.
[http://dx.doi.org/10.1021/jf302468h] [PMID: 23176387]

[95] Al-Jadi, A.M.; Kanyan Enchang, F.; Mohd Yusoff, K. The effect of Malaysian honey and its major components on the proliferation of cultured fibroblasts. *Turk. J. Med. Sci.,* **2014**, *44*(5), 733-740.
[http://dx.doi.org/10.3906/sag-1303-43] [PMID: 25539538]

[96] Tan, J.J.; Azmi, S.M.; Yong, Y.K.; Cheah, H.L.; Lim, V.; Sandai, D.; Shaharuddin, B. Tualang honey improves human corneal epithelial progenitor cell migration and cellular resistance to oxidative stress *in vitro. PLoS One,* **2014**, *9*(5), e96800.
[http://dx.doi.org/10.1371/journal.pone.0096800] [PMID: 24802273]

[97] Beretta, G.; Orioli, M.; Facino, R.M. Antioxidant and radical scavenging activity of honey in

endothelial cell cultures (EA.hy926). *Planta Med.,* **2007**, *73*(11), 1182-1189.
[http://dx.doi.org/10.1055/s-2007-981598] [PMID: 17823875]

[98] Schramm, D.D.; Karim, M.; Schrader, H.R.; Holt, R.R.; Cardetti, M.; Keen, C.L. Honey with high levels of antioxidants can provide protection to healthy human subjects. *J. Agric. Food Chem.,* **2003**, *51*(6), 1732-1735.
[http://dx.doi.org/10.1021/jf025928k] [PMID: 12617614]

[99] Jubri, Z.; Narayanan, N.; Karim, N.; Wan Ngah, W. Antiproliferative Activity and Apoptosis Induction by Gelam Honey on Liver Cancer Cell. *Int. J. Appl. Sci. Technol.,* **2012**, *2*, 135-141.

[100] Al-Waili, N.S.; Boni, N.S. Natural honey lowers plasma prostaglandin concentrations in normal individuals. *J. Med. Food,* **2003**, *6*(2), 129-133.
[http://dx.doi.org/10.1089/109662003322233530] [PMID: 12935324]

[101] Gheldof, N.; Wang, X.H.; Engeseth, N.J. Buckwheat honey increases serum antioxidant capacity in humans. *J. Agric. Food Chem.,* **2003**, *51*(5), 1500-1505.
[http://dx.doi.org/10.1021/jf025897t] [PMID: 12590505]

[102] Shin, H.; Ustunol, Z. Influence of honey-containing marinades on heterocyclic aromatic amine formation and overall mutagenicity in fried beef steak and chicken breast. *J. Food Sci.,* **2004**, *69*, C147-C153.

[103] Shin, H.; Strasburg, G.; Ustunol, Z. Influence of different unifloral honeys on heterocyclic aromatic amine formation and overall mutagenicity in fried ground-beef patties. *J. Food Sci.,* **2003**, *68*, 810-815.
[http://dx.doi.org/10.1111/j.1365-2621.2003.tb08247.x]

[104] Afroz, R.; Tanvir, E.; Hossain, F.; Gan, S.; Parvez, M.; Islam, A.; Khalil, I. Protective Effect of Sundarban Honey against Acetaminophen-Induced Acute Hepatonephrotoxicity in Rats. *Evidence-based Complementary and Alternative Medicine,* **2014**, *2014*, 1-6.
[http://dx.doi.org/10.1155/2014/143782]

[105] Sahhugi, Z.; Hasenan, S.; Jubri, Z. Protective Effects of Gelam Honey against Oxidative Damage in Young and Aged Rats. *Oxid. Med. Cell. Long.,* **2014**, *2014*, 1-8.
[http://dx.doi.org/10.1155/2014/673628]

[106] Sibel, S.; Enis, Y.M.; Hüseyin, S.; Timucin, A.A.; Duran, O. Analysis of grayanatoxin in Rhododendron honey and effect on antioxidant parameters in rats. *J. Ethnopharmacol.,* **2014**, *156*, 155-161.
[http://dx.doi.org/10.1016/j.jep.2014.08.027] [PMID: 25193682]

[107] Anarkooli, I.; Ganji, H.; Pourheidar, M. The Protective Effects of Insulin and Natural Honey against Hippocampal Cell Death in Streptozotocin-Induced Diabetic Rats. *J. Diabetes Res.,* **2014**, *2014*, 1-8.
[http://dx.doi.org/10.1155/2014/491571]

[108] Nasrolahi, O.; Khaneshi, F.; Rahmani, F.; Razi, M. Honey and metformin ameliorated diabetes-induced damages in testes of rat; correlation with hormonal changes. *Iran. J. Reprod. Med.,* **2013**, *11*(12), 1013-1020.
[PMID: 24639728]

[109] Khalil, I.; Moniruzzaman, M.; Boukraâ, L.; Benhanifia, M.; Islam, A.; Islam, N.; Sulaiman, S.A.; Gan,

S.H. Physicochemical and antioxidant properties of Algerian honey. *Molecules,* **2012**, *17*(9), 11199-11215.
[http://dx.doi.org/10.3390/molecules170911199] [PMID: 22996344]

[110] Moniruzzaman, M.; Khalil, M.I.; Sulaiman, S.A.; Gan, S.H. Physicochemical and antioxidant properties of Malaysian honeys produced by Apis cerana, Apis dorsata and Apis mellifera. *BMC Complement. Altern. Med.,* **2013**, *13*, 43-54.
[http://dx.doi.org/10.1186/1472-6882-13-43] [PMID: 23433009]

[111] Blasa, M.; Candiracci, M.; Accorsi, A.; Piacentini, M.; Albertini, M.; Piatti, E. Raw Millefiori honey is packed full of antioxidants. *Food Chem.,* **2006**, *97*, 217-222.
[http://dx.doi.org/10.1016/j.foodchem.2005.03.039]

[112] Moniruzzaman, M.; Sulaiman, S.A.; Azlan, S.A.; Gan, S.H. Two-year variations of phenolics, flavonoids and antioxidant contents in acacia honey. *Molecules,* **2013**, *18*(12), 14694-14710.
[http://dx.doi.org/10.3390/molecules181214694] [PMID: 24287998]

[113] Saxena, S.; Gautam, S.; Sharma, A. Physical, biochemical and antioxidant properties of some Indian honeys. *Food Chem.,* **2010**, *118*, 391-397.
[http://dx.doi.org/10.1016/j.foodchem.2009.05.001]

CHAPTER 8(II)

Antioxidant Properties of Bee Products of Plant-Origin Part 2. Propolis and Pollen

Pedro A.R. Fernandes[1,2], **Sónia S. Ferreira**[1,2], **Alice Fonte**[2], **Dulcineia F. Wessel**[2,3], **Susana M. Cardoso**[1,4,*]

[1] *QOPNA, Department of Chemistry, University of Aveiro, Campus Universitário de Santiago, 3810-193 Aveiro, Portugal*

[2] *School of Agriculture, Polytechnic Institute of Viseu, 3500-606 Viseu, Portugal*

[3] *CI&DETS, Polytechnic Institute of Viseu, 3500-606 Viseu, Portugal*

[4] *CERNAS, School of Agriculture, Polytechnic Institute of Coimbra, Bencanta, 3045-601 Coimbra, Portugal*

Abstract: Over the last years, the hive products such as propolis and pollen have been highlighted due to their potential health benefits, including antioxidant abilities that have been correlated with their content in phenolic compounds. Regardless of the several factors that may affect propolis and pollen antioxidant activity, these products have been shown to possess, either through the use of *in vitro* or *in vivo* models, important features concerning the modulation of cellular oxidative stress caused by environmental factors (*e.g.* UV-light), metals, pesticides and other xenobiotics. This modulatory effect focus not only on the capture of radicals that these elements might eventually generate, but also by the activation of cellular antioxidant mechanisms such as enzymatic antioxidants or by modifying gene expression patterns. Although the mechanisms behind these responses are not fully known, it has been showed that caffeic acid phenethyl ester, pinocembrin and chrisin are some of the compounds responsible for some of these responses. Taking into account the gathered results, propolis and pollen can be viewed as potential agents in the re-stabilization of cellular oxidative imbalance and in the prevention of oxidative stress related diseases.

Keywords: Antioxidant activity, Antioxidant defenses, Antioxidant enzymes,

* **Correspondence to S.M. Cardoso:** Department of Chemistry, University of Aveiro, Campus Universitário de Santiago, 3810-193 Aveiro, Portugal; Tel: +351 234 370360; Fax: +351 234 370084; Email: susanacardoso@ua.pt.

CAPE catalase, Chrysin, DPPH, Glutathione, Glutathione peroxidase, Hive, Lipid oxidation, Oxidative stress, Phenolic compounds, Pinocembrin, Pollen, Propolis, ROS.

1. ANTIOXIDANT PROPERTIES OF PROPOLIS

Propolis *i.e.*, the sticky dark-colored substance that bees produce from the collected resins of plants, has been the focus of many studies over the past decades with regard to biological activities, in particular that of antioxidant. In general, authors have associated this biological property to phenolic compounds present in propolis samples [1 - 7]. Obviously, since the phenolic content and profile of propolis is dependent on several factors (*e.g.* botanical origin [1, 8] and geographical location [2]), variations in antioxidant abilities of samples are also expected.

Following, relevant work on antioxidant properties of worldwide propolis is presented. Overall, this capacity has been mainly assessed in chemical models that, as known, do not mimic an *in vivo* environment and should only be used as a first approach. Still, over the last years, there has been also a considerable number of works dealing with the evaluation of antioxidant properties of propolis in biological models, both in cellular and *in vivo*. Due to the numerous literature data on this theme, we will mainly focus on recently reported data.

1.1. Assessment using Chemical Models

The 2,2-diphenyl-1-picrylhydrazyl (DPPH) [1, 2, 4 - 18] assay, along with the 2, 2'-azino-bis(3-ethylbenzothiazoline-6-sulfonic acid) diammonium salt (ABTS) [1, 2, 12, 13, 17] and ferric reducing antioxidant power (FRAP) assays [4, 5, 7 - 8, 12, 14, 15, 17], are undoubtedly the most frequently used methods for assessing the antioxidant activities of propolis extracts (see Table **1**), as they are reliable and fast-executing methods [19]. This last also facilitates the obtaining of experimental data from more than one assay, as recommended, in a short time period [20].

There is a wide variation among the reported values of EC_{50} (most of the times defined as the concentration of extract needed to reduce half of the radical or the ferric ion concentration) when evaluated by DPPH, ABTS and FRAP assays and often, the antioxidant abilities are expressed in variable units, which clearly hamper the comparison of results.

Fluctuations in EC_{50} values (or antioxidant ability) of propolis samples have been closely correlated with their total phenolic content (TPC) [1, 2, 5 - 7, 13, 15, 17]. Still, there are exceptions to this *e.g.*, not always the sample with the highest TPC shows the highest scavenging activity or reducing power [8, 22]. Cases like that can be due to the distinct antioxidant ability of individual phenolics in the extracts and/or to the presence of other antioxidants besides phenolic compounds.

Table 1. Selected studies (2009-2015) of antioxidant activities of worldwide propolis extracts, as measured using chemical models.

Geographic location	Botanical Origin	Solvent	Assay	Results	Ref.
China (Shandong)	*Populus sp.*	Mix EtOH/H_2O	DPPH ABTS	DPPH: EC_{50} = 15.49±0.59 μg/ml Ext; ABTS: EC_{50} = 36.66±1.82 μg/ml Ext.	[1]
Turkey (Bengol, Rize, Tekirgard and Van)	N.D	Mix MeOH/H_2O/HCl	DPPH ABTS	DPPH: 409.6 to 503.7 mg trolox eq./g Ext; ABTS: 237.7 to 285.3 mg trolox eq./g Ext.	[2]
Italy (Venetia)	N.D	Mix H_2O/EtOH	DPPH LOPerox	DPPH: 100 to 150 mM catechin eq. for 150 μg/mL; LOPerox: 62% to 75% for 1.2 μg/mL Ext.	[3]
China (temperate, subtropical and tropical zones/26 regions)	N.D	HotH_2O	DPPH RP	DPPH: EC_{50} = 0.28 to 3.29 μg/mL Ext; RP: 0.20 to 3.47 %/μg/mL Ext.	[4]
Poland (Northern region)	N.D	EtOH	DPPH RP	DPPH: 60.78±10.12 % for 25 μL of liq. EtOH Ext (1 g propolis/10 mL EtOH); RP: 930.5±66.34 mmol Fe^{2+}/g Ext	[5]
Mexico (Sonora)	N.D	EtOH	DPPH	DPPH: 16 to 64.8% at 100 μg/mL Ext.	[6]

(Table 1) contd.....

Geographic location	Botanical Origin	Solvent	Assay	Results	Ref.
China (Anhui)	N.D	EtOH; Chloroform; Ethyl acetate; *n*-butanol	DPPH ABTS RP	DPPH: EC_{50}≈ 33, 38, 14 and 25 μg/mL for EtOH, chloroform, Ethyl acetate and n-butanol Ext, respectively; ABTS: EC_{50}≈ 42, 41, 23 and 33 μg/mL for EtOH, chloroform, Ethyl acetate and n-butanol Ext, respectively; RP: 1.39, 1.40, 2.30 and 1.80 mg trolox eq./mg for for EtOH, chloroform, Ethyl acetate and n-butanol Ext, respectively.	[7]
Portugal (BraganÇa e Leiria)	*Populus x Canadensis* hybrid and *Cistus ladanifer*	Mix H_2O/EtOH	DPPH RP	DPPH : EC_{50}= 18 to 23 μg/mL Ext; RP: ≈ 400 to 710 mg/g caffeic acid: galangin: pinocembrin (1:1:1).	[8]
France (Southern, Western and Central regions/24 locations)	*Populus* spp. *P. nigra* L.	H_2O; EtOH; Mix H_2O/EtOH; MeOH; DCM; Mix DCM/ MeOH/H_2O	DPPH ORAC Anti-AGEs	DPPH: 1731±28, 1605±26, 1650±149, 1386±171, 1437 ± 105 and 1964 ± 124 μmol trolox eq./g of H_2O, EtOH, Mix H_2O/EtOH, MeOH, DCM and Mix solvents Ext, respectively; ORAC: 9722±273, 8155±114, 9890±480, 7769±360, 9242±739 and 11278±11 μmol trolox eq./g of H_2O, EtOH, Mix H_2O/EtOH, MeOH, DCM and Mix solvents Ext, respectively; Anti-AGEs: EC_{50}= 0.34, 0.05, 0.03, 0.03, 0.03 and 0.04 mg/mL of H_2O, EtOH, Mix H_2O/EtOH, MeOH, DCM and Mix solvents Ext, respectively	[9]
Brasil (Mato Grosso do Sul)	N.D	Mix H_2O/EtOH	DPPH	DPPH: EC_{50}= 40±4.8μg/mL Ext.	[10]
Brasil (Brejo Grande)	N.D	Mix H_2O/EtOH	DPPH SOD-Like CAT-Like	DPPH: EC_{50}= 270.13±24.77 μg/mL Ext; SOD-Like: 466.90±12.40 of SOD like activity; CAT-Like: 13.13±2.65 mmol of H_2O_2 decomposed/min for 20 μL of 100 μg/mL Ext.	[11]

(Table 1) contd.....

Geographic location	Botanical Origin	Solvent	Assay	Results	Ref.
Turkey (Erzurum province)	N.D	HotH_2O	DPPH ABTS DMPD SORSC HPRSA MCA RP CUPRAC LOPerox	DPPH: EC_{50} = 31.81 µg/mL Ext; ABTS: EC_{50}= 14.29 µg/mL Ext; DMPD: EC_{50} =18.32 µg/mL Ext; SORSC: EC_{50}= 9.89 µg/mL Ext; HPRSA: EC_{50}=6.54 µg/mL Ext; MCA: EC_{50}= 12.04 µg/mL Ext; RP= 0.568 absorbance for 10 µg/mL Ext; CUPRAC= 0.814 absorbance for 10 µg/mL Ext; LOPerox= 93.2% of 10 µg/mL Ext.	[12]
Portugal (Algarve)	N.D	Mix H_2O/EtOH	DPPH ABTS SORSC MCA	DPPH: EC_{50}= 9 to 40 µg/mL Ext; ABTS: EC_{50}=11 to 26 µg/mL Ext; SORSC: EC_{50}= 13 to 47 µg/mL Ext; MCA: 32.3 to 53.8 %	[13]
Italy (Bologna)	N.D	EtOH; Acetone; Chloroform.	DPPH TBARS RP	DPPH: EC_{50}= 2.06±0.45, 1.97±0.44 and 0.752±0.36 Trolox eq./g of EtOH, Acetone and Chloroform Ext, respectively; TBARS: 0.017±0.005, 0.016±0.007 and 0.021±0.009 mmol TEP/g of EtOH, Acetone and Chloroform Ext, respectively; RP: 0.69±0.09, 0.67±0.1 and 0.47±0.07 TEs/g of EtOH, Acetone and Chloroform Ext respectively.	[14]
Brasil (Minas Gerais, São Paulo, Rio Grande do Sul, Paraná and Santa Catarina)	N.D	Mix H_2O/EtOH	DPPH RP	DPPH: EC_{50}= 33.36 ±.22 µg/mL Ext; RP: EC_{50}= 270±3.27 µg/mL Ext.	[15]
Brasil (Paraná)	3 botanical species	Mix H_2O/EtOH	DPPH	DPPH: EC_{50}= 43 ±22 µg/ mL Ext.	[16]
Poland (Southern, Central and Northern zones/9 regions)	N.D	EtOH	DPPH ABTS RP	DPPH: 1.92 to 2.69 mM Trolox eq./g Ext; ABTS: 3.96 to 4.98 mM Trolox eq./g Ext; RP: 6.23 to 9.19 mM Fe^{+2}/g Ext.	[17]
India (Coimbatore)	N.D	Mix EtOH/H_2O	DPPH HRSA TBARS	DPPH: EC_{50}= 16.20 µg/mL Ext; HRSA: EC_{50}= 34.33 µg/mL Ext; TBARS: EC_{50}= 55.56 µg/mL Ext	[18]
Uruguay (Southern region)	N.D	Mix H_2O/EtOH	ORAC	ORAC: 1.8 to 9.0 µmol Trolox eq./mg propolis.	[21]

(Table 1) contd.....

Geographic location	Botanical Origin	Solvent	Assay	Results	Ref.
Croatia (Zagreb)	*Populus sp.*	EtOH; Mix H_2O/EtOH	BCB DPPH RP MCA	BCB: 87.3 to 96.81 %; DPPH: EC_{50}= 27.82 to 55.97 μg/mL Ext; RP: 0.53 to 0.56 slope/mg; MCA: EC_{50}= 47.81	[22]

ABTS - 2,2'-azino-bis(3-ethylbenzothiazoline-6-sulfonic acid) diammonium salt ; Anti-AGEs – inhibition of advanced glycation products formation assay; BCB - β-carotene Bleaching assay; CAT – Catalase; CUPRAC – Cupric ion reducing antioxidant capacity assay; DCM – Dichloromethane; DMPD - N,N-dimethyl-p-phenylenediamine; DPPH - 2,2-diphenyl-1-picrylhydrazyl; EtOH – Ethanol; HotH_2O – hot water; HPRSA – hydrogen peroxide radical scavenging ability; HRSA – Hydroxyl radical scavenging ability; LOPerox – linoleic acid peroxidation; MCA – metal chelating activity; MeOH – Methanol; N.D – not determined; ORAC - Oxygen radical absorbance capacity; RP – Reducing Power; SOD – Superoxide dismutase; SORSC – superoxide radical scavenging capacity; TEP - 1,1,3,3-tetraethoxypropane.

In addition to these popular spectrophotometric chemical assays, the antioxidant activity of propolis extracts have been evaluated by other chemical methods, which include the oxygen radical absorbance capacity (ORAC) assay, the monitoring of the ability to scavenge radicals such as the superoxide anion (O_2^-), hydroxyl ($OH^•$) or hydroperoxide (H_2O_2) or the ability to inhibit lipid oxidation (*e.g.* thiobarbituric acid reactive substances (TBARS) assay, and β-carotene bleaching assay) or even the assessment of antioxidant enzyme activities (Table **1**). Despite variations in results according to standard reference and/or propolis botanical origin are observable, results also confirm the relevant antioxidant activity of propolis extracts.

1.2. Assessment in Cellular Models

As more complex models, *in vitro* cellular assays take into account the availability factor of the propolis constituents, as well as their target cellular mechanism of action. Notably, the ability of propolis extracts of distinct botanical and geographical origins in decreasing the content of cellular reactive oxygen species (ROS) have been reported by numerous authors and for distinct cellular models (see Table **2**). In addition, distinct polar propolis extracts, mostly ethanolic ones, have been proved to effectively reduce biomolecular oxidation in cellular models (Table **2**), such as human erythrocytes, and spermatozoa under oxidative stress situations [10, 23]. Moreover, propolis extracts at non-toxic concentrations could reduce 8-oxodeoxyguanosine, a biomarker for oxidative damaged of DNA, in microglia subjected to hypoxia [24 - 25] and in a primary culture of rat hepatocytes exposed to 2,3,7,8-tetrachlorodibenzo-p-dioxin [26]. It is possible that the decreased oxidation of biomolecules is due to the ability of propolis in reducing the ROS levels [25, 27 - 34], or to its iron chelation properties [29].

Table 2. Antioxidant properties of selected propolis extracts, as demonstrated in cellular models.

Geographic Location	Model	Treatment Conditions	Effects	Ref.
Inhibition of ROS levels				
Brazilian green propolis (Yamada Apiculture Center, Inc Ltd., Japan)	Microglial, MG6, cell line exposed to normoxia or hypoxia	50 μg/mL EtOH Ext	↑viability; ↓TNF-α, IL-1β, IL-6, and mitochondrial ROS; ↓NF-κB activation	[25]
Portuguese propolis (Northeast region)	primary cortical neurons	0.01 to 10 μg/ml mix H_2O/EtOH for 1 h prior 24 h with STS or H_2O_2	↓ROS production and caspase-3 activation	[28]
Portuguese propolis (Bornes and Fundão)	heparinised blood of spherocytosis and healthy individuals	10 μg/mL MeOH Ext prior to NaCl osmotic stress or H_2O_2	↑ resistance to osmotic stress with/without exposure to H_2O_2	[29]
Brazilian green propolis (Minas gerais)	661W cell line	10-30 μg/mL Aq Ext 1h prior to white fluorescent light or UVA exposure	↑cell viablity; ↓phosphorylation of p38 with UVA exposure; ↓ ROS	[30]
Brazilian green propolis (Cajuru)	rabbit PMNs	25 μg/mL mix H_2O/EtOH Ext prior to 1 mg/mL OZ	↓ROS production; extracts from May, June, and August present higher antioxidant activity than those collected in other months	[31]
Brazilian green propolis (Nihon Natural Foods Co., Ltd., Japan)	J774A.1 cell line	0.1% (v/v) EtOH Ext for 24 h with IFN-γ and LPS	↓ROS and RNS, NO, IL-1α, IL-1β, IL-4, IL-6, IL-12p40, IL-13, TNF-α, G-CSF, GMCSF, MCP-1, MIP-1α, MIP-1β, and RANTES	[32]
Brazilian green propolis (Minas Gerais)	HUVECs deprived of serum and FGF-2	12.5, 25 and 50 μg/ml EtOH Ext	↓expression of integrin β4, p53 and ROS levels; ↑integrin β4, p53 and ROS; ↓mitochondrial membrane potential.	[33]

(Table 2) contd.....

Geographic Location	Model	Treatment Conditions	Effects	Ref.
Chinese red propolis (Shandong)	HUVECs deprived of serum and FGF-2	6.25, 12.5, and 25μg/mL EtOH Ext	↓PC-PLC activity, p53 and ROS levels; ↓mitochondrial membrane potential for the higher concentration.	[34]
Reduction of Biomolecular Oxidation				
Brazilian propolis (Mato Grosso do Sul)	human erythrocytes	50 to 125 μg/mL Mix H_2O/EtOH Ext prior to AAPH	↓lipid oxidation levels and hemolysis	[10]
Chilean and propolis (NATUR-ANDES-CHILE, San Vincente de Tágua)	human spermatozoa	12 and 25 μg/ml Mix H_2O/EtOH Ext after 1 h with benzo[α]pyrene; H_2O_2 ; or ADP, H_2O_2 and $FeSO_4$	↓Intracellular oxidants, lipid oxidation, and DNA damage	[23]
Portuguese propolis (Bornes and Fundão)	human erythrocytes	5 to 40 μg/mL MeOH Ext prior to AAPH	↓lipid oxidation and hemolysis	[24]
Turkish propolis (Erzurum)	hepatocytes from male Sprague–Dawley rat	25, 50 and 100 μM EtOH Ext with TCDD	↑viability and TAA; ↓8-oxo-2-dG adducts	[26]
Expression of antioxidant enzymes				
Uruguayan brown propolis (Southern region)	BAE cells	3.2 to 5.3 mg/mL Mix H_2O/EtOH Ext	↑eNOS protein expression; ↓Nox activity; ↓Nox4 mRNA expression	[21]
Chinese propolis (Shandong)	HUVECS	0.5, 15 and 30 μg/mL EtOH Ext with oxLDL	↑viability; ↓apoptotic cells, caspase-3 activity, and ox-LDL uptake; ↓LOX-1 protein and mRNA expression; ↓ROS generation and lipid oxidation; ↑ SOD and CAT; ↓ NADPH oxidase activity	[27]
Brazilian green propolis (Minas Gerais)	NB1-RGB cells	30 μg/mL Aq Ext 1h prior to UVA exposure	↑cell viablity; ↓phosphorylation of p38 and ERK; ↓ROS	[30]
Chinese and Brazilian red propolis(Api Co., Ltd., Japan)	COS7 Cells	0–20 μg/mL EtOH 1h prior to cadmium treatment	↓intracellular ROS accumulation; ↑cellular viability; ↑mRNA of HO-1 with involvement of HIF-1α	[35]

(Table 2) contd.....

Geographic Location	Model	Treatment Conditions	Effects	Ref.
Chinese propolis (Shandong)	RAW264.7 cell line	1.25-5 μg/mL EtOH Ext	↓ROS; ↑mRNA of GCLM, GCLC, HO-1, TrxR1,	[36]
Chinese red propolis (Shandong) and Brazilian green propolis(Minas Gerais)	HUVECs	12.5 μg/ml EtOH Ext with ox-LDL for 24 h	↑cell viability; ↓apoptosis; ↓PC-PLC activity; ↑Annexin a7 level; ↓NF-κB p65 level and translocation of NF-κB p65 from cytoplasm to nucleus; ↓ROS; ↑mitochondrial membrane potential	[37]
Brazilian propolis (Guarapari)	*Saccharomyces cerevisiae* BY4741	25 μg/mL EtOH Ext prior to menadione	↑cell survival; ↑SOD activity; ↓lipid oxidation levels	[38]
Brazilian propolis (Universal in Natura Products Ltda., Brazil)	A549 cells	0.015-0.15% (v/v) Mix H_2O/EtOH 6h prior to TGF-β1	↓N-cadherin levels, stress fiber formation, and cellular migration; ↓Smad and AKT phosphorylation and Snail expression; ↑PPARγ protein; ↓TGF-β1-induced EMT; ↓ROS	[39]

AAPH - 2,2-Azobis-(2-amidinopropane) dihydrochlorid; 8-oxo-2-dG - 8-oxo-2-deoxyguanosine; ADP - Adenosine diphosphate; AKT - Protein kinase B; Aq – Aqueous; BAE - Bovine aortic endothelial; CAT - Catalase; EMT - Epithelial–mesenchymal transition; eNOS - Endothelial nitric oxide synthase; ERK - Extracellular-regulated protein kinase; EtOH - Ethanol; Ext - Extract; FGF-2 - basic fibroblast growth factor; GCLC - Glutamate-cysteine ligase catalytic unit; GCLM - Glutamate-cysteine ligase modified unit; G-CSF - Granulocyte colony-stimulating factor; GMCSF - Granulocyte-macrophage colony stimulating factor; HIF-1α - Hypoxia inducible factor-1α; HO-1 - Heme oxygenase-1; HUVECs - Human umbilical vein endothelial cells; IFN - Interferon; IL - Interleukin; LOX-1 - Lectin-likeoxidized low-density lipoprotein receptor-1; LPS - Lipopolysaccharide; MCP-1 - Monocyte chemotactic protein 1; MeOH - Methanol; MG6 - c-my--immortalized mouse microglialcellline; MIP-1α - macrophage inflammatory protein 1; NADPH - Nicotinamide adenine dinucleotide phosphate; NB1-RGB - Normal human skin fibroblast cells; NF-κB - nuclear factor kappa-light-chain-enhancer of activated B cells; Nox - mono-nitrogen oxides; oxLDL - oxidized low-density lipoprotein; OZ - opsonized zymosan; PC-PLC - phosphatidylcholine-specific phospholipase C; PMN - polymorphonuclear cells; PPARγ - Peroxisome proliferator-activated receptor gamma; RANTES - regulated upon activation normal T cell expressed and secreted; RNS - reactive nitrogen species; SOD - Superoxide dismutase; STS - staurosporine; TAA - total antioxidant activity; TCDD - 2,3,7,8-tetrachlorodibenzo-p-dioxin; TGF-β1 - Transforming growth factor beta 1; TNF-α - Tumor necrosis factor alpha; TrxR1 - Thioredoxin reductase 1.

Besides the previous mentioned effects, several *in vitro* studies suggested that propolis antioxidant abilities are also exerted through modulation of antioxidant enzymes and/or genes. In particular, the decrease of ROS levels in kidney tubule COS7 cells induced by ethanolic extracts of Brazilian and Chinese propolis has been suggested to be associated with an increase of superoxide dismutase (SOD) activity [35]. In addition, Zhang *et al.* [36] proposed that the decrement of ROS levels in RAW 264.7 macrophages induced by the exposure to a Chinese propolis extract, might be associated to the increased expression of mRNA subunits of glutamate-cysteine ligase (GCL) (an enzyme linked to the synthesis of reduced glutathione (GSH). In their study, the authors also observed the translocation of nuclear factor (erythroid-derived 2)-like 2 (Nrf2) to the nucleus, as well the concomitant stimulation of the expression of *heme oxygenase 1* (HO-1), *i.e.* an antioxidant enzyme that limits the breakdown of heme into carbon monoxide, iron, and bilirubin, therefore, suggesting the elimination of ROS *via* Nrf2/HO-1 pathway.

Moreover, Xuan *et al.* [37] related the reduction of ROS, as registered for human umbilical vein endothelial cells (HUVECs) treated with ethanolic extracts of both Chinese and Brazilian propolis, with a decrease of nuclear factor kappa-ligh--chain-enhancer of activated B cells (NF-κB) induced by oxidized low-density lipoprotein (ox-LDL)-induced injury. In the same cellular model, Fang *et al.* [27] also reported that Chinese propolis extract ameliorated the suppression of SOD and catalase (CAT) activities and enhanced the inhibition of the endothelial nicotinamide adenine dinucleotide phosphate (NADPH) oxidase activity. The modulation enzymatic activities by propolis was also registered when using *Saccharomyces cerevisiae* as a model [38], where the reduction of ROS and lipid oxidation was associated with the activation of the antioxidant enzyme Cu/Zn-SOD.

1.3. Assessment using *In vivo* Models

The capacity of propolis in modulating antioxidant defenses was also shown *in vivo*, through mitigation of metals and xenobiotics toxicity, by alleviation of oxidative stress and/or restoring antioxidant defenses (Table **3**). *E.g.*, Garoui *et al.* [40] have shown that an ethanolic extract of Tunisian propolis exerted protective effects against cobalt-induced oxidative damages in the kidney of pregnant Wistar rats. Overall, the oral administration of the propolis extract decreased malondialdehyde (MDA) levels and restored the antioxidant defenses, increasing CAT, SOD, and glutathione peroxidase (GPx) activities and GSH levels in kidney [40]. Similar results were observed in several xenobiotics-induced oxidative stress models, such as those of oxytetracycline [41], malathion [42] and chlorpyrifos [43] (Table **3**).

Importantly, the protection of propolis extracts to metals and xenobiotics induced-toxicity has also been registered for renal and/or hepatic function markers. Garoui *et al.* [40] observed that the supplementation of Tunisian propolis extract in cobalt-treated rats could partially restore the levels of markers of renal function, like creatinine, urea, and uric acid. In turn, Nirala and Bhadauria [44] reported hepatorenal protective effects of an ethanolic extract of Indian propolis, as assessed in acetaminophen (APAP)-treated female Sprague Dawley rats. The authors observed that a single oral administration of propolis extract could restore the antioxidant status of liver and kidney tissues, as well as the blood markers of renal and hepatic function, namely aspartate aminotransferase (AST), alanine aminotransferase (ALT), serum alkaline phosphatase (SALP) and lactate dehydrogenase (LDH) activities.

Table 3. Antioxidant properties of selected propolis extracts, as demonstrated *in vivo* models.

Geographic Location	Model	Treatment Conditions	Effects	Ref.
Protection from metals				
Tunisian propolis (Mahares)	Cobalt-treated pregnant Wistar rats (kidney, plasma, and urine)	1 g/100 gdiet of Mix H_2O/EtOH Ext for 28 days, oral administration	↓lipid oxidation; ↑CAT, SOD, GPx, and GSH; ↓ Creatinine, urea, and LDH in plasma; ↑Creatinine, urea, and LDH in kidney; ↑Uric acid in plasma; ↓Uric Acid in urine; ↑ Creatinine clearance	[40]
Turkish propolis (Kocaavsar)	Arsenic-exposed carp (*Cyprinus carpio*) (liver, gill, and muscle)	10 mg/L of Mix H_2O/EtOH Ext in tank water, for 7 days	↓lipid oxidation; ↑CAT	[48]
Turkish propolis (Kayseri province)	Chromium-exposed carp (*Cyprinus carpio*) (blood, liver, kidney, spleen, and gill)	5 and 10 mg/kg/day of mix Mix H_2O/EtOH Ext for 28 days, oral administration	↓lipid oxidation and SOD; ↑ CAT, GPx, and GSH	[51]

(Table 3) contd.....

Geographic Location	Model	Treatment Conditions	Effects	Ref.
Protection from xenobiotics				
Turkish propolis (Kayseri province)	Oxytetracycline treated rainbow trout (*Oncorhynchus mykiss*) (blood, liver, kidney, spleen, and heart)	50 mg/kg/day of Mix H_2O/EtOH Ext for 14 days before, with, or after oxytetracycline, oral administration	↓lipid oxidation; ↑GSH, CAT, GPx, SOD, WBC, TI, and Phagocytic activity	[41]
Turkish propolis (Kayseri)	Malathion-exposed carp (*Cyprinus carpio carpio*) (liver, kidney, and gill)	10 mg/kg/day of Mix H_2O/EtOH Ext for 10 days, oral administration	↓lipid oxidation, SOD and CAT; ↑GSH and GPx; ↑erythrocyte counts; ↑Hemoglobin.	[42]
Turkish propolis (Kaiseri province)	Chlorpyrifos -exposed carp (*Cyprinus carpio carpio*) (liver, kidney, and gill)	10 mg/kg/day of Mix H_2O/EtOH Ext for 10 days, oral administration	↓lipid oxidation, SOD, and WBC; ↑CAT, GPx, erythrocyte counts, and Hemoglobin.	[43]
Indian propolis (N.D.)	APAP-treated female Sprague Dawley rats (blood, liver, and kidney)	100 and 200 mg/kg of EtOH Ext single dose with APAP, oral administration	↑Hemoglobin, glucose, bilirubin, TP, GSH, glycogen, CYP-AH, CYP-AND, ALP and ATPase; ↓ MDA, TC, ACP, AST, ALT, SALP, and LDH	[44]
Croatian propolis (N.D)	Skin lesion on male Swiss albino mice (skin)	30 μL, 5% of Mix H_2O/EtOH Ext for 5 days with HXS or PPD, topical administration	↓macrophage spreading index and number of cells in peritoneal cavity; ↓MDA; ↑GSH	[45]
Turkish propolis (Kocaavsar)	Cypermethrin-treated rainbow trout (*Oncorhynchus mykiss*) (brain)	10 ppm of Mix H_2O/EtOH Ext in tank water for 4 days	↓lipid oxidation; ↑CAT	[46]
Egyptian brown propolis (Beni-Suef)	Cisplatin-injected males mice *Mus musculus* (liver, kidney, and testis)	8.4 mg/kg/day of H_2O Ext, oral administration	↓Lipid oxidation; ↑CAT and GSH	[47]
Egyptian propolis (Dakah-lia Governorate)	Doxorubicin treated male Wistar rats (testis and serum)	200 mg/kg/day of Mix H_2O/EtOH Ext for 5 days a week for 21 days, oral administration	↑LDH, SDH, ACP, ALP, G6PD, IL-4, and GSH; ↓ALT, AST, MPO, TNF-α, lipid oxidation, and testosterone increase	[50]
Brazilian green propolis (Coapi, Brazil)	CS-exposed male C57BL/6 mice (lungs and BAL)	200 mg/kg/day of Mix H_2O/EtOH Ext after CS exposure for 5 days, oral gavage	↓Alveolar macrophage and neutrophils; ↓NO and MPO; ↓MDA, SOD, CAT, and GPx; ↑GSH/GSSG	[52]

(Table 3) contd.....

Geographic Location	Model	Treatment Conditions	Effects	Ref.
Protection from radiation				
Romanian propolis (Transylvania, Cluj County)	UVB irradiated female Swiss mice (skin)	1.5 and 3.0 mg/cm^2 of Mix H_2O/EtOH Ext 3 times in 24 h prior to or after UVB, topical administration	↓lipid oxidation, IL-6, neutrophil infiltration; caspase 3, skin cell death; epidermal CPD formation	[49]
Brazilian brown propolis extract (APIS FLORA, Brazil) and green propolis (Bioessens Ltda., Brazil)	UVB irradiated HRS/J mice (skin)	100 mg/kg of Mix H_2O/EtOH Ext 18h before and 30 min after UV, oral administration	↑GSH.	[53]
Protection from oxidative stress in model of disease				
Croatian brown propolis (Zagreb)	Alloxan-induced diabetic Swiss albino mice (kidney and liver)	50 mg/kg/day of Mix H_2O/EtOH or H_2O Ext for 8 days, intraperitoneal injection	↓lipid oxidation; ↓hepatorenal damages	[22]
Brazilian green propolis (Yamada Apiculture Center, Inc Ltd., Jpana)	Hypoxia-exposed mice (somatosensory cortex)	8.33 mg/kg of EtOH Ext 2 times/day for 7 days, intraperitoneal injection	↓TNF-α, IL-1β, IL-6, and 8-oxo-dG	[25]
Brazilian green propolis and Chinese red propolis (Hangzhou BEE-WORDS Apiculture Ltd., China)	Streptozotocin induced diabetic male Sprague Dawley rats (serum, liver, and kidney)	100 mg/kg/day of EtOH Ext for 56 days, oral administration	↓NOS, lipid oxidation, AST, UAER, MDA, and GPx; ↑SOD, GPx, and CAT	[54]
Brazilian green propolis (Minas Gerais)	WIRS-hepatic damage on male Wistar rats (serum and liver)	10,50 or 100 mg/kg of EtOH Ext, single oral administration	↓lipid oxidation, ALT, AST, NOx, and MPO; ↑SOD, NPSH, and ascorbic acid.	[55]
Brazilian green propolis (Minas Gerais)	WIRS-induced gastric mucosal lesions male Wistar rats (gastric mucosa)	50 mg/kg/day of EtOH Ext for 6 days or 1 day before WIRS, oral administration	↓gastric mucosal lesion; ↑NPSH, AA, and VE; ↓lipid oxidation, NOx, XO, and MPO	[56]

(Table 3) contd.....

Geographic Location	Model	Treatment Conditions	Effects	Ref.
Brazilian brown propolis (Santa Flora)	HSV-2-inoculated female adult BALB/c mice (blood, vaginal tissues)	50 mg/kg/day of Mix H_2O/EtOH Ext for 5 days prior to HSV-2 intravaginally inoculation, and 5 days after, orally by gavage	↓WBC, neutrophils RS, tyrosine nitration, MPO and ascorbic acid; ↑CAT in vaginal tissues	[57]
Chinese propolis (Dalion Garo International Trade Co., Ltda., China)	High cholesterol-fed male New-Zealand white rabbits (serum)	75 mg/kg/day of EtOH Ext for 28 days, oral administration	↓TC, LDL-C, TGs, and lipid oxidation; ↑HDL-C and GSH	[58]
Spanish propolis (Verbiotech I+D+I S.L., Spain)	Male Wistar albino rats with 22 months (plasma)	2 g/100 g of diet, for 90 days, oral administration	↓Glucose and TC; ↓lipid and protein oxidation	[59]

8-oxo-dG - 8-oxo-deoxyguanosine; ACP - acid phosphatase; ALP - alkaline phosphatase; ALT - alanine transaminase; APAP - acetaminophen; AST - aspartate transaminase; ATPase - adenosine triphosphatase; BAL - broncoalveolar lavage; CAT - catalase; CPD - cyclobutane pyrimidine dimer; CS - cigarette smoke; CYP - cytochrome P450; CYP-AH - aniline hydroxylase; CYP-AND - amidopyrine N-demethylase; Ext - extract; EtOH - ethanol; G6PD - glucose-6-phosphate dehydrogenase; GPx - glutathione peroxidase; GSH - reduced glutathione; HbAlc - glycated hemoglobina; HDL-C - high density lipoproteincholesterol; HRS/J - sex-matched hairless mice; HSV-2 - herpes simplex virus type 2; HXS - n-Hexyl salycilate; IL - interleukin; LDH - lactate dehydrogenase; LDL-C - low density lipoprotein-cholesterol; MPO - myeloperoxidase activity; N.D. – not determined; NO - nitric oxide; NOS - nitric synthetase; NOx - mono-nitrogen oxides; NPSH - non-protein thiols content; PPD - Di-n-Propyl Disulfide; ROS - reactive oxygen species; RS - reactive species; SALP - serum alkaline phosphatase; SDH - sorbitol dehydrogenase; SOD - superoxide dismutase; TAC - total antioxidant capacity; TC - total cholesterol; TGs - triglycerides; TI - Total immunoglobulin; TNF-α - tumor necrosis factor-α; TP - total protein; UAER - urinary albumin excretion rate; VE - vitamin E; WBC - white blood cell; WIRS - water-immersion restraint stress exposure.

Nonetheless, the protective effect of propolis extracts against toxic agents was also observed in lungs [44], skin [45], brain [46], testis [47], muscle [48], spleen [41], heart tissues [41] and for haematological parameters [41 - 43]. *E.g.* the oral administration of an ethanolic extract of Brazilian propolis has been reported to protect mice lungs when exposed to cigarette smoke. The treatment could normalize all biochemical parameters, namely the activities of antioxidant enzymes and reduced GSH/oxidized GSH ratio and lipid oxidation, when compared with those of the control group [44]. In turn, Orsolic and coworkers [45] had registered protective effects from a Croatian propolis extract against skin-lesions by topical administration, while Bolfa *et al.* [49], when investigating the effects of a topically-administered Romanian propolis extract (either prior to or after UV-light exposure) in a Swiss mouse model, suggested that the protective effect of propolis extracts can be explained in part by their inhibitory effect in inflammation, especially by reducing sources of free radicals in skin [49].

In addition, Kakoolaki *et al.* [46] had observed a reversed tendency on cypermethrin-induced increased lipid oxidation and reduced CAT activity, by Turkish propolis extract, in rainbow trout brain tissue [46]. Moreover, the oral administration of an ethanolic extract of Egyptian propolis prevented doxorubicin (Dox)-induced increase of lipid oxidation and reduction of GSH content in testis [50]. Tohamy *et al.* [47] also observed protective antioxidant effects of a water extract from brown Brazilian propolis, when administrated by oral gavage, in the hepatic, renal and testicular functions as well as the genotoxicity in the testis of cisplatin (CDDP) in male albino mice. The exposure of arsenic-exposed carps to tank water with Turkish propolis ethanolic extract was also reported to reverse the increased MDA levels and the decreased activity of CAT in muscle tissue [48]. Positive effects were also observed in muscle tissue, as well as in heart and spleen tissues of arsenic-exposed carps. Furthermore, oral administration of Turkish propolis extracts neutralized the negative impact of chlorpyrifos, malathion, and oxytetracycline in haematological parameters [41 - 43].

Likewise, the antioxidant activity of propolis has been demonstrated in *in vivo* models of oxidative stress related diseases, confirming its health promoting potential. The effect of ethanolic extracts of Brazilian green propolis and Chinese poplar propolis on streptozotocin (STZ)-induced type-1 diabetes mellitus in Sprague Dawley rats was reported by Zhu *et al.* [54]. According to the authors, both propolis samples not only improved body and kidney weights, but also the serum glucose, lipid profile, as well as the renal function. Simultaneously, renal GPx and serum SOD activities were improved whereas serum and hepatorenal MDA levels were significantly reduced. In another diabetic mice model, propolis extracts could attenuate alloxan-induced hepatotoxicity damages and decrease MDA levels [22].

Propolis extracts might also have a protective effect against hypercholesterolemia and oxidative stress, associated to the development of atherosclerosis, the main cause of cardiovascular diseases. Nader *et al.* [58] studied the oral administration of an ethanolic extract of Chinese propolis in simultaneous with cholesterol-enriched diet to rabbits. They observed a reduction of MDA levels and an increase of GSH in the serum, in addition to reduction of serum total cholesterol (TC), low density lipoprotein (LDL)-cholesterol, triglycerides and an increase of high density lipoprotein (HDL)-cholesterol concentration, therefore indicating that propolis can diminish the risk of atherosclerosis lesions by hypercholesterolemia *via* an antioxidant mechanism.

Furthermore, a Brazilian green propolis extract has been demonstrated to prevent DNA damages in the brain of hypoxia-exposed mice after intraperitoneal injection for 7 days [25]. Along with the decrease of oxidative damages, propolis administration could decrease the levels of pro-inflammatory cytokines in the somatosensory cortex, namely tumor necrosis factor-α (TNF-α), interleukin-1β (IL-1β) and IL-6. This protection against neuroinflammation can be useful to protect against cognitive deficits in aged people and in neurodegenerative pathologies, such as Alzheimer's disease. In agreement with this, the adding of crude Spanish propolis in the diet of aged-Wistar albino rats resulted in a reduction of lipid and protein oxidation in the plasma, in addition to decreased levels of glucose and TC in the blood [59].

Using a model of gastric lesions, the water immersion restraint stress (WIRS) model, Nakamura *et al.* [56] also observed that the oral administration of an ethanolic extract of Brazilian propolis to Wistar rats significantly attenuated the gastric mucosal lesions and decreased the levels of lipid oxidation. Other examples of beneficial effects of propolis reported in disease models can be observed in Table **3**.

1.4. Correlation of Antioxidant Activities of Propolis with Phenolic Compounds

Many reported studies have pointed out a strong correlation between the propolis antioxidant activity and its content in total phenolics and/or total flavonoids or even with specific phenolic constituents [1, 2, 5 - 7, 13, 15, 17]. In this last topic, caffeic acid phenethyl ester (CAPE), chrysin and pinocembrin are the most cited ones. This section resumes the most recent works (2014-2015) for which these isolated propolis constituents were accessed regarding their antioxidant properties. More information about the antioxidant activity of these compounds and their respective role in oxidative stress can be found elsewhere [60 - 62].

As can be observed in Table **4**, CAPE, chrysin and pinocembrin, at a concentration range of 10 to 50 μM, have been shown to efficiently reduce the toxicity of H_2O_2 [63], as well as that of 1-methyl-4-phenylpyridinium (MPP^+) [64], not only through their antioxidant nature, but also by activating signaling pathways. Actually, it has been demonstrated that one of CAPE´s antioxidant mechanisms of action evolves the activation of the ERK-Nrf2 signaling pathway responsible for the induction of HO-1 activation, as shown in HepG2 cells [63]. Besides this pathway, other signaling mechanisms may be activated leading to increased activation of antioxidant proteins such as glutathione reductase (Gr), glutathione peroxidase (Gpx), SOD and CAT. Thus, lipid oxidation reactions, ROS levels and antioxidant molecules such as GSH are modulated by these compounds, when cells are exposed to xenobiotics [60].

Table 4. Selected studies of antioxidant activities of CAPE, chrysin and pinocembrin, as measured by biological models.

Compound	Model	Treatment Conditions	Effects	Ref
Cellular models				
CAPE	HepG2 cells	AAPH or H_2O_2 + 1 to 50 μM	↑ARE and HO-1 expression; ↑ Nrf2 levels; CAPE Induces HO-1 Expression through the ERK-Nrf2 Signaling Pathway; ↓intracellular ROS	[63]
Pinocembrin	SH-SY5Y Cells	Pre-incubation with pinocembrin for 4h, and in the medium for an additional 24 h from 10 to 50 μM + MPP^+	↑cell viability; ↓ROS generation, apoptotic rate and caspase-3 activation; ↑membrane potential, decreased Bcl-2/Bax ratio; ↓cytochrome c release	[64]
Chrysin	PASMC under hypoxia	1 hour pre-treatment of 1 to 100 μM	↓cell proliferation; ↓collagen I and collagen III protein expression; ↓NOX4 expression; ↓lipid oxidation and ROS formation;	[65]
CAPE	retinal pigment epithelial cells (ARPE-19) under hypoxia	N.D*	↓VEGF ↓ROS levels; ↓PI3K/AKT and HIF-1α expression	[66]

(Table 4) contd.....

Compound	Model	Treatment Conditions	Effects	Ref
CAPE	mouse fibroblast cell lines L929	777.76 cGy/ Min + 50μM CAPE and/or trolox	↑cell viability; ↓ROS formation; ↓lipid oxidation; ↑GSH/GSSG ratio; ↑Nrf2, Gr, Gpx and SOD expression;	[73]
In vivo models				
Chrysin	Male Wistar rats (Serum and Liver)	Methotrexate + 40 and 80 mg/kg bw/day, oral administration	↓lipid oxidation; ↓ALT, AST and LDH activity; ↑GSH content; ↑Gr, Gpx, SOD and CAT activity; hepatoprotective effect by histopathological assay; ↓p53, Bax, and caspase-3 expression	[67]
Chrysin	Wistar rats (kidneys)	5-Fluorouracil + pre- and co-treatment of 50 and 100 mg/kg bw/day, oral administration	↓lipid oxidation; ↑Gpx, Gr, CAT and SOD activity; ↑GSH levels; ↓creatinine, blood urea nitrogen and KIM-1 levels; ↓LDH activity; ↓p53, Bax, and caspase-3 expression; renal protective effect by histopathological assay	[68]
Chrysin	Wistar rats (Serum and liver)	Cisplatin + pre-treatment of 25 and 50 mg/kg bw/day, oral administration	↑GSH levels; ↑Gpx, Gr, GST, G6-PD, CAT, SOD and QR activity; ↓XO activity; ↓lipid oxidation; ↓AST, ALT, LDH and GGT levels; ↓NOx, NF-kB, TNF-α levels; ↓iNOS and COX-2 expression; hepatoprotective effect by histopathological assay	[69]
Chrysin	albino rats (heart)	Doxorubicin + pre-treatment of 25 and 50 mg/kg bw/day, oral administration	↓CK-MB and LDH activity; cardiac protective effect by histopathological assay; ↓lipid oxidation; ↑GSH levels and CAT and SOD activity; ↓iNOS and COX-2 expression; ↓NOx and TNF-α and Bax levels; ↓Ccl2 levels	[70]
Chrysin	Rats (blood)	TCDD + 50 mg/kg/day, oral administration	↓lipid oxidation; ↓TNF-α levels; ↑IFN-γ	[71]
CAPE	Sprague- Dawley rats (brain)	Isoniazid and/or ethambutol + 10 mol/kg bw/day, intraperitoneal injection	↑total antioxidant activity; ↓total oxidant status; ↓lipid oxidation; ↑SOD activity; ↑PON-1 levels. Cerebral protective effect by histopathological assay	[72]

(Table 4) contd.....

Compound	Model	Treatment Conditions	Effects	Ref
Pinocembrin	Induced ischemia–reperfusion Male Wistar rats (hippocampus)	10 mg/kg bw/day, oral adminsitration	↓infarct size; ↓LDH and MPO activity; ↓NF-κB, TNF-α and IL-6; ↓lipid oxidation and NOx; ↑GSH content; ↓caspase-3 activity and cytochrome c release.	[74]
Chrysin	Freund's complete adjuvant induced arthritis in Wistar rats	50 mg/kg bw/day, oral administration	↓paw edema; testis protective effect by histopathological assay; ↑gonadosomatic index and testosterone levels; ↓LH and FSH; ↑StAR gene expression; ↓MPO activity, TNF-α levels, COX-2 and iNOS protein expression; ↑IL-10 level; ↓lipid oxidation and NOx; ↓FasL expression and caspase 3 activity	[75]
Chrysin	Cerebral ischemia induced in C57BL/6 mice (brain)	Pre-treatment of 25 to 100 mg/kg bw/day, oral administration	↓infarct volume; ↓lipid oxidation; ↑SOD activity; ↓iNOS, COX-2 and NF-κB expression; ↓GFAP and Iba-1 expression; ↓IL-1β, IL-6, IL-12, IL-1α, IL-17A, IFN-γ and TNF-α levels	[76]
CAPE	Wistar albino rats (lungs, plasma and bronchoal-veolar lavage fluid)	Pneumoperitoneum + 10 µmol/kg bw, intraperitoneal injection	↑total antioxidant activity in all samples ↓total antioxidant status in all samples; ↑PON-1 activity except for bronchoalveolar lavage fluid; ↓TNF-α and IL-6 levels; Lung protective effect by histopathological assay	[77]

AAPH - 2,2-Azobis-(2-amidinopropane) dihydrochlorid; ALT- alanine aminotransferase; AKT - Protein kinase B; ARE - Antioxidant Response Element; AST - aspartate aminotransferase; Bax - bcl-2-like protein 4; Bcl-2 - B-cell lymphoma 2; CAT - Catalase; CK-MB - Creatine kinase isoenzyme-MB; COX-2 - Cyclooxygenase-2; FasL - apoptosis stimulating fragment ligand; FSH - follicle-stimulating hormone; G6-PD - glucose-6-phosphate dehydrogenase; GFAP - glial fibrillary acidic protein; GGT - γ-glutamyl transpeptidase; Gpx - glutathione peroxidase; Gr - glutathione reductase; GSH - glutathione; GSSG - Glutathione disulfide; GST - glutathione-S-transferase; HO-1 - Heme Oxygenase-1; Iba-1 - ionized calcium binding adapter molecule; IFN-γ - interferon-γ; IL - interleukine; iNOS - inducible nitric oxide synthase; KIM-1 - Kidney Injury Molecule-1; LDH - lactate dehydrogenase; LH - luteinizing Hormone; MPO - myeloperoxidase; MPP+ - 1-methyl-4-phenylpyridinium; N.D – not determined; NF-κB - nuclear factor kappa B; Nox - mono-nitrogen oxides; NOX4 - NADPH oxidase 4; Nrf2 - nuclear factor (erythroid-derived 2)-like 2; PASMCs - pulmonary artery smooth muscle cells; PI3K - phosphoinositide 3-kinase; PON-1 - Serum paraoxonase/arylesterase 1; QR - quinone reductase; ROS - reactive oxugen species; SOD - superoxide dismutase; StAR - steroidogenic acute regulatory protein; TCDD - 2,3,7,8-tetrachlorodibenzo-p-dioxin; TNF-α - tumor necrosis factor-alpha; VEGF - Vascular endothelial growth factor; XO - xanthine oxidase.

*it was not possible for the authors to access the article to clarify these conditions.

In agreement with that, several mechanisms were described to be modulated by CAPE when inhibiting radiation-induced oxidative stress, as well as for this compound and for chrysin, when protecting from hypoxia conditions *in vitro* studies [65 - 66]. Similarly, factors associated with cell viability, as apoptotic index and caspase-3 activation, were already been shown to be reduced in SH-SY5Y line cells exposed to MPP^+ and treated with pinocembrin, and also in *in vivo* models using rats that were administered, either by orally or by intraperiotenial injection, with chrysin at doses ranging from 50 to 100 mg/kg bw/day [67 - 68]. In these and other studies, the co-administration of xenobiotics with CAPE and chrysin lead to the increase of antioxidant enzymes activity, reduced levels of ROS and inhibition of lipid oxidation reactions, similar to that observed in *in vitro* tests [67 - 72].

In vivo protection of CAPE, chrysin and pinocembrin have also been shown in models of oxidative stress-associated disorders, including cerebral ischemia [74, 76], arthritis [75] and pneumoperitoneum [77]. Besides reducing the characteristic organ damages, as demonstrated by histopathological assays, the compounds generally inhibited lipid oxidation and promoted antioxidant defenses, and commonly diminished cell-death [64, 74] and anti-inflammatory markers [70, 75 - 76].

2. ANTIOXIDANT PROPERTIES OF POLLEN

2.1. Assessment using Chemical Models

Unlike propolis, pollen is only modestly explored concerning its antioxidant properties, either using chemical or biological models. However, similarly to propolis, this hive product has been used in humans´ diet for centuries due to its well-recognized and unique composition in essential amino acids, fatty acids, minerals, vitamins and phytochemicals [78]. Simultaneously, pollen has been widely present in folk medicine as a result from its claimed beneficial effects. The latter have been partially associated to the presence of polyphenols in pollen samples, leading in the last few years to the development of several strategies to obtain polyphenolic enriched extracts from pollen matrices.

One of the main approaches is focused on the use of various polar solvents for obtaining enriched-phenolic extracts [79 - 88]. Indeed, some studies have shown that the prevalence of phenolic compounds might be strongly correlated with their antioxidant activity [80, 89, 90]. These extracts proved to be able to scavenge H_2O_2, $O_2^{\cdot-}$ and $OH^{\cdot}$ [91 - 94], showing a protective ability against ROS that can induce lipid oxidation reactions, as shown in β-carotene bleaching assay and TBARS assay (Table **5**). Nevertheless, as represented in the same Table, within the various methods that allow to evaluate the antioxidant activity, the 2,2-diphenyl-1-picrylhydrazyl (DPPH) assay has been the most frequently used, and therefore, the most suitable for comparisons between different studies regarding bee pollen extracts.

Table 5. Selected studies of antioxidant activities of worldwide pollen extracts, as measured by chemical models.

Geographic location	Botanical Origin	Solvent	Assay	Results	Ref.
Brazil (Paraíba)	*Mimosa gemmulata* and *Fabaceae*	EtOH Hexane Ethyl acetate	DPPH	DPPH: 104.5 to 106.1, 212.0 to 236.5 and 41.9 to 43.7 μg/mL Ext for EtOH, Hexane and Ethyl acetate	[79]
USA (Sonoram Desert)	6 MultiF	MeOH	DPPH RP	DPPH: EC_{50}= 20 to 145 μg/mL; RP: 0.0096 to 0,1136 mM trolox in liquid Ext	[80]
Romania (Transylvania)	12 MultiF	MeOH	DPPH RP ABTS	DPPH: 0.135 to 2.615 mmol trolox/g Ext; RP: 0.255 to 5.355 mmol Fe^{2+}/g Ext; ABTS: 0.546 to 6.838 mmol trolox/g Ext	[81]
Portugal (five national parks)	4 MonoF /1 MultiF	MeOH	DPPH BCB	DPPH: EC_{50}= 2160 to 8870 μg/mL Ext ;BCB: EC_{50}= 3110 to 6520 μg/mL Ext	[82]
Serbia (Belgrade)	*Zea mays L.* ≠genotypes	dissolved in the reaction mix	ABTS	ABTS: 79.94 to 104.38 mmol trolox/kg pollen	[83]
Brazil (Paraná and Alagoas)	N.D	Mix H_2O/EtOH	BCB	BCB: 8.8 to 23% inhib. for 50 μL of liquid Ext	[84]
Portugal and New Zeland	10 MonF	Mix H_2O/EtOH	DPPH	DPPH: EC_{50}= 40 to over 500 μg/mL Ext	[85]

(Table 5) contd.....

Geographic location	Botanical Origin	Solvent	Assay	Results	Ref.
Greece (Peloponnesus)	MultiF	MeOH fraction from H_2O Ext	DPPH	DPPH: EC_{50}= 250 µg/mL Ext	[86]
China (Jilin)	*Crataegus pinnatifida*	H_2O	DPPH RP MCA	DPPH: 10.68 to 37.89% at 100 to 400 µg/mL Ext; 82.06 ± 1.07 µmol Asc. Acid/g Ext; RP: 72.88 ± 0.93 µmol trolox/g Ext; 42.13 ± 1.14 µmol trolox/g of pollen; MCA: 14.48 ± 0.21 mg NA_2EDTA eq/g Ext.	[87]
China	*Schisandra chinensis*	Mix H_2O/EtOH	DPPH RP MCA	DPPH: inhibition of 60% at 75 µg/mL Ext; RP: 0.6 OD at 77.2 µg/mL Ext; MCA: 23.24 ± 0.79 mg NA_2EDTA eq/g Ext.	[88]
Portugal (Douro)	22 MonoF	MeOH	DPPH BCB	DPPH: EC_{50}= 300 to 450 µg/mL Ext; BCB: EC_{50}= 250 to 450 µg/mL Ext	[89]
Brasil (Bahia)	22 MonoF /3 MultiF	Fraction purified by Ethyl acetate	DPPH ABTS MCA	DPPH: EC_{50}= 10.7 to 209.1 µg/mL Ext; ABTS: 6.0 to 97.2 µg/mL Ext; MCA: EC_{50}= 171.9 to 1507.0 µg/mL Ext.	[90]
Korea (Kangwondo)	*Pinus densiflora*	Mix H_2O/EtOH	DPPH HPRSA LOP TBARS PO; RP	DPPH: EC_{50}≈ 500 µg/mL Ext; HPRSA: 22% inhib. at 500 µg/mL Ext; LOP: 35% inhib. at 2000 µg/mL Ext; TBARS: 35% inhib. at 2000 µg/mL Ext; PO: 13% inhib. at 2000 µg/mL Ext; RP: 0.75 OD at 2000 µg/mL Ext.	[91]
Poland (Kraków)	12 MonoF	Mix H_2O/MeOH	DPPH LPO HRSA	DPPH: 8.6 to 91% inhib. at 830 µg/mL Ext; LPO: 6.8 to 86.4 % at 1670 µg/mL Ext; HRSA: 10.5 to 98% at 1000 µg/mL Ext.	[92]
Japan	*Cistus ladaniferus*	HotH_2O H_2O EtOH	DPPH LO HRSC SORSC	DPPH: 55, 60 and 75 % inhib. for HotH_2O, H_2O and EtOH for 300µL of liq. Ext (1 g of pollen for 10 mL of solvent); LO: 98, 52 and 98% for HotH_2O, H_2O and EtOH for 200µL of liq. Ext; HRSC: 60, 60 and 90% inhibition for HotH_2O, H_2O and EtOH for 75 µL of liq. Ext; SORSC: 98,8%, 49,4% and 99% for HotH_2O, H_2O and EtOH for 20 µL of liq. Ext.	[93]

(Table 5) contd.....

Geographic location	Botanical Origin	Solvent	Assay	Results	Ref.
Spain (Extremadura)	*Echium plantagineum L.*	Mix H_2O/MeOH	SORSC NORSC	SORSC: EC_{25} = 800µg/mL Ext; NORSC: EC_{25} = 1900µg/mL Ext.	[94]
Thailand (Nan)	*Zea mays L*	Mix H_2O/MeOH	DPPH	DPPH: EC_{50} = 428.6 ± 29.0 µg/mL Ext.	[95]
India (west Bengal)	*Brassica juncea*;	EtOH	DPPH	DPPH: EC_{50} = 48 µg/mL Ext.	[96]
Turkey (Anzer)	N.D	MeOH	DPPH RP CUPRAC	DPPH: 0.65 to 8.98 mg/mL Ext; RP: 11.77 to 105.06 µmol trolox/g of pollen; CUPRAC: 33.1 to 91.8 µmol trolox/g of pollen.	[97]
Portugal and Spain	8 MonF	MeOH	DPPH TBARS	DPPH: EC_{50} = 10.7 to 209.1 µg/mL Ext; TBARS: 350 to 3700 µg/mL Ext.	[98]
Eslovakia (Nitra region)	3 MonoF	EtOH	DPPH RP	DPPH: inhib. of 47.97 to 86.25% for 100 µL of liq. Ext (1 g of pollen for 10 mL of EtOH); RP: 3055,8 to 4944,5 µg eq. Asc. Acid/mL for 100 µL of liq. Ext.	[99]
Bulgary (Ajtos)	N.D	H_2O	DPPH	DPPH: 28% inhib for 50 µg/mL Ext.	[100]
Brasil (Southeast Brazil)	3 MonF/ 4 MultiF	Mix H_2O/MeOH	DPPH	DPPH: 75.9 to 94.1% inhibition for 25 µg/mL Ext.	[101]

ABTS - 2,2'-azinobis(3-ethylbenzothiazoline-6-sulfonic acid) diammonium salt; BCB - β-carotene Bleaching assay; CUPRAC - Cupric ion reducing antioxidant capacity assay; DPPH - 2,2-diphenyl-1-picrylhydrazyl; EtOH - Ethanol; Ext - Extract; FRAP - ferric reducing antioxidant power; HotH_2O - Hot aqueous; HPRSA - hydrogen peroxide radical scavenging ability; HRSA - Hydroxyl radical scavenging ability; HRSC - Hydroxyl radical scavenging capability; LOP - linoleic acid peroxidation; MCA - metal chelating activity; MeOH - Methanol; MonF - Monofloral; MultiF - Multifloral; NORSC - nitric oxide radical scavenging capacity; PO - protein oxidation; RP - Reducing power; SORSC - superoxide radical scavenging capacity; TBARS - thiobarbituric acid reactive substances.

In general, the EC_{50} of bee pollen extracts for DPPH assay range from 2.16 to over 500 µg of extract/mL [79, 80, 82, 85, 86, 89, 90, 95 - 98], being the geographic origin [82, 84], the species that constitute pollen [80 - 82, 92, 99], the genotype [83], the seasonality [90], the pre-treatment before extraction [100 - 101] as well as the solvent used for extraction [79, 93, 97, 98] determinant factors in the extracts′ antioxidant activity. For instance, the antioxidant activity of water and hot water extract from Japanese *Cistus ladaniferus* pollen was lower than that observed in ethanol extracts, as shown in the work of Nagay *et al.* [93]. These variations result in part from the different solubility between families of polyphenolic compounds in the solvents used, as well as from the solubility of other macromolecules which can affect the extract purity and also its antioxidant activity. In fact, a study developed by Campos *et al.* [85] pointed out that non-phenolic material, possibly proteins, was responsible for part of pollen′s antioxidant activity. In addition to this, the presence of vitamins, like vitamin E, may affect the pollens′ antioxidant activity, leaving far from absolute the correlation between the total phenolic and/or flavonoid content of pollen with its antioxidant activity. However it is clear that phenolic compounds and flavonoids found in bee pollen like quercitrin, 8-methoxyherbacetin, 7-methoxyherbacetin, 7-methoxyherbacetin-3-O-sophoroside, quercetin-3-O-sophoroside, luteolin, tricetin and myricetin and phenylpropanoids largely contribute to the antioxidant activity, since these compounds present free radical scavenging properties [85, 92, 101].

2.2. Assessment using *In vitro* and *In vivo* Biological Models

Although it is not possible to directly transpose the beneficial effects of pollen extracts only by the evaluation of their antioxidant activity using chemical models, it is a fact that the application of pollen in folk medicine goes back several centuries. As a result, increasing tries to examine the mechanisms behind the therapeutic action of pollen have been developed. One of pollens′ effects may be related to the inhibition of ROS formation and the consequent phenomena (Table **6**). As show in *in vitro* studies, pollen extracts obtained with DMSO, methanol and hydromethanol solvents from a multifloral Turkish sample and from Spanish *Echium plantagineum L* and Korean *Typhaangustata* botanical origins, have been shown to reduce the intracellular ROS production [102 - 104] at concentration ranges of 0.3125 to 50 mg/mL. Therefore, cellular structures and molecules like DNA can be protected, as proved by Cheng *et al.* [87] for *Crataegus pinnatifida* pollen extracts within the aforementioned concentration ranges. Simultaneously, lipid and protein oxidation reactions have been shown to be inhibited by pollen hydroethanolic extracts, favoring cells viability, as reported by Lee *et al.* [104]. These protective effects were also observed *in vivo* models, when supplementing the mice diet with pollen [105, 106].

Besides the direct effects on ROS production at cellular level, pollen has also been suggested to modulate enzymatic antioxidant defenses (SOD, CAT, GPx) and expression of certain genes, even when using *in vivo* models (Table **6**). For instance, it has been reported that the content of GSH, SH groups and activity of SOD, CAT and GPx can be significantly increased in erythrocytes, upon administration of apis flower pollen or pollen from Croatia to the animals [105, 106]. At the same time, the expression of crucial proteins might be reduced as in the case of Heat-shock protein (HSPa9a) and tumor necrosis factor (ligand) superfamily, member 6 (Tnfsf6) which are involved in the stress and apoptosis signaling pathways. Secondary antioxidant defense mechanisms such as the proteossome, responsible for the turnover of normal and damaged intracellular proteins, may also be activated upon administration of pollen extracts, as demonstrated in HFL-1 embryonic fibroblast cell lines treated with aqueous extracts from Greek multifloral pollen [86].

Taking into account that living organisms are constantly subjected to multiple stress factors, several studies have attempted to evaluate the protective effect of crude pollen or of their extracts, especially at the level of xenobiotics (Table **6**). In an *in vitro* study, developed by Sousa *et al.* [107], the increased cellular ROS levels and cell mortality induced by *tert*-butyl hydroperoxide were demonstrated to be impaired by a Spanish *Echium plantagineum L* purified methanolic extract. Similar observations were seen in other studies using *in vivo* biological models. This was the case for propoxur (2-isopropoxyphenyl methylcarbamate) and for carbaryl, carbamate pesticides which are shown slightly toxic to humans and animals. When given to mice at doses of 225 mg/kg bw/day (for carbaryl) or 20 mg/kg bw/day (for propoxur), these agents negatively affected both the antioxidant enzymes (SOD, CAT, GPx), and oxidative stability in erythrocytes, liver, kidneys, brain and heart [108, 109]. These effects could still be partially inhibited by the simultaneous administration of toxins with *Brassica napus L.* pollen [108, 109]. A similar protective effect was observed for mice administered with an Egyptian pollen aqueous extracts by intraperiotaneal injection plus cisplatin [47]. Turkish multifloral pollen and Chinese *Schisandra chinensis* pollen also shown the ability to inhibit the effects of carbon tetrachloride (an hepatotoxic compound that yields trichloromethyl and trichloromethyl peroxyl radicals as a result of its metabolization in the liver) [11, 34]. Additionally, pollen extracts from Egyptian *Phoenix dactylifera L.* pollen were registered to protect against the toxicity of cadmium [110]. These compounds, as well as other reactive chemicals may additionally induce mutagenesis processes or DNA lesions like double strand breaks, that if not properly repaired may give rise to carcinogenesis processes.

Table 6. Selected studies of antioxidant activities of worldwide pollen extracts, as measured by *in vitro* and *in vivo* biological models.

Geographic Location	Model	Treatment Conditions	Effects	Ref
Cellular models				
Greece (MultiF pollen)	HFL-1 human embryonic fibroblasts	0.5 to 10 μg/mL of Aq Ext	↑CT-L proteossome activity; ↑β2 and β5 proteossome subunits.	[86]
China (*Crataegus pinnatifida*)	Mouse lymphocytes	100 μM H_2O_2 + 4 and 8 mg/mL of Aq Ext	↓DNA damages in a dependent concentration manner	[87]
Turkey (MultiF pollen)	K-562 cells and mononuclear cells	0 to 50 mg/mL of DMSO Ext	↓the production of intracellular ROS induced in K-562 cells	[102]
Spain (*Jara pringosa* and *Jara blanca*)	(RGC-5)	Induced stress H_2O_2, O_2^- and $OH^·$ + 300 μg/mL of EtOH Ext	↓intracellular ROS, presenting an EC_{50} of 9.99, 8.44 and 57.6 μg/mL for intracellular H_2O_2, O_2^- and $OH^·$.	[103]
Korea (*Typhaangustata*)	Murine osteoblastic MC3T3-E1 cell	0.3 mM H_2O_2 + 2 to 10 μg/mL of mix H_2O/EtOH Ext	↑viability; ↑production of collagen, ALP activity and mineralization; ↓TNFSF11 production;. ↓ lipid and protein oxidation.	[104]
Spain (*Echium plantagineum L)*	Caco-2 cells	0.3125 to 20 mg/mL of purified MeOH Ext for 24h + 4-6h 150 μM of t-BHP after Ext removal	↑viability; ↓ROS at low tested concentrations.	[107]
In vivo models				

(Table 6) contd.....

Geographic Location	Model	Treatment Conditions	Effects	Ref
Egypt (N.D)	Mice *Mus musculus* (kidney, liver and testis)	cisplatin twice a week for three weeks + 140 mg/kg bw/day of Aq Ext starting at the 2nd week, intraperiotaneal injection	↓lipid oxidation all the tissues; ↑CAT activity and GSH content; ↓clastogenetic changes; hepatoprotective effect by histopathological assay.	[47]
China (*Schisandra chinensis*)	Kunming mice (liver)	CCl4 on day 42 + 10, 20 and 40 g /kg bw/day of Mix H_2O/EtOH Ext for 42 days, oral administation	↓lipid oxidation; ↑SOD and GPx activity; ↓ALT and AST activity; hepatoprotective effect by histopathological assay	[88]
Ukraine (N.D)	Rats (erythrocytes)	N.D*, oral administration	↓lipid oxidation; ↑GSH, total SH groups. ↑Gpx and Gr activity.	[105]
Croatia (MultiF pollen - *Quercus* ilex - 47.2 %)	CBA/Hr mice (erythrocytes, liver and brain)	100 mg/kg bw/day of pollen, oral administration	↓lipid oxidation in the liver; ↓SOD activity in the liver and brain; ↑CAT activity in the liver; ↓expression of Hspa9a and Tnfsf6 in the liver; ↓expression of Caspase 1 and Ccl21c in the brain.	[106]
Turkey (*Brassica napus L*)	Wistar albino rats (erythrocytes, liver, brain, kidneys and heart)	carbaryl + 50 or 100 mg/kg bw/day of pollen, oral administation	↓lipid oxidation in all tissues except for the heart; ↑SOD, CAT e Gpx activity in all the tissues	[108]
Turkey (*Brassica napus L*)	Wistar rats (plasma, erythrocytes, liver, brain, kidneys and heart)	propoxur + 100 mg/kg bw/day of Aq Ext, oral administation	↓lipid oxidation in all tissues; ↑SOD, CAT, GPx activity; ↓ALP and AST activity.	[109]
Egypt (*Phoenix dactylifera L.*)	Wistar rats (testis)	$CaCl_2$ in the 1st day + 40 mg/kg bw of Mix H_2O/EtOH Ext for 56 days, intaperiotaneal injection	↓lipid oxidation; ↑GSH; ↑the sperm counts and mobility, testosterone levels.	[110]

(Table 6) contd.....

Geographic Location	Model	Treatment Conditions	Effects	Ref
Turkey (MF pollen - *Castanea sativa* L.- >45%)	Sprague- Dawley rats (plasma, erythrocytes and liver)	CCl_4 + 200 and 400 mg /kg bw/day of pollen, oral administration	↓lipid oxidation ↑SOD activity in all tissues; ↓AST and ALT activity and apoptotic index of hepatocytes; hepatoprotective effect by histopathological assay	[111]
Korea (*Pinus sylvestris L.*)	Freund's complete adjuvant induced arthritis in ICR mice (blood/sérum)	100 and 200 mg/kg bw/day of Mix H_2O/EtOH Ext, oral administration	↓paw edema in mice; ↓TNF-α, IL-1β and IL-6; ↓CIA**, rheumatoid factor and anti-type II collagen anti-body; ↓lipid oxidation, protein oxidation and advanced glycation products	[112]

ALP - Alkaline phosphatase; ALT - alanine aminotransferase; Aq – Aqueous; AST - aspartate aminotransferase; CAT – Catalase; Ccl21c - chemokine (C-C motif) ligand 21C (leucine); DMSO - Dimethyl sulfoxide; DNA - Deoxyribonucleic acid; EtOH – Ethanol; Ext – Extract; Gpx – glutathione peroxidase; Gr – glutathione reductase; GSH – glutathione; HSP – heat shock protein; IL – interleukin; MeOH – methanol; MultiF – Multifloral; N.D – not determined; RGC-5 - Retinal ganglion cells; ROS - reactive oxugen species; SOD – superoxide dismutase; t-BHP - *tert*-Butyl hydroperoxide; TNF-α – tumour necrosis factor;TNFSF - tumor necrosis factor ligand superfamily member.

*it was not possible for the authors to assess to the whole article. ** a T cell-dependent, Ab-mediated autoimmune condition induced by type II collagen.

As far as we are aware, published data concerning the application of pollen extracts in the inhibition of oxidative stress-related diseases is nearly nonexistent, being only found one study developed by Lee *et al.* [112], who evaluated the oral administration of an hydroethanolic extracts of Korean *Pinus sylvestris L.* pollen in arthritis-induced mice. The authors observed that lipid and protein oxidation, as well as advanced glycation products could be inhibited in mice supplemented with pollen, with additional effects in inflammatory mediators such as the TNF-α, IL-1β and IL-6.

Although pollen demonstrates a great potential in preventing oxidative stress, especially due to its rich composition in phenolic compounds that are able to quench radicals or neutralize xenobiotics, the data concerning its applications, benefits, mechanisms of action and absorption through the different cell tissues is still scarce. Therefore as the interest in pollen antioxidant properties increased almost exponentially in the past 5 years, it is expected that in the upcoming years further studies by different research areas will continue to rise, uncovering all these fundamental topics allowing to address the different mechanisms of action around this hive product.

CONCLUSION

Since ancient times, propolis and pollen have been associated to health beneficial effects. For the last years, many of these health effects have been related to their potential antioxidant activities, which in turn have been closely associated to their phenolic constituents. Several chemical assays have already shown that pollen and propolis antioxidant activity results from their radical scavenging activity and reducing power effect. When transposing to biological models, these hive products have, beyond a protective ability against ROS and inhibition of biomolecular oxidation in several cellular models, an effect around the enzymatic antioxidant defenses (SOD, CAT, GPx) and expression of specific genes. Moreover, these antioxidant mechanisms may be responsible for the observed reduction of oxidative stress in multiple organs. Taking into account the different approaches that have been used and the gathered results, propolis and pollen can be viewed as potential agents in the re-stabilization of cellular oxidative imbalance and in the prevention of oxidative stress related diseases. To confirm these results more *in vivo* studies should be conducted to determine their respective mechanisms of action, opening the door to develop potential pharmaceutical applications and/or functional foods.

CONFLICT OF INTEREST

The authors declare no conflict of interest regarding this publication.

ACKNOWLEDGEMENTS

The authors gratefully acknowledge the Portuguese Foundation for Science and Technology (FCT), European Union, QREN, FEDER, COMPETE, for funding the Organic Chemistry Research Unit (QOPNA) (project PEst-C/QUI/UI0062/2013; FCOMP-01-0124-FEDER- 037296) (Project PEst-OE/AGR/UI0681/2011), CI&DETS of Viseu (project PEst-OE/CED/UI4016/2014) and CERNAS (project PEst-OE/AGR/UI0681/2014).

REFERENCES

[1] Wang, K.; Zhang, J.; Ping, S.; Ma, Q.; Chen, X.; Xuan, H.; Shi, J.; Zhang, C.; Hu, F. Anti-inflammatory effects of ethanol extracts of Chinese propolis and buds from poplar (Populus×canadensis). *J. Ethnopharmacol,* **2014**, *155*(1), 300-311. [http://dx.doi.org/10.1016/j.jep.2014.05.037] [PMID: 24882729]

[2] Erdogan, S.; Ates, B.; Durmaz, G.; Yilmaz, I.; Seckin, T. Pressurized liquid extraction of phenolic compounds from Anatolia propolis and their radical scavenging capacities. *Food Chem. Toxicol,* **2011**, *49*(7), 1592-1597. [http://dx.doi.org/10.1016/j.fct.2011.04.006] [PMID: 21530603]

[3] Gregoris, E.; Stevanato, R. Correlations between polyphenolic composition and antioxidant activity of Venetian propolis. *Food Chem. Toxicol,* **2010**, *48*(1), 76-82. [http://dx.doi.org/10.1016/j.fct.2009.09.018] [PMID: 19766694]

[4] Guo, X.; Chen, B.; Luo, L.; Zhang, X.; Dai, X.; Gong, S. Chemical compositions and antioxidant activities of water extracts of Chinese propolis. *J. Agric. Food Chem,* **2011**, *59*(23), 12610-12616. [http://dx.doi.org/10.1021/jf202818p] [PMID: 22026502]

[5] Tyszka-Czochara, M.; Paśko, P.; Reczyński, W.; Szlósarczyk, M.; Bystrowska, B.; Opoka, W. Zinc and propolis reduces cytotoxicity and proliferation in skin fibroblast cell culture: total polyphenol content and antioxidant capacity of propolis. *Biol. Trace Elem. Res,* **2014**, *160*(1), 123-131. [http://dx.doi.org/10.1007/s12011-014-0019-3] [PMID: 24913100]

[6] Vargas-Sánchez, R.D.; Torrescano-Urrutia, G.R.; Acedo-Félix, E.; Carvajal-Millán, E.; González-Córdova, A.F.; Vallejo-Galland, B.; Torres-Llanez, M.J.; Sánchez-Escalante, A. Antioxidant and antimicrobial activity of commercial propolis extract in beef patties. *J. Food Sci,* **2014**, *79*(8), C1499-C1504. [http://dx.doi.org/10.1111/1750-3841.12533] [PMID: 25039743]

[7] Yang, H.; Dong, Y.; Du, H.; Shi, H.; Peng, Y.; Li, X. Antioxidant compounds from propolis collected in Anhui, China. *Molecules,* **2011**, *16*(4), 3444-3455. [http://dx.doi.org/10.3390/molecules16043444] [PMID: 21512452]

[8] Falcão, S.I.; Vale, N.; Cos, P.; Gomes, P.; Freire, C.; Maes, L.; Vilas-Boas, M. *In vitro* evaluation of Portuguese propolis and floral sources for antiprotozoal, antibacterial and antifungal activity. *Phytother. Res,* **2014**, *28*(3), 437-443. [http://dx.doi.org/10.1002/ptr.5013] [PMID: 23722631]

[9] Boisard, S.; Le Ray, A.M.; Gatto, J.; Aumond, M.C.; Blanchard, P.; Derbré, S.; Flurin, C.; Richomme, P. Chemical composition, antioxidant and anti-AGEs activities of a French poplar type propolis. *J. Agric. Food Chem,* **2014**, *62*(6), 1344-1351. [http://dx.doi.org/10.1021/jf4053397] [PMID: 24443994]

[10] Campos, J.F.; dos Santos, U.P.; Macorini, L.F.; de Melo, A.M.; Balestieri, J.B.; Paredes-Gamero, E.J.; Cardoso, C.A.; de Picoli Souza, K.; dos Santos, E.L. Antimicrobial, antioxidant and cytotoxic activities of propolis from Melipona orbignyi (Hymenoptera, Apidae). *Food Chem. Toxicol,* **2014**, *65*, 374-380. [http://dx.doi.org/10.1016/j.fct.2014.01.008] [PMID: 24412556]

[11] Frozza, C.O.; Garcia, C.S.; Gambato, G.; de Souza, M.D.; Salvador, M.; Moura, S.; Padilha, F.F.; Seixas, F.K.; Collares, T.; Borsuk, S.; Dellagostin, O.A.; Henriques, J.A.; Roesch-Ely, M. Chemical characterization, antioxidant and cytotoxic activities of Brazilian red propolis. *Food Chem. Toxicol,* **2013**, *52*, 137-142.
[http://dx.doi.org/10.1016/j.fct.2012.11.013] [PMID: 23174518]

[12] Gülçin, I.; Bursal, E.; Sehitoğlu, M.H.; Bilsel, M.; Gören, A.C. Polyphenol contents and antioxidant activity of lyophilized aqueous extract of propolis from Erzurum, Turkey. *Food Chem. Toxicol,* **2010**, *48*(8-9), 2227-2238.
[http://dx.doi.org/10.1016/j.fct.2010.05.053] [PMID: 20685228]

[13] Miguel, M.G.; Nunes, S.; Dandlen, S.A.; Cavaco, A.M.; Antunes, M.D. Phenols and antioxidant activity of hydro-alcoholic extracts of propolis from Algarve, South of Portugal. *Food Chem. Toxicol,* **2010**, *48*(12), 3418-3423.
[http://dx.doi.org/10.1016/j.fct.2010.09.014] [PMID: 20849908]

[14] Papotti, G.; Bertelli, D.; Bortolotti, L.; Plessi, M. Chemical and functional characterization of Italian propolis obtained by different harvesting methods. *J. Agric. Food Chem,* **2012**, *60*(11), 2852-2862.
[http://dx.doi.org/10.1021/jf205179d] [PMID: 22360702]

[15] Rocha, B.A.; Bueno, P.C.; Vaz, M.M.; Nascimento, A.P.; Ferreira, N.U.; Moreno Gde, P.; Rodrigues, M.R.; Costa-Machado, A.R.; Barizon, E.A.; Campos, J.C.; de Oliveira, P.F.; Acesio Nde, O.; Martins Sde, P.; Tavares, D.C.; Berretta, A.A. Evaluation of a propolis water extract using a reliable RP-HPLC methodology and *In vitro* and *In vivo* efficacy and safety characterisation. *Evid. Based Complement. Alternat. Med. 2013,* **2013**, 670451.

[16] Schmidt, E.M.; Stock, D.; Chada, F.J.; Finger, D.; Christine Helena Frankland Sawaya, A.; Eberlin, M.N.; Felsner, M.L.; Quinaia, S.P.; Monteiro, M.C.; Torres, Y.R. A Comparison between Characterization and Biological Properties of Brazilian Fresh and Aged Propolis *Bio.Med ,* **2014**, *2014*, 10.
[http://dx.doi.org/10.1155/2014/257617] [PMID: 257617]

[17] Socha, R.; Gałkowska, D.; Bugaj, M.; Juszczak, L. Phenolic composition and antioxidant activity of propolis from various regions of Poland. *Nat. Prod. Res,* **2015**, *29*(5), 416-422.
[http://dx.doi.org/10.1080/14786419.2014.949705] [PMID: 25185953]

[18] Thirugnanasampandan, R.; Raveendran, S.B.; Jayakumar, R. Analysis of chemical composition and bioactive property evaluation of Indian propolis. *Asian Pac. J. Trop. Biomed,* **2012**, *2*(8), 651-654.
[http://dx.doi.org/10.1016/S2221-1691(12)60114-2] [PMID: 23569988]

[19] Krishnaiah, D.; Sarbatly, R.; Nithyanandam, R. A review of the antioxidant potential of medicinal plant species. *Food Bioprod. Process,* **2011**, *89*(3), 217-233.
[http://dx.doi.org/10.1016/j.fbp.2010.04.008]

[20] Moon, J-K.; Shibamoto, T. Antioxidant assays for plant and food components. *J. Agric. Food Chem,* **2009**, *57*(5), 1655-1666.
[http://dx.doi.org/10.1021/jf803537k] [PMID: 19182948]

[21] Silva, V.; Genta, G.; Möller, M.N.; Masner, M.; Thomson, L.; Romero, N.; Radi, R.; Fernandes, D.C.; Laurindo, F.R.; Heinzen, H.; Fierro, W.; Denicola, A. Antioxidant activity of uruguayan propolis. *In vitro* and cellular assays. *J. Agric. Food Chem,* **2011**, *59*(12), 6430-6437.

[http://dx.doi.org/10.1021/jf201032y] [PMID: 21563839]

[22] Oršolić, N.; Sirovina, D.; Končić, M.Z.; Lacković, G.; Gregorović, G. Effect of Croatian propolis on diabetic nephropathy and liver toxicity in mice. *BMC Complement. Altern. Med,* **2012**, *12*, 117. [http://dx.doi.org/10.1186/1472-6882-12-117] [PMID: 22866906]

[23] Russo, A.; Troncoso, N.; Sanchez, F.; Garbarino, J.A.; Vanella, A. Propolis protects human spermatozoa from DNA damage caused by benzo[a]pyrene and exogenous reactive oxygen species. *Life Sci,* **2006**, *78*(13), 1401-1406. [http://dx.doi.org/10.1016/j.lfs.2004.10.085] [PMID: 16457855]

[24] Valente, M.J.; Baltazar, A.F.; Henrique, R.; Estevinho, L.; Carvalho, M. Biological activities of Portuguese propolis: protection against free radical-induced erythrocyte damage and inhibition of human renal cancer cell growth *in vitro. Food Chem. Toxicol,* **2011**, *49*(1), 86-92. [http://dx.doi.org/10.1016/j.fct.2010.10.001] [PMID: 20934479]

[25] Wu, Z.; Zhu, A.; Takayama, F.; Okada, R.; Liu, Y.; Harada, Y.; Wu, S.; Nakanishi, H. Brazilian green propolis suppresses the hypoxia-induced neuroinflammatory responses by inhibiting NF-kappaB activation in microglia. *Oxid. Med. Cell Longev. 2013,* **2013**.

[26] Türkez, H.; Yousef, M.I.; Geyikoglu, F. Propolis protects against 2,3,7,8-tetrachlorodibenzo-p-dioin-induced toxicity in rat hepatocytes. *Food Chem. Toxicol,* **2012**, *50*(6), 2142-2148. [http://dx.doi.org/10.1016/j.fct.2011.09.018] [PMID: 21963951]

[27] Fang, Y.; Li, J.; Ding, M.; Xu, X.; Zhang, J.; Jiao, P.; Han, P.; Wang, J.; Yao, S. Ethanol extract of propolis protects endothelial cells from oxidized low density lipoprotein-induced injury by inhibiting lectin-like oxidized low density lipoprotein receptor-1-mediated oxidative stress. *Exp. Biol. Med. (Maywood),* **2014**, *239*(12), 1678-1687. [http://dx.doi.org/10.1177/1535370214541911] [PMID: 24962173]

[28] Cardoso, S.M.; Ribeiro, M.; Ferreira, I.L.; Cristina Rego, A. Northeast Portuguese propolis protects against staurosporine and hydrogen peroxide-induced neurotoxicity in primary cortical neurons. *Food Chem. Toxicol,* **2011**, *49*(11), 2862-2868. [http://dx.doi.org/10.1016/j.fct.2011.08.010] [PMID: 21864632]

[29] Moreira, L.L.; Dias, T.; Dias, L.G.; Rogão, M.; Da Silva, J.P.; Estevinho, L.M. Propolis influence on erythrocyte membrane disorder (hereditary spherocytosis): a first approach. *Food Chem. Toxicol,* **2011**, *49*(2), 520-526. [http://dx.doi.org/10.1016/j.fct.2010.11.042] [PMID: 21130830]

[30] Murase, H.; Shimazawa, M.; Kakino, M.; Ichihara, K.; Tsuruma, K.; Hara, H. The Effects of brazilian green propolis against excessive light-induced cell damage in retina and fibroblast cells. *Evid. Based Complement. Alternat. Med., 2013,* **2013**. [http://dx.doi.org/10.1155/2013/238279]

[31] Simões-Ambrosio, L.M.; Gregório, L.E.; Sousa, J.P.; Figueiredo-Rinhel, A.S.; Azzolini, A.E.; Bastos, J.K.; Lucisano-Valim, Y.M. The role of seasonality on the inhibitory effect of Brazilian green propolis on the oxidative metabolism of neutrophils. *Fitoterapia,* **2010**, *81*(8), 1102-1108. [http://dx.doi.org/10.1016/j.fitote.2010.07.008] [PMID: 20637843]

[32] Szliszka, E.; Kucharska, A.Z.; Sokol-Letowska, A.; Mertas, A.; Czuba, Z.P.; Krol, W. Chemical composition and anti-inflammatory effect of ethanolic extract of brazilian green propolis on activated

J774A.1 macrophages. *Evid. Based Complement. Alternat. Med., 2013,* **2013**, 976415.

[33] Xuan, H.; Zhao, J.; Miao, J.; Li, Y.; Chu, Y.; Hu, F. Effect of Brazilian propolis on human umbilical vein endothelial cell apoptosis. *Food Chem. Toxicol,* **2011**, *49*(1), 78-85. [http://dx.doi.org/10.1016/j.fct.2010.09.034] [PMID: 20933568]

[34] Xuan, H.; Zhu, R.; Li, Y.; Hu, F. Inhibitory effect of chinese propolis on phosphatidylcholine-specific phospholipase C activity in vascular endothelial cells. *Evid. Based Complement. Alternat. Med., 2011,* **2011**, 985278. [http://dx.doi.org/10.1155/2011/985278]

[35] Kamiya, T.; Izumi, M.; Hara, H.; Adachi, T. Propolis suppresses CdCl(2)-induced cytotoxicity of COS7 cells through the prevention of intracellular reactive oxygen species accumulation. *Biol. Pharm. Bull,* **2012**, *35*(7), 1126-1131. [http://dx.doi.org/10.1248/bpb.b12-00062] [PMID: 22791161]

[36] Zhang, J.; Cao, X.; Ping, S.; Wang, K.; Shi, J.; Zhang, C.; Zheng, H.; Hu, F. Comparisons of ethanol extracts of chinese propolis (poplar type) and poplar gums based on the antioxidant activities and molecular mechanism. *Evid. Based Complement. Alternat. Med., 2015,* **2015**.

[37] Xuan, H.; Li, Z.; Wang, J.; Wang, K.; Fu, C.; Yuan, J.; Hu, F. Propolis reduces phosphatidylcholine-specific phospholipase C activity and increases annexin a7 level in oxidized-LDL-Stimulated human umbilical vein endothelial cells. *Evid. Based Complement. Alternat. Med., 2014,* **2014**, 307594. [http://dx.doi.org/10.1155/2014/465383]

[38] de Sá, R.A.; de Castro, F.A.; Eleutherio, E.C.; de Souza, R.M.; da Silva, J.F.; Pereira, M.D. Brazilian propolis protects Saccharomyces cerevisiae cells against oxidative stress. *Braz. J. Microbiol,* **2013**, *44*(3), 993-1000. [http://dx.doi.org/10.1590/S1517-83822013005000062] [PMID: 24516431]

[39] Kao, H-F.; Chang-Chien, P-W.; Chang, W-T.; Yeh, T-M.; Wang, J-Y. Propolis inhibits TGF-β--induced epithelial-mesenchymal transition in human alveolar epithelial cells *via* PPARγ activation. *Int. Immunopharmacol,* **2013**, *15*(3), 565-574. [http://dx.doi.org/10.1016/j.intimp.2012.12.018] [PMID: 23328619]

[40] Garoui, M.; Troudi, A.; Fetoui, H.; Soudani, N.; Boudawara, T.; Zeghal, N. Propolis attenuates cobalt induced-nephrotoxicity in adult rats and their progeny. *Exp. Toxicol. Pathol,* **2012**, *64*(7-8), 837-846. [http://dx.doi.org/10.1016/j.etp.2011.03.004] [PMID: 21507616]

[41] Enis Yonar, M.; Mişe Yonar, S.; Silici, S. Protective effect of propolis against oxidative stress and immunosuppression induced by oxytetracycline in rainbow trout (Oncorhynchus mykiss, W.). *Fish Shellfish Immunol,* **2011**, *31*(2), 318-325. [http://dx.doi.org/10.1016/j.fsi.2011.05.019] [PMID: 21642000]

[42] Yonar, S.M.; Ural, M.S.; Silici, S.; Yonar, M.E. Malathion-induced changes in the haematological profile, the immune response, and the oxidative/antioxidant status of Cyprinus carpio carpio: protective role of propolis. *Ecotoxicol. Environ. Saf,* **2014**, *102*, 202-209. [http://dx.doi.org/10.1016/j.ecoenv.2014.01.007] [PMID: 24480596]

[43] Enis Yonar, M.; Yonar, S.M.; Ural, M.S.; Silici, S.; Düşükcan, M. Protective role of propolis in chlorpyrifos-induced changes in the haematological parameters and the oxidative/antioxidative status of Cyprinus carpio carpio. *Food Chem. Toxicol,* **2012**, *50*(8), 2703-2708.

[http://dx.doi.org/10.1016/j.fct.2012.05.032] [PMID: 22634289]

[44] Nirala, S.K.; Bhadauria, M. Propolis reverses acetaminophen induced acute hepatorenal alterations: a biochemical and histopathological approach. *Arch. Pharm. Res,* **2008**, *31*(4), 451-461. [http://dx.doi.org/10.1007/s12272-001-1178-5] [PMID: 18449502]

[45] Oršolić, N.; Skurić, J.; Dikić, D.; Stanić, G. Inhibitory effect of a propolis on di-n-propyl disulfide or n-hexyl salycilate-induced skin irritation, oxidative stress and inflammatory responses in mice. *Fitoterapia,* **2014**, *93*, 18-30. [http://dx.doi.org/10.1016/j.fitote.2013.12.007] [PMID: 24370661]

[46] Kakoolaki, S.; Talas, Z.S.; Cakir, O.; Ciftci, O.; Ozdemir, I. Role of propolis on oxidative stress in fish brain. *Basic Clin. Neurosci,* **2013**, *4*(2), 153-158. [PMID: 25337342]

[47] Tohamy, A.A.; Abdella, E.M.; Ahmed, R.R.; Ahmed, Y.K. Assessment of anti-mutagenic, anti-histopathologic and antioxidant capacities of Egyptian bee pollen and propolis extracts. *Cytotechnology,* **2014**, *66*(2), 283-297. [http://dx.doi.org/10.1007/s10616-013-9568-0] [PMID: 23677589]

[48] Talas, Z.S.; Gulhan, M.F.; Erdogan, K.; Orun, I. Antioxidant effects of propolis on carp Cyprinus carpio exposed to arsenic: biochemical and histopathologic findings. *Dis. Aquat. Organ,* **2014**, *108*(3), 241-249. [http://dx.doi.org/10.3354/dao02714] [PMID: 24695237]

[49] Bolfa, P.; Vidrighinescu, R.; Petruta, A.; Dezmirean, D.; Stan, L.; Vlase, L.; Damian, G.; Catoi, C.; Filip, A.; Clichici, S. Photoprotective effects of Romanian propolis on skin of mice exposed to UVB irradiation. *Food Chem. Toxicol,* **2013**, *62*, 329-342. [http://dx.doi.org/10.1016/j.fct.2013.08.078] [PMID: 24013074]

[50] Rizk, S.M.; Zaki, H.F.; Mina, M.A. Propolis attenuates doxorubicin-induced testicular toxicity in rats. *Food Chem. Toxicol,* **2014**, *67*, 176-186. [http://dx.doi.org/10.1016/j.fct.2014.02.031] [PMID: 24593989]

[51] Yonar, M.E.; Yonar, S.M.; Çoban, M.Z.; Eroğlu, M. Antioxidant effect of propolis against exposure to chromium in Cyprinus carpio. *Environ. Toxicol,* **2014**, *29*(2), 155-164. [http://dx.doi.org/10.1002/tox.20782] [PMID: 22052654]

[52] Lopes, A.A.; Ferreira, T.S.; Nesi, R.T.; Lanzetti, M.; Pires, K.M.; Silva, A.M.; Borges, R.M.; Silva, A.J.; Valença, S.S.; Porto, L.C. Antioxidant action of propolis on mouse lungs exposed to short-term cigarette smoke. *Bioorg. Med. Chem,* **2013**, *21*(24), 7570-7577. [http://dx.doi.org/10.1016/j.bmc.2013.10.044] [PMID: 24262889]

[53] Fonseca, Y.M.; Marquele-Oliveira, F.; Vicentini, F.T.; Furtado, N.A.; Sousa, J.P.; Lucisano-Valim, Y.M.; Fonseca, M.J. Evaluation of the potential of brazilian propolis against UV-induced oxidative stress. *Evid. Based Complement. Alternat. Med., 2011,* **2011**. [http://dx.doi.org/10.1155/2011/863917]

[54] Zhu, W.; Li, Y.H.; Chen, M.L.; Hu, F.L. Protective effects of Chinese and Brazilian propolis treatment against hepatorenal lesion in diabetic rats. *Hum. Exp. Toxicol,* **2011**, *30*(9), 1246-1255. [http://dx.doi.org/10.1177/0960327110387456] [PMID: 20956460]

[55] Nakamura, T.; Ohta, Y.; Ohashi, K.; Ikeno, K.; Watanabe, R.; Tokunaga, K.; Harada, N. Protective effect of Brazilian propolis against hepatic oxidative damage in rats with water-immersion restraint stress. *Phytother. Res,* **2012**, *26*(10), 1482-1489.
[PMID: 22298415]

[56] Nakamura, T.; Ohta, Y.; Ikeno, K.; Ohashi, K.; Ikeno, T. Protective effect of repeatedly preadministered brazilian propolis ethanol extract against Stress-induced gastric mucosal lesions in rats. *Evid. Based Complement. Alternat. Med., 2014,* , 383482.**2014**
[http://dx.doi.org/10.1155/2014/383482]

[57] Sartori, G.; Pesarico, A.P.; Pinton, S.; Dobrachinski, F.; Roman, S.S.; Pauletto, F.; Rodrigues, L.C., Jr; Prigol, M. Protective effect of brown Brazilian propolis against acute vaginal lesions caused by herpes simplex virus type 2 in mice: involvement of antioxidant and anti-inflammatory mechanisms. *Cell Biochem. Funct,* **2012**, *30*(1), 1-10.
[http://dx.doi.org/10.1002/cbf.1810] [PMID: 22025285]

[58] Nader, M.A.; el-Agamy, D.S.; Suddek, G.M. Protective effects of propolis and thymoquinone on development of atherosclerosis in cholesterol-fed rabbits. *Arch. Pharm. Res,* **2010**, *33*(4), 637-643.
[http://dx.doi.org/10.1007/s12272-010-0420-1] [PMID: 20422375]

[59] Lisbona, C.; Díaz-Castro, J.; Alférez, M.J.; Guisado, I.M.; Guisado, R.; López-Aliaga, I. Positive influence of a natural product as propolis on antioxidant status and lipid peroxidation in senescent rats. *J. Physiol. Biochem,* **2013**, *69*(4), 919-925.
[http://dx.doi.org/10.1007/s13105-013-0270-7] [PMID: 23813443]

[60] Tolba, M.F.; Azab, S.S.; Khalifa, A.E.; Abdel-Rahman, S.Z.; Abdel-Naim, A.B. Caffeic acid phenethyl ester, a promising component of propolis with a plethora of biological activities: a review on its anti-inflammatory, neuroprotective, hepatoprotective, and cardioprotective effects. *IUBMB Life,* **2013**, *65*(8), 699-709.
[http://dx.doi.org/10.1002/iub.1189] [PMID: 23847089]

[61] Kasala, E.R.; Bodduluru, L.N.; Madana, R.M.; v, A.K.; Gogoi, R.; Barua, C.C. Chemopreventive and therapeutic potential of chrysin in cancer: mechanistic perspectives. *Toxicol. Lett,* **2015**, *233*(2), 214-225.
[http://dx.doi.org/10.1016/j.toxlet.2015.01.008] [PMID: 25596314]

[62] Rasul, A.; Millimouno, F.M.; Ali Eltayb, W.; Ali, M.; Li, J.; Li, X. Pinocembrin: A novel natural compound with versatile pharmacological and biological activities. *Biomed. Res. Int., 2013,* **2013**, *9*
[http://dx.doi.org/10.1155/2013/379850]

[63] Kim, J-K.; Jang, H-D. Nrf2-mediated HO-1 induction coupled with the ERK signaling pathway contributes to indirect antioxidant capacity of caffeic acid phenethyl ester in HepG2 cells. *Int. J. Mol. Sci,* **2014**, *15*(7), 12149-12165.
[http://dx.doi.org/10.3390/ijms150712149] [PMID: 25007817]

[64] Wang, Y.; Gao, J.; Miao, Y.; Cui, Q.; Zhao, W.; Zhang, J.; Wang, H. Pinocembrin protects SH-SY5Y cells against MPP+-induced neurotoxicity through the mitochondrial apoptotic pathway. *J. Mol. Neurosci,* **2014**, *53*(4), 537-545.
[http://dx.doi.org/10.1007/s12031-013-0219-x] [PMID: 24395092]

[65] Li, X-W.; Wang, X-M.; Li, S.; Yang, J-R. Effects of chrysin (5,7-dihydroxyflavone) on vascular

remodeling in hypoxia-induced pulmonary hypertension in rats. *Chin. Med,* **2015**, *10*(1), 4. [http://dx.doi.org/10.1186/s13020-015-0032-2] [PMID: 25722740]

[66] Paeng, S.H.; Jung, W.K.; Park, W.S.; Lee, D.S.; Kim, G.Y.; Choi, Y.H.; Seo, S.K.; Jang, W.H.; Choi, J.S.; Lee, Y.M.; Park, S.; Choi, I.W. Caffeic acid phenethyl ester reduces the secretion of vascular endothelial growth factor through the inhibition of the ROS, PI3K and HIF-1α signaling pathways in human retinal pigment epithelial cells under hypoxic conditions. *Int. J. Mol. Med,* **2015**, *35*(5), 1419-1426.
[PMID: 25738890]

[67] Ali, N.; Rashid, S.; Nafees, S.; Hasan, S.K.; Sultana, S. Beneficial effects of Chrysin against Methotrexate-induced hepatotoxicity *via* attenuation of oxidative stress and apoptosis. *Mol. Cell. Biochem,* **2014**, *385*(1-2), 215-223.
[http://dx.doi.org/10.1007/s11010-013-1830-4] [PMID: 24154663]

[68] Rashid, S.; Ali, N.; Nafees, S.; Hasan, S.K.; Sultana, S. Mitigation of 5-Fluorouracil induced renal toxicity by chrysin *via* targeting oxidative stress and apoptosis in wistar rats. *Food Chem. Toxicol,* **2014**, *66*(0), 185-193.
[http://dx.doi.org/10.1016/j.fct.2014.01.026] [PMID: 24486618]

[69] Rehman, M.U.; Ali, N.; Rashid, S.; Jain, T.; Nafees, S.; Tahir, M.; Khan, A.Q.; Lateef, A.; Khan, R.; Hamiza, O.O.; Kazim, S.; Qamar, W.; Sultana, S. Alleviation of hepatic injury by chrysin in cisplatin administered rats: probable role of oxidative and inflammatory markers. *Pharmacol. Rep,* **2014**, *66*(6), 1050-1059.
[http://dx.doi.org/10.1016/j.pharep.2014.06.004] [PMID: 25443734]

[70] Mantawy, E.M.; El-Bakly, W.M.; Esmat, A.; Badr, A.M.; El-Demerdash, E. Chrysin alleviates acute doxorubicin cardiotoxicity in rats *via* suppression of oxidative stress, inflammation and apoptosis. *Eur. J. Pharmacol,* **2014**, *728*(0), 107-118.
[http://dx.doi.org/10.1016/j.ejphar.2014.01.065] [PMID: 24509133]

[71] Ciftci, O.; Ozdemir, I. Protective effects of quercetin and chrysin against 2,3,7,8-tetrachlorodibenzo-p-dioxin (TCDD) induced oxidative stress, body wasting and altered cytokine productions in rats. *Immunopharmacol. Immunotoxicol,* **2011**, *33*(3), 504-508.
[http://dx.doi.org/10.3109/08923973.2010.543686] [PMID: 21214421]

[72] Uzar, E.; Varol, S.; Acar, A.; Firat, U.; Basarslan, S.K.; Evliyaoglu, O.; Yucel, Y.; Alp, H.; Gökalp, O. Assesment the role of oxidative stress and efficacy of caffeic acid phenethyl ester (CAPE) on neurotoxicity induced by isoniazid and ethambutol in a rat model. *Eur. Rev. Med. Pharmacol. Sci,* **2014**, *18*(19), 2953-2959.
[PMID: 25339492]

[73] Bai, H.; Liu, R.; Chen, H-L.; Zhang, W.; Wang, X.; Zhang, X-D.; Li, W-L.; Hai, C-X. Enhanced antioxidant effect of caffeic acid phenethyl ester and Trolox in combination against radiation induced-oxidative stress. *Chem. Biol. Interact,* **2014**, *207*(0), 7-15.
[http://dx.doi.org/10.1016/j.cbi.2013.10.022] [PMID: 24211618]

[74] Saad, M.A.; Abdel Salam, R.M.; Kenawy, S.A.; Attia, A.S. Pinocembrin attenuates hippocampal inflammation, oxidative perturbations and apoptosis in a rat model of global cerebral ischemia reperfusion. *Pharmacol. Rep,* **2015**, *67*(1), 115-122.
[http://dx.doi.org/10.1016/j.pharep.2014.08.014] [PMID: 25560584]

[75] Darwish, H.A.; Arab, H.H.; Abdelsalam, R.M. Chrysin alleviates testicular dysfunction in adjuvant arthritic rats *via* suppression of inflammation and apoptosis: comparison with celecoxib. *Toxicol. Appl. Pharmacol,* **2014**, *279*(2), 129-140.
[http://dx.doi.org/10.1016/j.taap.2014.05.018] [PMID: 24932515]

[76] Yao, Y.; Chen, L.; Xiao, J.; Wang, C.; Jiang, W.; Zhang, R.; Hao, J. Chrysin protects against focal cerebral ischemia/reperfusion injury in mice through attenuation of oxidative stress and inflammation. *Int. J. Mol. Sci,* **2014**, *15*(11), 20913-20926.
[http://dx.doi.org/10.3390/ijms151120913] [PMID: 25402649]

[77] Davarci, I.; Alp, H.; Ozgur, T.; Karcioglu, M.; Tuzcu, K.; Evliyaoglu, O.; Motor, S.; Durgun Yetim, T. Ameliorating effects of CAPE on oxidative damage caused by pneumoperitoneum in rat lung tissue. *Int. J. Clin. Exp. Med,* **2014**, *7*(7), 1698-1705.
[PMID: 25126167]

[78] Yang, K.; Wu, D.; Ye, X.; Liu, D.; Chen, J.; Sun, P. Characterization of chemical composition of bee pollen in China. *J. Agric. Food Chem,* **2013**, *61*(3), 708-718.
[http://dx.doi.org/10.1021/jf304056b] [PMID: 23265625]

[79] Silva, T.M.; Camara, C.A.; da Silva Lins, A.C.; Maria Barbosa-Filho, J.; da Silva, E.M.; Freitas, B.M.; de Assis Ribeiro dos Santos, F. Chemical composition and free radical scavenging activity of pollen loads from stingless bee Melipona subnitida Ducke. *J. Food Compos. Anal,* **2006**, *19*(6–7), 507-511.
[http://dx.doi.org/10.1016/j.jfca.2005.12.011]

[80] LeBlanc, B.W.; Davis, O.K.; Boue, S.; DeLucca, A.; Deeby, T. Antioxidant activity of Sonoran Desert bee pollen. *Food Chem,* **2009**, *115*(4), 1299-1305.
[http://dx.doi.org/10.1016/j.foodchem.2009.01.055]

[81] Mărghitaş, L.A.; Stanciu, O.G.; Dezmirean, D.S.; Bobiş, O.; Popescu, O.; Bogdanov, S.; Campos, M.G. *In vitro* antioxidant capacity of honeybee-collected pollen of selected floral origin harvested from Romania. *Food Chem,* **2009**, *115*(3), 878-883.
[http://dx.doi.org/10.1016/j.foodchem.2009.01.014]

[82] Morais, M.; Moreira, L.; Feás, X.; Estevinho, L.M. Honeybee-collected pollen from five Portuguese Natural Parks: palynological origin, phenolic content, antioxidant properties and antimicrobial activity. *Food Chem. Toxicol,* **2011**, *49*(5), 1096-1101.
[http://dx.doi.org/10.1016/j.fct.2011.01.020] [PMID: 21291944]

[83] Žilić, S.; Vančetović, J.; Janković, M.; Maksimović, V. Chemical composition, bioactive compounds, antioxidant capacity and stability of floral maize (*Zea mays* L.) pollen. *J. Funct. Foods,* **2014**, *10*, 65-74.
[http://dx.doi.org/10.1016/j.jff.2014.05.007]

[84] Carpes, S.T.; Begnini, R.; Alencar, S.M.; Masson, M.L. Study of preparations of bee pollen extracts, antioxidant and antibacterial activity. *Cienc. Agrotec,* **2007**, *31*, 1818-1825.
[http://dx.doi.org/10.1590/S1413-70542007000600032]

[85] Campos, M.G.; Webby, R.F.; Markham, K.R.; Mitchell, K.A.; Da Cunha, A.P. Age-induced diminution of free radical scavenging capacity in bee pollens and the contribution of constituent flavonoids. *J. Agric. Food Chem,* **2003**, *51*(3), 742-745.
[http://dx.doi.org/10.1021/jf0206466] [PMID: 12537451]

[86] Graikou, K.; Kapeta, S.; Aligiannis, N.; Sotiroudis, G.; Chondrogianni, N.; Gonos, E.; Chinou, I. Chemical analysis of Greek pollen - Antioxidant, antimicrobial and proteasome activation properties. *Chem. Cent. J,* **2011**, *5*(1), 33.
[http://dx.doi.org/10.1186/1752-153X-5-33] [PMID: 21699688]

[87] Cheng, N.; Wang, Y.; Gao, H.; Yuan, J.; Feng, F.; Cao, W.; Zheng, J. Protective effect of extract of Crataegus pinnatifida pollen on DNA damage response to oxidative stress. *Food Chem. Toxicol,* **2013**, *59*, 709-714.
[http://dx.doi.org/10.1016/j.fct.2013.07.015] [PMID: 23871827]

[88] Cheng, N.; Ren, N.; Gao, H.; Lei, X.; Zheng, J.; Cao, W. Antioxidant and hepatoprotective effects of Schisandra chinensis pollen extract on CCl4-induced acute liver damage in mice. *Food Chem. Toxicol,* **2013**, *55*(0), 234-240.
[http://dx.doi.org/10.1016/j.fct.2012.11.022] [PMID: 23201450]

[89] Feás, X.; Vázquez-Tato, M.P.; Estevinho, L.; Seijas, J.A.; Iglesias, A. Organic bee pollen: botanical origin, nutritional value, bioactive compounds, antioxidant activity and microbiological quality. *Molecules,* **2012**, *17*(7), 8359-8377.
[http://dx.doi.org/10.3390/molecules17078359] [PMID: 22785265]

[90] Freire, K.R.; Lins, A.C.; Dórea, M.C.; Santos, F.A.; Camara, C.A.; Silva, T.M. Palynological origin, phenolic content, and antioxidant properties of honeybee-collected pollen from Bahia, Brazil. *Molecules,* **2012**, *17*(2), 1652-1664.
[http://dx.doi.org/10.3390/molecules17021652] [PMID: 22314384]

[91] Lee, K-H.; Kim, A-J.; Choi, E-M. Antioxidant and antiinflammatory activity of pine pollen extract *in vitro*. *Phytother. Res,* **2009**, *23*(1), 41-48.
[http://dx.doi.org/10.1002/ptr.2525] [PMID: 19107823]

[92] Leja, M.; Mareczek, A.; Wyżgolik, G.; Klepacz-Baniak, J.; Czekońska, K. Antioxidative properties of bee pollen in selected plant species. *Food Chem,* **2007**, *100*(1), 237-240.
[http://dx.doi.org/10.1016/j.foodchem.2005.09.047]

[93] Nagai, T.; Inoue, R.; Inoue, H.; Suzuki, N. Scavenging capacities of pollen extracts from cistus ladaniferus on autoxidation, superoxide radicals, hydroxyl radicals, and DPPH radicals. *Nutr. Res,* **2002**, *22*(4), 519-526.
[http://dx.doi.org/10.1016/S0271-5317(01)00400-6]

[94] Moita, E.; Gil-Izquierdo, A.; Sousa, C.; Ferreres, F.; Silva, L.R.; Valentão, P.; Domínguez-Perles, R.; Baenas, N.; Andrade, P.B. Integrated analysis of COX-2 and iNOS derived inflammatory mediators in LPS-stimulated RAW macrophages pre-exposed to *Echium plantagineum* L. bee pollen extract. *PLoS One,* **2013**, *8*(3), e59131.
[http://dx.doi.org/10.1371/journal.pone.0059131] [PMID: 23520554]

[95] Chantarudee, A.; Phuwapraisirisan, P.; Kimura, K.; Okuyama, M.; Mori, H.; Kimura, A.; Chanchao, C. Chemical constituents and free radical scavenging activity of corn pollen collected from Apis mellifera hives compared to floral corn pollen at Nan, Thailand. *BMC Complement. Altern. Med,* **2012**, *12*(1), 45.
[http://dx.doi.org/10.1186/1472-6882-12-45] [PMID: 22513008]

[96] Ketkar, S.S.; Rathore, A.S.; Lohidasan, S.; Rao, L.; Paradkar, A.R.; Mahadik, K.R. Investigation of the

nutraceutical potential of monofloral Indian mustard bee pollen. *J. Integr. Med,* **2014**, *12*(4), 379-389. [http://dx.doi.org/10.1016/S2095-4964(14)60033-9] [PMID: 25074888]

[97] Ulusoy, E.; Kolayli, S. Phenolic composition and antioxidant properties of anzer bee pollen. *J. Food Biochem,* **2014**, *38*(1), 73-82. [http://dx.doi.org/10.1111/jfbc.12027]

[98] Pascoal, A.; Rodrigues, S.; Teixeira, A.; Feás, X.; Estevinho, L.M. Biological activities of commercial bee pollens: antimicrobial, antimutagenic, antioxidant and anti-inflammatory. *Food Chem. Toxicol,* **2014**, *63*, 233-239. [http://dx.doi.org/10.1016/j.fct.2013.11.010] [PMID: 24262487]

[99] Fatrcová-Šramková, K.; Nôžková, J.; Kačániová, M.; Máriássyová, M.; Rovná, K.; Stričík, M. Antioxidant and antimicrobial properties of monofloral bee pollen. *J. Environ. Sci. Health B,* **2013**, *48*(2), 133-138. [http://dx.doi.org/10.1080/03601234.2013.727664] [PMID: 23305281]

[100] Marinova, M.D.; Tchorbanov, B.P. Preparation of antioxidant enzymatic hydrolysates from honeybee-collected pollen using plant enzymes. *Enzyme Res., 2010,* **2010**. [http://dx.doi.org/10.4061/2010/415949]

[101] Negri, G.; Teixeira, E.W.; Alves, M.L.; Moreti, A.C.; Otsuk, I.P.; Borguini, R.G.; Salatino, A. Hydroxycinnamic acid amide derivatives, phenolic compounds and antioxidant activities of extracts of pollen samples from Southeast Brazil. *J. Agric. Food Chem,* **2011**, *59*(10), 5516-5522. [http://dx.doi.org/10.1021/jf200602k] [PMID: 21500799]

[102] Aliyazicioglu, Y.; Deger, O.; Ovali, E.; Barlak, Y.; Hosver, I.; Tekelioglu, Y.; Karahan, S.C. Effects of Turkish pollen and propolis extracts on respiratory burst for K-562 cell lines. *Int. Immunopharmacol,* **2005**, *5*(11), 1652-1657. [http://dx.doi.org/10.1016/j.intimp.2005.04.005] [PMID: 16039555]

[103] Nakajima, Y.; Tsuruma, K.; Shimazawa, M.; Mishima, S.; Hara, H. Comparison of bee products based on assays of antioxidant capacities. *BMC Complement. Altern. Med,* **2009**, *9*(1), 4. [http://dx.doi.org/10.1186/1472-6882-9-4] [PMID: 19243635]

[104] Lee, Y.S.; Choi, E.M. Effect of pollen from Typha angustata on hydrogen peroxide induced toxicity in osteoblastic MC3T3-E1 cells. *J. Oral Pathol. Med,* **2012**, *41*(2), 171-177. [http://dx.doi.org/10.1111/j.1600-0714.2011.01077.x] [PMID: 21958211]

[105] Dudov, I.A.; Starodub, N.F. [Antioxidant system of rat erythrocytes under conditions of prolonged intake of honeybee flower pollen load]. *Ukr. Biokhim. Zh,* **1994**, *66*(6), 94-96. [Antioxidant system of rat erythrocytes under conditions of prolonged intake of honeybee flower pollen load]. [PMID: 7785093]

[106] Sarić, A.; Balog, T.; Sobočanec, S.; Kusić, B.; Šverko, V.; Rusak, G.; Likić, S.; Bubalo, D.; Pinto, B.; Reali, D.; Marotti, T. Antioxidant effects of flavonoid from Croatian *Cystus incanus* L. rich bee pollen. *Food Chem. Toxicol,* **2009**, *47*(3), 547-554. [http://dx.doi.org/10.1016/j.fct.2008.12.007] [PMID: 19124059]

[107] Sousa, C.; Moita, E.; Valentão, P.; Fernandes, F.; Monteiro, P.; Andrade, P.B. Effects of colored and noncolored phenolics of *Echium plantagineum* L. bee pollen in Caco-2 cells under oxidative stress induced by tert-butyl hydroperoxide. *J. Agric. Food Chem,* **2015**, *63*(7), 2083-2091.

[http://dx.doi.org/10.1021/jf505568h] [PMID: 25642978]

[108] Eraslan, G.; Kanbur, M.; Silici, S. Effect of carbaryl on some biochemical changes in rats: the ameliorative effect of bee pollen. *Food Chem. Toxicol,* **2009**, *47*(1), 86-91. [http://dx.doi.org/10.1016/j.fct.2008.10.013] [PMID: 18996165]

[109] Eraslan, G.; Kanbur, M.; Silici, S.; Cem Liman, B.; Altınordulu, S.; Soyer Sarica, Z. Evaluation of protective effect of bee pollen against propoxur toxicity in rat. *Ecotoxicol. Environ. Saf,* **2009**, *72*(3), 931-937. [http://dx.doi.org/10.1016/j.ecoenv.2008.06.008] [PMID: 18707757]

[110] El-Neweshy, M.S.; El-Maddawy, Z.K.; El-Sayed, Y.S. Therapeutic effects of date palm (*Phoenix dactylifera* L.) pollen extract on cadmium-induced testicular toxicity. *Andrologia,* **2013**, *45*(6), 369-378. [http://dx.doi.org/10.1111/and.12025] [PMID: 22998418]

[111] Yilddiz, O.; Can, Z.; Saral, Ö.; Yulug, E.; Öztürk, F.; Aliyazicioilu, R.; Canpolat, S.; Kolayli, S. Hepatoprotective potential of chestnut bee pollen on carbon tetrachloride-induced hepatic damages in rats. *Evid. Based Complement. Alternat. Med., 2013,,* **2013**, *9*

[112] Lee, K-H.; Choi, E-M. Effect of pine pollen extract on experimental chronic arthritis. *Phytother. Res,* **2009**, *23*(5), 651-657. [http://dx.doi.org/10.1002/ptr.2526] [PMID: 19148883]

CHAPTER 9

Anti-Inflammatory Activity of the Honeybee Plant-Derived Products Honey, Pollen and Propolis

Joana Liberal[1,2], Isabel V. Ferreira[1], Eliza O. Cardoso[3], Ana Silva[1], Ariane R. Bartolomeu[3], João Martins[1,2], Karina B. Santiago[3], Bruno J. Conti[3], Bruno M. Neves[1,2,4], Maria T. Batista[1,2], José M. Sforcin[3,*], Maria T. Cruz[1,2,*]

[1] *Center for Neuroscience and Cell Biology, University of Coimbra 3004-517, Coimbra, Portugal*

[2] *Faculty of Pharmacy, University of Coimbra 3000-548, Coimbra, Portugal*

[3] *Biosciences Institute, UNESP, 18618-000, Botucatu, SP, Brazil*

[4] *Department of Chemistry, Mass Spectrometry Center, QOPNA, University of Aveiro, Campus Universitário de Santiago, 3810-193 Aveiro, Portugal*

Abstract: This chapter aims to discuss the effects of honeybee plant-derived products in inflammatory processes, with particular focus on honey, pollen and propolis. Honey is mainly composed by fructose and glucose, containing also minerals, proteins, free amino acids, vitamins and polyphenols and has long been used by humans not only for nutritional purposes but also as a medicine. The biological properties of honey can be ascribed to its polyphenolic content which, in turn, is usually associated to its anti-inflammatory activity, as well as antioxidant, antiproliferative and antimicrobial benefits. Bee pollen results from the agglutination of flower pollens with nectar and salivary substances of the honeybees. Due to its optimal nutritional balance, it has been considered as a perfect food all around the world and also used as a therapeutical agent. However, there is a lack of scientific support addressing the biological activities of bee pollen. Propolis is produced by bees from secretions of trees, trunks, buds, leaves and pollen, adding wax and substances secreted by bee glands.

* **Address correspondence to José Maurício Sforcin:** Department of Microbiology and Immunology, Biosciences Institute, UNESP, 18618-970, Botucatu, SP, Brazil; Tel: +55 14 3880-0445; Fax: +55 14 3815-3744; Email: sforcin@ibb.unesp.br and *__Maria Teresa Cruz:__ Faculty of Pharmacy and Center for Neurosciences and Cell Biology, Pólo das Ciências da Saúde, Azinhaga de Santa Comba, 3000-548 Coimbra, Portugal; Tel: +351 239 480209; Fax: +351 239 487362; Email: trosete@ff.uc.pt

Susana M. Cardoso (Ed)

The large and diverse number of chemicals in propolis may justify their biological activities, namely anti-inflammatory properties. Herein we emphasize the anti-inflammatory potential of the honeybee plant-derived products propolis, honey and pollen. Whenever possible we also disclose the action mechanisms and the principal compounds responsible for the biological activity. The intracellular signaling targets of propolis, honey and pollen are highlighted and summarized in Fig. (**1**). Overall, the production of inflammatory mediators, *i.e.* nitric oxide (NO) and prostaglandins, are inhibited by the three products partially due to the inhibition of nuclear factor kappa B (NF-κB) and mitogen activated protein kinases (MAPKs) signaling pathways.

Keywords: Bee pollen, Chemokines, Cytokines, Flavonoids, Honey, Honeybee, Immune cells, Immune system, Inflammation, Intracellular signaling pathways, Lipopolysaccharide, Macrophages, Mechanism of action, Nitric oxide, Nuclear factor kappa B, Polyphenols, Propolis, Prostaglandins.

1. INFLAMMATORY PROCESS: AN INTRODUCTION

The innate immune system provides a first line of defense and consists of cells and proteins that are always present and ready to mobilize immediately or within hours of an antigen's appearance in the body. The main components of the innate immune system include physical epithelial barriers, dendritic cells, phagocytic leukocytes, natural killer (NK) cells and plasma proteins. Besides its crucial role in the clearance of the antigen, cells of the innate immune system also take part in the initiation and subsequent development of adaptive immune responses as well as in the elimination of antigens that have been targeted by an adaptive immune response. The adaptive immune system is mediated by lymphocytes that once activated, proliferate and generate potent mechanisms for neutralizing or eliminating the antigen, also providing increased protection against a subsequent attack by the same antigen. The two types of adaptive immune responses include: humoral immunity, mediated by antibodies produced by B lymphocytes, and cell-mediated immunity, mediated by T lymphocytes [1]. The activation of both types of immunity initiates inflammation.

Inflammation is part of the body's immune response that may be triggered by an infection, burn, or other injuries, as an attempt of self-protection. Two stages of inflammation exist, acute and chronic inflammation. Acute inflammation is an initial stage (innate immunity) that usually persists for a short period and is generally beneficial for the host. If inflammation lasts for a longer time period, the second stage of inflammation, or chronic inflammation, become self-perpetuating over time predisposing the host to various chronic illnesses, including rheumatoid arthritis [2], asthma [3], inflammatory bowel diseases [4], atherosclerosis [5] and cancer [6]. Several cell types and chemical mediators are involved in the inflammatory response. Briefly, immune cells present receptors on their surfaces named pattern recognition receptors (PRRs), such as Toll-like receptors (TLRs), which recognize molecules highly conserved and shared among pathogens, collectively referred to as pathogen-associated molecular patterns (PAMPs). Macrophages have a key role in providing an early and immediate defense against foreign agents. Upon recognition by TLR4 of the gram negative cell wall component, lipopolysaccharide (LPS), macrophages become activated and produce a variety of pro-inflammatory mediators, including prostaglandins, nitric oxide (NO), tumor necrosis factor (TNF)-α, interleukin (IL)-1β, IL-6 and IL-10. These cytokines, which may be also produced by other immune cells, play a major role in the induction and regulation of inflammation, hematopoiesis, and immune response.

TNF-α is an endogenous pyrogen displaying systemic effects in the acute phase reaction and able to induce IL-1β and IL-6 production. IL-1β is a pleiotropic pro-inflammatory cytokine that stimulates systemic and local responses to infection. It induces the expression of adhesion molecules and chemokines on endothelial cells, leading to the traffic and infiltration of inflammatory and immunocompetent cells at the site of injury [7]. In addition, IL-1β causes vasodilatation and hypotension, fever and enhances pain sensitivity. IL-6 is a multifunctional regulator of immune response, hematopoiesis, and acute phase reactions [8]. In contrast, the anti-inflammatory cytokine IL-10 decreases the production of pro-inflammatory cytokines TNF-α and IL-1β [9].

Other important inflammatory mediators are prostaglandin E2 (PGE2), which is synthesized by the rate limiting enzyme cyclooxygenase (COX), and NO that is synthesized by nitric oxide synthase (NOS). The high-output of NO by the inducible form of NOS (iNOS) contributes to the pathogenesis of septic shock and inflammatory diseases. Therefore, the selective inhibition of COX-2 and iNOS in macrophages has been used to disclose new anti-inflammatory drugs, namely for honeybee plant-derived products screening [10].

The expression of pro-inflammatory molecules in macrophages is tightly regulated by several transcription factors and signaling pathways. Among these pathways, mitogen activated proteins kinases (MAPKs) play critical roles in the regulation of inflammation. The MAPKs pathways include p38 MAPK, c-jun NH_2-terminal kinase (JNK) and extracellular signal-regulated kinase (ERK), being these cascades involved on LPS-induced COX-2 and iNOS expression in macrophages. Furthermore, many inflammatory stimuli like LPS, cytokines and oxidants activate nuclear factor kappa B (NF-κB) through several signaling pathways that lead to the phosphorylation of inhibitory protein κB (IκB) by IκB kinase (Iκκ). This phosphorylation event triggers IκB ubiquitination and subsequent degradation by proteasome. IκB degradation exposes the nuclear localization motif of NF-κB that in turn is rapidly translocated to the nucleus, where it activates the transcription of target genes [10].

Currently, anti-inflammatory drugs are still present with limitations, mainly regarding their effectiveness, responsiveness, safety and cost of manufacture. Natural products are important sources of new drugs and notably, it has been widely shown that many bee-derived products have potential therapeutic effects, mainly through decreasing inflammation and modulating immune responses. In fact, therapies involving bees and their products, called "apitherapy", have been used for centuries in traditional medicine. Herein we describe the effects of the honeybee plant-derived products honey, pollen and propolis, on several inflammatory-associated signaling pathways and on the modulation of a inflammatory mediators' production.

2. HONEY

Honey is produced by honeybees from the nectar of flowers, plant secretions and excretions of plant-sucking insects. The chemical composition of the honey depends on its geographical origin, botanical sources and environmental conditions. The dominance of one particular plant classifies the honey as monofloral; if there are several botanical sources and none of them is predominant it is labeled multifloral (also known as polyfloral). Honey is a high concentrated solution of sugars, mainly fructose and glucose, but also contains other bioactive constituents as polyphenols, organic acids, carotenoids, vitamins and proteins [11].

Honey is one of the oldest known medicines and its use can be traced back to prehistoric times, as represented in Stone Age paintings. Besides its nutritional purposes, honey has been used to overcome liver, cardiovascular and gastrointestinal ailments, as well as for wound healing. However, due to the lack of scientific support, the use of honey was limited to folk medicine. Since a few decades ago, honey was subjected to laboratory and clinical investigations and it has found a place in modern medicine (reviewed in [12 - 14]). For instance, honey is now reintroduced in modern medical care for the treatment of wounds and burns.

Effectiveness of topical administration of honey is demonstrated by animal experiments and clinical trials, allowing to conclude that in addition to its high acidity, its favorable effect on wound regeneration was due to anti-infectious, anti-inflammatory and antioxidant properties [15,16]. Histological and clinical studies of wound healing were performed in two groups of 50 randomly allocated patients with fresh partial-thickness burns treated with honey dressing or with the antibiotic mafenide acetate. The results obtained clearly demonstrated an early subsidence of acute inflammatory changes, better control of infection, and quicker wound healing in honey-dressed wounds, while in mafenide acetate treated wounds the inflammatory reaction was sustained even on epithelialization [17]. Moreover, a clinical randomized study aimed to compare the abscess wounds healing dressed with either crude undiluted honey or a standard approach. Honey-treated wounds had quicker healing, thus allowing shorter length of hospital stay, relatively to those treated with the standard approach [18]. Recently, Makhdoom [19] and colleagues showed excellent results after diabetic wounds treatment with dressings soaked with natural honey. Indeed, the debility of diabetic foot patients was minimized by decreasing the frequency of leg or foot amputations, thus improving the quality and productivity of patients life. All the effects were partially ascribed to honey ability to lower prostaglandin levels and elevate NO end products [20].

Additionally, the ability of honey to modulate production and quenching of free radicals may contribute to the resolution of the inflammation in chronic wounds [21, 22]. More recently Majtan and colleagues [23] reported a new action mechanism of honey in wound healing through modulation of matrix metalloproteinase-9 (MMP-9), at both mRNA and protein levels, in human keratinocytes. MMP-9 is a protease with a crucial role in the degradation of matrix and growth-promoting agents in chronic wounds. The authors detected that a honey aqueous extract and its flavonoids apigenin and kaempferol inhibited TNF-α-induced production of MMP-9 in keratinocytes in a dose-dependent manner, thus preventing proteolytic activity in cutaneous inflammation.

Besides honey therapeutic value in wound healing, recently several *in vivo* (summarized in Table **1**) and *in vitro* (summarized in Table **2**) studies point out other benefits for different pathological conditions, which are strongly related to honey's anti-inflammatory properties. For instance, Majtanova and colleagues [24] reported the complementary use of honey for the treatment of contact lens-induced corneal ulcer. The treatment with topical levofloxacin and 25% (w/v) γ-irradiated honeydew honey solution was active and the patient attained final best corrected visual acuity of the damaged eye. The positive clinical outcome was associated to the antibacterial, anti-biofilm and anti-inflammatory properties of honey. The use of topical honey in the treatment of corneal abrasions and endotoxin-induced keratitis was also addressed in Lewis rats. Topical application of honey to injured corneas resulted in faster epithelial healing and decreased expression of vascular endothelial growth factor (VEGF), transforming growth factor (TGF)-β, interferon (IFN)-γ, IL-12 and TNF-α, thus corroborating the above exposed results performed in humans [24].

The authors also established that honey treatment reduced the inflammation in endotoxin-induced keratitis by reducing the levels of the angiogenic factors (VEGF and TGF-β), inflammatory cytokines (IL-12) and chemokines [25]. In another study [26], the anti-inflammatory and antioxidant effects of tualang honey in alkali injury were evaluated. Bashkaran and colleagues [26] demonstrated that honey treatment was as good as the conventional approach in conjunctival hyperemia, corneal edema and epithelial healing.

Table 1. *In vivo* Anti-inflammatory properties of honeybee plant-derived products.

	Type/Origin	*In vivo* model	Route of administration	Anti-inflammatory effects	Ref.
HONEY	Unprocessed	Fresh-thickness burns from 50 patients	Topical Honey dressing	Early subsidence of acute inflammatory changes Better control of infection Earlier woud-healing	[17]
	Unprocessed	Children pyomyositis abscesses	Topical Honey dressing	Faster healing Earlier granulation and re-epithelization (in comparison with chlorinated lime and boric acid solution)	[18]
	Unprocessed	Diabetic foot patients	Topical Honey dressing	↓ Leg or foot amputation	[19]
	Unprocessed	Lewis rats with corneal abrasions or endotoxin-induced keratitis	Topical	↓VEGF,TGF-β, IFN-γ, IL-12, TNF-α and CCR-5 levels	[25]
	Tualang	Rabbits with alkali injury on cornea	Topical and oral	Similar effects to those observed in conventional therapy	[26]
	Sterilised honeydew	One patient with persistent gluteofemoral fistulas	Topical	↓Inflammation, pain and induration	[27]
	Manuka Manuka + sulfasalazine	TNBS-induced rat colitis model	Intrarectal	Restored lipid peroxidation ↑Antioxidant parameters ↓Colonic inflammation	[28, 29]
	Honey	TNBS-induced rat colitis model	Intrarectal	Prevented free radical formation on inflamed tissues	[30]
	Monofloral and multifloral	Necrotising-induced lesions in rats	Oral	Total protection against ethanol and acidified aspirin-induced lesions, and positive protection against indomethacin-induced gastric lesions	[31]
	Unprocessed	Acetylsalicylic acid induced gastric ulcer in rats	Oral	The reduction in ulcers number was as effective as in the cimetidine (anti-ulcer drug) treated rats	[32]

(Table 1) contd.....

	Type/Origin	*In vivo* model	Route of administration	Anti-inflammatory effects	Ref.
HONEY	Tualang	Ovalbumin-induced asthma in rabbits	Aerosolized	↓Inflammatory cells ↓Goblet cell hyperplasia ↓Infiltration of inflammatory cells in the peribronchial region	[33]
	Gelam	Endotoxemia induced by LPS in rats	Intravenous	↓TNF-α, IL1-β, IL6, HMGB1 and NO serum levels ↑Increase of HO-1 serum levels	[34]
	Gelam	Endotoxemia induced by LPS in rabbits	Intravenous	↑Organ functions and survival rate ↓PMNs infiltration and MPO activity	[35]
	Gelam	Endotoxemia induced by LPS in rats	Intravenous	↓Peroxynitrite production	[45]
	Mimosa scabrella	12-*O*-tetradecanoylphorbol-13-acet-te-induced ear edema model on mice	Topical	↓Ear edema ↓MPO activity ↓Leucocyte infiltration ↓ROS	[36]
	Mixture (unprocessed honey, olive oil and beeswax)	Patients with atopic dermatitis or psoriasis	Topical	↓Pruritus, erythema, scaling and oozing	[37]
	Unprocessed	Formalin-induced paw licking CIPE	Oral	↓Paw edema	[38]
	Unprocessed	CIPE, Cotton pellet-induced granuloma and formaldehyde-induced arthritis	Oral	↓NO release	[39]
	Sterilized Gelam	CIPE	Oral	↓p65, p50 and IκBα gene expression ↓Translocation and activation of NF-κB ↓Cytosolic degradation of IκBα ↓COX-2, TNF-α expression	[40]
	Sterilized Gelam	CIPE	Oral	↓Paw edema size ↓NO, PGE2,TNF-α and IL-6 production ↓NOS, COX-2, TNF-α, and IL-6 protein and RNA levels	[41]
	Gelam	LPS-induced paw edema and CIPE	Intraperitoneal	↓Paw edema size ↓NO and PGE2 production	[42]
	Rewarewa	Arachidonic acid-induced ear oedema in mice	Topical	↓Leukocyte infiltration	[43]
	Unprocessed	Rat air pouch model of inflammation	Subcutaneous	↓Granulation tissue weight and angiogenesis ↓PGE2 and VEGF	[44]

(Table 1) contd.....

	Type/Origin	*In vivo* model	Route of administration	Anti-inflammatory effects	Ref.
PROPOLIS	Ethanolic extract Lageado Farm (FMVZ - UNESP)	BALB/c male mice treated with IFN-γ	Intraperitoneal	↓H_2O_2 production ↓NO production (higher concentrations)	[80]
	Brazilian green propolis Ethanolic extract Yamada Apiculture Center, Okayama, Japan	C57B/6N mice exposed to hypoxia	Intraperitoneal	↓IL-1*β*, TNF-α, IL-6, and 8-oxo- deoxyguanosine	[82]
	Propolis (WSD) Bulgaria	ICR mice infected with bacteria and fungus	Oral and parenteral	Improved survival Prevented cyclosphamide-induced imunosupression	[86]
	Ethanolic extract Lageado Farm, UNESP	Spleen cells from BALB/c male mice treated with concanavalin A	Intraperitoneal	↓ Splenocyte proliferation both in the presence or abscence of ConA	[87]
	Ethanolic extract Beekeeping Section, UNESP	C57BL/6 mice inoculated with B16F10 cells	Oral	↓Th1 (IFN-γ and IL-2) and Th2 (IL-10) cytokine expression/production	[89]
	Brazilian propolis ethanolic extract Yamada Apiculture Center, Okayama Japan	CIA in mice	Oral	↓Clinical arthritis scores numbers of IL-1- -producing cells	[90]
	Aqueous extract, Propharma (Stenlose, Denmark)	Patients with mild to moderate asthma	Oral	↓Incidence and severity of nocturnal attacks Improvement of ventilator functions ↓PGE_2, $PGF_{2\alpha}$, LTD-4 ↓TNF-α, IL-6, IL-8, ICAM-1 ↑IL-10	[93]

(Table 1) contd.....

	Type/Origin	*In vivo* model	Route of administration	Anti-inflammatory effects	Ref.
POLLEN	Cernilton N (pollen extract)	Patients with chronic prostatitis syndrome	Oral	42% improved significantly and 36% were cured of signs and symptoms	[99]
	Cistus sp. (ethanolic extract)	CIPE	Oral	↓Swelling thickness	[100]
	Honey-bee pollen mix (HMB)	CIPE	Oral	↓Swelling thickness	[101]

CCR5, chemokine receptor 5; CFs, patients without associated complicating factors; CIPE, carrageenan induced paw edema; CIA, collagen-induced arthritis; ConA, concanavalin A; COX-2, cyclooxygenase 2; HMGB1, high mobility Group Box 1; HO-1, heme oxygenase-1; ICAM-1, intercellular adhesion molecule 1; IFNγ, interferon gamma; IκBα, inhibitor of κBα; IL, interleukin; LTD-4, leukotriene D4; MPO, myeloperoxidase; NF-κB, nuclear factor κB; NO, nitric oxide; NOS, nitric oxide synthase; PGE, prostaglandin; PGF2α, prostaglandin F2α; PMNs, polymorphonuclear leukocytes; ROS, reactive oxygen species; TGF-β, transforming growth factor beta; TNBS, 2,4,6-trinitrobenzenesulfonic acid; TNF-α, tumor necrosis alpha; VEGF, vascular endothelial growth factor.

Another inflammatory condition for which honey has revealed beneficial results is inflammatory bowel disease. For instance, Vlcekova and colleagues [27] reported a patient with persistent fistulas in whom conventional medical and surgical therapy were unsuccessful. Importantly, most of the gluteal and femoral fistulas were healed and closed after 6 months of honey treatment. Honey also reduced inflammation, pain and induration of the affected region [27]. Corroborating these results, Khanduja and collaborators [28, 29] demonstrated that Manuka honey offers protection in experimentally induced inflammatory bowel disease in rats. These studies noticed that different doses of honey restored lipid peroxidation as well as improved antioxidant parameters. Morphological and histological scores as well as colonic inflammation were significantly reduced. Combination therapy (Manuka honey + sulfasalazine) also reduced colonic inflammation [28, 29]. In the same inflammatory model of colitis, intrarectal honey administration demonstrated potential therapeutic value, probably through preventing the formation of free radicals in inflamed tissues [30].

The gastric cytoprotective properties of natural honey (monofloral and polyfloral specimens) were also evaluated in rats using absolute ethanol, indomethacin and acidified acetylsalicylic acid as necrotising agents. Both types of honey elicited almost total protection against ethanol and acidified acetylsalicylic acid-induced lesions. Honey protection after indomethacin-induced gastric lesions was lower but still remained positive [31]. These results were later confirmed by the study of Bukhari and colleagues [32] that assessed the protective effects of natural honey in acetylsalicylic acid induced gastric ulcer. The authors concluded that the positive healing effects were almost the same in the honey and cimetidine (reference drug) groups [32].

Honey has shown beneficial therapeutic effects in inflammatory-related pathologies such as asthma, endotoxemia and skin inflammation. For instance, in a rabbit model of ovalbumin-induced asthma the effect of aerosolized honey on airway tissues was investigated. Histopathological analyses revealed that aerosolized honey promoted structural changes of the epithelium, mucosa, and submucosal regions, decreased the number of airway inflammatory cells present in bronchoalveolar lavage fluid and inhibited goblet cell hyperplasia [33]. Concerning endotoxemia, the works of Kassim and colleagues [34, 35] clearly proved that gelam honey ensured a protective effect against LPS-induced organ failure in both rabbits and rats. The protective effect on organs was related to a reduction in neutrophils infiltration as well as to a decrease in cytokines (TNF-α, IL-1β, and IL-10), high-motility group protein (HMGB)-1 and NO levels [34, 35].

The potential anti-inflammatory activity of the monofloral honey of *Mimosa scabrella,* produced by *Melipona marginata,* on mice skin was also investigated using a 12-*O*-tetradecanoylphorbol-13-acetate-induced ear edema model of inflammation [36]. Topical application of honey extract was able to reduce ear edema, decrease myeloperoxidase activity and diminish leucocyte infiltration and reactive oxygen species (ROS) production. This anti-inflammatory activity could be due to a synergic effect of the phenolic compounds identified in the honey sample, namely kaempferol and caffeic acid [36]. In an attempt to demonstrate the therapeutic value of honey in human skin disorders, Al-Waili showed that topical application of honey in patients suffering from atopic dermatitis and psoriasis promoted a significant improvement after 2 weeks of treatment [37].

In addition to these studies, other authors used more classic models of inflammation in an attempt to better explore the anti-inflammatory potential of honey and its molecular targets. Several authors evaluated the anti-inflammatory effect of honey using the carrageenan paw edema method of inflammation [38 - 41]. The proposed action mechanism involved inhibition of NF-κB translocation to the nucleus through the reduction of IκBα degradation, with subsequent decrease of iNOS, COX-2 and the pro-inflammatory cytokines TNF-α and IL-6 expression [40 - 42]. In a similar experimental model, the arachidonic acid-induced ear edema, other authors reported the anti-inflammatory effect of honey through suppression of leukocyte (monocyte and neutrophil) infiltration [43]. Furthermore, the anti-angiogenic effect of honey and its impact on inflammatory mediators was also disclosed in the rat air pouch model of inflammation. Honey was able to reduce granulation tissue weight and angiogenesis, which was correlated to its potent inhibitory activities against PGE2 and VEGF [44].

In a mechanistic point of view several *in vitro* studies (Table **2**) deeply explored putative molecular targets of honey. Kassim and colleagues [45] addressed the effect of gelam honey in LPS/IFN-γ-treated macrophages (RAW 264.7). The authors concluded that honey improved the viability of cells and inhibited NO production in a similar way to the inhibitor of iNOS - 1400W. Furthermore, honey, in contrast to 1400W, inhibited peroxynitrite production from the synthetic substrate 3-morpholinosydnonimine (SIN-1) and prevented the peroxynitrite-mediated conversion of dihydrorhodamine 123 into its fluorescent oxidation product rhodamine 123. In addition, honey also repressed peroxynitrite synthesis in LPS-treated rats [45].

Corroborating the inhibitory effects of honey in the production of pro-inflammatory mediators in other cell types, Ahmad and colleagues [46] demonstrated that tualang honey inhibited UVB-induced IκBα degradation and NF-κB activation, in a murine keratinocyte cell line. The treatment with tualang honey also inhibited UVB-triggered inflammatory cytokines, iNOS and COX-2 expression and also PGE production [46]. Additionally, the anti-inflammatory effect of 4 crude extracts from manuka (*Leptospermum scoparium*), kanuka (*Kunzea ericoides*), clover (*Trifolium spp.*), and a manuka/kanuka blend of honeys were examined.

Table 2. *In vitro* Anti-inflammatory properties of Honeybee plant-derived products.

	Type/Origin	Model	Effect/Mechanism of action	Ref.
HONEY	Heat-processed (commercial), Manuka and Pasture	Cell free assay	↓Free-radicals quenching activities	[21]
	Unifloral and multifloral	Cell free assay - antioxidant activity	↓Superoxide anion scavenging	[22]
		Zymosan-activated human neutrophils (PMNs)	↓ROS	
		Complement modulation assay with human serum and sheep erythrocytes	↓Human complement activity	
	Honeydew	TNF-α- stimulated HACAT	↓MMP9 mRNA and protein levels ↓MMP9 activity	[23]
	Gelam (methanolic and ethyl acetate extracts)	LPS-treated macrophages TNF-α-induced cytotoxicity in L929 cells	↓NO production ↓TNF- αtoxicity	[42]
	Gelam	LPS/IFNγ-treated macrophages (Raw 264.7 cells)	↑Cell viability ↓NO levels ↓Peroxynitrite scavenging	[45]
	Tualang	PAM212 mouse keratinocyte cell line exposed to UVB radiation	↓Translocation of NF-κB, and degradation of IκBα ↓IL6, TNF-α, IL-1β and iNOS protein expression ↓COX-2 expression and PGE2 production	[46]
	Kanuka and Manuka	HEK-Blue™-2, HEK-Blue™-4 or NOD2-WT cells	↓Inflammatory process through TLR1/TLR2 signaling pathway	[47]
	Multifloral (flavonoid extract)	LPS-stimulated N13 microglia	↓TNF-α, IL-1β and iNOS mRNA expression ↓iNOS protein expression ↓ROS production	[52]

(Table 2) contd.....

	Type/Origin	Model	Effect/Mechanism of action	Ref.
PROPOLIS	Ethanolic extract Beekeeping Section, UNESP, Campus of Botucatu	LPS-stimulated RAW 264.7 macrophages	↓NO levels ↓p38 MAPK, JNK1/2 and NF-κB expression	[53]
	Ethanolic extract Lageado Farm, UNESP, Campus of Botucatu *Baccharis dracunculifolia*	LPS-stimulated peritoneal macrophages from BALB/c mice	↓IL-6 production	[76]
	Poplar-type propolis Shandong province, North of China	LPS-stimulated RAW 264.7 macrophages	↓NO, IL1 and IL6 production ↓IκBα and AP-1 phosphorylation ↓iNOS, IL-6 and IL-1 mRNA expression	[81]
	Brazilian green propolis ethanolic extract Yamada Apiculture Center, Okayama, Japan	MG6 mouse microglial cells exposed to hypoxia	↓ROS levels ↓Activation of NF-κB ↓IL-1*β*, TNF-α, and IL-6 expression	[82]
	Ethanolic extract Beekeeping Section, UNESP, Campus de Botucatu	PBMC	↑IL-10 production	[84]
	Propolis (WSD) Bulgaria	Macrophages from IRC mice infected with bacteria and fungus (with/without immunosuppressors)	Improved survival Prevented cyclosphamide-induced immunosuppression	[86]
	Brazilian propolis Ethanolic extract Yamada Apiculture Center, Okayama, Japan	Spleen cells of normal mice stimulated with PMA	↓IL17 expression (dose dependent)	[90]
	Extracts Honig-Mehler, Neichen, Germany	PBMC Purified T lymphocytes	↑TGF-β ↓IL-1β, IL-12, IL-2 and IL-4 ↓ERK-2 ↓Phytohemagglutinin (PHA)-induced DNA synthesis of PBMC and T cells	[91]

(Table 2) contd.....

	Type/Origin	Model	Effect/Mechanism of action	Ref.
POLLEN	Comercial	Cell free Assay	↓Hyaluronidase activity	[95]
	Echium plantagineum L.	LPS-stimulated RAW 264.7 macrophages	↓NO, L-citruline levels ↓PGE production (low concentrations)	[97]
	Cistus sp.	LPS-stimulated RAW 264.7 macrophages	↓COX-1 and COX-2 activities ↓NO levels	[100]

AP-1, activator protein 1; COX, cyclooxygenase; ERK, extracellular signal-regulated kinase; IFNγ, interferon gamma; IκBα, inhibitor of κBα; IL, interleukin; iNOS, inducible nitric oxide synthase; JNK, c-Jun N-terminal kinase; LPS, lipopolysaccharide; MAPK, mitogen-activated protein kinases; MMP9, Matrix metallopeptidase 9; NF-κB, nuclear factor κB; NO, nitric oxide; PBMC, peripheral blood mononuclear cell; PGE, prostaglandin; PMNs, polymorphonuclear leukocytes; ROS, reactive oxygen species; TGF-β, transforming growth factor beta; TLR, toll-like receptor; TNF-*α*, tumor necrosis alpha; WSD, water-soluble derivative of propolis.

Anti-inflammatory assays were conducted in HEK-Blue™-2, HEK-Blue™-4, and nucleotide oligomerization domain (NOD)2-Wild Type (NOD2-WT) cell lines, to evaluate their inflammatory properties and their specificity towards different signaling pathways. Kanuka honey, and to a lesser extent, manuka honey, produced a powerful anti-inflammatory effect related to their phenolic content. The effect was observed in HEK-Blue™-2 cells using the synthetic tripalmitoylated lipopeptide Pam3CysSerLys4 (Pam3CSK4) ligand, suggesting that honey acts specifically through the TLR1/TLR2 signaling pathway. The manuka/kanuka blend and clover honeys had no anti-inflammatory effect in any cell line [47].

It is reasonable to speculate that honey's anti-inflammatory potential could be partially ascribed to its polyphenolic content. Indeed, epidemiological studies indicated that consumption of phenolic compounds-rich foods reduces the incidence of chronic inflammatory diseases. Actually, phenolic compounds are recognized for their therapeutic properties, such as antioxidant, anti-inflammatory, antiallergic, antiviral, antibacterial and antitumoral activities. Their anti-inflammatory properties are not only related to their antioxidant activity, but also to the modulation of signaling transduction pathways involved in the inflammatory response [10]. Accordingly, Kassim and colleagues [48] hypothesized that the different anti-inflammatory activities of honey were related to the varying amounts of phenolic compounds within the extracts tested.

Subsequently, honey methanol and honey ethyl acetate extracts were tested *in vitro* for their effect on NO production in stimulated macrophages. The major phenolic compounds in the extracts were ellagic, gallic, ferulic, chlorogenic and caffeic acids and the flavonol, myricetin. Other compounds found in lower concentrations were hesperetin, *p*-coumaric acid, chrysin, quercetin, luteolin and kaempferol. The authors concluded that both honey extracts exhibited anti-inflammatory potential that may be attributed, at least in part, to the phenolic compounds [48]. Several studies emphasize the anti-inflammatory potential of phenolic compounds presented in honey. For instance, chrysin suppressed LPS-stimulated pro-inflammatory response by blocking NF-κB and JNK activation, as well as COX-2, iNOS and cytokines production, in macrophages and microglia cells [49, 50]. Ahad and colleagues reported similar effects in an *in vivo* model [51]. The protective effect of a honey flavonoid extract on the production of pro-inflammatory mediators by LPS-stimulated N13 microglia was also addressed. The results showed that the honey flavonoid extract significantly inhibited the release of pro-inflammatory cytokines such as TNF-α and IL-1β. The expression of iNOS and ROS production were also significantly inhibited, which could be useful in preventive-therapeutic strategies for neurodegenerative diseases involving neuroinflammation [52]. Previously, our group also demonstrated that caffeic acid and luteolin suppress LPS-stimulated pro-inflammatory response by blocking NF-κB and MAPK activation in macrophages [53, 54].

In conclusion, the extensive data in the literature reinforces the role of honey as a therapeutic approach for the management of inflammatory diseases. This biological activity seems to be related to honey phenolic content and due to blockage of signaling pathways involved in the expression of several pro-inflammatory mediators.

3. PROPOLIS

Propolis (bee glue) has been used for medicinal purposes since ancient times due to its several therapeutic properties [55]. Propolis is produced by bees from secretions of trees, trunks, buds, leaves, pollen and substances secreted by bee glands, such as wax. The large number of chemical components in propolis may justify its several biological activities [56]. As for honey, propolis' chemical composition is highly variable and depends on the local flora [57] and hence, pharmacological variability among propolis samples is expected, thus hampering a systematical comparison between studies [58 - 60]. Therefore, propolis' biological properties should be linked to its chemical composition and botanical sources, but universal and reliable criteria for chemical standardization of different propolis types are still missing [61, 62].

Since propolis contains a huge amount of compounds, it is possible to hypothesize that its complex composition would lead to damage in the organism [63]. Nevertheless, it seems to be safe to humans, and no side effects were related after propolis administration to rats [56, 63 - 65]. Indeed, propolis has been used in some countries to treat a wide range of human ailments, such as respiratory infections, cardiovascular and blood systems disorder (anemia), dental care, dermatology (tissue regeneration, ulcers, healing of burn wounds, mycosis), cancer management, immune system support and enhancement, digestive tract conditions (ulcers and infections), liver protection, as well as an anti-inflammatory agent [66 - 70].

Propolis is undoubtedly the most studied of the three botanical products regarding the inflammatory potential with different approaches being used *in vitro* and *in vivo* models (see Tables **2** and **1**, respectively), in order to understand propolis' mechanism of action [71 - 74]. However, the greatest problem to carry out biological assays is both propolis standardization and the design of experimental protocols, since different concentrations of propolis in *in vivo* and *in vitro* studies, as well as different extracts, intake period and routes of administration have been used differently by researchers [75]. Moreover, biological assays should include well-established positive and/or negative controls, in order to compare propolis efficiency.

3.1. Propolis Effects on Innate Immune Cells

Recent studies have provided information of propolis influence on the immune/inflammatory response [62,75]. Previously, we treated macrophages with propolis or its phenolic acids before or after LPS addition to the cell culture, in order to respectively evaluate a preventive role of the natural products in inhibiting LPS pro-inflammatory action or alternatively a possible treatment of inflammatory diseases. This study was carried out with a propolis sample produced in southeast Brazil, which main vegetal sources were *Baccharis dracunculifolia* DC., followed by *Eucalyptus citriodora* Hook and *Araucaria angustifolia* (Bert.) O. Kuntze. Notably, the treatment with propolis or phenolic acids prevented LPS action before and after its challenge, and IL-6 production was lower than that induced by LPS alone [76], thus suggesting that these natural products have both preventive and protective effects. Since elevated levels of IL-6 are related to chronic inflammation, cardiac diseases and depression [77], propolis administration may represent a new option of treatment for such diseases. Propolis also inhibited IL-10 production either before or after macrophages incubation with LPS, whereas cinnamic and coumaric acids counteracted significantly LPS action only when added after LPS treatment [76]. Note that IL-10 is important for maintaining homeostasis of the host due to its immunoregulatory action [78] and high levels of this anti-inflammatory cytokine are found in immunosuppressed patients [79]. Hence, propolis inhibitory action on IL-10 production may be useful to prevent infections.

The process of phagocytosis is complex and involves the binding of the target to the surface of macrophages and ingestion, which usually triggers the so-called oxidative burst. Notably, Brazilian propolis has been described to increase hydrogen peroxide (H_2O_2) generation by peritoneal macrophages, favoring microorganisms killing [80]. Brazilian propolis and caffeic acid have also been reported to inhibit NO production in macrophages, at non-cytotoxic concentrations. Furthermore, both propolis and caffeic acid suppressed LPS-induced signaling pathways, namely MAPKs such as p38 MAPK and JNK and NF-κB. The ERK1/2 was not affected by propolis extract and caffeic acid. These findings suggest its use as a natural source of safe anti-inflammatory drugs [53]. In addition, Wang and colleagues [81] verified that poplar (*Populus* sp.) Chinese propolis inhibited the production of NO, IL-1 and IL-6 in LPS-stimulated RAW 264.7 cells and suppressed mRNA expression of iNOS, IL-1 and IL-6 in a time- and dose-dependent manner. The authors also demonstrated that propolis suppressed the phosphorylation of IκBα and activator protein (AP)-1 without affecting IκBα degradation [81].

Wu and colleagues [82] reported that systemic treatment with Brazilian propolis for 7 days inhibited microglial expression of pro-inflammatory cytokines and 8-oxo-deoxyguanosine, a biomarker for oxidative damaged DNA, in the somatosensory cortex of mice subjected to hypoxia exposure. These observations indicate that propolis overwhelms the hypoxia-induced neuroinflammatory responses through inhibition of NF-κB activation in microglia, suggesting that this might be beneficial in preventing hypoxia-induced neuroinflammation [82]. Additionally, Ivanovska and colleagues [83] detected a significant reduction of acute inflammation after oral application of water soluble derivative (WSD) of propolis in an *in vivo* mouse model of inflammation [83].

Brazilian propolis increased the expression of TLR-4 and CD80 in human monocytes, which are important for the recognition of microorganism and stimulation of T cells, respectively. Concerning TNF-α and IL-10 production by human monocytes, both cytokines were stimulated by low concentrations of propolis and inhibited by higher ones. Surprisingly, the fungicidal activity was increased using high concentrations of propolis, and no correlation with cytokine production was found, which occurred using different concentrations [84].

It is not known which compounds of propolis are responsible for its activities, although literature suggests that there may be a synergistic effect amongst its components. Besides cafeic acid, and cinnamic acid, other bioactive components of propolis have also proven to possess anti-inflammatory effects. For instance, Szliszka and colaborators [85] investigated the effects of artepillin C (3,5-diprenyl-4-hydroxycinnamic acid *i.e.*, the main bioactive constituent of Brazilian green propolis) in RAW 264.7 macrophages for further understanding its anti-inflammatory mechanisms of action. The authors observed that artepillin C inhibited the production of oxygen and nitrogen reactive species, as well as that of cytokines (IL-1, IL-3, IL-4, IL-5, IL-9, IL-12p40, IL-13, IL-17, TNF-α, granulocyte colony stimulating factor (G-CSF), granulocyte macrophage colony-stimulating factor (GM-CSF), monocyte chemoattractant protein (MCP)-1, macrophage inflammatory protein (MIP)-1, RANTES and chemokine C-X-C motif ligand (CXL)-1). Artepillin C also blocked NF-κB expression in macrophages [85]. Therefore, it is reasonable to conclude that propolis exerts anti-inflammatory effects probably due to inhibition of NO and cytokine production.

3.2. Propolis Effects on Adaptive Immune Cells

Adaptive immunity is important to the recognition of "non-self" antigens, in order to eliminate specific pathogens or pathogen infected cells, followed by the development of immunological memory, so that each pathogen is "remembered" after the first exposure and quickly eliminated on a latter contact. When T helper cells (Th) interact with antigen presenting cells (APC), the following events are necessary to their activation: 1) TCR-CD3 complex on Th cell binds to the antigen-major histocompatibility complex (MHC) on the APC surface, and 2) CD28 on Th cell binds to CD80 (B7.1) or CD86 (B7.2) on APC. After these two signals, Th cells produce IL-2, which acts in an autocrine way to induce their proliferation. Afterwards, Th cells develop into effector cells and may differentiate into two subtypes: Th1 and Th2 cells. Th1 cells produce IFN-γ and IL-2, among others, and active cell-mediated immunity, favoring the defense against intracellular pathogens, by activating macrophages, CD8+ T cells and NK cells. On the other hand, Th2 cells produce predominantly IL-4, IL-5, IL-6, IL-10 and IL-13, which promote humoral immunity, inducing antibody production against extracellular microorganisms.

Propolis action was thought to be limited mainly to macrophages, with no influence on lymphocyte proliferation [86]. However, more recently, lymphocyte polyclonal activation on propolis-treated mice was monitored and an inhibitory effect of Brazilian propolis on splenocyte proliferation was observed [87]. Moreover, previous studies have shown that flavonoids have an immunossupressor effect on the lymphoproliferative response [88]. Since propolis contains flavonoids in its chemical composition, they could be responsible for this effect. Baseline proliferation of splenocytes was not affected after mice treatment with propolis for 3 days. However, concanavalin (Con)A-stimulated cells of propolis-treated animals had a significantly reduction in proliferation, whereas control mice showed a normal proliferative response to this mitogen [87]. One explanation for these results could be the production of cytokines with antiproliferative effect on responding T cells or also the induction of macrophages' mediators that could decrease the proliferative response.

Furthermore, propolis administration for 14 days to C57BL/6 mice led to inhibition of IL-1β, IL-6, IFN-γ, IL-2 and IL-10 production by spleen cells. These effects may account for its anti-inflammatory activity since it is well established that cytokines orchestrate and perpetuate the chronic inflammatory response associated with several diseases [89]. In order to verify whether propolis would be able to moderate the severity of rheumatoid arthritis, Ansorge and colleagues evaluated the effect of a Brazilian propolis ethanolic extract on the pathogenesis of collagen-induced arthritis (CIA) in mice. Propolis inhibited the differentiation of Th17 cells from murine splenocytes in a concentration-dependent way, providing new insights on the potential mechanism behind its immunosuppressive and anti-inflammatory effects [90]. Another study, reported that propolis inhibited IL-12, IL-2, IL-4 and IL-10 production, whereas the production of TGF-β1 by T regulatory cells was increased in human peripheral blood mononuclear cells (PBMC) or T cell cultures. TGF-β1 and IL-10 may be produced by T regulatory cells. Since propolis increases TGF-β1 production, this cytokine could also influence cell division, as well as decrease the production of other pro-inflammatory cytokines. IL-12 is thought to drive differentiation of T cells towards Th1 type cell. Since propolis was able to inhibit the production of IL-12, IL-2 and IL-4, it was suggested that propolis and some of its constituents could inhibit Th1 and Th2 cells. In addition propolis strongly suppressed DNA synthesis on human PBMC and purified T cells in a dose-dependent way. These effects were, at least in part, mediated by some of its constituents, namely caffeic acid phenethyl ester (CAPE), and the flavonoids quercetin and hesperidin [91].

In order to understand the molecular mechanisms behind the negative regulation of cellular growth by propolis, Ansorge and colleagues [91] studied the MAPK signal pathway, measuring the induction of ERK-2 mRNA expression that is involved in the regulation of several transcription factors, which in turn control the regulation of critical genes of lymphocytes, including IL-2. In this study, the authors showed that ERK-2 was strongly suppressed in propolis-stimulated PBMC, clearly suggesting that one way of signaling triggered by propolis is mediated by the MAPK ERK-2 [91]. Another explanation for propolis inhibitory effects on lymphoproliferation derives from the observation that CAPE has inhibitory effects on both transcription factors NF-κB and Nuclear factor of activated T-cells (NFAT) [92]. Consequently, CAPE inhibited IL-2 gene transcription, IL-2R (CD25) expression, and proliferation of human T cells, providing new insights into the molecular mechanisms involved in the anti-inflammatory and immunomodulatory activities of this natural compound.

Khayyal and collaborators [93] investigated the effects of a 13% aqueous extract of propolis as an adjuvant to therapy of patients with mild to moderate asthma, after its administration daily for 2 months [93]. At the end of the study, patients receiving propolis showed a marked reduction in the incidence and severity of nocturnal attacks and improvement of ventilation functions, which were associated with decreases of PGE_2 and $F_{2\alpha}$ ($PGF_{2\alpha}$), leukotriene D4 (LTD-4) pro-inflammatory cytokines (TNF-α, IL-6, IL-8) and increased IL-10. Chan and colleagues [94] reported that two main immunopotent chemicals have been identified: CAPE and artepillin C. In addition, Brazilian propolis, CAPE, and artepillin C have been shown to exert an immunosuppressive action on T lymphocyte subsets but paradoxically activated macrophage function [94].

Overall, the reported data suggests that propolis may modulate both innate and adaptive responses, showing pro- and anti-inflammatory activities, depending on concentration, intake period and experimental conditions. Thus, it all should be taken into consideration to obtain the expected results, since inhibitory or stimulatory activities might be expected. Little is known concerning propolis clinical efficiency, although there have been a great number of publications lately considering the immunomodulatory and antitumor action of propolis and its constituents, which indicates their potential for the development of new pharmaceutical agents.

4. BEE POLLEN

Since memorable times, bee pollen is considered a good source of energy and nourishing substances. Pollen is claimed the "only perfectly complete food" since it contains carbohydrates, crude fibers, lipids, vitamins, minerals and all the essential amino acids needed for the human organism. It also contains phenolic compounds such as flavonoids. Although several therapeutic properties have been recently highlighted, such as antimicrobial, antifungal, antioxidant, hepatoprotective and chemoprotective activities, studies addressing the imunomodulatory potential of bee pollen are much scarcer than those reported for propolis and/or honey. For instance, Pascoal and colleagues [95] studied the anti-inflammatory activity of commercial bee pollen through measurement of the inhibitory effect on the reactions catalyzed by hyaluronidase in a cell-free assay [95]. The authors reported anti-inflammatory properties for pollen, although lower than that of propolis [96], what might be related with the lower amount of polyphenols in pollen. The anti-inflammatory potential of an *Echium plantagineum* L.bee pollen hydromethanolic extract (mainly constituted by kaempferol heterosides) was also explored by Moita and colleagues [97] through assessment of COX-2 and iNOS derived inflammatory mediators in LPS-stimulated RAW 264.7 macrophages. The results demonstrated that the *Echiumplantagineum* bee pollen extract dose-dependently decreased NO and L-citrulline production in LPS-stimulated cells at non-toxic concentrations. Additionally, the levels of prostaglandins, their metabolites and isoprostanes also decreased with low extract concentrations [97].

Besides those *in vitro* studies (Table **2**), other reports also addressed the *in vivo* anti-inflammatory potential of bee pollen (Table **1**). Accordingly, extracts of bee pollen are thought to be effective in prostatic disorders because of their presumed anti-inflammatory and anti-androgen effects [98]. Indeed, in an open-label study, 90 patients received one tablet of the pollen extract Cernilton N, 3 times a day for 6 months [99]. Patients with complicating anatomic factors, with minimal responses to treatment, were excluded. The 36% of the remaining patients were cured of their symptoms while another 42% saw their symptoms improved [99]. Other *in vivo* study investigated the anti-inflammatory effect of bee pollen from *Cistus* sp. of Spanish origin in carrageenan-induced paw edema (CIPE) in rats [100]. Additionally, the anti-inflammatory mechanism and the compounds responsible for bee pollen bioactivity were also disclosed [100]. The authors demonstrated that the ethanol extract of bee pollen had a potent anti-inflammatory activity through inhibition of NO production, besides the inhibition of COX-2 activity [100]. Some flavonoids of bee pollen (quercetin-7-rhamnoside, kaempferol-3-glucoside, isorhamnetin, kaempferol and quercetin) could partly contribute for the anti-inflammatory action [100].

The *in vivo* anti-inflammatory effect of pure honey and a honey-bee pollen mix (HBM) formulation was also recently evaluated [101]. The authors concluded that HBM displayed significant anti-inflammatory activity without inducing any acute toxicity or gastric damage. On the other hand, pure honey did not exert any remarkable anti-inflammatory activity. Results clearly demonstrated that mixing pure honey with bee pollen significantly increased the healing potential of honey (partially due to phenolic and flavonoid contents), thus providing additional support for its traditional use.

CONCLUSION

While much progress has been made in the treatment of inflammatory diseases, it still remains a therapeutic area with immense unmet medical needs. Given the limitations of conventional pharmaceuticals, there is a clear call for new molecules. Natural products are important sources of new drugs, however it is crucial to understand the molecular mechanisms behind the healing properties of the products used in traditional medicine. The anti-inflammatory properties of honeybee plant-derived products are reported in ethnopharmacological studies and have been validated in several *in vitro* and *in vivo* models of inflammation. The intracellular signaling targets of pollen, propolis and honey are summarized in Fig. (**1**). Overall, the production of inflammatory mediators, *i.e.* NO and prostaglandins, were inhibited by the three products. These effects are probably due to the inhibition of NF-κB and MAPKs signaling pathways. Regarding the honeybee products reviewed, the molecular targets of the pollen were the less investigated. However, considering its rich nutritional composition and traditional uses, the mechanisms behind its anti-inflammatory properties should be deeply explored.

Importantly, it is necessary to take into account that the different origins of these products change their composition and consequently their efficacy. Additionally, other factors could modulate their biological activity, namely the concentrations used, the period of ingestion and the type of preparation. Accordingly, it is crucial to identify the constituents responsible for those effects and also to investigate the synergism/antagonism of the different molecules present in honey, pollen and propolis in order to maximize their therapeutic anti-inflammatory potential.

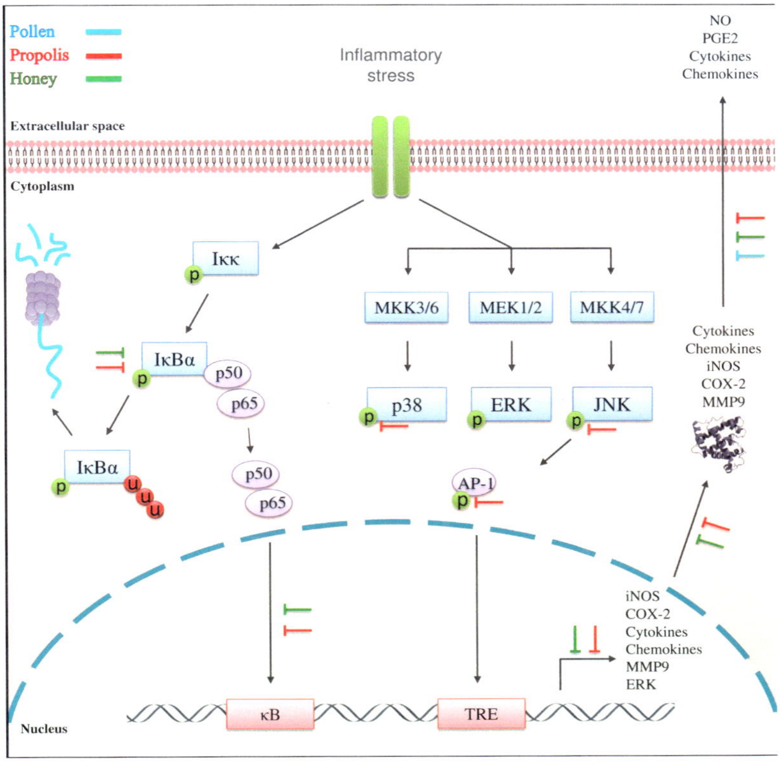

Fig. (1). Inflammatory intracellular signaling pathways modulated by honeybee plant-derived products: pollen, propolis and honey. AP-1, activator protein 1; COX-2, cyclooxygenase-2; ERK, extracellular signal-regulated kinase; IκBα, inhibitor of κBα; IKK, IκBα kinase; iNOS, inducible nitric oxide synthase; JNK, c-Jun N-terminal kinase; MEK and MKK, mitogen-activated protein kinase kinase; MMP9, Matrix metallopeptidase 9; NO, nitric oxide; PGE, prostaglandins; TRE, tetradecanoylphorbol-13-acetate responsive element.

CONFLICT OF INTEREST

The author confirms that author has no conflict of interest to declare for this publication.

ACKNOWLEDGE.MENTS

This work was supported by Fundação para a Ciência e Tecnologia, Fundo Comunitário Europeu (FEDER), Centro de Neurociências e Biologia Celular [PEst-2014], and SFRH/BD/72918/2010 and SFRH/BD/73065/2010 (J.L. and J.M. Ph.D. fellowships).

REFERENCES

[1] Delves, P.J.; Martin, S.J.; Burton, D.R.; Roitt, I.M. *Roitt's Essential Immunology,* 12; John Wiley & Sons: Oxford, UK, **2011**.

[2] Goronzy, J.J.; Weyand, C.M. Developments in the scientific understanding of rheumatoid arthritis. *Arthritis Res. Ther.,* **2009**, *11*(5), 249.
[http://dx.doi.org/10.1186/ar2758] [PMID: 19835638]

[3] Murdoch, J.R.; Lloyd, C.M. Chronic inflammation and asthma. *Mutat. Res.,* **2010**, *690*(1-2), 24-39.
[http://dx.doi.org/10.1016/j.mrfmmm.2009.09.005] [PMID: 19769993]

[4] Wallace, K.L.; Zheng, L-B.; Kanazawa, Y.; Shih, D.Q. Immunopathology of inflammatory bowel disease. *World J. Gastroenterol.,* **2014**, *20*(1), 6-21.
[http://dx.doi.org/10.3748/wjg.v20.i1.6] [PMID: 24415853]

[5] Wong, B.W.; Meredith, A.; Lin, D.; McManus, B.M. The biological role of inflammation in atherosclerosis. *Can. J. Cardiol.,* **2012**, *28*(6), 631-641.
[http://dx.doi.org/10.1016/j.cjca.2012.06.023] [PMID: 22985787]

[6] Mantovani, A. Molecular pathways linking inflammation and cancer. *Curr. Mol. Med.,* **2010**, *10*(4), 369-373.
[http://dx.doi.org/10.2174/156652410791316968] [PMID: 20455855]

[7] Dinarello, C.A. Immunological and inflammatory functions of the interleukin-1 family. *Annu. Rev. Immunol.,* **2009**, *27*, 519-550.
[http://dx.doi.org/10.1146/annurev.immunol.021908.132612] [PMID: 19302047]

[8] Badache, A.; Hynes, N.E. Interleukin 6 inhibits proliferation and, in cooperation with an epidermal growth factor receptor autocrine loop, increases migration of T47D breast cancer cells. *Cancer Res.,* **2001**, *61*(1), 383-391.
[PMID: 11196191]

[9] O'Shea, J.J.; Murray, P.J. Cytokine signaling modules in inflammatory responses. *Immunity,* **2008**, *28*(4), 477-487.
[http://dx.doi.org/10.1016/j.immuni.2008.03.002] [PMID: 18400190]

[10] Costa, G.; Francisco, V.; Lopes, M.C.; Cruz, M.T.; Batista, M.T. Intracellular signaling pathways modulated by phenolic compounds: application for new anti-inflammatory drugs discovery. *Curr. Med. Chem.,* **2012**, *19*(18), 2876-2900.
[http://dx.doi.org/10.2174/092986712800672049] [PMID: 22519398]

[11] Wang, J.; Li, Q.X. Chemical composition, characterization, and differentiation of honey botanical and geographical origins. *Adv. Food Nutr. Res.,* **2011**, *62*, 89-137.
[http://dx.doi.org/10.1016/B978-0-12-385989-1.00003-X] [PMID: 21504822]

[12] Alvarez-Suarez, J.M.; Giampieri, F.; Battino, M. Honey as a source of dietary antioxidants: structures, bioavailability and evidence of protective effects against human chronic diseases. *Curr. Med. Chem.,* **2013**, *20*(5), 621-638.
[http://dx.doi.org/10.2174/092986713804999358] [PMID: 23298140]

[13] Eteraf-Oskouei, T.; Najafi, M. Traditional and modern uses of natural honey in human diseases: a review. *Iran. J. Basic Med. Sci.,* **2013**, *16*(6), 731-742.

[PMID: 23997898]

[14] Al-Waili, N.; Salom, K.; Al-Ghamdi, A.; Ansari, M.J.; Al-Waili, A.; Al-Waili, T. Honey and cardiovascular risk factors, in normal individuals and in patients with diabetes mellitus or dyslipidemia. *J. Med. Food,* **2013**, *16*(12), 1063-1078. [http://dx.doi.org/10.1089/jmf.2012.0285] [PMID: 24328699]

[15] Yaghoobi, R.; Kazerouni, A.; Kazerouni, O. Evidence for clinical use of honey in wound healing as an anti-bacterial, anti-inflammatory, anti-oxidant and anti-viral Agent: A review. *Jundishapur J. Nat. Pharm. Prod.,* **2013**, *8*(3), 100-104. [http://dx.doi.org/10.17795/jjnpp-9487] [PMID: 24624197]

[16] Zbuchea, A. Up-to-date use of honey for burns treatment. *Ann. Burns Fire Disasters,* **2014**, *27*(1), 22-30. [PMID: 25249844]

[17] Maghsoudi, H.; Salehi, F.; Khosrowshahi, M.K.; Baghaei, M.; Nasirzadeh, M.; Shams, R. Comparison between topical honey and mafenide acetate in treatment of burn wounds. *Ann. Burns Fire Disasters,* **2011**, *24*(3), 132-137. [PMID: 22396671]

[18] Okeniyi, J.A.; Olubanjo, O.O.; Ogunlesi, T.A.; Oyelami, O.A. Comparison of healing of incised abscess wounds with honey and EUSOL dressing. *J. Altern. Complement. Med.,* **2005**, *11*(3), 511-513. [http://dx.doi.org/10.1089/acm.2005.11.511] [PMID: 15992237]

[19] Makhdoom, A.; Khan, M.S.; Lagahari, M.A.; Rahopoto, M.Q.; Tahir, S.M.; Siddiqui, K.A. Management of diabetic foot by natural honey. *J. Ayub Med. Coll. Abbottabad,* **2009**, *21*(1), 103-105. [PMID: 20364754]

[20] Al-Waili, N.; Salom, K.; Al-Ghamdi, A.A. Honey for wound healing, ulcers, and burns; data supporting its use in clinical practice. *ScientificWorldJournal,* **2011**, *11*, 766-787. [http://dx.doi.org/10.1100/tsw.2011.78] [PMID: 21479349]

[21] Henriques, A.; Jackson, S.; Cooper, R.; Burton, N. Free radical production and quenching in honeys with wound healing potential. *J. Antimicrob. Chemother.,* **2006**, *58*(4), 773-777. [http://dx.doi.org/10.1093/jac/dkl336] [PMID: 16963460]

[22] van den Berg, A.J.; van den Worm, E.; van Ufford, H.C.; Halkes, S.B.; Hoekstra, M.J.; Beukelman, C.J. An *in vitro* examination of the antioxidant and anti-inflammatory properties of buckwheat honey. *J. Wound Care,* **2008**, *17*(4), 172-174, 176-178. [http://dx.doi.org/10.12968/jowc.2008.17.4.28839] [PMID: 18494436]

[23] Majtan, J.; Bohova, J.; Garcia-Villalba, R.; Tomas-Barberan, F.A.; Madakova, Z.; Majtan, T.; Majtan, V.; Klaudiny, J. Fir honeydew honey flavonoids inhibit TNF-α-induced MMP-9 expression in human keratinocytes: a new action of honey in wound healing. *Arch. Dermatol. Res.,* **2013**, *305*(7), 619-627. [http://dx.doi.org/10.1007/s00403-013-1385-y] [PMID: 23812412]

[24] Majtanova, N.; Vodrazkova, E.; Kurilova, V.; Horniackova, M.; Cernak, M.; Cernak, A.; Majtan, J. Complementary treatment of contact lens-induced corneal ulcer using honey: a case report. *Cont. Lens Anterior Eye,* **2015**, *38*(1), 61-63. [http://dx.doi.org/10.1016/j.clae.2014.09.004] [PMID: 25278429]

[25] Uwaydat, S.; Jha, P.; Tytarenko, R.; Brown, H.; Wiggins, M.; Bora, P.S.; Bora, N.S. The use of topical honey in the treatment of corneal abrasions and endotoxin-induced keratitis in an animal model. *Curr. Eye Res.,* **2011**, *36*(9), 787-796.
[http://dx.doi.org/10.3109/02713683.2010.544441] [PMID: 21812661]

[26] Bashkaran, K.; Zunaina, E.; Bakiah, S.; Sulaiman, S.A.; Sirajudeen, K.; Naik, V. Anti-inflammatory and antioxidant effects of Tualang honey in alkali injury on the eyes of rabbits: experimental animal study. *BMC Complement. Altern. Med.,* **2011**, *11*, 90.
[http://dx.doi.org/10.1186/1472-6882-11-90] [PMID: 21982267]

[27] Vlcekova, P.; Krutakova, B.; Takac, P.; Kozanek, M.; Salus, J.; Majtan, J. Alternative treatment of gluteofemoral fistulas using honey: a case report. *Int. Wound J.,* **2012**, *9*(1), 100-103.
[http://dx.doi.org/10.1111/j.1742-481X.2011.00844.x] [PMID: 21977900]

[28] Medhi, B.; Prakash, A.; Avti, P.K.; Saikia, U.N.; Pandhi, P.; Khanduja, K.L. Effect of Manuka honey and sulfasalazine in combination to promote antioxidant defense system in experimentally induced ulcerative colitis model in rats. *Indian J. Exp. Biol.,* **2008**, *46*(8), 583-590.
[PMID: 18814487]

[29] Prakash, A.; Medhi, B.; Avti, P.K.; Saikia, U.N.; Pandhi, P.; Khanduja, K.L. Effect of different doses of Manuka honey in experimentally induced inflammatory bowel disease in rats. *Phytother. Res.,* **2008**, *22*(11), 1511-1519.
[http://dx.doi.org/10.1002/ptr.2523] [PMID: 18688794]

[30] Bilsel, Y.; Bugra, D.; Yamaner, S.; Bulut, T.; Cevikbas, U.; Turkoglu, U. Could honey have a place in colitis therapy? Effects of honey, prednisolone, and disulfiram on inflammation, nitric oxide, and free radical formation. *Dig. Surg.,* **2002**, *19*(4), 306-311.
[http://dx.doi.org/10.1159/000064580] [PMID: 12207075]

[31] Gharzouli, K.; Amira, S.; Gharzouli, A.; Khennouf, S. Gastroprotective effects of honey and glucose-fructose-sucrose-maltose mixture against ethanol-, indomethacin-, and acidified aspirin-induced lesions in the rat. *Exp. Toxicol. Pathol.,* **2002**, *54*(3), 217-221.
[http://dx.doi.org/10.1078/0940-2993-00255] [PMID: 12484559]

[32] Bukhari, M.H.; Khalil, J.; Qamar, S.; Qamar, Z.; Zahid, M.; Ansari, N.; Bakhshi, I.M. Comparative gastroprotective effects of natural honey, Nigella sativa and cimetidine against acetylsalicylic acid induced gastric ulcer in albino rats. *J. Coll. Physicians Surg. Pak.,* **2011**, *21*(3), 151-156.
[PMID: 21419021]

[33] Kamaruzaman, N.A.; Sulaiman, S.A.; Kaur, G.; Yahaya, B. Inhalation of honey reduces airway inflammation and histopathological changes in a rabbit model of ovalbumin-induced chronic asthma. *BMC Complement. Altern. Med.,* **2014**, *14*, 176.
[http://dx.doi.org/10.1186/1472-6882-14-176] [PMID: 24886260]

[34] Kassim, M.; Yusoff, K.M.; Ong, G.; Sekaran, S.; Yusof, M.Y.; Mansor, M. Gelam honey inhibits lipopolysaccharide-induced endotoxemia in rats through the induction of heme oxygenase-1 and the inhibition of cytokines, nitric oxide, and high-mobility group protein B1. *Fitoterapia,* **2012**, *83*(6), 1054-1059.
[http://dx.doi.org/10.1016/j.fitote.2012.05.008] [PMID: 22626749]

[35] Kassim, M.; Mansor, M.; Al-Abd, N.; Yusoff, K.M. Gelam honey has a protective effect against

lipopolysaccharide (LPS)-induced organ failure. *Int. J. Mol. Sci.,* **2012**, *13*(5), 6370-6381. [http://dx.doi.org/10.3390/ijms13056370] [PMID: 22754370]

[36] Borsato, D.M.; Prudente, A.S.; Döll-Boscardin, P.M.; Borsato, A.V.; Luz, C.F.; Maia, B.H.; Cabrini, D.A.; Otuki, M.F.; Miguel, M.D.; Farago, P.V.; Miguel, O.G. Topical anti-inflammatory activity of a monofloral honey of Mimosa scabrella provided by Melipona marginata during winter in southern Brazil. *J. Med. Food,* **2014**, *17*(7), 817-825. [http://dx.doi.org/10.1089/jmf.2013.0024] [PMID: 24650139]

[37] Al-Waili, N.S. Topical application of natural honey, beeswax and olive oil mixture for atopic dermatitis or psoriasis: partially controlled, single-blinded study. *Complement. Ther. Med.,* **2003**, *11*(4), 226-234. [http://dx.doi.org/10.1016/S0965-2299(03)00120-1] [PMID: 15022655]

[38] Owoyele, B.V.; Oladejo, R.O.; Ajomale, K.; Ahmed, R.O.; Mustapha, A. Analgesic and anti-inflammatory effects of honey: the involvement of autonomic receptors. *Metab. Brain Dis.,* **2014**, *29*(1), 167-173. [http://dx.doi.org/10.1007/s11011-013-9458-3] [PMID: 24318481]

[39] Owoyele, B.V.; Adenekan, O.T.; Soladoye, A.O. Effects of honey on inflammation and nitric oxide production in Wistar rats. *Zhong Xi Yi Jie He Xue Bao,* **2011**, *9*(4), 447-452. [http://dx.doi.org/10.3736/jcim20110415] [PMID: 21486559]

[40] Hussein, S.Z.; Mohd Yusoff, K.; Makpol, S.; Mohd Yusof, Y.A. Gelam honey attenuates carrageenan-induced rat paw inflammation *via* NF-κB pathway. *PLoS One,* **2013**, *8*(8), e72365. [http://dx.doi.org/10.1371/journal.pone.0072365] [PMID: 24015236]

[41] Hussein, S.Z.; Mohd Yusoff, K.; Makpol, S.; Mohd Yusof, Y.A. Gelam honey inhibits the production of proinflammatory, mediators nO, PGE(2), TNF-α, and IL-6 in carrageenan-induced acute paw edema in rats. *Evid. Based Complement. Alternat. Med.,* **2012**, *2012*, 109636. [http://dx.doi.org/10.1155/2012/109636] [PMID: 22919407]

[42] Kassim, M.; Achoui, M.; Mansor, M.; Yusoff, K.M. The inhibitory effects of Gelam honey and its extracts on nitric oxide and prostaglandin E(2) in inflammatory tissues. *Fitoterapia,* **2010**, *81*(8), 1196-1201. [http://dx.doi.org/10.1016/j.fitote.2010.07.024] [PMID: 20708657]

[43] Leong, A.G.; Herst, P.M.; Harper, J.L. Indigenous New Zealand honeys exhibit multiple anti-inflammatory activities. *Innate Immun.,* **2012**, *18*(3), 459-466. [http://dx.doi.org/10.1177/1753425911422263] [PMID: 21978989]

[44] Eteraf-Oskouei, T.; Najafi, M.; Gharehbagheri, A. Natural honey: a new and potent anti-angiogenic agent in the air-pouch model of inflammation. *Drug Res. (Stuttg.),* **2014**, *64*(10), 530-536. [PMID: 24357137]

[45] Kassim, M.; Mansor, M.; Suhaimi, A.; Ong, G.; Yusoff, K.M. Gelam honey scavenges peroxynitrite during the immune response. *Int. J. Mol. Sci.,* **2012**, *13*(9), 12113-12129. [http://dx.doi.org/10.3390/ijms130912113] [PMID: 23109904]

[46] Ahmad, I.; Jimenez, H.; Yaacob, N.S.; Yusuf, N. Tualang honey protects keratinocytes from ultraviolet radiation-induced inflammation and DNA damage. *Photochem. Photobiol.,* **2012**, *88*(5), 1198-1204.

[http://dx.doi.org/10.1111/j.1751-1097.2012.01100.x] [PMID: 22276569]

[47] Tomblin, V.; Ferguson, L.R.; Han, D.Y.; Murray, P.; Schlothauer, R. Potential pathway of anti-inflammatory effect by New Zealand honeys. *Int. J. Gen. Med.,* **2014**, *7*, 149-158. [http://dx.doi.org/10.2147/IJGM.S45839] [PMID: 24623989]

[48] Kassim, M.; Achoui, M.; Mustafa, M.R.; Mohd, M.A.; Yusoff, K.M. Ellagic acid, phenolic acids, and flavonoids in Malaysian honey extracts demonstrate *in vitro* anti-inflammatory activity. *Nutr. Res.,* **2010**, *30*(9), 650-659. [http://dx.doi.org/10.1016/j.nutres.2010.08.008] [PMID: 20934607]

[49] Woo, K.J.; Jeong, Y-J.; Inoue, H.; Park, J-W.; Kwon, T.K. Chrysin suppresses lipopolysaccharide-induced cyclooxygenase-2 expression through the inhibition of nuclear factor for IL-6 (NF-IL6) DNA-binding activity. *FEBS Lett.,* **2005**, *579*(3), 705-711. [http://dx.doi.org/10.1016/j.febslet.2004.12.048] [PMID: 15670832]

[50] Ha, S.K.; Moon, E.; Kim, S.Y. Chrysin suppresses LPS-stimulated proinflammatory responses by blocking NF-κB and JNK activations in microglia cells. *Neurosci. Lett.,* **2010**, *485*(3), 143-147. [http://dx.doi.org/10.1016/j.neulet.2010.08.064] [PMID: 20813161]

[51] Ahad, A.; Ganai, A.A.; Mujeeb, M.; Siddiqui, W.A. Chrysin, an anti-inflammatory molecule, abrogates renal dysfunction in type 2 diabetic rats. *Toxicol. Appl. Pharmacol.,* **2014**, *279*(1), 1-7. [http://dx.doi.org/10.1016/j.taap.2014.05.007] [PMID: 24848621]

[52] Candiracci, M.; Piatti, E.; Dominguez-Barragán, M.; García-Antrás, D.; Morgado, B.; Ruano, D.; Gutiérrez, J.F.; Parrado, J.; Castaño, A. Anti-inflammatory activity of a honey flavonoid extract on lipopolysaccharide-activated N13 microglial cells. *J. Agric. Food Chem.,* **2012**, *60*(50), 12304-12311. [http://dx.doi.org/10.1021/jf302468h] [PMID: 23176387]

[53] Búfalo, M.C.; Ferreira, I.; Costa, G.; Francisco, V.; Liberal, J.; Cruz, M.T.; Lopes, M.C.; Batista, M.T.; Sforcin, J.M. Propolis and its constituent caffeic acid suppress LPS-stimulated pro-inflammatory response by blocking NF-κB and MAPK activation in macrophages. *J. Ethnopharmacol.,* **2013**, *149*(1), 84-92. [http://dx.doi.org/10.1016/j.jep.2013.06.004] [PMID: 23770030]

[54] Francisco, V.; Figueirinha, A.; Costa, G.; Liberal, J.; Lopes, M.C.; García-Rodríguez, C.; Geraldes, C.F.; Cruz, M.T.; Batista, M.T. Chemical characterization and anti-inflammatory activity of luteolin glycosides isolated from lemongrass. *J. Funct. Foods,* **2014**, *10*, 436-443. [http://dx.doi.org/10.1016/j.jff.2014.07.003]

[55] Ghisalberti, E.L. Propolis: A review - international bee research association. *Bee World,* **1979**, *27*, 519-550.

[56] Burdock, G.A. Review of the biological properties and toxicity of bee propolis (propolis). *Food Chem. Toxicol.,* **1998**, *36*(4), 347-363. [http://dx.doi.org/10.1016/S0278-6915(97)00145-2] [PMID: 9651052]

[57] Bankova, V.S.; de Castro, S.L.; Marcucci, M.C. Propolis: recent advances in chemistry and plant origin. *Apidologie (Celle),* **2000**, *31*, 3-15. [http://dx.doi.org/10.1051/apido:2000102]

[58] Bankova, V. Chemical diversity of propolis and the problem of standardization. *J. Ethnopharmacol.,*

2005, *100*(1-2), 114-117.
[http://dx.doi.org/10.1016/j.jep.2005.05.004] [PMID: 15993016]

[59] Rubio, O.C.; Cuellar Cuellar, A.; Rojas, N.; Castro, H.V.; Rastrelli, L.; Aquino, R. A polyisoprenylated benzophenone from Cuban propolis. *J. Nat. Prod.,* **1999**, *62*(7), 1013-1015.
[http://dx.doi.org/10.1021/np980339n] [PMID: 10425129]

[60] Montenegro, G.; Avila, G.; Peña, R.C. Botanical origin and seasonal production of propolis in hives of central chile. *Bol. Botânica,* **2001**, *19*, 1.

[61] Bankova, V. Recent trends and important developments in propolis research. *Evid. Based Complement. Alternat. Med.,* **2005**, *2*(1), 29-32.
[http://dx.doi.org/10.1093/ecam/neh059] [PMID: 15841275]

[62] Sforcin, J.M.; Bankova, V. Propolis: is there a potential for the development of new drugs? *J. Ethnopharmacol.,* **2011**, *133*(2), 253-260.
[http://dx.doi.org/10.1016/j.jep.2010.10.032] [PMID: 20970490]

[63] Sforcin, J.M.; Novelli, E.L.; Funari, S.R. Seasonal effect of Brazilian propolis on seric biochemical variables. *J. Venom. Anim. Toxins,* **2002**, *8*, 244-254.
[http://dx.doi.org/10.1590/S0104-79302002000200005]

[64] Mani, F.; Damasceno, H.C.; Novelli, E.L.; Martins, E.A.; Sforcin, J.M. Propolis: Effect of different concentrations, extracts and intake period on seric biochemical variables. *J. Ethnopharmacol.,* **2006**, *105*(1-2), 95-98.
[http://dx.doi.org/10.1016/j.jep.2005.10.011] [PMID: 16293383]

[65] Mani, F.; Damasceno, H.C.; Novelli, E.L.; Sforcin, J.M. Biochemical determinations of propolis-treated rats: Effects of different concentrations, extracts and intake period. *Biosaúde,* **2008**, *10*, 3-16.

[66] Grunberger, D.; Banerjee, R.; Eisinger, K.; Oltz, E.M.; Efros, L.; Caldwell, M.; Estevez, V.; Nakanishi, K. Preferential cytotoxicity on tumor cells by caffeic acid phenethyl ester isolated from propolis. *Experientia,* **1988**, *44*(3), 230-232.
[http://dx.doi.org/10.1007/BF01941717] [PMID: 2450776]

[67] Hausen, B.M.; Evers, P.; Stüwe, H.T.; König, W.A.; Wollenweber, E. Propolis allergy (IV). Studies with further sensitizers from propolis and constituents common to propolis, poplar buds and balsam of Peru. *Contact Dermat.,* **1992**, *26*(1), 34-44.
[http://dx.doi.org/10.1111/j.1600-0536.1992.tb00865.x] [PMID: 1600736]

[68] Ikeno, K.; Ikeno, T.; Miyazawa, C. Effects of propolis on dental caries in rats. *Caries Res.,* **1991**, *25*(5), 347-351.
[http://dx.doi.org/10.1159/000261390] [PMID: 1836157]

[69] Scheller, S.; Gazda, G.; Krol, W.; Czuba, Z.; Zajusz, A.; Gabrys, J.; Shani, J. The ability of ethanolic extract of propolis (EEP) to protect mice against gamma irradiation. *Z. Naturforsch. C.,* **1989**, *44*(11-12), 1049-1052.
[PMID: 2698623]

[70] Scheller, S.; Krol, W.; Swiacik, J.; Owczarek, S.; Gabrys, J.; Shani, J. Antitumoral property of ethanolic extract of propolis in mice-bearing Ehrlich carcinoma, as compared to bleomycin. *Z. Naturforsch., C, J. Biosci.,* **1989**, *44*(11-12), 1063-1065.

[PMID: 2483616]

[71] Khayyal, M.T.; el-Ghazaly, M.A.; el-Khatib, A.S. Mechanisms involved in the antiinflammatory effect of propolis extract. *Drugs Exp. Clin. Res.,* **1993**, *19*(5), 197-203. [PMID: 7513636]

[72] Miyataka, H.; Nishiki, M.; Matsumoto, H.; Fujimoto, T.; Matsuka, M.; Satoh, T. Evaluation of propolis. I. Evaluation of Brazilian and Chinese propolis by enzymatic and physico-chemical methods. *Biol. Pharm. Bull.,* **1997**, *20*(5), 496-501. [http://dx.doi.org/10.1248/bpb.20.496] [PMID: 9178928]

[73] Hu, F.; Hepburn, H.R.; Li, Y.; Chen, M.; Radloff, S.E.; Daya, S. Effects of ethanol and water extracts of propolis (bee glue) on acute inflammatory animal models. *J. Ethnopharmacol.,* **2005**, *100*(3), 276-283. [http://dx.doi.org/10.1016/j.jep.2005.02.044] [PMID: 15899563]

[74] Paulino, N.; Teixeira, C.; Martins, R.; Scremin, A.; Dirsch, V.M.; Vollmar, A.M.; Abreu, S.R.; de Castro, S.L.; Marcucci, M.C. Evaluation of the analgesic and anti-inflammatory effects of a Brazilian green propolis. *Planta Med.,* **2006**, *72*(10), 899-906. [http://dx.doi.org/10.1055/s-2006-947185] [PMID: 16902858]

[75] Sforcin, J.M. Propolis and the immune system: a review. *J. Ethnopharmacol.,* **2007**, *113*(1), 1-14. [http://dx.doi.org/10.1016/j.jep.2007.05.012] [PMID: 17580109]

[76] Bachiega, T.F.; Orsatti, C.L.; Pagliarone, A.C.; Sforcin, J.M. The effects of propolis and its isolated compounds on cytokine production by murine macrophages. *Phytother. Res.,* **2012**, *26*(9), 1308-1313. [http://dx.doi.org/10.1002/ptr.3731] [PMID: 22275284]

[77] Howren, M.B.; Lamkin, D.M.; Suls, J. Associations of depression with C-reactive protein, IL-1, and IL-6: a meta-analysis. *Psychosom. Med.,* **2009**, *71*(2), 171-186. [http://dx.doi.org/10.1097/PSY.0b013e3181907c1b] [PMID: 19188531]

[78] Mosser, D.M. The many faces of macrophage activation. *J. Leukoc. Biol.,* **2003**, *73*(2), 209-212. [http://dx.doi.org/10.1189/jlb.0602325] [PMID: 12554797]

[79] Alter, G.; Kavanagh, D.; Rihn, S.; Luteijn, R.; Brooks, D.; Oldstone, M.; van Lunzen, J.; Altfeld, M. IL-10 induces aberrant deletion of dendritic cells by natural killer cells in the context of HIV infection. *J. Clin. Invest.,* **2010**, *120*(6), 1905-1913. [http://dx.doi.org/10.1172/JCI40913] [PMID: 20440075]

[80] Orsi, R.O.; Funari, S.R.; Soares, A.M.; Calvi, S.A.; Oliveira, S.L.; Sforcin, J.M.; Bankova, V. Immunomodulatory action of propolis on macrophage activation. *J. Venom. Anim. Toxins,* **2000**, *6*, 205-219. [http://dx.doi.org/10.1590/S0104-79302000000200006]

[81] Wang, K.; Ping, S.; Huang, S.; Hu, L.; Xuan, H.; Zhang, C.; Hu, F. Molecular mechanisms underlying the *in vitro* anti-inflammatory effects of a flavonoid-rich ethanol extract from chinese propolis (poplar type). *Evid. Based Complement. Alternat. Med.,* **2013**, *2013*, 127672. [http://dx.doi.org/10.1155/2013/127672] [PMID: 23401705]

[82] Wu, Z.; Zhu, A.; Takayama, F.; Okada, R.; Liu, Y.; Harada, Y.; Wu, S.; Nakanishi, H. Brazilian green propolis suppresses the hypoxia-induced neuroinflammatory responses by inhibiting NF-κB activation

in microglia. *Oxid. Med. Cell. Longev.,* **2013**, *2013*, 906726.
[http://dx.doi.org/10.1155/2013/906726] [PMID: 23983903]

[83] Ivanovska, N.D.; Dimov, V.B.; Bankova, V.S.; Popov, S.S. Immunomodulatory action of propolis. VI. Influence of a water soluble derivative on complement activity *in vivo*. *J. Ethnopharmacol.,* **1995**, *47*(3), 145-147.
[http://dx.doi.org/10.1016/0378-8741(95)01272-F] [PMID: 8569238]

[84] Búfalo, M.C.; Bordon-Graciani, A.P.; Conti, B.J.; de Assis Golim, M.; Sforcin, J.M. The immunomodulatory effect of propolis on receptors expression, cytokine production and fungicidal activity of human monocytes. *J. Pharm. Pharmacol.,* **2014**, *66*(10), 1497-1504.
[http://dx.doi.org/10.1111/jphp.12279] [PMID: 24934761]

[85] Szliszka, E.; Mertas, A.; Czuba, Z.P.; Król, W. Inhibition of inflammatory response by artepillin C in activated rAW264.7 macrophages. *Evid. Based Complement. Alternat. Med.,* **2013**, *2013*, 735176.
[http://dx.doi.org/10.1155/2013/735176] [PMID: 23781267]

[86] Dimov, V.; Ivanovska, N.; Manolova, N.; Bankova, V.; Nikolov, N.; Popov, S. Immunomodulatory action of propolis. Influence on anti-infectious protection and macrophage function. *Apidologie (Celle),* **1991**, *22*, 155-162.
[http://dx.doi.org/10.1051/apido:19910208]

[87] Sá-Nunes, A.; Faccioli, L.H.; Sforcin, J.M. Propolis: lymphocyte proliferation and IFN-gamma production. *J. Ethnopharmacol.,* **2003**, *87*(1), 93-97.
[http://dx.doi.org/10.1016/S0378-8741(03)00121-1] [PMID: 12787960]

[88] You, K.M.; Son, K.H.; Chang, H.W.; Kang, S.S.; Kim, H.P. Vitexicarpin, a flavonoid from the fruits of Vitex rotundifolia, inhibits mouse lymphocyte proliferation and growth of cell lines *in vitro*. *Planta Med.,* **1998**, *64*(6), 546-550.
[http://dx.doi.org/10.1055/s-2006-957511] [PMID: 9741302]

[89] Missima, F.; Pagliarone, A.C.; Orsatti, C.L.; Araújo, J.P., Jr; Sforcin, J.M. The Effect of propolis on Th1/Th2 cytokine expression and production by melanoma-bearing mice submitted to stress. *Phytother. Res.,* **2010**, *24*(10), 1501-1507.
[http://dx.doi.org/10.1002/ptr.3142] [PMID: 20878701]

[90] Tanaka, M.; Okamoto, Y.; Fukui, T.; Masuzawa, T. Suppression of interleukin 17 production by Brazilian propolis in mice with collagen-induced arthritis. *Inflammopharmacology,* **2012**, *20*(1), 19-26.
[http://dx.doi.org/10.1007/s10787-011-0088-2] [PMID: 21861090]

[91] Ansorge, S.; Reinhold, D.; Lendeckel, U. Propolis and some of its constituents down-regulate DNA synthesis and inflammatory cytokine production but induce TGF-beta1 production of human immune cells. *Z. Naturforsch., C, J. Biosci.,* **2003**, *58*(7-8), 580-589.
[http://dx.doi.org/10.1515/znc-2003-7-823] [PMID: 12939048]

[92] Márquez, N.; Sancho, R.; Macho, A.; Calzado, M.A.; Fiebich, B.L.; Muñoz, E. Caffeic acid phenethyl ester inhibits T-cell activation by targeting both nuclear factor of activated T-cells and NF-kappaB transcription factors. *J. Pharmacol. Exp. Ther.,* **2004**, *308*(3), 993-1001.
[http://dx.doi.org/10.1124/jpet.103.060673] [PMID: 14617683]

[93] Khayyal, M.T.; el-Ghazaly, M.A.; el-Khatib, A.S.; Hatem, A.M.; de Vries, P.J.; el-Shafei, S.; Khattab,

M.M. A clinical pharmacological study of the potential beneficial effects of a propolis food product as an adjuvant in asthmatic patients. *Fundam. Clin. Pharmacol.,* **2003**, *17*(1), 93-102. [http://dx.doi.org/10.1046/j.1472-8206.2003.00117.x] [PMID: 12588635]

[94] Chan, G.C.; Cheung, K-W.; Sze, D.M. The immunomodulatory and anticancer properties of propolis. *Clin. Rev. Allergy Immunol.,* **2013**, *44*(3), 262-273. [http://dx.doi.org/10.1007/s12016-012-8322-2] [PMID: 22707327]

[95] Pascoal, A.; Rodrigues, S.; Teixeira, A.; Feás, X.; Estevinho, L.M. Biological activities of commercial bee pollens: antimicrobial, antimutagenic, antioxidant and anti-inflammatory. *Food Chem. Toxicol.,* **2014**, *63*, 233-239. [http://dx.doi.org/10.1016/j.fct.2013.11.010] [PMID: 24262487]

[96] Silva, J.C.; Rodrigues, S.; Feás, X.; Estevinho, L.M. Antimicrobial activity, phenolic profile and role in the inflammation of propolis. *Food Chem. Toxicol.,* **2012**, *50*(5), 1790-1795. [http://dx.doi.org/10.1016/j.fct.2012.02.097] [PMID: 22425940]

[97] Moita, E.; Gil-Izquierdo, A.; Sousa, C.; Ferreres, F.; Silva, L.R.; Valentão, P.; Domínguez-Perles, R.; Baenas, N.; Andrade, P.B. Integrated analysis of COX-2 and iNOS derived inflammatory mediators in LPS-stimulated RAW macrophages pre-exposed to Echium plantagineum L. bee pollen extract. *PLoS One,* **2013**, *8*(3), e59131. [http://dx.doi.org/10.1371/journal.pone.0059131] [PMID: 23520554]

[98] Duclos, A.J.; Lee, C-T.; Shoskes, D.A. Current treatment options in the management of chronic prostatitis. *Ther. Clin. Risk Manag.,* **2007**, *3*(4), 507-512. [PMID: 18472971]

[99] Rugendorff, E.W.; Weidner, W.; Ebeling, L.; Buck, A.C. Results of treatment with pollen extract (Cernilton N) in chronic prostatitis and prostatodynia. *Br. J. Urol.,* **1993**, *71*(4), 433-438. [http://dx.doi.org/10.1111/j.1464-410X.1993.tb15988.x] [PMID: 8499988]

[100] Maruyama, H.; Sakamoto, T.; Araki, Y.; Hara, H. Anti-inflammatory effect of bee pollen ethanol extract from *Cistus* sp. of Spanish on carrageenan-induced rat hind paw edema. *BMC Complement. Altern. Med.,* **2010**, *10*, 30. [http://dx.doi.org/10.1186/1472-6882-10-30] [PMID: 20573205]

[101] Küpeli Akkol, E.; Orhan, D.D.; Gürbüz, I.; Yesilada, E. *In vivo* activity assessment of a "honey-bee pollen mix" formulation. *Pharm. Biol.,* **2010**, *48*(3), 253-259. [http://dx.doi.org/10.3109/13880200903085482] [PMID: 20645809]

CHAPTER 10

Antitumor Properties of Honeybee Plant-Derived Products: Honey, Propolis and Pollen

Cristina Almeida-Aguiar[1,2,*], **Ricardo Silva-Carvalho**[3,4,#], **Fátima Baltazar**[3,4]

[1] *CITAB, Centre for the Research and Technology of Agro-Environmental and Biological Sciences, AgroBioPlant Group, University of Minho, Braga, Portugal*

[2] *Biology Department, School of Sciences, University of Minho, Campus de Gualtar, 4710-057 Braga, Portugal*

[3] *Life and Health Sciences Research Institute (ICVS), School of Health Sciences, University of Minho, Braga, Portugal;*

[4] *ICVS/3B's - PT Government Associate Laboratory, Braga/Guimarães, Portugal*

Abstract: The majority of cancers have no curable treatment and the main available therapies have serious side effects, justifying the need for development of new antitumor agents. Several efforts have been made to identify natural products useful in the cancer setting. This area has emerged as an important research field, providing the possibility to both identify novel potentially useful agents and to study the mechanisms of antitumor action. Honeybee plant-derived products have shown anti-cancer activity in a series of experimental and clinical studies with cell lines, animals and humans. Honey, the viscous, golden and sweet liquid produced by bees from the nectar of flowering plants has proven to display antiproliferative and apoptotic effects, along with other activities that contribute for its antitumor properties. Propolis, a special substance made by honeybees through mixing tree saps with salivary secretions, is used to seal fissures and openings in the hive, strength combs, seal brood cells and protect the hive from infections. Propolis contains phytonutrients that may be useful in different pathological conditions, including cancer.

[*] **Address correspondence to C. Almeida Aguiar:** Biology Department, School of Sciences, University of Minho, Campus de Gualtar, 4710-057 Braga, Portugal; Tel: +351 253601513; Fax: +351 253604319; Email: cristina.aguiar@bio.uminho.pt and [#]**Ricardo Silva-Carvalho:** Present address: Instituto de Investigação e Inovação em Saúde, Universidade do Porto, Portugal. INEB - Instituto de Engenharia Biomédica, Universidade do Porto, Rua do Campo Alegre 823, 4150-180 Porto, Portugal

Susana M. Cardoso (Ed)

Bee pollen, the bees´ primary food source, is plant pollen collected from a variety of plants and processed by honeybees. Demand for this natural product is rising since it has effects on a variety of biological functions, which contribute to the fight and prevention of cancer. This review focuses on the antitumor properties of honey, propolis and bee pollen as well as on the potential use of these honeybee plant-derived products to develop new therapeutic approaches for patients with different types of tumors.

Keywords: Angiogenesis, Apoptosis, Autophagy, Bioactivities, Cancer, Flavonoids, Honey, Honeybees, Immortality, Invasion, Metastasis, Natural products, Oncogenes, Plants, Pollen, Polyphenols, Proliferation, Propolis, Standardization, Tumor suppressors.

1. CANCER

Cancer can be defined as a disease where cells suffered alterations and acquired an abnormal capacity of proliferation, invasion to adjacent tissues or metastasis to distant organs, through blood and lymphatic vessels. Two main types of genes are involved in cancer development: oncogenes and tumor suppressor genes. Oncogenes are genes that contribute to the development of cancer, while tumor suppressor genes have the opposite effect. The development of many cancers is often related to alterations in these genes, either by promoting the activity of oncogenes or decreasing the activity of tumor suppressors.

There are several types of cancer, with diverse characteristics and sensitivity to therapy, therefore with different mortality rates. The worldwide incidence and mortality of the main types of cancer are presented in Fig. (**1**) [1].

Despite advances of therapeutic regimens over the years, there are still some cancers which do not respond well to current therapy and consequently have high mortality rates. Thus, new therapeutic approaches are urgently needed.

1.1. Hallmarks of Cancer

In the publication "The hallmarks of cancer" [2], which was recently updated [3], Hanahan and Weinberg detail the main alterations - "hallmarks" - acquired by human cancer cells during the development of malignant tumors, which include the sustained capacity of proliferation, insensitivity to anti-growth signaling, resistance to cell death, replicative immortality, angiogenesis induction and the mechanisms of invasion and metastasis (Fig. **2**). In their most recent publication, the authors propose additional hallmarks, known as enabling characteristics and emerging hallmarks [3].

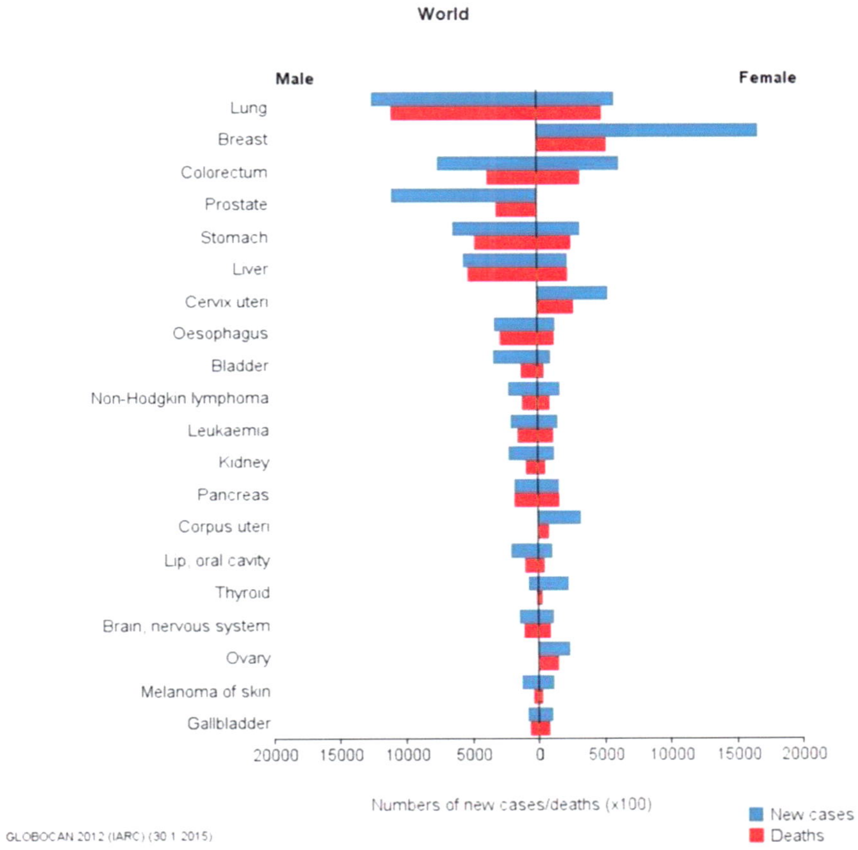

Fig. (1). Worldwide incidence and mortality of the main cancers [1].

1.1.1. Sustaining Proliferative Signaling

The capacity of chronic proliferation is probably the most important alteration of cancer cells. While the release of growth factors is tightly regulated in normal tissues, cancer cells present deregulation of this process, with consequent uncontrolled growth. Cell proliferation is mainly regulated by growth factors that activate cell surface receptors.

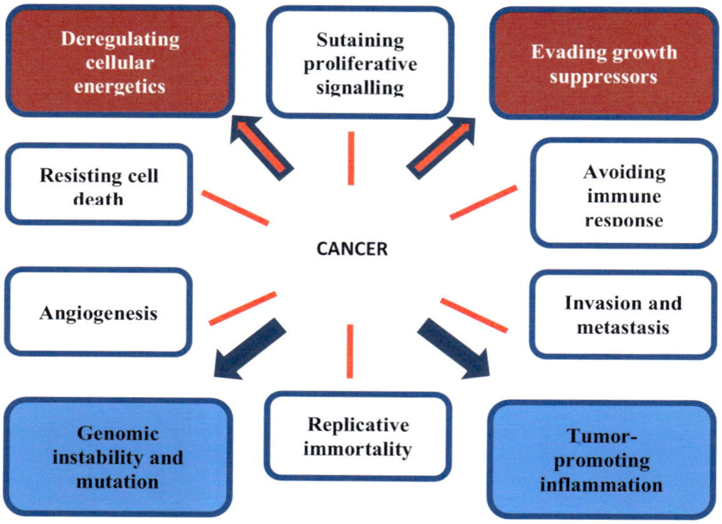

Fig. (2). Hallmarks of cancer - the new generation. White boxes represent the essential hallmarks recognized by Hanahan e Weinberg, blue boxes are the enabling characteristics and red boxes correspond to the emergent hallmarks (adapted from [3]).

These growth factor receptors with tyrosine kinase activity (RTKs) have a similar structure: an extracellular domain which contains the ligand-binding region, a transmembrane domain, and a cytoplasmic domain containing the protein tyrosine kinase (TK) region [4]. Ligand binding leads to receptor activation by dimerization, with activates a series of intracellular cascades, involved in cell cycle progression and growth, including the PI3K (PI3 kinase)/Akt (Protein kinase B), PLC (phospholipase C)/PKC (Protein kinase C), Src (sarcoma) family kinase pathways, Ras (rat sarcoma)/Raf (rapidly accelerated fibrosarcoma) and JAK (Janus-activated kinase)/STAT (signal transducer and activator of transcription) pathways. This intracellular signaling is also involved in other cell functions, such as escape from apoptosis (programmed cell death), induction of angiogenesis, alteration in cell metabolism, resistance to chemotherapy and invasion.

Alterations in the expression or structure of growth receptors, as well as in downstream signaling molecules, are the main contributors to sustained cancer cell proliferation. Common alterations include activating B-Raf mutations in melanomas and PI3K mutations in several tumors. Also, while mechanisms of negative feedback control the growth of normal tissues, cancer cells exhibit defects in these mechanisms, contributing to their increased proliferative capacity. This is the case of disruption of the phosphatase activity of the multifunctional tumor suppressor PTEN (phosphatase and tensin homolog), seen in several tumors, which counteracts PI3K activity, promoting tumorigenesis.

1.1.2. Insensitivity to Anti-growth Signaling

Beyond stimulating their own proliferation, cancer cells can also escape the signaling that affects proliferation negatively, and which mostly depends on the activity of tumor suppressor molecules. Two of the most important tumor suppressor genes are the ones which codify the proteins retinoblastoma (Rb) and tumor protein 53 (TP53 or p53) [5]. Rb protein regulates cell cycle and division, working as a guardian of the cell cycle progression. Thus, cells with defects in Rb present continuous cells proliferation. P53 is involved in cell response to stress, such as genome damage, leading to cell cycle arrest until the situation is controlled. If cell damage is too severe, p53 can induce apoptosis. The p53 coding gene is found mutated in about 50% of human tumors.

1.1.3. Resistance to Cell Death

Induction of cell death as a result of several *stimuli* constitutes a barrier to the development of cancer. However, this function is compromised in cancers that progress to more advanced stages of malignancy. Apoptosis or programmed cell death is regulated by two main circuits, the intrinsic pathway, that processes signals of intracellular origin, and the extrinsic pathway (involving Fas ligand/Fas receptor) that deals with extracellular *stimuli* [6]. Each of these pathways leads ultimately to the activation of caspases (cysteine-aspartic proteases), which initiate a proteolytic cascade, leading to cell death. Apoptosis is controlled by pro-apoptotic and anti-apoptotic molecules, which belong to the Bcl-2 (B-cell lymphoma 2) family. Bcl-2 and its related proteins Bcl-xL, Bcl-w, Mcl-1 (myeloid leukemia cell differentiation protein) are inhibitors of apoptosis. The proteins Bax (Bcl-2 associated X protein) and Bak (Bcl-2 antagonist/killer protein

1) are apoptosis promoters which compromise the integrity of the mitochondrial membrane and mediate the liberation of pro-apoptotic molecules, being cytochrome c the most important [6].

Autophagy is another type of cell response to stress, being connected for example to lack of nutrients. In this process, organelles such as ribosomes and mitochondria are broken down into catabolites, which will be recycled for cell biosynthesis and energetic metabolism. This process is mediated by the formation of intracellular vesicles, which surround the organelles and fuse with the lysosomal membrane, leading to organelle degradation. Autophagy may have opposite effects in cancer, acting as a tumor suppressor in the initial phase of tumorigenesis and as oncogene in a later phase of tumor progression [7].

Another form of cell death is necrosis, in which cells increase in size and literally explode, releasing their content to the microenvironment. Similarly to the other processes, recent evidence points to necrosis as a genetically controlled process [8]. Necrosis could also play a dual role in cancer, either having an anti-proliferative effect or promoting the development of cancer, by recruiting inflammatory cells, stimulating angiogenesis, proliferation and invasion.

1.1.4. Replicative Immortality

Another important alteration in cancer is the capacity of unlimited proliferation, in opposition to normal cells, which enter senescence after a certain number of cell divisions. Here telomeres play an essential role, by protecting the ends of the chromosomes. The size of the telomeres is progressively reduced when cells divide, until they lose the capacity to protect chromosomes from fusion, threatening cell viability. Telomerases are DNA polymerases that add segments to the telomeric DNA terminations, which expression is high in immortalized cells, including cancer [9]. There are tumors in which there is extensive telomere shortening due to low telomerase activity. In these cases, lack of p53 could allow these neoplasias to survive and acquire a number of chromosome alterations, leading to mutations in oncogenes and tumor suppressors. On the other hand, some tumors acquire telomerase activity latter in development, leading to maintenance of the mutated genome. Recently, additional functions of telomerase in tumorigenesis independent from telomere maintenance have been identified, including in cell proliferation [9].

1.1.5. Induction of Angiogenesis

Tumor vasculature has an important role in the maintenance of tumors, by providing the necessary nutrients and oxygen, as well as for secreting the products of metabolism and carbon dioxide. In normal situations in the adult organism, such as cicatrisation and menstrual cycle, development of blood vessels is a transient process, while in cancer angiogenesis is almost always activated and stays throughout tumor progression to support tumor expansion.

The angiogenic switch is controlled by signaling proteins that bind to receptors at the surface of endothelial cells, which either repress or stimulate angiogenesis. The main stimulator molecules are VEGF-A (vascular endothelial growth factor A) and TSP-1 (thrombospondin 1). Signaling through VEGF is done by means of three receptors with tyrosine kinase activity, VEGFR-1, VEGFR-2 e VEGFR-3. On the other hand, TSP-1 signaling recruits anti-angiogenic factors that counteract pro-angiogenic stimuli. Since angiogenesis in cancer is an uncontrolled process (blood vessels formed are typically aberrant, tortuous and blood flux is distorted), the entry of chemotherapy inside the tumor and its efficacy are compromised [10].

1.1.6. Mechanisms of Invasion and Metastasis

A main first step in cancer epithelial cell invasion is loss of cell-cell adhesion, as well as adhesion to the extracellular matrix. Cadherins (calcium-dependent cell adhesion molecules) are transmembrane proteins with a crucial role in cell adhesion, being responsible for cohesion of tissues [11]. There are three main types of cadherins, E-cadherins (present in epithelia), N-cadherins (in neurons) and P-cadherins (in placenta), being the alterations in E-cadherin the best studied. High expression of E-cadherin functions as an anti-invasion and anti-metastasis mechanism, while its down-regulation stimulates these mechanisms. Down-regulation and inactivating mutations of E-cadherin are commonly found in human carcinomas (cancer of epithelial origin).

The invasion and metastasis process involves a series of steps, which starts with local invasion, followed by entry of the cancer cells into blood of lymphatic vessels, extravasation to distant tissues, formation of micrometastasis and then growth into macroscopic tumors. The epithelial-to-mesenchymal transition (EMT), in which cancer cells acquire the capacity to invade, resist to apoptosis and disseminate, has also been associated with invasion and metastasis [12].

1.2. Enabling Cancer Characteristics

Different mechanisms are in the basis of the carcinogenesis process, and likely the most important is genetic instability, which leads to development of genetic alterations, namely mutations and chromosome rearrangements. Some mutated genomes have proliferation advantage and dominant phenotype. The accumulation of mutations could be facilitated by the compromise of surveillance systems such as the one of p53, involved in the recognition of genetic damage, promoting cell death. Beyond mutations, also epigenetic alterations, mediated by gene methylation and histone modifications, could be responsible for the inactivation of tumor suppressor genes.

Another important factor is the inflammatory stage of the malignant lesion, mediated by inflammatory cells, which can have pro-tumoral properties, contributing to tumor progression [13]. Inflammatory cells secrete bioactive molecules to the microenvironment, such as growth and survival factors, pro-angiogenic factors, enzymes that modify the extra-cellular matrix, promoting angiogenesis, invasion and metastasis. On the other hand, inflammatory cells also release reactive oxygen species (ROS), which are mutagenic and contribute to genetic instability and malignancy.

1.3. Emerging Cancer Hallmarks

Aberrant cell proliferation also includes alterations in the cell energetic metabolism. In contrast to normal cells, which adjust their metabolism in function of oxygen availability, most cancer cells use glycolysis as the main source of energy, and convert pyruvate into lactate, which is exported to the microenvironment. In order to cope with this less efficient energetic pathway, cancer cells increase the rates of glycolysis, by upregulating glucose transporters (*e.g.* GLUT1, GLUT3), glycolytic enzymes, lactate dehydrogenase (LDH) and a series of pH regulators and lactate transporters (MCTs (monocarboxylate transporters)) [14]. This metabolic phenotype provides several advantages to cancer cells, including survival in conditions of low oxygen tension and deviation of intermediates to anabolic reactions for new cell formation. Besides, the resulting low extracellular pH is associated with escape to the immune system, invasion, metastasis and promotion of angiogenesis. This metabolic switch has been associated with genetic alterations in oncogenes (*e.g.* Myc (myelocytomatosis), Ras) and tumor suppressors (*e.g.* p53) and is also stimulated by the presence of hypoxia which is characteristic of many tumors [15].

The role of the immunologic system in cancer is still not completely understood. It is assumed that the immunologic cells are responsible for the recognition and elimination of cancer cells in the initial stages of a malignant tumor. Thus, successful tumors were able to escape the immune system in some way, either by avoiding recognition or destruction. On the other hand, high immunogenic cells can also escape destruction by the immune system through activation of immunossupressor systems, which were designed to destroy them.

1.4. Tumor Microenvironment

Nowadays, it is known that malignant tumors are not just a homogeneous group of cancer cells but very complex organs instead. For example, beyond cancer cells, carcinomas also contain stem cells, endothelial cells, pericytes, inflammatory cells, fibroblasts and stromal cells [3]. This tumor complexity constitutes a challenge to cancer therapy. For example, cancer stem cells are recognized to be more resistant to cancer therapy and could be responsible for cancer relapse, after apparent successful treatment. Besides blood vessels, lymphatic vessels also play a role in tumor maintenance. As stated above, intra-tumor vessels often collapse and are non-functional; however, in the periphery of the tumor they proliferate actively and contribute to dissemination of cancer cells. Also, immune cells are part of the tumors and can have opposite roles, on one hand fighting against cancer cells and on the other hand promoting cancer progression. Cancer associated fibroblasts can be of two different types, the ones which support epithelial tissues and myofibroblasts. It is known that recruitment of fibroblasts promotes cancer cell proliferation, angiogenesis, invasion and metastasis.

2. HONEYBEE PLANT-DERIVED PRODUCTS

Natural products have long been used as sources of active agents with pharmacological activity and vast therapeutic possibilities. In fact, they have provided a great number of compounds with applications in medicine, pharmacy and biology. In particular, there are a large number of anticancer drugs obtained from natural sources [16] and many studies focusing on the use of natural products for cancer prevention and treatment are being performed worldwide.

The scientific interest in modified plant products by animals, such as the honeybee products, which normally have been largely ignored and wasted, has also been increased throughout the years [17] and some have been extensively studied more recently [18 - 20]. Here we report and discuss the current findings and potential of some honeybee plant-derived products for the development of anticancer drugs and strategies, following the suggestions drawn from several studies reporting that some natural bee products can inhibit tumor cell growth as well as metastasis and may induce apoptosis of cancer cells. Likewise, the majority of studies regarding the mode of action of honeybee products against cancer cells suggest mediation by apoptosis, necrosis and lysis of the tumor cells [21]. Phenolic compounds, in particular the flavonoids of honeybee products seem to be the main bioactive compounds responsible for anti-cancer effects and have been described to have an important role as chemopreventive agents in a plethora of studies [22 - 25]. Overall, flavonoids display antitumor properties due to their antioxidant activity and the related capacity to interfere with several signaling pathways, as the stimulation of tumor necrosis factor alpha (TNF-a), the inhibition of cell proliferation, the induction of apoptosis and cell cycle arrest [26 - 28].

2.1. Honey

Honey, frequently the main desired product of the hive, is a sweet viscous solution of variable color that results from flowers' nectars that various species of bees collect, mix with salivary secretions, and regurgitate repeatedly. During nectar collection, pollen can also be incorporated into honey through several ways. In the hive, the mixture is deposited and stored in honeycombs that are sealed with wax. Bees use their wings to fan honey combs, thereby evaporating most of honey water and avoiding its fermentation. Inside the sealed comb the honey matures and can last for many years.

In antiquity raw honey was as appreciated on the table as on the battlefield and its medicinal and nutritional value were long acknowledged. Democritus (500 BC) consumed honey daily, both for longevity and for fertility. Hippocrates and other physicians considered honey very important to strength the body and improve health [29]. The traditional use of honey is also documented in ancient Egypt, as embalming material, and in the texts of Old Testament. Recognized as one of the purest and most natural medicines, honey has been used for eras to treat an extensive range of medical problems, like wounds, burns, and scrapes. More recently, the manufacture of a range of honey-based products for medical treatment, specifically for first and second-degree burns and traumatic and surgical wounds, has received clearance by the Food and Drug Administration (FDA). Such goods are made with a monofloral honey produced in New Zealand and Australia from the nectar of *Leptospermum scoparium*, the manuka tree [30].

As this work is focused on the anti-cancer effects of honeybee plant-derived products, the following lines will highlight the current knowledge concerning such effects and the possible mechanisms underlying antitumor action. The clinical value of honey and its role in cancer have been reviewed in recent studies [23, 31 - 33] which highlighted its effectiveness in chemotherapy and radiotherapy-induced skin reactions, radiation-induced oral mucositis, stomatitis, periodontal gum disease, external surgical wounds, malignant ulcers and infected lesions in oncology patients [34 - 36]. In general, honey can reduce tumor proliferation either by inhibiting cell proliferation and/or inducing apoptosis [23, 37], blocking cell cycle [28], affecting the immune system [38, 39] and modulating signaling pathways [26].

Table **1** summarizes relevant reported data for the anti-cancer effects of honey and its compounds.

Table 1. ***In vitro*** **and** ***in vivo*** **antitumor effects of honey and its compounds.**

Type/Origin/ Isolated compound	Model	Route of administration	Antitumor effects	Refs.
Multifloral honey (4 leading Indian brands)	HCT 15 and HT 9 colon cancer cell lines Breast and liver cancer cells	-	Induce apoptosis arresting cells at subG1 phase, ↓MMP and ↑ROS, up-regulates p53, modulates expression of pro and anti-apoptotic proteins ↑caspase-3	[23, 27, 37] [24]
Tualang honey	Human MCF-7 and MDA-MB-231 breast adenocarcinoma cell lines Oral squamous and osteosarcoma cell lines	-	Promotes anticancer activity of TAM. Induces apoptosis with activation of caspase-3/7, -8 and -9 ↑depolarization of mitochondrial membrane when used with TAM dose-dependent antiproliferative effect by inducing early apoptosis	[40] [41]
Ethyl acetate extracts of Greek honeys (thyme, pine and fir)	Breast (MCF-7), endometrial (Ishikawa) and prostate (PC-3) cancer cell lines	-	Antiestrogenic and a weak estrogenic effect at low and high concentration, respectively, in MCF-7 cells. Thyme honey ↓viability of Ishikawa and PC-3 cells, fir honey ↑viability of MCF-7	[42]
Pure unfractionated honey purchased from Manitoba (Tokyo, Japan)	Human bladder cancer (T24, 253J and RT4), murine bladder cancer cell line (MBT-2)	-	Significant ↓of T24 and MBT-2 cell lines proliferation by 1-25% honey and of RT4 and 253J cell lines by 6-25% honey	[45]
Manuka honey	murine melanoma (B16.F1), colorectal carcinoma (CT26) and human breast cancer (MCF-7) cells	-	Potent time- and dose-dependent anti-proliferative effect, mediated by activation of a caspase 9-dependent apoptotic pathway, with ↑caspase 3, ↓Bcl-2 expression, DNA fragmentation and cell death	[55]
Honey	HepG2 cell line	-	↓cell viability, ↑cell antioxidant status and induces apoptotic death	[59]
Honey	Patients with head and neck cancer under radiation	Topical	Significant ↓ in the symptomatic grade 3/4 mucositis	[34, 43]
Jungle honey	Mice	Intra-peritoneal	Chemotactic activity for neutrophils. ↓tumor incidence and weight and ↑ROS producing cells	[39]
Life-Mel Honey	Cancer patients under chemotherapy for primary or metastatic disease and with grade 4 neutropenia	Oral	Prevention of chemotherapy-induced neutropenia 1/3 patients reported ↑quality of life and no side effects during honey intake	[54]

(Table 1) contd.....

Type/Origin/ Isolated compound	Model	Route of administration	Antitumor effects	Refs.
HoneySoft®	Adult females with breast neoplasms under radiotherapy	Topical Honey dressing	Faster healing of skin reactions. More patient satisfaction with honey than with paraffin gauzes	[45]
Medihoney™	Patient with mantle cell lymphoma under chemotherapy	Topical	Improvement of healing rates of complex wound	[44]
Pure unfractionated honey purchased (Manitoba - Tokyo, Japan)	Mice with MBT-2 bladder cancer implantation model	Intralesional injection Oral ingestion	Intralesional injection of honey (6 and 12%) as well as oral ingestion significantly ↓tumor growth	[45]
Honey	BALB/c strain mice inoculated with Ehrlich ascites tumor	Topical Honey dressing	↓ tumor implantation	[51]
Honey	murine tumor models (mammary and colon carcinomas)	Oral	When applied before tumor-cell inoculation leads to a pronounced anti-metastatic effect	[52]
Manuka honey	mouse melanoma model	intravenously	33% inhibition tumor growth, with ↑tumor apoptosis Dramatic ↑host survival in the co-treatment group, alleviating chemotherapy-induced toxicity	[55]
Quercetin	Mice MCF-7 mammary carcinoma model	-	↓the tumor, cell cycle arrest at G2/M phase	[25]
Chrysin	U937 leukemia cells	-	Induces apoptosis through caspase activation and inactivation of Akt signaling	[26]
	B16-F1 and A375 melanoma cells		Induces apoptosis through caspase-dependenrt mechanisms, activation of p38 MAP kinases and down-regulation of ERK	[28]
Eugenol	Colon carcinoma cell lines HCT-15 and HT-29	-	↑intracellular non-protein thiols, ↑in the earlier lipid layer break, dissipation of MMP and ↑ROS, activation of p53 and caspase-3	[46]
Caffeic acid	HCT 15 colon cancer cells	-	↑ cells accumulation at sub-G1 phase. Induction of apoptosis in dose- and time-dependent way. ↑ROS and ↓MMP	[47]
p-Coumaric acid	Human colorectal carcinoma (HCT-15) cells	-	↓proliferation and cells colony forming ability. ↑ROS, ↑lipid layer breaks and ↓MMP	[48]

MMP - mitochondrial membrane potential, ROS – reactive oxygen species, TAM - tamoxifen.

2.1.1. Honey Effects on Cancer Cell Growth and Proliferation

Research evidence for honey inhibition of cell growth comes from studies with cell and tissue cultures [37, 40 - 42], animal models [25, 39, 41] and also clinical trials [34, 35, 43, 44]. The anti-proliferative effect of honey revealed to be dependent on dose, time [37] and also on the type of cell line. Swellam *et al.* [45] reported a significant *in vitro* inhibition of the proliferation of bladder cancer cell lines T24, MBT-2, RT4 and 253J and observed a significant arrest in the sub-G1 phase of cells treated with 3% honey. Honey and/or its phenolics components were also reported to cause perturbation of growth and proliferation in colon cancer [37, 46 - 48], glioma [49], and melanoma [28] cell lines blocking cell cycle in G0/G1 phase. Jungle honey – a honey made by wild honeybees living in the tropical forest of Nigeria – has antitumor activity against human oral, breast, cervical cancer and osteosarcoma derived cell lines [40, 41]. Inhibition of the estrogenic activity and of cell viability of breast (MCF-7), endometrial (Ishikawa) and prostate (PC-3) cancer cells was also detected upon treatment with different Greek honey extracts [42].

Results from studies using animal models indicate that honey displays moderate antitumor as well as antimetastatic effects against renal cell carcinoma [38, 45] and potentiates the activity of chemotherapeutic medications like 5-fluorouracil or cyclophosphamide [50]. Honey as well as its bioactive compounds caffeic acid, CAPE (caffeic acid phenethyl ester) and flavonoid glycones proved to have an inhibitory effect on tumor cell proliferation, blocking cell cycle by down-regulating many cellular pathways *via* tyrosine cyclooxygenase, ornithine decarboxylase, and kinase [23, 28, 49]. Honey was also shown to be involved in modulation of p53 regulation [23], which is involved in tumor suppression. Hamzaoglu and co-workers [51] verified that the protective covering of surgical injuries with honey prevents tumor implantation in mice and proposed that honey could possibly inhibit proteolysis of the basal membrane that occurs during tumor cell invasion and act as a barrier. Lung tumor incidence and weight decreased in mice injected with Jungle honey but this antitumor activity was associated to ROS production by infiltrated neutrophils into tumor tissue [39].

Honey can lead to promotion of tumor growth too, as it contains a mixture of amino acids, vitamins and minerals plus large amounts of glucose. Furthermore, honey high osmolarity can lead to an outflow of lymph which in turn enhances nutrification and oxygenation while its acidity favors oxygen release, from haemoglobin, in the capillaries of adjacent tissues. In accordance, in murine and rat tumor models the antitumor effect was only observed when honey was given prior to tumor cell inoculation (and development) - seeming that honey polyphenolic components stimulate host antitumor defenses only before tumor, whereas its nutritive constituents prevail in tumor´s presence [52].

One of the first clinical studies concerning the use of honey in oncology patients, dated from 1970, reported the effectiveness of honey topical application in the treatment of wounds of urogenital carcinomas in 12 patients, which showed clearance of infection within 3-6 days [32]. Topical application of natural honey showed to be effective and proved to be a simple and economical treatment too, particularly in the prevention of oral mucositis, a condition where patients (in particular head and neck cancer patients) are more susceptible to ulcerations and infections due to the damage of epithelial cells by chemotherapy and/or radiotherapy [34, 43]. Honey seems to be also useful in the treatment of wounds of other cancers like breast [35] or mantle cell lymphoma [44], even if infected with methicillin-resistant *Staphylococcus aureus* (MRSA) [36]. Medihoney™, the medical-grade honey from New Zealand, and other therapeutic honeys have been proving to successfully treat diverse wounds in different oncologic patients [30, 35, 36, 53, 54]. Honey may also be used as adjuvant in cancer therapy: one third of a group of cancer patients receiving chemotherapy reported improvement of quality of life and no side effects during honey intake [54].

2.1.2. Honey Effect on Cancer Cell Apoptosis

Many chemotherapeutics currently in use are apoptosis inducers and some honey anti-cancer effects have shown to be mediated by apoptosis processes, making honey a potential natural anti-cancer agent. Antiproliferative and early apoptotic effects of different samples of honey were detected in human osteosarcoma cell lines [41] as well as in oral squamous cell carcinoma [41], HCT 15 and HT 29 colon cancer cell lines [23, 37], human MCF-7 breast cancer cells [40] and cervical cancer cell lines [40]. The inhibition of bladder cancer cell lines involved a cell cycle arrest [45] and honey induced apoptosis was associated with the activation of caspase-3 by CAPE [49]. Honey apoptotic effects can also occur *via* depolarization of mitochondrial membrane [40], increased expression of caspase-3, induction of p53, and pro-apoptotic protein Bax [23, 27], down-regulation of the expression of the antiapoptotic protein Bcl-2 [26] and ROS generation, which in turn activates p53 and p53 modulates the expression of both pro- and anti-apoptotic proteins like Bax and Bcl-2 [27]. Manuka honey induces apoptosis in cancer cells through the induction of caspase-9 and consequent activation of the executor caspase-3, induction of DNA fragmentation, and loss of Bcl-2 expression [55].

2.1.3. Other Effects of Honey

Beyond antiproliferative and apoptotic effects, honey displays antioxidant, anti-inflammatory, estrogenic and immunomodulatory activities which are implicated in its antitumor properties. Antioxidant and anti-inflammatory properties of honey and other honeybee products are covered elsewhere in this book and will only be referred briefly here. Honey reduces chronic inflammation, a well known risk factor for cancer pathogenesis, improves the immune status and stimulates the healing of chronic ulcers and wounds by enabling an increase in lymphocytes and phagocytes, and helping monocytes to release cytokines and interleukins (*e.g.* TNF-α, NF-κB (nuclear factor kappa-light-chain-enhancer of activated B cells), IL(interleukin)-1β, IL-6, also involved in honey-induced apoptosis [56].

ROS and reactive nitrogen species (RNS) are oxidative stress agents which damage cell biomolecules and play a crucial role in human cancer development [57]. Antioxidants may avoid or delay the onset of certain types of cancer mainly through its action as free radical scavengers. In honey, phenolic acids and flavonoids are responsible for its well-established antioxidant activity and may inhibit the cancer process *in vivo* [58]. An enhanced antioxidant *status* with apoptosis has been observed in hepatocellular carcinoma cells [59].

Mutagenicity is interlinked with carcinogenicity. As honey has been shown to act as a strong antimutagenic agent, this can be associated to its anticarcinogenic properties [60]. Honey also displays an estrogenic modulator activity through an antagonistic action, which may be relevant in estrogen-dependent cancers such as endometrial and breast [42]. Estrogen receptors tie to estrogens and then translocate into the nuclei to bind to estrogen-response elements, resulting in expression of the estrogenic effect in the targeted tissue. Some studies report modulation of estrogen receptor activity by some honeys, an effect that was again linked to honey phenolic contents [42]. For example, honey extracts prepared with different Greek honeys showed dual estrogen effects: agonistic at high concentrations and antagonistic at low concentrations.

A role of honey as a probiotic agent has also been suggested to contribute for its antitumor properties, due to a stimulating effect on colonic probiotic bacteria [61]. Probiotics have multiple modes of action that are particularly important to protect the damaged immune system after chemo- and radiotherapy [32].

2.2. Propolis

Propolis, also called bee glue, is a complex natural gummy product produced by bees from the resins collected from parts of plants, buds, and exudates, which is combined with secretions from salivary glands rich in enzymes, pollen, beeswax and probably with other compounds of bee metabolism [21, 62].

This complex natural resinous product has been extensively employed by man in many parts of the world since ancient times, at least to 300 BC, especially in folk medicine to treat several diseases [17, 18, 63 - 66]. The use of propolis continues today, despite not being considered a therapeutic agent in conventional medicine. It is still used in many regions of the world, including Japan and the European Union. Actually, it is commercially available in the form of capsules, extracts, mouthwash solutions, throat lozenges, creams, powder and also in more purified products from which the wax was removed. It was expected that the biological properties of propolis would be different regarding its chemical composition variation in different geographic origins. Nevertheless, in many cases, samples of different origins can display identical biological activity [67], being flavonoids, aromatic acids, diterpenic acids and phenolic compounds the principal constituents responsible for the biological activities [49, 62]. In the last years, many studies using different worldwide samples of propolis have demonstrated antibacterial and antifungal [17, 68 - 72], antiviral [68, 69, 73], anti-inflammatory [74], antioxidant [75, 76], immunomodulatory [17 , 77] and antitumor activities [18 , 76 - 80].

2.2.1. Propolis Effect on Cancer

Propolis is a complex natural product which is relatively non-toxic in mammals, contains a large diversity of compounds that can act through different pathways in the oncogenic network, leading to the end of carcinogenesis [81]. Recently, using a variety of propolis samples from different geographic origins, independent *in vitro* and *in vivo* studies have shown that this product can decrease cell proliferation and growth, increase apoptosis, promote antiangiogenic effects and modulate the tumor microenvironment [82 , 83], which in turn leads to cessation of tumorigenesis. Table **2** summarizes relevant reported data for the anti-cancer effects of propolis and its compounds.

Table 2. ***In vitro*** **and** ***in vivo*** **antitumor effects of propolis and its compounds.**

Type/Origin/ Isolated compound	Model	Route of administration	Antitumor effects	Ref
Hexane extract/ Thailand	SW620 (colon), BT474 (breast), Hep-G2 (hepatic), Chago (lung), Kato-III (stomach) cancer cells; CH-liver and HS-27 (fibroblast) normal cells	-	Antiproliferative activity against all cancer cell lines; low cytotoxic activity on the normal cell lines	[84]
Ethanol extract/ Portugal	MDA-MB-231, MDA-MB-468 and MCF7 (breast), DU145 and 22RV1 (prostate), U251 and SW1088 (glioblastoma) cancer cells. HB4a (breast), PNT-1 (prostate) and hDFb190 (fibroblast) normal cells	-	↓viability of all cancer cell lines, with lowest effect in the normal cell ↓MDA-MB-231 and DU145 cell proliferation and affects cell cycle	[78]
	CAM.	-	↓vessel sprouting	
Methanol extract/ Portugal	Normal and cancerous renal cells derived from human renal cell carcinoma patients and A-498 (human renal carcinoma) cells	-	Selective toxicity against malignant cells compared to normal cells ↓tumor cells growth	[76]
Ethanol extract/ Turkey	MCF-7 (breast) cancer cells		↓cell growth and promotes apoptosis by the activation of caspase-8, 9 and 6	[92]

(Table 2) contd.....

Type/Origin/ Isolated compound	Model	Route of administration	Antitumor effects	Ref
Ethanol extract/ India	MCF-7 (breast), HT-29 (colon), Caco-2 (colorectal) and B16F1 (murine melanoma) cancer cells	-	↓cells viability and promotes DNA fragmentation	[94]
Ethanol extract/ China	Breast cancer cells (MCF-7 and MDA-MB-231) and human umbilical vein endothelial cells (HUVECs)	-	Induces apoptosis by regulating annexin A7 (ANXA7) and p53 protein expression, inhibiting NF-kB and regulating ROS levels and MMP	[95]
Ethanol extract/Poland	LNCaP (prostate) cancer cells	-	Promotes TRAIL-induced apoptosis by ↑TRAIL-R2 expression, ↓activity of NF-κB, ↑disruption of membrane potential	[97 - 99]
Hydro-alcoholic extract/ Brazil	CAM and yolk-sac membranes of chick embryos	-	↓angiogenesis and vasculogenesis	[106]
Ethanol extract/ Brazil	5637 (bladder) cancer cells	-	Induces AIF, p53, Bax, and Bcl-2 activation, promoting apoptosis	[96]
Ethanol extract/ Brazil	DU145 and PC-3 (prostate) cancer cells and primary malignant tumor (RC58T/h/SA#4)-derived human prostate cancer	-	↓proliferation of prostate cancer cells *via* regulation of cyclin D1, B1, CDK and p21 protein expression	[87]
	Neuro2a (mouse neuroblastoma) cancer cells	-	Promotes cell death and growth retardation *via* histone deacetylase inhibitory activity	[86]
Ethanol extract/ Brazil	Female nu/nu mice	Intraperitoneal injection	Block PAK1 signaling and suppress almost completely the growth of human neurofibromatosis tumor xenografts	[88]
Artepillin C				
Ethanol extract/Brazil	MCF-7 (breast) cancer cells	-	Promotes apoptosis through induction of mitochondrial dysfunction, DNA fragmentation and caspase-3 activity	[10, 21]
Ethanol extract/ Korean	CAM	-	↓ the number of newly formed vessels	[107]
Ethanol extract/ Brazil	HEK293 (embryonic kidney) cells	-	Inhibits expression of HIF-1α protein and its downstream target genes (GLUT1, hexokinase 2, and VEGF-A)	[105]
Galangin	WSU-HN4, HN6 and HN13 (head and neck squamous cell carcinoma) cancer cells	-	↓proliferation and colony formation of the cancer cells and promotes G0/G1 phase cell cycle arrest, accompanied by ↓cyclin D1, CDK4 and CDK6 expression, ↓Rb phosphorylation	[91]

(Table 2) contd.....

Type/Origin/ Isolated compound	Model	Route of administration	Antitumor effects	Ref
7-epi-nemorosone	LNCaP (prostate) cancer cells	-	Induces cell cycle disruption (↑sub-G0/G1, G1, and ↓S phase populations) and apoptosis by targeting the protein kinase MEK1/2 and Akt/PKB	[93]
Chrysin	U937 (leukemia) cancer cells	-	Induces apoptosis by Akt dephosphorylation and down-regulation of XIAP	[26]
Bacharin, beturetol, kaempferide, isosakuranetin and drupanin	CAM	-	↓number of newly formed vessels	[105]
	HEK293 (embryonic kidney) cells	-	Inhibits expression of HIF-1α protein and its downstream target genes (GLUT1, hexokinase 2 and VEGF-A)	[105]
CAPE	Tw2.6 (human oral squamous cell carcinoma) cancer cells	-	Suppress cell proliferation and colony formation *via* inhibition of Akt signaling, ↓G1 phase cell population, ↑G2/M phase cell population and induces apoptosis	[88]
	MDA-MB-231, MCF-7, and SKBR3 (breast) cancer cells		Promotes accumulation of acetylated histone proteins suggesting histone deacetylase inhibitory properties ↓estrogen receptor (ER) and progesterone receptor, EGFR and Her2 protein	[89]
	MDA-MB-231, MCF-7, MCF-10A and MCF-12A (breast) cancer cells		Suppress the growth of MCF-7 and MDA-MB-231 cells without much effect on normal cells, promotes cell cycle arrest and apoptosis by down-regulation Bcl-2 proteins, ↓NF-κB and mdr-1 gene expression	[90]
	MCF-7 (breast) cancer cells	-	Promotes apoptosis by Bax protein induction, activation of caspases and MAPK family proteins p38 and JNK.	[101]
			Promotes apoptosis through induction of caspase-3 activity, mitochondrial dysfunction and DNA fragmentation	[102]
	HT1080 (fibrosarcoma) cancer cells	-	↑Invasion, cell migration, motility and colony formation, associated to ↓MMP TIMP-2 mRNA levels, down-regulation of MMP-2 and MMP-9 expression, ↓MMP-2 activity	[103]
	Hep3B (hepatocellular carcinoma) cells	-	↓MMP-9 activity	[104]

(Table 2) contd.....

Type/Origin/ Isolated compound	Model	Route of administration	Antitumor effects	Ref
CAPE	C57BL/6 and BALB/c mice infected with LLC (lewis lung carcinoma), B16F10 (melanoma) and CT26 (colon) cancer cells.	Intraperitoneal injection	Inhibit the growth and neovascularization of primary tumor cells	[109]

CAM - chicken chorioallantoic membrane; CDK - cyclin dependent kinase; ER - estrogen receptor; HIF-1α - hypoxia-inducible factor 1-alpha; MMP - the mitochondrial membrane potential ; Rb -retinoblastoma protein; TIMP-2 - tissue inhibitor metalloproteinase-2; VEGF-A - vascular endothelial growth factor A.

2.2.2. Propolis Effect on Cancer Cell Growth and Proliferation

As mentioned before, cancer cells present uncontrolled growth and proliferation, which is a result of the abnormal function of various genes. In fact, sustained proliferation is the most fundamental characteristic of cancer cells [2 , 3]. It was reported that an hexane extract of Thailand propolis provides an high antiproliferative activity against cancer cell lines of distinct organ origin (Chago, BT474, Hep-G2, KATO-III and SW620) while being less toxic against normal fibroblast and liver cells [84]. Recently, our group [78] also showed that an ethanol extract of poplar Portuguese propolis decreased the viability of seven tested cancer cell lines (MDA-MB-231, MDA-MB-468, MCF7, DU145, 22RV1, U251 and SW1088), with minor effects observed for the two normal cell lines (HB4a and PNT-1) and fibroblasts (hDFb190). Additionally, MDA-MB-231 and DU145 cell proliferation decreased, with cell cycle changes. This selective toxicity against malignant cells compared to normal cells was observed in another study using poplar Portuguese propolis [76].

Other studies showed the antitumor activity of Brazilian propolis [85 - 87]. It was described that ethanol extracts of Brazilian propolis can affect cell proliferation by regulating the expression of cyclins (D1 and B1), proteins responsible to control the progression of cells through the cell cycle by triggering cyclin-dependent kinase (CDK) enzymes as well as p21 [86], the cyclin-dependent kinase inhibitor 1 or CDK-interacting protein 1. Ishiai *et al.*[85] showed that ethanol extracts of Brazilian green propolis can retard cancer cell growth *via* histone deacetylase inhibitory activity. Additionally, ethanol extracts of green propolis together with artepillin C, when administered to mice with tumors associated with neurofibromatosis, blocked PAK1 signaling, which is required for tumor growth.

The gene PAK1 codes for a serine/threonine p21-activated kinase that regulates cell motility and morphology. As a result, the tumor was suppressed almost completely [87]. Interestingly, a similar inhibitory effect can be induced by CAPE as well.

Kuo *et al.*[88] showed that CAPE blocked cell proliferation of human oral squamous cell carcinoma (TW2.6) by decreasing G1 phase cell population, increasing G2/M phase cell population. Also, it decreased the abundance of serine/threonine-specific protein kinases Akt, Akt1, Akt2, Akt3, phospho-Akt Ser473, phospho-Akt Thr 308, glycogen synthase kinase 3 β (GSK3β), members of the class O of forkhead box transcription factors (FOXO) family such as FOXO1, phospho-FOXO1 FoxO1a (Thr24), FOXO3a, phospho-FoxO3a FoxO3a (Thr32), and also NF-κB, phospho-NF-κB Ser536, Rb, phospho-Rb Ser807/811, S-phase kinase-associated protein 2 (Skp2), and cyclin D1, but increased the cell cycle inhibitor $p27^{Kip1}$[88]. CAPE has also a role as a histone deacetylase inhibitor. Results show that CAPE promotes an accumulation of acetylated histone proteins in MCF-7 (ER+) and MDA-MB-231 (ER−/PR−/Her2-) cells with associated decrease in estrogen (ER) and progesterone receptors (ER and PR, respectively) in MCF-7 cells, and up-regulation of ER and decrease in EGFR in MDA-MB-231 cells [89]. Additionally, it was shown that CAPE effect in different cancer cell lines was mediated through inhibition of NF-κB [90]. The flavonoid galangin, present in many types of propolis, inhibits the proliferation and colony formation of human head and neck squamous cell carcinoma and promoted cell cycle arrest of the tumor cells at the G0/G1 phase, accompanied by decreased expression of cyclin D1, CDK4, CDK6 and phosphorylation of Rb [91].

2.2.3. Propolis Effect on Cancer Cell Apoptosis

The ability to promote apoptosis in cancer cells is a positive effect of the anticancer therapy. Recently, some *in vitro* and *in vivo* studies showed that propolis extracts can induce apoptosis in diverse tumor cells. Different extracts of Turkish propolis induced apoptosis in the MCF-7 human breast cell line, effect that was associated with an increase in the expression of caspases [92]. Diaz-Carballo *et al.* [93] showed that the polyisoprenylated benzophenone 7-*epi*-nemorosone induces cell cycle disruption and apoptosis by targeting mitogen-activated protein kinase MEK1/2 and Akt/PKB. Another study showed that an ethanol extract of Indian stingless bee propolis promoted DNA fragmentation in different cancer cell lines [94]. Xuan *et al.* [95] showed that an ethanolic extract of Chinese propolis and its polyphenolic/flavonoids compounds induce apoptosis in breast cancer cells by regulating annexin A7 (ANXA7) and p53 protein expression, inhibiting NF-κB and regulating the levels of ROS and the mitochondrial membrane potential. Begnini *et al.* [96] showed that Brazilian red propolis may trigger apoptosis or apoptosis-like PCD induction through p53, Bax, and Bcl-2 activation. In other studies it was shown that an ethanolic extract of Brazilian green propolis as well as its main constituents (artepillin C, kaempferol, *p*-coumaric acid and quercetin) sensitize prostate cancer cells to TRAIL-induced death [97 - 99]. Additionally, a potent apoptosis effect of artepilin C, baccharin and drupanin (propolis cinnamic acid derivatives) has been demonstrated in colon cancer cells, with stimulation of both intrinsic and extrinsic apoptosis pathways. Especially for baccharin and drupanin, the mechanism involved transduction TRAIL/DR4/5 and/or FasL/Fas death-signaling loops and miR-143, with decreased expression of MAPK/Erk5 and its downstream target c-Myc [100].

CAPE has been identified as one of the major active compounds in propolis with apoptotic effect on different cancer cell lines [49]. In breast MCF-7 cancer cells, CAPE activated caspases, induced p53-regulated Bax protein and activated Fas *via* a Fas ligand (Fas-L)-independent mechanism. Therefore, Fas activation seems to have a role in NF-κB inhibition-induced apoptosis [101]. In addition, Kamiya and co-workers [102] revealed that either a Brazilian red propolis ethanol extract or CAPE reduce MCF-7 cell viability through induction of caspase-3 activity, DNA fragmentation and mitochondrial dysfunction. Also, an increased expression in CHOP (CCAAT/enhancer-binding protein homologous protein) was seen, which only occurs in the presence of endoplasmic reticulum stress. So, MCF-7 cell apoptosis occurs, at least partially, by ER stress-related signaling.

Chrysin, another bioactive compound of propolis, has been shown to have the capacity to influence the apoptotic process in leukemia cells by the activation of caspases, Akt dephosphorylation and down-regulation of XIAP, the X-linked inhibitor of apoptosis protein [26].

2.2.4. Propolis Can Exert Antiangiogenic Effects and Modulate the Tumor Microenvironment

Similar to normal tissues, tumors need to have means of capturing nutrients and oxygen and excreting the products of their metabolism and carbon dioxide to the extracellular medium. As stated above, the tumor-associated vasculature and angiogenesis have a great importance in the maintenance of these functions [3]. Additionally, it is known that cancer microenvironment is very important for cancer development and angiogenesis. Many studies showed that propolis and/or its components can interfere with the symbiosis cancer cells/microenvironment and inhibit angiogenesis, which in turn impair tumorigenesis.

Hwang *et al.* [103] showed that CAPE suppressed gene expression of the matrix metalloproteinases MMP-2, MMP-9 and MT1-MMP and the metallopeptidase inhibitor 2 (TIMP-2) in human HT1080 fibrosarcoma cells. Additionally, it inhibited the activated MMP-2 activity, cell migration, motility, invasiveness and colony formation of tumor cells. CAPE also inhibits MMP-9 activity in hepatocellular carcinoma [104]. Another study that demonstrates that cancers and their tumor microenvironment can be countered by natural products has been performed recently [105]. The authors examined components which are derived from Brazilian green propolis - bacharin, beturetol, kaempferide, isosakuranetin and drupanin - and found that some components significantly inhibited the expression of hypoxia-inducible factor 1-alpha (HIF-1α) protein and its downstream target genes such as GLUT1, hexokinase 2, and VEGF-A [105]. On the other hand, Silva-Carvalho *et al.* [78] showed that an ethanol extract of Portuguese propolis increased expression of HIF-1α, pyruvate dehydrogenase kinase, GLUT1, LDH and carbonic anhydrase in breast cancer cells.

Regarding the effect of propolis on angiogenesis, it was shown using HUVEC (human umbilical vein endothelial cells) that hydro-alcoholic propolis extract from Brazil decreased cell viability, proliferation and migration and also inhibited the tube-like structure formation on matrigel. Additionally, *in vivo* propolis administration inhibited both angiogenesis and vasculogenesis in the chorioallantoic and yolk-sac membranes of chick embryos [106]. Silva-Carvalho *et al.* [78] also showed that the ethanol extract of Portuguese propolis decreased

HBMEC (human brain microvascular endothelial cells) total biomass and proliferation and decreased vessel sprouting in the chicken chorioallantoic membrane (CAM). Another study reported that ethanol extracts of Korean propolis suppressed the proliferation and inhibited the tube formation of HUVECs. Also, it deacreased the number of newly formed vessels in the CAM [107], effect that was also observed using Brazilian green propolis [105].

Daleprane *et al.* [108] characterized the molecular mechanism that explains the antiangiogenic action of red propolis polyphenols on endothelial cells. This study provides evidence that such polyphenols display antiangiogenic effects *in vitro* and *ex* and *in vivo*, which result from an effective inhibition of VEGF gene expression. Under hypoxic conditions, this inhibition of VEGF gene expression was attributed to the destabilization of HIF-1α protein, which is achieved *via* an increase in the von Hippel-Lindau (pVHL)-dependent proteasomal degradation. This increase only occurs as a consequence of reduced Cdc42 protein expression. Recently, it was shown that CAPE suppressed VEGF-induced proliferation, migration, tube formation, loss of VE-cadherin at cell-cell contacts in endothelial cells and the formation of actin stress fibers. *In vivo* it reduced vascular permeability in mouse skin capillaries, blocked VEGF-stimulated neovascularization and inhibited the growth and neovascularization of primary tumor cells in C57BL/6 and BALB/c mice [109].

2.3. Pollen

Pollen is the set of small, fine, powdery grains which are the male reproduction units or gametophytes formed in the anthers of flowers. Commercially traded pollen is bee pollen, resulting from the agglutination of flower pollen - collected by the honeybee *Apis mellifera* from a variety of plants - with nectar (and/or honey) plus bee salivary secretions [110]. Collected pollen grains are packed as pellets into pollen baskets on honeybees′ rear legs with the aid of special combs and hairs [111] and usually mixed with nectar or regurgitated honey which makes it stick together and adhere to the legs. In this way, it is easily transported to the hive where it will feed larvae in the early stages of development. Bee pollen is recovered by humans at the entrance of the hive and is therefore a wild product, produced without manipulation [110].

Besides being the bee's primary food source, bee pollen is seen as one of the richest natural foods ever discovered and one of the nature's most complete foods. Its nutritional and medicinal value has been recognized for thousands of years. Human consumption of bee pollen is reported both in the Bible and other religious books, ancient Chinese texts and Egyptian papyri where it was described as "a life-giving dust". Initial mentions to its medical uses are found in books by Arab and Jewish physicians in Islamic Spain and some recommendations included its use as a tranquilizing tonic, a aphrodisiac, or to cure stomach swellings produced by certain foods [112 , 113]. Traditional health practitioners prescribed pollen to their patients to alleviate or to treat several conditions such as allergic reactions, anemia, colitis, colds, enteritis, flu, premature ageing and ulcers. In humans pollen was shown to be easily absorbed in the digestive tract, or partly digested, and passes directly from the stomach into the blood stream [114 , 115].

Several research studies have been trying to document and support both the nutritional claims and the therapeutic efficacy and safety of bee pollen. However, research with cell lines and animals is still scarce and human pre-clinical trials are almost absent. Furthermore, many of the studies were conducted with flower pollen and most medical applications of pollen are pollen preparations of flower pollen as well because only the use of specific pollen can assure a constant concentration of the active ingredient(s).

The most promising results regarding pollen's potential and the most outstanding therapeutic action of clinical value is pollen effect in prostatitis, or prostate inflammation. Chronic prostatitis is a disease of unknown etiology affecting a large number of men, in particular the elderly, and might be related to age and to hormone changes. Benign prostatic hyperplasia is characterized by an enlarged prostate which results from small non-cancerous growths inside the prostate. In placebo-controlled, double-blind clinical trials, researchers found that intake of a bee pollen extract by patients with benign prostatic hyperplasia significantly decrease the symptoms of the disease [116 , 117], being an effective treatment for prostate enlargement and prostatitis. Quercetin, one of the main flavonoids in bee pollen, is thought to be related with this antiprostatitic activity [118]. Quercetin inhibits permanently androgen-independent PC-3 cancer cells blocking cell cycle by inhibiting the expression of several specific genes, up-regulates the expression of various tumor suppressor genes and down-regulates oncogene expression [119]. Other flavonoids detected in bee pollen (kaempferol, apigenin) have also shown to inhibit growth of prostate tissue and to reduce inflammation, pain, and the risk of prostate cancer [120]. The phytochemicals β-sitosterol and other phytosterols of bee pollen might be implicated in the antiprostatitis action too [121], as well as the carotenes β-carotene and lycopene, which have been associated with a decrease of the risk for some prostate carcinoma [122].

But pollen is thought to have other health benefits, including antitumor effects [113 , 123 - 125]. Some findings report that extracts of Turkish bee pollen inhibit respiratory burst of K-562 cancer cells [126], a Brassica bee pollen extract shows anticancer activity by increasing apoptosis of human PC-3 prostate cancer cells [127], bee pollen extracts inhibit the proliferation of HUVECs [128] and bee pollen polysaccharides of *Rosa rugosa* show significant anti-proliferative activity *in vitro*, inhibiting the proliferation of human colon cancer HT-29 and HCT116 cells in a dose-dependent manner [129]. Other findings suggest that pollen can protect animals and humans against the adverse effects or x-ray radiation treatments and cytotoxic damage of chemotherapy, supporting its use as chemoprotective/chemopreventive agent [130 , 131]. Table **3** summarizes relevant reported data for the anti-cancer effects of bee pollen and its compounds.

Table 3. ***In vitro*** **and** ***in vivo*** **antitumor effects of bee pollen and its compounds.**

Type/Origin/ Isolated compound	Model	Route of administration	Antitumor effects	Ref.
Hexane extract/ Thailand	SW620 (colon), BT474 (breast), Hep-G2 (hepatic), Chago (lung), Kato-III (stomach) cancer cells; CH-liver and HS-27 (fibroblast) normal cells	-	Antiproliferative activity against all cancer cell lines; low cytotoxic activity on the normal cell lines	[84]
Apigenin	Mouse skin derived cell lines, C50 and 308 Human HL-60 cells		Induces reversible G_2/M arrest, partly due to inhibition of $p34^{cdc2}$ mitotic kinase activity oand perturbation of cyclin B1 levels	[120]
β-carotene	1439 men subsequently diagnosed with cancer over 12 years of follow-up	Oral	Men with plasma β-carotene at baseline and assigned at random to β-carotene supplementation showed a possible but non-significant ↓in overall cancer risk, primarily due to a significant ↓risk of prostate carcinoma	[122]

(Table 3) contd.....

Type/Origin/ Isolated compound	Model	Route of administration	Antitumor effects	Ref.
Water soluble fraction (Cernitin T60) of pollen extract (product of SAPEC S.A., Barbengo, Switzerland)	Syngeneic mice with implanted Lewis lung carcinoma		Prolonged the life-span of mice carrying the tumor without any apparent side effects Cernitin T60 seems a potent immunostimulator of macrophages	[123]
Polysaccharide LBPP, extracted and isolated from the pollen of *Brassica napus*L	Sarcoma 180-bearing mice and B16 melanoma-bearing mice	Oral	Significant ↓in tumor formation, significant ↑natural killer cell activity, phagocytic function of monocyte, lymphocyte proliferation, serum hemolysis antibody and a significant ↑peripheral blood abnormality and anemia	[124]
Pollen extract prepared by DMSO/ Turkish	K-562 cell cultures and mononuclear cell (MNC) cultures		Inhibit respiratory burst in cancer cell lines, probably by an antioxidant potential	[126]
Chloroform extract from bee pollen of *Brassica campestris*(steroid fraction)	PC-3 (prostate), MCF-7 (estrogen-responsive breast adenocarcinoma), LNCaP (androgen-sensitive prostate), Hela (cérvix), BCG-823 (gastric), BEL-7402 (hepatocarcinoma), KB (squamous), A549 (lung) and HO8910 (ovarian) cancer cell lines	-	Strongest cytotoxicity in PC-3 cell line Induces apoptosis in PC-3 cells, with ↑caspase-3 activity, time-dependent ↓Bcl-2 expression.	[127]
Ethanol extract of bee pollen/China	Human umbilical vein endothelial cells (HUVECs).	-	Suppresses VEGF-induced *in vitro*tube formation suppression of HUVEC proliferation	[128]
Bee pollen polysaccharides from *Rosa rugosa*	Colon cancer cell lines HT-29 and HCT116	-	Anti-proliferative activity Sub-fractions with significant synergistic effect	[129]
Extracts of bee pollen from *Salix alba*L. and *Cystus incanus*L	*Saccharomyces cerevisiae*RMY326 and human stimulated lymphocytes	-	Bee pollens able to ↓chromosome damage induced by three cancer drugs, supporting their use as chemoprotective/ chemopreventive agents	[131]

As already mentioned, flavonoids, phytosterols and carotenoids are important pollen components associated with its main biological activities [132]. The pollen flavonoids quercetin, rutin and chyrisin have shown a chemopreventive activity by increasing apoptosis, thus acting in cancer prevention [28 , 133]. As already mentioned, phytosterols are other important group of bee pollen bioactive compounds which main effects are the blood cholesterol-lowering effect due to a partial inhibition of intestinal cholesterol absorption and an immune stimulating and anti-inflammatory actions, which have been mainly attributed to β-sitosterol.

Recent evidence strongly suggests that plant sterols may be valuable against the development of different cancers like colorectal, breast and prostate cancers [134].

CONCLUSION

Worldwide interest in honeybee plant-derived products has recently developed as more information on their chemical composition and physiological effects has become available and highlighted. In particular, an increasing number of studies provided valuable insights about the anti-cancer effects of honeybee plant-derived products, reporting the inhibition of tumor cell growth, metastasis, as well as induction of apoptosis. Honey, often the main desired product of beekeeping and the most well-known natural product obtained from the hive, has been extensively researched. Although not yet completely understood, multifactorial processes seem to be involved in its antitumor properties: antiproliferative and apoptotic, antioxidant, anti-inflammatory, estrogenic and immunomodulatory activities. Propolis, probably the most important bees′ "chemical weapon" against pathogenic microorganisms, and still one of the most frequently used remedies in some parts of the world, has been found to contain phytonutrients that may have cancer-preventing and antitumor properties in humans. Bee pollen is an exquisite functional food with many health-enhancing effects (antioxidant, anti-inflammatory and/or antimicrobial activities) and having as main biologically active compounds flavonoids and phytosterols. It is an excellent source of antioxidant polyphenols and it has a clinically proven effect against prostatitis.

The results obtained so far, in particular the ones obtained with worldwide samples of honey, propolis and/or the main bioactive compounds – polyphenols and phenolic acids, in particular flavonoids - bring hope of improved treatment for human tumors. However, clinical trials are scarce and there is still a long way until bee products could take a place in modern medicine. More randomized experiments, with improved grading criteria and reporting standards, are essential to validate research and clinical evidence, further contributing to the knowledge of the anticancer nature of honeybee plant-derived products and also to endorse them as potential candidates in the war against cancer.

Additionally, the exact chemical composition of honey, pollen and propolis, as natural products that are gathered rather than secreted, depends on the plants the worker bees visit and can vary even in the same apiary. Bioactivities show also some variation with time and local of collection of the bee product. Thus, quality assurance of honeybee plant-derived products is of utmost importance and there is a need to overcome the problem of chemical standardization and to fully understand the conditions under which such natural products may promote health.

CONFLICT OF INTEREST

The authors declare no conflict of interest regarding this publication.

ACKNOWLEDGEMENTS

The authors are grateful to FCT - Portuguese Foundation for Science and Technology for financial support.

REFERENCES

[1] Ferlay, J.; Soerjomataram, I.; Ervik, M.; Dikshit, R.; Eser, S.; Mathers, C.; Rebelo, M.; Parkin, D.M.; Forman, D.; Bray, F. *GLOBOCAN 2012 v1.0, Cancer Incidence and Mortality Worldwide: IARC CancerBase No. 11 [Internet]*; International Agency for Research on Cancer: Lyon, France, **2013**. Available from: http://globocan.iarc.fr, accessed on 31/January/2015

[2] Hanahan, D.; Weinberg, R.A. The hallmarks of cancer. *Cell,* **2000**, *100*(1), 57-70. [http://dx.doi.org/10.1016/S0092-8674(00)81683-9] [PMID: 10647931]

[3] Hanahan, D.; Weinberg, R.A. Hallmarks of cancer: the next generation. *Cell,* **2011**, *144*(5), 646-674. [http://dx.doi.org/10.1016/j.cell.2011.02.013] [PMID: 21376230]

[4] Lemmon, M.A.; Schlessinger, J. Cell signaling by receptor tyrosine kinases. *Cell,* **2010**, *141*(7), 1117-1134. [http://dx.doi.org/10.1016/j.cell.2010.06.011] [PMID: 20602996]

[5] Sherr, C.J.; McCormick, F. The RB and p53 pathways in cancer. *Cancer Cell,* **2002**, *2*(2), 103-112. [http://dx.doi.org/10.1016/S1535-6108(02)00102-2] [PMID: 12204530]

[6] Jackson, S.P.; Bartek, J. The DNA-damage response in human biology and disease. *Nature,* **2009**, *461*(7267), 1071-1078. [http://dx.doi.org/10.1038/nature08467] [PMID: 19847258]

[7] Apel, A.; Zentgraf, H.; Büchler, M.W.; Herr, I. Autophagy-A double-edged sword in oncology. *Int. J. Cancer,* **2009**, *125*(5), 991-995. [http://dx.doi.org/10.1002/ijc.24500] [PMID: 19452527]

[8] Galluzzi, L.; Kroemer, G. Necroptosis: a specialized pathway of programmed necrosis. *Cell,* **2008**, *135*(7), 1161-1163. [http://dx.doi.org/10.1016/j.cell.2008.12.004] [PMID: 19109884]

[9] Artandi, S.E.; DePinho, R.A. Telomeres and telomerase in cancer. *Carcinogenesis,* **2010**, *31*(1), 9-18. [http://dx.doi.org/10.1093/carcin/bgp268] [PMID: 19887512]

[10] Nagy, J.A.; Chang, S.H.; Shih, S.C.; Dvorak, A.M.; Dvorak, H.F. Heterogeneity of the tumor vasculature. *Semin. Thromb. Hemost.,* **2010**, *36*(3), 321-331. [http://dx.doi.org/10.1055/s-0030-1253454] [PMID: 20490982]

[11] Berx, G.; van Roy, F. Involvement of members of the cadherin superfamily in cancer. *Cold Spring Harb. Perspect. Biol.,* **2009**, *1*(6), a003129. [http://dx.doi.org/10.1101/cshperspect.a003129] [PMID: 20457567]

[12] Klymkowsky, M.W.; Savagner, P. Epithelial-mesenchymal transition: a cancer researcher's conceptual friend and foe. *Am. J. Pathol.,* **2009**, *174*(5), 1588-1593. [http://dx.doi.org/10.2353/ajpath.2009.080545] [PMID: 19342369]

[13] Grivennikov, S.I.; Greten, F.R.; Karin, M. Immunity, inflammation, and cancer. *Cell,* **2010**, *140*(6), 883-899. [http://dx.doi.org/10.1016/j.cell.2010.01.025] [PMID: 20303878]

[14] Semenza, G.L. Tumor metabolism: cancer cells give and take lactate. *J. Clin. Invest.,* **2008**, *118*(12), 3835-3837. [PMID: 19033652]

[15] DeBerardinis, R.J.; Lum, J.J.; Hatzivassiliou, G.; Thompson, C.B. The biology of cancer: metabolic reprogramming fuels cell growth and proliferation. *Cell Metab.,* **2008**, *7*(1), 11-20. [http://dx.doi.org/10.1016/j.cmet.2007.10.002] [PMID: 18177721]

[16] Gordaliza, M. Natural products as leads to anticancer drugs. *Clin. Transl. Oncol.,* **2007**, *9*(12), 767-776. [http://dx.doi.org/10.1007/s12094-007-0138-9] [PMID: 18158980]

[17] Sforcin, J.M.; Bankova, V. Propolis: is there a potential for the development of new drugs? *J. Ethnopharmacol.,* **2011**, *133*(2), 253-260. [http://dx.doi.org/10.1016/j.jep.2010.10.032] [PMID: 20970490]

[18] Khalil, M.L. Biological activity of bee propolis in health and disease. *Asian Pac. J. Cancer Prev.,* **2006**, *7*(1), 22-31. [PMID: 16629510]

[19] Alvarez-Suarez, J.M.; Giampieri, F.; Battino, M. Honey as a source of dietary antioxidants: structures, bioavailability and evidence of protective effects against human chronic diseases. *Curr. Med. Chem.,* **2013**, *20*(5), 621-638. [http://dx.doi.org/10.2174/092986713804999358] [PMID: 23298140]

[20] Viuda-Martos, M.; Ruiz-Navajas, Y.; Fernández-López, J.; Pérez-Alvarez, J.A. Functional properties of honey, propolis, and royal jelly. *J. Food Sci.,* **2008**, *73*(9), R117-R124. [http://dx.doi.org/10.1111/j.1750-3841.2008.00966.x] [PMID: 19021816]

[21] Premratanachai, P.; Chanchao, C. Review of the anticancer activities of bee products. *Asian Pac. J. Trop. Biomed.,* **2014**, *4*(5), 337-344. [http://dx.doi.org/10.12980/APJTB.4.2014C1262] [PMID: 25182716]

[22] Fresco, P.; Borges, F.; Diniz, C.; Marques, M.P. New insights on the anticancer properties of dietary polyphenols. *Med. Res. Rev.,* **2006**, *26*(6), 747-766. [http://dx.doi.org/10.1002/med.20060] [PMID: 16710860]

[23] Jaganathan, S.K.; Mandal, M. Antiproliferative effects of honey and of its polyphenols: A Review. *J. Biomed. Biotechnol,* **2009**, *2009*, 1-13. [http://dx.doi.org/10.1155/2009/830616]

[24] Jafari, S.; Saeidnia, S.; Abdollahi, M. Role of natural phenolic compounds in cancer chemoprevention *via* regulation of the cell cycle. *Curr. Pharm. Biotechnol.,* **2014**, *15*(4), 409-421. [http://dx.doi.org/10.2174/1389201015666140813124832] [PMID: 25312621]

[25] Dai, J.; Mumper, R.J. Plant phenolics: extraction, analysis and their antioxidant and anticancer properties. *Molecules,* **2010**, *15*(10), 7313-7352. [http://dx.doi.org/10.3390/molecules15107313] [PMID: 20966876]

[26] Woo, K.J.; Jeong, Y.J.; Park, J.W.; Kwon, T.K. Chrysin-induced apoptosis is mediated through caspase activation and Akt inactivation in U937 leukemia cells. *Biochem. Biophys. Res. Commun.,* **2004**, *325*(4), 1215-1222. [http://dx.doi.org/10.1016/j.bbrc.2004.09.225] [PMID: 15555556]

[27] Jaganathan, S.K.; Mandal, M. Involvement of non-protein thiols, mitochondrial dysfunction, reactive oxygen species and p53 in honey-induced apoptosis. *Invest. New Drugs,* **2010**, *28*(5), 624-633. [http://dx.doi.org/10.1007/s10637-009-9302-0] [PMID: 19705065]

[28] Pichichero, E.; Cicconi, R.; Mattei, M.; Canini, A. Chrysin-induced apoptosis is mediated through p38 and Bax activation in B16-F1 and A375 melanoma cells. *Int. J. Oncol.,* **2011**, *38*(2), 473-483. [PMID: 21170506]

[29] Skiadas, P.K.; Lascaratos, J.G. Dietetics in ancient Greek philosophy: Plato's concepts of healthy diet. *Eur. J. Clin. Nutr.,* **2001**, *55*(7), 532-537. [http://dx.doi.org/10.1038/sj.ejcn.1601179] [PMID: 11464226]

[30] Lusby, P.E.; Coombes, A.; Wilkinson, J.M. Honey: a potent agent for wound healing? *J. Wound Ostomy Continence Nurs.,* **2002**, *29*(6), 295-300. [PMID: 12439453]

[31] Bardy, J.; Slevin, N.J.; Mais, K.L.; Molassiotis, A. A systematic review of honey uses and its potential value within oncology care. *J. Clin. Nurs.,* **2008**, *17*(19), 2604-2623. [http://dx.doi.org/10.1111/j.1365-2702.2008.02304.x] [PMID: 18808626]

[32] Oršolić, N. Bee honey and cancer. *JAAS,* **2009**, *1*(4), 93-103.

[33] Ahmed, S.; Othman, N.H. Honey as a Potential Natural Anticancer Agent: A Review of Its Mechanisms. *Evid. Based Complement. Alternat. Med,* **2013**, *2013*, 829070 . [http://dx.doi.org/10.1155/2013/829070]

[34] Biswal, B.M.; Zakaria, A.; Ahmad, N.M. Topical application of honey in the management of radiation mucositis: a preliminary study. *Support. Care Cancer,* **2003**, *11*(4), 242-248. [PMID: 12673463]

[35] Moolenaar, M.; Poorter, R.L.; van der Toorn, P.P.; Lenderink, A.W.; Poortmans, P.; Egberts, A.C. The effect of honey compared to conventional treatment on healing of radiotherapy-induced skin toxicity in

breast cancer patients. *Acta Oncol.,* **2006**, *45*(5), 623-624.
[http://dx.doi.org/10.1080/02841860600781799] [PMID: 16864180]

[36] Simon, A.; Traynor, K.; Santos, K.; Blaser, G.; Bode, U.; Molan, P. Medical honey for wound care--still the 'latest resort'? *Evid. Based Complement. Alternat. Med.,* **2009**, *6*(2), 165-173.
[http://dx.doi.org/10.1093/ecam/nem175] [PMID: 18955301]

[37] Jaganathan, S.K.; Mandal, M. Honey constituents and their apoptotic effect in colon cancer cells. *JAAS,* **2009**, *1*(2), 29-36.

[38] Al-Waili, N.S. Effects of daily consumption of honey solution on hematological indices and blood levels of minerals and enzymes in normal individuals. *J. Med. Food,* **2003**, *6*(2), 135-140.
[http://dx.doi.org/10.1089/109662003322233549] [PMID: 12935325]

[39] Fukuda, M.; Kobayashi, K.; Hirono, Y.; Miyagawa, M.; Ishida, T.; Ejiogu, E.C.; Sawai, M.; Pinkerton, K.E.; Takeuchi, M. Jungle Honey Enhances Immune Function and Antitumor Activity. *Jungle Honey Evid. Based Complement. Alternat. Med.,* **2009**, *2011*, 908743.
[http://dx.doi.org/10.1093/ecam/nen086]

[40] Fauzi, A.N.; Norazmi, M.N.; Yaacob, N.S. Tualang honey induces apoptosis and disrupts the mitochondrial membrane potential of human breast and cervical cancer cell lines. *Food Chem. Toxicol.,* **2011**, *49*(4), 871-878.
[http://dx.doi.org/10.1016/j.fct.2010.12.010] [PMID: 21167897]

[41] Ghashm, A.A.; Othman, N.H.; Khattak, M.N.; Ismail, N.M.; Saini, R. Antiproliferative effect of Tualang honey on oral squamous cell carcinoma and osteosarcoma cell lines. *BMC Complement. Altern. Med.,* **2010**, *10*(1), 49.
[PMID: 20840769]

[42] Tsiapara, A.V.; Jaakkola, M.; Chinou, I.; Graikou, K.; Tolonen, T.; Virtanen, V.; Moutsatsou, P. Bioactivity of Greek honey extracts on breast cancer (MCF-7), prostate cancer (PC-3) and endometrial cancer (Ishikawa) cells: Profile analysis of extracts. *Food Chem.,* **2009**, *116*(3), 702-708.
[http://dx.doi.org/10.1016/j.foodchem.2009.03.024]

[43] Motallebnejad, M.; Akram, S.; Moghadamnia, A.; Moulana, Z.; Omidi, S. The effect of topical application of pure honey on radiation-induced mucositis: A randomised clinical trial. *J. Contemp. Dent. Pract.,* **2008**, *9*(3), 1-9.
[PMID: 18335114]

[44] Dunford, C.E. Treatment of a wound infection in a patient with mantle cell lymphoma. *Br. J. Nurs.,* **2001**, *10*(16), 1058-1062, 1060, 1062, 1064-1065.
[http://dx.doi.org/10.12968/bjon.2001.10.16.9375] [PMID: 11907456]

[45] Swellam, T.; Miyanaga, N.; Onozawa, M.; Hattori, K.; Kawai, K.; Shimazui, T.; Akaza, H. Antineoplastic activity of honey in an experimental bladder cancer implantation model: *in vivo* and *in vitro* studies. *Int. J. Urol.,* **2003**, *10*(4), 213-219.
[http://dx.doi.org/10.1046/j.0919-8172.2003.00602.x] [PMID: 12657101]

[46] Jaganathan, S.K.; Mazumdar, A.; Mondhe, D.; Mandal, M. Apoptotic effect of eugenol in human colon cancer cell lines. *Cell Biol. Int.,* **2011**, *35*(6), 607-615.
[http://dx.doi.org/10.1042/CBI20100118] [PMID: 21044050]

[47] Jaganathan, S.K. Growth inhibition by caffeic acid, one of the phenolic constituents of honey, in HCT 15 colon cancer cells. *Sci. World J.,* **2012**, *2012*, 372345.

[48] Jaganathan, S.K.; Supriyanto, E.; Mandal, M. Events associated with apoptotic effect of p-Coumaric acid in HCT-15 colon cancer cells. *World J. Gastroenterol.,* **2013**, *19*(43), 7726-7734. [http://dx.doi.org/10.3748/wjg.v19.i43.7726] [PMID: 24282361]

[49] Lee, Y.J.; Kuo, H.C.; Chu, C.Y.; Wang, C.J.; Lin, W.C.; Tseng, T.H. Involvement of tumor suppressor protein p53 and p38 MAPK in caffeic acid phenethyl ester-induced apoptosis of C6 glioma cells. *Biochem. Pharmacol.,* **2003**, *66*(12), 2281-2289. [http://dx.doi.org/10.1016/j.bcp.2003.07.014] [PMID: 14637186]

[50] Wattenberg, L.W. *Chemoprevention of cancer by naturally occurring and synthetic compounds*; Wattenberg, L.W.; Lipkin, M.; Boone, C.W.; Kellof, G.J., Eds.; CRC Press: Cancer chemoprevention. Boca Ranton, **1992**, pp. 19-40.

[51] Hamzaoglu, I.; Saribeyoglu, K.; Durak, H.; Karahasanoglu, T.; Bayrak, I.; Altug, T.; Sirin, F.; Sariyar, M. Protective covering of surgical wounds with honey impedes tumor implantation. *Arch. Surg.,* **2000**, *135*(12), 1414-1417. [http://dx.doi.org/10.1001/archsurg.135.12.1414] [PMID: 11115344]

[52] Orsolić, N.; Knezević, A.; Šver, L.; Terzić, S.; Hackenberger, B.K.; Basić, I. Influence of honey bee products on transplantable murine tumours. *Vet. Comp. Oncol.,* **2003**, *1*(4), 216-226. [http://dx.doi.org/10.1111/j.1476-5810.2003.00029.x] [PMID: 19379183]

[53] Blaser, G.; Santos, K.; Bode, U.; Vetter, H.; Simon, A. Treatment of MRSA-colonized or infected wounds with MedihoneyTM antibacterial honey products. *J. Wound Care,* **2007**, *16*, 325-328. [http://dx.doi.org/10.12968/jowc.2007.16.8.27851] [PMID: 17927079]

[54] Zidan, J.; Shetver, L.; Gershuny, A.; Abzah, A.; Tamam, S.; Stein, M.; Friedman, E. Prevention of chemotherapy-induced neutropenia by special honey intake. *Med. Oncol.,* **2006**, *23*(4), 549-552. [http://dx.doi.org/10.1385/MO:23:4:549] [PMID: 17303914]

[55] Fernandez-Cabezudo, M.J.; El-Kharrag, R.; Torab, F.; Bashir, G.; George, J.A.; El-Taji, H.; al-Ramadi, B.K. Intravenous administration of manuka honey inhibits tumor growth and improves host survival when used in combination with chemotherapy in a melanoma mouse model. *PLoS One,* **2013**, *8*(2), e55993. [http://dx.doi.org/10.1371/journal.pone.0055993] [PMID: 23409104]

[56] Oršolić, N.; Bašić, I. Honey as cancer-preventive agent. *Period. Biol.,* **2004**, *106*, 397-401.

[57] Orrenius, S.; Gogvadze, V.; Zhivotovsky, B. Mitochondrial oxidative stress: implications for cell death. *Annu. Rev. Pharmacol. Toxicol.,* **2007**, *47*, 143-183. [http://dx.doi.org/10.1146/annurev.pharmtox.47.120505.105122] [PMID: 17029566]

[58] Nagai, T.; Sakai, M.; Inoue, R.; Inoue, H.; Suzuki, N. Antioxidative activities of some commercially honeys, royal jelly, and propolis. *Food Chem.,* **2001**, *75*(2), 237-240. [http://dx.doi.org/10.1016/S0308-8146(01)00193-5]

[59] Hassan, M.I.; Mabrouk, G.M.; Shehata, H.H.; Aboelhussein, M.M. Antineoplastic effects of bee honey and Nigella sativa on hepatocellular carcinoma cells. *Integr. Cancer Ther.,* **2012**, *11*(4), 354-363. [http://dx.doi.org/10.1177/1534735410387422] [PMID: 21147814]

[60] Saxena, S.; Gautam, S.; Maru, G.; Kawle, D.; Sharma, A. Suppression of error prone pathway is responsible for antimutagenic activity of honey. *Food Chem. Toxicol.,* **2012**, *50*(3-4), 625-633. [http://dx.doi.org/10.1016/j.fct.2012.01.003] [PMID: 22269905]

[61] Ezz El-Arab, A.M.; Girgis, S.M.; Hegazy, E.M.; Abd El-Khalek, A.B. Effects of dietary honey on intestinal microflora and toxicity of mycotoxins in mice. *BMC Complement. Altern. Med.,* **2006**, *6*(6), 1-13.
[PMID: 16412227]

[62] Fokt, H.; Pereira, A.; Ferreira, A.M.; Cunha, A.; Aguiar, C. How do bees prevent hive infections? The antimicrobial properties of propolis. In: *Current Research, Technology and Education Topics in Applied Microbiology and Microbial Biotechnology. Microbiology Book Series, Mendez-Vilas*; , A., Ed.; , **2010**; 21, pp. 481-493.

[63] Burdock, G.A. Review of the biological properties and toxicity of bee propolis (propolis). *Food Chem. Toxicol.,* **1998**, *36*(4), 347-363.
[http://dx.doi.org/10.1016/S0278-6915(97)00145-2] [PMID: 9651052]

[64] Castaldo, S.; Capasso, F. Propolis, an old remedy used in modern medicine. *Fitoterapia,* **2002**, *73* Suppl. 1, S1-S6.
[http://dx.doi.org/10.1016/S0367-326X(02)00185-5] [PMID: 12495704]

[65] Kuropatnicki, A.K.; Szliszka, E.; Krol, W. Historical aspects of propolis research in modern times. *Evid. Based Complement. Alternat. Med.,* **2013**, *2013*, 964149.
[http://dx.doi.org/10.1155/2013/964149] [PMID: 23710243]

[66] Sulaiman, G.M.; Ad'hiah, A.H.; Al-Sammarrae, K.W.; Bagnati, R.; Frapolli, R.; Bello, E.; Uboldi, S.; Romano, M.; Panini, N.; Scanziani, E.; Pezzolato, M.; Erba, E.; D'Incalci, M. Assessing the anti-tumour properties of Iraqi propolis in vitro and in vivo. *Food Chem. Toxicol.,* **2012**, *50*(5), 1632-1641. [http://dx.doi.org/10.1016/j.fct.2012.01.022] [PMID: 22306915]

[67] Bankova, V. Recent trends and important developments in propolis research. *Evid. Based Complement. Alternat. Med.,* **2005**, *2*(1), 29-32.
[http://dx.doi.org/10.1093/ecam/neh059] [PMID: 15841275]

[68] Marcucci, M.C. Propolis: chemical composition, biological properties and therapeutic activity. *Apidologie (Celle),* **1995**, *26*, 83-99.
[http://dx.doi.org/10.1051/apido:19950202]

[69] Kujumgiev, A.; Tsvetkova, I.; Serkedjieva, Y.; Bankova, V.; Christov, R.; Popov, S. Antibacterial, antifungal and antiviral activity of propolis of different geographic origin. *J. Ethnopharmacol.,* **1999**, *64*(3), 235-240.
[http://dx.doi.org/10.1016/S0378-8741(98)00131-7] [PMID: 10363838]

[70] Ordóñez, R.M.; Zampini, I.C.; Moreno, M.I.; Isla, M.I. Potential application of Northern Argentine propolis to control some phytopathogenic bacteria. *Microbiol. Res.,* **2011**, *166*(7), 578-584. [http://dx.doi.org/10.1016/j.micres.2010.11.006] [PMID: 21237629]

[71] Scazzocchio, F.; D'Auria, F.D.; Alessandrini, D.; Pantanella, F. Multifactorial aspects of antimicrobial activity of propolis. *Microbiol. Res.,* **2006**, *161*(4), 327-333.
[http://dx.doi.org/10.1016/j.micres.2005.12.003] [PMID: 16427259]

[72] Silva, J.C.; Rodrigues, S.; Feás, X.; Estevinho, L.M. Antimicrobial activity, phenolic profile and role in the inflammation of propolis. *Food Chem. Toxicol.,* **2012**, *50*(5), 1790-1795. [http://dx.doi.org/10.1016/j.fct.2012.02.097] [PMID: 22425940]

[73] Amoros, M.; Simões, C.M.; Girre, L.; Sauvager, F.; Cormier, M. Synergistic effect of flavones and flavonols against herpes simplex virus type 1 in cell culture. Comparison with the antiviral activity of propolis. *J. Nat. Prod.,* **1992**, *55*(12), 1732-1740. [http://dx.doi.org/10.1021/np50090a003] [PMID: 1338212]

[74] Hu, F.; Hepburn, H.R.; Li, Y.; Chen, M.; Radloff, S.E.; Daya, S. Effects of ethanol and water extracts of propolis (bee glue) on acute inflammatory animal models. *J. Ethnopharmacol.,* **2005**, *100*(3), 276-283. [http://dx.doi.org/10.1016/j.jep.2005.02.044] [PMID: 15899563]

[75] Miguel, M.G.; Nunes, S.; Dandlen, S.A.; Cavaco, A.M.; Antunes, M.D. Phenols and antioxidant activity of hydro-alcoholic extracts of propolis from Algarve, South of Portugal. *Food Chem. Toxicol.,* **2010**, *48*(12), 3418-3423. [http://dx.doi.org/10.1016/j.fct.2010.09.014] [PMID: 20849908]

[76] Valente, M.J.; Baltazar, A.F.; Henrique, R.; Estevinho, L.; Carvalho, M. Biological activities of Portuguese propolis: protection against free radical-induced erythrocyte damage and inhibition of human renal cancer cell growth *in vitro. Food Chem. Toxicol.,* **2011**, *49*(1), 86-92. [http://dx.doi.org/10.1016/j.fct.2010.10.001] [PMID: 20934479]

[77] Chan, G.C.; Cheung, K.W.; Sze, D.M. The immunomodulatory and anticancer properties of propolis. *Clin. Rev. Allergy Immunol.,* **2013**, *44*(3), 262-273. [http://dx.doi.org/10.1007/s12016-012-8322-2] [PMID: 22707327]

[78] Silva-Carvalho, R.; Miranda-Gonçalves, V.; Ferreira, A.M.; Cardoso, S.M.; Sobral, A.J.; Almeida-Aguiar, C.; Baltazar, F. Antitumoral and antiangiogenic activity of Portuguese propolis in *in vitro* and *in vivo* models. *J. Funct. Foods,* **2014**, *11*, 160-171. [http://dx.doi.org/10.1016/j.jff.2014.09.009]

[79] Valença, I.; Morais-Santos, F.; Miranda-Gonçalves, V.; Ferreira, A.M.; Almeida-Aguiar, C.; Baltazar, F. Portuguese propolis disturbs glycolytic metabolism of human colorectal cancer *in vitro. BMC Complement. Altern. Med.,* **2013**, *13*, 184. [http://dx.doi.org/10.1186/1472-6882-13-184] [PMID: 23870175]

[80] Barlak, Y.; Değer, O.; Colak, M.; Karataylı, S.C.; Bozdayı, A.M.; Yücesan, F. Effect of Turkish propolis extracts on proteome of prostate cancer cell line. *Proteome Sci.,* **2011**, *9*, 74. [http://dx.doi.org/10.1186/1477-5956-9-74] [PMID: 22152088]

[81] Watanabe, M.A.; Amarante, M.K.; Conti, B.J.; Sforcin, J.M. Cytotoxic constituents of propolis inducing anticancer effects: a review. *J. Pharm. Pharmacol.,* **2011**, *63*(11), 1378-1386. [http://dx.doi.org/10.1111/j.2042-7158.2011.01331.x] [PMID: 21988419]

[82] Araújo, J.R.; Gonçalves, P.; Martel, F. Chemopreventive effect of dietary polyphenols in colorectal cancer cell lines. *Nutr. Res.,* **2011**, *31*(2), 77-87. [http://dx.doi.org/10.1016/j.nutres.2011.01.006] [PMID: 21419311]

[83] Sawicka, D.; Car, H.; Borawska, M.H.; Nikliński, J. The anticancer activity of propolis. *Folia*

Histochem. Cytobiol., **2012**, *50*(1), 25-37.
[http://dx.doi.org/10.5603/FHC.2012.0004] [PMID: 22532133]

[84] Umthong, S.; Phuwapraisirisan, P.; Puthong, S.; Chanchao, C. *In vitro* antiproliferative activity of partially purified Trigona laeviceps propolis from Thailand on human cancer cell lines. *BMC Complement. Altern. Med.,* **2011**, *11*, 37.
[http://dx.doi.org/10.1186/1472-6882-11-37] [PMID: 21548933]

[85] Ishiai, S.; Tahara, W.; Yamamoto, E.; Yamamoto, R.; Nagai, K. Histone deacetylase inhibitory effect of Brazilian propolis and its association with the antitumor effect in Neuro2a cells. *Food Sci. Nutr.,* **2014**, *2*(5), 565-570.
[http://dx.doi.org/10.1002/fsn3.131] [PMID: 25473514]

[86] Li, H.; Kapur, A.; Yang, J.X.; Srivastava, S.; McLeod, D.G.; Paredes-Guzman, J.F.; Daugsch, A.; Park, Y.K.; Rhim, J.S. Antiproliferation of human prostate cancer cells by ethanolic extracts of Brazilian propolis and its botanical origin. *Int. J. Oncol.,* **2007**, *31*(3), 601-606.
[PMID: 17671687]

[87] Messerli, S.M.; Ahn, M.R.; Kunimasa, K.; Yanagihara, M.; Tatefuji, T.; Hashimoto, K.; Mautner, V.; Uto, Y.; Hori, H.; Kumazawa, S.; Kaji, K.; Ohta, T.; Maruta, H.; Artepillin, C. Artepillin C (ARC) in Brazilian green propolis selectively blocks oncogenic PAK1 signaling and suppresses the growth of NF tumors in mice. *Phytother. Res.,* **2009**, *23*(3), 423-427.
[http://dx.doi.org/10.1002/ptr.2658] [PMID: 19003952]

[88] Kuo, Y.Y.; Lin, H.P.; Huo, C.; Su, L.C.; Yang, J.; Hsiao, P.H.; Chiang, H.C.; Chung, C.J.; Wang, H.D.; Chang, J.Y.; Chen, Y.W.; Chuu, C.P. Caffeic acid phenethyl ester suppresses proliferation and survival of TW2.6 human oral cancer cells *via* inhibition of Akt signaling. *Int. J. Mol. Sci.,* **2013**, *14*(5), 8801-8817.
[http://dx.doi.org/10.3390/ijms14058801] [PMID: 23615471]

[89] Omene, C.; Kalac, M.; Wu, J.; Marchi, E.; Frenkel, K.; O'Connor, O.A. Propolis and its Active Component, Caffeic Acid Phenethyl Ester (CAPE), Modulate Breast Cancer Therapeutic Targets *via* an Epigenetically Mediated Mechanism of Action. *J. Cancer Sci. Ther.,* **2013**, *5*(10), 334-342.
[PMID: 24466386]

[90] Wu, J.; Omene, C.; Karkoszka, J.; Bosland, M.; Eckard, J.; Klein, C.B.; Frenkel, K. Caffeic acid phenethyl ester (CAPE), derived from a honeybee product propolis, exhibits a diversity of anti-tumor effects in pre-clinical models of human breast cancer. *Cancer Lett.,* **2011**, *308*(1), 43-53.
[http://dx.doi.org/10.1016/j.canlet.2011.04.012] [PMID: 21570765]

[91] Zhu, L.; Luo, Q.; Bi, J.; Ding, J.; Ge, S.; Chen, F. Galangin inhibits growth of human head and neck squamous carcinoma cells *in vitro* and *in vivo*. *Chem. Biol. Interact.,* **2014**, *224C*, 149-156.
[http://dx.doi.org/10.1016/j.cbi.2014.10.027] [PMID: 25450235]

[92] Seda Vatansever, H.; Sorkun, K.; Ismet Deliloğlu Gurhan, S.; Ozdal-Kurt, F.; Turkoz, E.; Gencay, O.; Salih, B. Propolis from Turkey induces apoptosis through activating caspases in human breast carcinoma cell lines. *Acta Histochem.,* **2010**, *112*(6), 546-556.
[http://dx.doi.org/10.1016/j.acthis.2009.06.001] [PMID: 19733388]

[93] Díaz-Carballo, D.; Gustmann, S.; Acikelli, A.H.; Bardenheuer, W.; Buehler, H.; Jastrow, H.; Ergun, S.; Strumberg, D. 7-epi-nemorosone from Clusia rosea induces apoptosis, androgen receptor down-

regulation and dysregulation of PSA levels in LNCaP prostate carcinoma cells. *Phytomedicine,* **2012**, *19*(14), 1298-1306.
[http://dx.doi.org/10.1016/j.phymed.2012.08.004] [PMID: 22981203]

[94] Choudhari, M.K.; Haghniaz, R.; Rajwade, J.M.; Paknikar, K.M. Anticancer activity of Indian stingless bee propolis: an *in vitro* study. *Evid. Based Complement. Alternat. Med.,* **2013**, *2013*, 928280.
[http://dx.doi.org/10.1155/2013/928280] [PMID: 23762169]

[95] Xuan, H.; Li, Z.; Yan, H.; Sang, Q.; Wang, K.; He, Q.; Wang, Y.; Hu, F. Antitumor Activity of Chinese Propolis in Human Breast Cancer MCF-7 and MDA-MB-231 Cells. *Evid. Based Complement. Alternat. Med.,* **2014**, *2014*, 280120.
[http://dx.doi.org/10.1155/2014/280120] [PMID: 24963320]

[96] Begnini, K.R.; Moura de Leon, P.M.; Thurow, H.; Schultze, E.; Campos, V.F.; Martins Rodrigues, F.; Borsuk, S.; Dellagostin, O.A.; Savegnago, L.; Roesch-Ely, M.; Moura, S.; Padilha, F.F.; Collares, T.; Pêgas Henriques, J.A.; Seixas, F.K. Brazilian red propolis induces apoptosis-like cell death and decreases migration potential in bladder cancer cells. *Evid. Based Complement. Alternat. Med.,* **2014**, *2014*, 639856.
[http://dx.doi.org/10.1155/2014/639856] [PMID: 25530785]

[97] Szliszka, E.; Krol, W. Polyphenols Isolated from Propolis Augment TRAIL-Induced Apoptosis in Cancer Cells. *Evid. Based Complement. Alternat. Med.,* **2013**, *2013*, 731940.
[http://dx.doi.org/10.1155/2013/731940]

[98] Szliszka, E.; Sokol-Letowska, A.; Kucharska, A.Z.; Jaworska, D.; Czuba, Z.P.; Krol, W. Ethanolic Extract of Polish Propolis: Chemical Composition and TRAIL-R2 Death Receptor Targeting Apoptotic Activity against Prostate Cancer Cells. *Evid. Based Complement. Alternat. Med.,* **2013**, *2013*, 757628.

[99] Szliszka, E.; Zydowicz, G.; Janoszka, B.; Dobosz, C.; Kowalczyk-Ziomek, G.; Krol, W. Ethanolic extract of Brazilian green propolis sensitizes prostate cancer cells to TRAIL-induced apoptosis. *Int. J. Oncol.,* **2011**, *38*(4), 941-953. b
[PMID: 21286663]

[100] Kumazaki, M.; Shinohara, H.; Taniguchi, K.; Yamada, N.; Ohta, S.; Ichihara, K.; Akao, Y. Propolis cinnamic acid derivatives induce apoptosis through both extrinsic and intrinsic apoptosis signaling pathways and modulate of miRNA expression. *Phytomedicine,* **2014**, *21*(8-9), 1070-1077.
[http://dx.doi.org/10.1016/j.phymed.2014.04.006] [PMID: 24854570]

[101] Watabe, M.; Hishikawa, K.; Takayanagi, A.; Shimizu, N.; Nakaki, T. Caffeic acid phenethyl ester induces apoptosis by inhibition of NFkappaB and activation of Fas in human breast cancer MCF-7 cells. *J. Biol. Chem.,* **2004**, *279*(7), 6017-6026.
[http://dx.doi.org/10.1074/jbc.M306040200] [PMID: 14625298]

[102] Kamiya, T.; Nishihara, H.; Hara, H.; Adachi, T. Ethanol extract of Brazilian red propolis induces apoptosis in human breast cancer MCF-7 cells through endoplasmic reticulum stress. *J. Agric. Food Chem.,* **2012**, *60*(44), 11065-11070.
[http://dx.doi.org/10.1021/jf303004n] [PMID: 23066995]

[103] Hwang, H.J.; Park, H.J.; Chung, H.J.; Min, H.Y.; Park, E.J.; Hong, J.Y.; Lee, S.K. Inhibitory effects of caffeic acid phenethyl ester on cancer cell metastasis mediated by the down-regulation of matrix

metalloproteinase expression in human HT1080 fibrosarcoma cells. *J. Nutr. Biochem.,* **2006**, *17*(5), 356-362.
[http://dx.doi.org/10.1016/j.jnutbio.2005.08.009] [PMID: 16214327]

[104] Jin, U.H.; Chung, T.W.; Kang, S.K.; Suh, S.J.; Kim, J.K.; Chung, K.H.; Gu, Y.H.; Suzuki, I.; Kim, C.H. Caffeic acid phenyl ester in propolis is a strong inhibitor of matrix metalloproteinase-9 and invasion inhibitor: isolation and identification. *Clin. Chim. Acta,* **2005**, *362*(1-2), 57-64.
[http://dx.doi.org/10.1016/j.cccn.2005.05.009] [PMID: 16004979]

[105] Hattori, H.; Okuda, K.; Murase, T.; Shigetsura, Y.; Narise, K.; Semenza, G.L.; Nagasawa, H. Isolation, identification, and biological evaluation of HIF-1-modulating compounds from Brazilian green propolis. *Bioorg. Med. Chem.,* **2011**, *19*(18), 5392-5401.
[http://dx.doi.org/10.1016/j.bmc.2011.07.060] [PMID: 21865046]

[106] Meneghelli, C.; Joaquim, L.S.; Félix, G.L.; Somensi, A.; Tomazzoli, M.; da Silva, D.A.; Berti, F.V.; Veleirinho, M.B.; Recouvreux, Dde.O.; de Mattos Zeri, A.C.; Dias, P.F.; Maraschin, M. Southern Brazilian autumnal propolis shows anti-angiogenic activity: an *in vitro* and *in vivo* study. *Microvasc. Res.,* **2013**, *88*, 1-11.
[http://dx.doi.org/10.1016/j.mvr.2013.03.003] [PMID: 23538317]

[107] Park, S.I.; Ohta, T.; Kumazawa, S.; Jun, M.; Ahn, M.R. Korean propolis suppresses angiogenesis through inhibition of tube formation and endothelial cell proliferation. *Nat. Prod. Commun.,* **2014**, *9*(4), 555-560.
[PMID: 24868883]

[108] Daleprane, J.B.; Schmid, T.; Dehne, N.; Rudnicki, M.; Menrad, H.; Geis, T.; Ikegaki, M.; Ong, T.P.; Brüne, B.; Abdalla, D.S. Suppression of hypoxia-inducible factor-1α contributes to the antiangiogenic activity of red propolis polyphenols in human endothelial cells. *J. Nutr.,* **2012**, *142*(3), 441-447.
[http://dx.doi.org/10.3945/jn.111.150706] [PMID: 22279137]

[109] Chung, T.W.; Kim, S.J.; Choi, H.J.; Kwak, C.H.; Song, K.H.; Suh, S.J.; Kim, K.J.; Ha, K.T.; Park, Y.G.; Chang, Y.C.; Chang, H.W.; Lee, Y.C.; Kim, C.H. CAPE suppresses VEGFR-2 activation, and tumor neovascularization and growth. *J. Mol. Med.,* **2013**, *91*(2), 271-282.
[http://dx.doi.org/10.1007/s00109-012-0952-6] [PMID: 22935775]

[110] Campos, M.G.; Bogdanov, S.; Almeida-Muradian, L.B.; Szczesna, T.; Mancebo, Y.; Frigerio, C.; Ferreira, F. Pollen composition and standardisation of analytical methods. *J. Apic. Res.,* **2008**, *47*(2), 154-161.
[http://dx.doi.org/10.3896/IBRA.1.47.2.12]

[111] Krell, R. Value added products from beekeeping. *FAO Agricultural Services Bulletin,* **1996**, *124*, 87-113.

[112] Crane, E. *History of other products from bees The world history of beekeeping and honey hunting*; Gerald Duckworth & Co Ltd: London, **1999**, pp. 545-553.

[113] Bogdanov, S. Pollen: Production, Nutrition and Health: A Review. *Bee Product Science, Available online: http://www.bee-hexagon.net/files/file/fileE/Health/PollenBook2Review.pdf,* **2014**. (accessed on 20 Jannuary 2015).

[114] Jorde, W.; Linskens, H.F. Zur Persorption von Pollen und Sporen durch die intakte Darmschleimhaut. *Acta Allergol.,* **1974**, *29*(3), 165-175.

[http://dx.doi.org/10.1111/j.1398-9995.1974.tb01646.x] [PMID: 4408086]

[115] Franchi, G.G.; Franchi, G.; Corti, P.; Pompella, A. Microspectrophotometric evaluation of digestibility of pollen grains. *Plant Foods Hum. Nutr.,* **1997**, *50*(2), 115-126. [http://dx.doi.org/10.1007/BF02436031] [PMID: 9201746]

[116] Buck, A.C.; Cox, R.; Rees, R.W.; Ebeling, L.; John, A. Treatment of outflow tract obstruction due to benign prostatic hyperplasia with the pollen extract, cernilton. A double-blind, placebo-controlled study. *Br. J. Urol.,* **1990**, *66*(4), 398-404. [http://dx.doi.org/10.1111/j.1464-410X.1990.tb14962.x] [PMID: 1699628]

[117] Mupakami, M.; Tsukada, O.; Okihara, K.; Hasimoto, K.; Yamada, H.; Yamaguhi, H. Beneficial effect of honeybee-collected pollen lump extract on benign prostatic hyperplasia (BPH) - A double-blind, placebo-controlled clinical trial. *Food Sci. Technol. Res.,* **2008**, *14*(3), 306-310. [http://dx.doi.org/10.3136/fstr.14.306]

[118] Duclos, A.J.; Lee, C.T.; Shoskes, D.A. Current treatment options in the management of chronic prostatitis. *Ther. Clin. Risk Manag.,* **2007**, *3*(4), 507-512. [PMID: 18472971]

[119] Lima, B.; Tapia, A.; Luna, L.; Fabani, M.P.; Schmeda-Hirschmann, G.; Podio, N.S.; Wunderlin, D.A.; Feresin, G.E. Main flavonoids, DPPH activity, and metal content allow determination of the geographical origin of propolis from the Province of San Juan (Argentina). *J. Agric. Food Chem.,* **2009**, *57*(7), 2691-2698. [http://dx.doi.org/10.1021/jf803866t] [PMID: 19334753]

[120] Lepley, D.M.; Li, B.; Birt, D.F.; Pelling, J.C. The chemopreventive flavonoid apigenin induces G2/M arrest in keratinocytes. *Carcinogenesis,* **1996**, *17*(11), 2367-2375. [http://dx.doi.org/10.1093/carcin/17.11.2367] [PMID: 8968050]

[121] Klippel, K.F.; Hiltl, D.M.; Schipp, B. A multicentric, placebo-controlled, double-blind clinical trial of beta-sitosterol (phytosterol) for the treatment of benign prostatic hyperplasia. German BPH-Phyto Study group. *Br. J. Urol.,* **1997**, *80*(3), 427-432. [http://dx.doi.org/10.1046/j.1464-410X.1997.t01-1-00362.x] [PMID: 9313662]

[122] Cook, N.; Stampfer, M.; MA, J.; Manson, J.; Sacks, F; Buring, J.; Hennekens, C. Beta-carotene supplementation for patients with low baseline levels and decreased risks of total and prostate carcinoma. *Cancer,* **1999**, *86*, 1629-1631. [http://dx.doi.org/10.1002/(SICI)1097-0142(19991101)86:9<1783::AID-CNCR21>3.0.CO;2-N]

[123] Furusawa, E.; Chou, S.C.; Hirazumi, A. Antitumor potential of pollen extract on Lewis lung carcinoma implanted intraperitoneally in syngenic mice. *Phytother. Res.,* **1995**, *9*, 255-259. [http://dx.doi.org/10.1002/ptr.2650090405]

[124] Yang, X.; Guo, D.; Zhang, J.; Wu, M. Characterization and antitumor activity of pollen polysaccharide. *Int. Immunopharmacol.,* **2007**, *7*(4), 427-434. [http://dx.doi.org/10.1016/j.intimp.2006.10.003] [PMID: 17321465]

[125] Li, F.; Yuan, Q.; Rashid, F. Isolation, purification and immunobiological activity of a new water-soluble bee pollen polysaccharide from Crataegus pinnatifida *Bge. Carbohydr. Polym.,* **2009**, *78*, 80-88. [http://dx.doi.org/10.1016/j.carbpol.2009.04.005]

[126] Aliyazicioglu, Y.; Deger, O.; Ovali, E.; Barlak, Y.; Hosver, I.; Tekelioglu, Y.; Karahan, S.C. Effects of Turkish pollen and propolis extracts on respiratory burst for K-562 cell lines. *Int. Immunopharmacol.,* **2005**, *5*(11), 1652-1657.
[http://dx.doi.org/10.1016/j.intimp.2005.04.005] [PMID: 16039555]

[127] Wu, Y.D.; Lou, Y.J. A steroid fraction of chloroform extract from bee pollen of Brassica campestris induces apoptosis in human prostate cancer PC-3 cells. *Phytother. Res.,* **2007**, *21*(11), 1087-1091.
[http://dx.doi.org/10.1002/ptr.2235] [PMID: 17639562]

[128] Izuta, H.; Shimazawa, M.; Tsuruma, K.; Araki, Y.; Mishima, S.; Hara, H. Bee products prevent VEGF-induced angiogenesis in human umbilical vein endothelial cells. *BMC Complement. Altern. Med.,* **2009**, *9*, 45.
[http://dx.doi.org/10.1186/1472-6882-9-45] [PMID: 19917137]

[129] Wang, B.; Diao, Q.; Zhang, Z.; Liu, Y.; Gao, Q.; Zhou, Y.; Li, S. Antitumor activity of bee pollen polysaccharides from Rosa rugosa. *Mol. Med. Rep.,* **2013**, *7*(5), 1555-1558.
[PMID: 23525233]

[130] Gao, Y.; Hu, F.; Zhu, W.; Li, Y. Study on the antitumor activity of propolis and bee pollen and royal jelly. *J. Bee,* **2003**, *7*, 3-4.

[131] Pinto, B.; Caciagli, F.; Riccio, E.; Reali, D.; Sarić, A.; Balog, T.; Likić, S.; Scarpato, R. Antiestrogenic and antigenotoxic activity of bee pollen from Cystus incanus and Salix alba as evaluated by the yeast estrogen screen and the micronucleus assay in human lymphocytes. *Eur. J. Med. Chem.,* **2010**, *45*(9), 4122-4128.
[http://dx.doi.org/10.1016/j.ejmech.2010.06.001] [PMID: 20598400]

[132] Han, X.; Shen, T.; Lou, H. Dietary Polyphenols and Their Biological Significance. *Int. J. Mol. Sci.,* **2007**, *8*, 950-988.
[http://dx.doi.org/10.3390/i8090950]

[133] Khan, N.; Afaq, F.; Mukhtar, H. Apoptosis by dietary factors: the suicide solution for delaying cancer growth. *Carcinogenesis,* **2007**, *28*(2), 233-239.
[http://dx.doi.org/10.1093/carcin/bgl243] [PMID: 17151090]

[134] Trautwein, E.; Demonty, I. Phytosterols: natural compounds with established and emerging health benefits. *OCL,* **2007**, *14*, 259-266.
[http://dx.doi.org/10.1051/ocl.2007.0145]

CHAPTER 11

Antimicrobial Activity of Honeybee Plant-Derived Products

Miroslava Kačániová[1,*], **Cristina Almeida-Aguiar**[2,3,*]

[1] *Department of Microbiology, Faculty of Biotechnology and Food Sciences, Slovak University of Agriculture in Nitra, Tr. A. Hlinku 2, 949 76 Nitra, Slovakia*

[2] *CITAB, Centre for the Research and Technology of Agro-Environmental and Biological Sciences, AgroBioPlant Group, University of Minho, Braga, Portugal*

[3] *Biology Department, School of Sciences, University of Minho, Campus de Gualtar, 4710-057 Braga, Portugal*

Abstract: Bee products have always been used as foods and/or as therapeutic agents against a number of diseases in alternative medicine since almost immemorial times. The millennial track record of the health and nutritional benefits of such natural bee products is being supported more recently by scientific research. The antimicrobial activity of bee products is considered one of its widespread and most important bioactivities, and highlights the potential of these natural products as promising antimicrobial agents for clinical and biotechnological applications. Honey and propolis fulfill all the criteria of ideal candidates for treatment of non-healing wounds and other diseases caused by microorganisms. These natural products find application against resistant pathogenic microorganisms without the risk of antimicrobial resistance acquisition and prevent the formation or distortion of biofilms. Honey is regarded as a pure natural and functional product of high nutritional value. Honey factors that contribute to its antimicrobial activity are diverse but researchers consider both the high sugar concentration and low pH as well as the presence of hydrogen peroxide, methylglyoxal and bee defensin-1 (an antimicrobial peptide) as the most relevant in different honeys.

* **Equally contributing and correspondent authors:Miroslava Kačániová**, Department of Microbiology, Faculty of Biotechnology and Food Sciences, Slovak University of Agriculture in Nitra, Tr. A. Hlinku 2, 949 76 Nitra, Slovakia; Tel: +421376414494; Email: miroslava.kacaniova@gmail.com and **Cristina Almeida-Aguiar,** Biology Department, School of Sciences, University of Minho, Campus de Gualtar, 4710-057 Braga, Portugal; Tel: +351253601513; Email: cristina.aguiar@bio.uminho.pt

Susana M. Cardoso (Ed)

Propolis is a sticky resin produced by bees, especially from coniferous, having plant-derived and bee-released compounds, but with a chemical composition difficult to standardize due to its dependence on vegetation, season and environmental conditions at the collection site. The mechanism of propolis antimicrobial activity is complex, probably relying in a synergistic activity between different phenolic compounds such as flavonoids and phenolic acids along with other components. Pollen is a source of phytochemicals and nutrients, extremely rich in carotenoids, flavonoids and phytosterols. Although less studied for this bioactivity, bee pollen has also been proven to possess antimicrobial activity.

Keywords: Antibacterial activity, Antifungal activity, Antiviral activity, Bacteria, Bees, Biofilm, Concentration, Nectar, Pathogenic microorganisms, Peroxidase activity, Phenolic compounds, Pollen, Propolis, Resistance, Synergism, Yeasts.

1. INTRODUCTION

Antimicrobial agents are essential to ensure human/animal health and welfare, as well as food security. Overall, antimicrobials are vital in the reduction of the global burden of infectious diseases. However, the increasing number of immunocompromised individuals - patients suffering from AIDS (acquired immune deficiency syndrome), with organ transplantation and undergoing chemotherapy regimens for cancer treatment - increases the number of opportunistic infections and has been leading to the development and spread of resistant pathogens, diminishing the efficacy of antimicrobial drugs. The evolution of resistant strains is a natural phenomenon, occurring when microorganisms replicate erroneously or when strains exchange resistant traits, but the use and misuse of antimicrobial drugs speed up the appearance of such drug-resistant strains, further amplified by inadequate infection control practices, poor sanitary conditions and inappropriate food-handling [1].

Microbial resistance to antimicrobial agents is a global threat and is considered a serious public health problem by the World Health Organization (WHO), making the discovery of new antimicrobial agents an urgent need. In 2014, the WHO reported the results of a first global surveillance on antimicrobial resistance, with data from 114 countries. This WHO's report reveals that antibiotic resistance is not a vision but a frightening reality instead, which is happening here and now, threatening the capacity to treat common infections and minor injuries, considered treatable for decades [1]. Resistance to antimicrobial agents, in particular to antibiotics, including the major last-resort drugs, continues rising across the world. Even more worryingly is the fact that very few new antibiotics are under development and few new therapies are on the horizon. Alternative antimicrobial strategies and new antimicrobial agents are therefore urgently demanded and have been leading to reconsider the use of old traditional therapeutic resources [2]. In the actual trend towards the use of natural and renewable resources, natural compounds with antimicrobial properties and potential biomedical applications - particularly found in plants and plant products, including honey and other bee products - have been extensively researched recently [3 - 5] and are of highest interest. Such products and/or its compounds might be used either directly or as leads for the development of novel antimicrobials agents.

In this work we aim to report and discuss the potential of some honeybee plant-derived products for the development of new antimicrobial drugs, supported by the results of several studies regarding the assessment of antimicrobial properties in honeybee plant-derived products against diverse microorganisms. Likewise, the action mechanisms of the best known compounds responsible for the antimicrobial activity of honey, propolis and pollen will be described.

2. ANTIMICROBIAL PROPERTIES OF HONEY

Honey has long been used and prescribed as a remedy in traditional medicine of all civilizations, either for oral intake or topical application. Widely seen as a food source of high nutritional value, honey is being increasingly recognized by its healing properties. Since its first documented use, in Sumerian writings (2100 to 2000 BC), as medication and ointment, several scientific works have reported its use in the treatment of ulcers, wounds, bed sores and other skin infections resultant from burns and wounds [5 - 7]. This healing property of honey is mainly attributed to its antibacterial activity, further helped by the maintenance of a moist wound condition and the supply of a protective barrier by honey´s high viscosity, which helps preventing infection [2]. A number of medical grade honeys with standardized levels of antibacterial activity are currently marketed, and display an effective *in vitro* bactericidal activity against antibiotic-resistant bacteria, including species causing several life-threatening infections to humans [2].

The antimicrobial activity of bee honey is one of its most studied biological properties. The specificity of this activity, as well as of the others honey´s bioactivities such as antitumor, anti-inflammatory, antioxidant and antiviral properties, depends on honey´s components, which vary according to its floral, geographical and entomological origin [8]. Differences in the antimicrobial activity and other biological properties among flower honeys have been proved in different studies [9 - 2]. In a study with honey obtained from eight different floral sources, for instance, honey showed bacteriostatic effect for *Bacillus cereus*, *Micrococcus flavus* and *Sarcina lutea* [13]. Yet, one out of the eight types of honey did not inhibit *Bacillus subtilis* growth.

Antimicrobial activity of honey may vary widely: MIC (minimum inhibitory concentration) values were reported from < 3% to 50% and highest [14 - 16]. The growth of fungi and bacteria from sludge, soil, tap water or air can be prevented by using solutions containing 25% of honey, while solutions containing 20% of honey prevent the growth of air contaminants. Overall, many species of bacteria are sensitive to honey but studies generally focus on microorganisms that cause human infections (Table **1**). Diseases of microbial origin which are susceptible to treatment with honey include: urinary tract infections (*Proteus* spp., *Pseudomonas aeruginosa*), cholera (*Vibrio cholerae*), nosocomial infections (*Staphylococcus aureus*), septicemia (*Escherichia coli*), bloodstream infections (*Stenotrophomonas maltophilia*), infections in burn wounds (*Micrococcus luteus, Cellulosimicrobium cellulans, Listonella anguillarum*), gastritis and gastric neoplasias (*Helicobacter pylori*) and tuberculosis (*Mycobacterium tuberculosis*) [8, 17 - 22].

Table 1. Minimum inhibitory concentration of different types of honey against several microorganisms.

Type of honey	MIC range (%)	Microorganisms	Clinical importance	Reference
Apis mellifera Manuka (*Leptospermum scoparium*)	3.4- 20.0	*S. pyogenes*, MRSA, *S. aureus, S. maltophilia, A. baumannii, E. coli, P. aeruginosa, S. typhi*	Clinical practice	[2, 17]
Apis mellifera pasture	3.6± 0.7	*Coagulase-negative Staphylococci*	Clinical practice	[14]

(Table 1) contd.....

Type of honey	MIC range (%)	Microorganisms	Clinical importance	Reference
Apis mellifera Egyptian clover (*Trifolium alexandrinum*)	35.0 - 40.0	*A. schubertii, A. baumannii, H.paraphrohaemlyticus, M. luteus, C. cellulans, L. anguillarum*	Clinical practice	[18]
Apis mellifera (different flora)	2.5 - 50.0	*E. coli, E. faecalis, P. aeruginosa, S. aureus, C. albicans*	Urinary tract infection (UTI), diarrhoea, septicaemia, wound infections, community acquired and nosocomial infections	[23 - 27]
Apis dorsata Tualang (*Koompassia excelsa*)	8.75 - 25.0	*S. pyogenes, Staphylococci,* MRSA, *S. aureus, S. maltophilia, A. baumannii, E. coli, P. aeruginosa, S. typhi, E. cloacae*	Chronic rhinosinusitis, clinical practice	[19 - 20]
Apis dorsata Nilgiri	25.0, 35.0 and 40.0	*S. aureus, P. aeruginosa, E. coli*	Community acquired and nosocomial infection, UTI, diarrhoea, septicaemia, wound infection , diabetic foot ulcer	[22, 25 - 28]
Melipona beecheii flora unknown	4.0 - 5.0	*S. aureus, P. aeruginosa, E. coli*	Community acquired and nosocomial infections, UTI, septicaemia, wound infections, diarrhea, diabetic foot ulcer	[25 - 29]
Tetragonisca angustula flora unknown	2.5 - 10.0	*B. cereus, S. cerevisiae*	Seborrheic dermatitis, atopic dermatitis, psoriasis	[30]
Stingless bees flora unknown	4.0 - 16.0	*Candida spp.*	Seborrheic dermatitis, atopic dermatitis, psoriasis	[31, 32]
Stingless bees flora unknown	32.0	*C. albicans, C. glabrata*	Seborrheic dermatitis, atopic dermatitis, psoriasis	[31, 32]
Medihoney and different potencies of Surgihoney	32.0 - 256.0	*S.aureus,* MRSA, β-haemolytic streptococci, *Enterococcus spp., E. faecium, E. coli, K. pneumoniae, S. marcescens, P. aeruginosa, A. lwoffii, P. acnes, B. fragilis, C. albicans, C. glabrata, A. fumigatus*	Community acquired and nosocomial infections, UTI, diarrhea, septicaemia, wound infections, diabetic foot ulcer, psoriasis, seborrheic and atopic dermatitis	[26 - 28, 33]

(Table 1) contd.....

Type of honey	MIC range (%)	Microorganisms	Clinical importance	Reference
Garlic and tazma honey	6.25 - 66.6	*Salmonella sp., S. aureus, L. monocytogenes*	Enteric fever	[26, 34 - 36]
Stingless bee *Trigona* spp.	5.0 - 50.0	*B. subtilis, M. luteus, B. megaterium, B. brevis , E. coli, P. syringae*	Ringworm, cellulitis, impetigo, and necrotizing fasciitis	[37]
Manuka honey	57.9 - 78.8	*S. aureus*		[27, 38]
Northwestern Argentina honey	0.10 - 0.25	*S. aureus, E. coli, E. faecalis, K. pneumoniae*	Surgical wounds and conjunctiva	[10, 39]
Manuka honey	5.3 - 8.2	*S. mutans, S. sobrinus, L. rhamnosus, A. viscosus, P. gingivalis, F. nucleatum*	Surgical wounds and conjunctiva	[10, 40]
South African honey	0.15 - 5.0	*H. pylori*	Disease, gastric adenocarcinomas and mucosa-associated lymphoid tissue (MALT) lymphoma, peptic ulcer, chronic gastritis, gastric malignancies	[41, 42]
Greek honey	0.35 - 3.2	*S. aureus, S. epidermidis, E. coli, E. cloacae, K. pneumoniae, P. aeruginosa, C. albicans, C. tropicalis, C. glabrata, Streptococcus mutans, S. viridans*	Chronic wounds associated either with poly or monomicrobial biofilm, Chronic rhinosinusitis	[20, 43, 44]
Talah honey, Dhahian honey, Sumra-1 honey, Sidr honey, Sumra-2 honey	40.5 - 44.0	*E. coli, B. subtillis, P. aeruginosa, K. pneumonia, S. aureus, S. typhimurum, M. luteus, S. epidermidis, B. cereus, A. nidulans, S. marcescens, E. aerogenes*	Chronic rhinosinusitis, Haematology and cardiology wards	[10, 21, 44]
Algeria honey	56.0 - 96.0	*E. coli, P. aeruginosa*	Atopic dermatitis, wound infection, diabetic foot ulcer, UTI	[26, 27, 45]
Nortest Portugal honey	4.0 - 40.0	*B. subtilis, E. coli, K. pneumoniae, P. aeruginosa, S. aureus, S. lentus*	Wound infection , diabetic foot ulcer, UTI, Surgical wounds and conjuntive	[10, 26, 27, 46]

Acinetobacter baumannii - A. baumannii, Acinetobacter lwoffii- A. lwoffii, Actinomyces viscosus- A. viscosus, Aeromonas schubertii- A. schubertii, Aspergillus fumigatus- A. fumigatus, Aspergillus nidulans- A. nidulans,

Bacillus cereus- B. cereus, Bacillus subtilis- B. subtilis, Bacillus brevis- B. brevis, Bacillus megaterium- B. megaterium, Bacteroides fragilis- B. fragilis, Candida albicans- C. albicans, Candida glabrata - C. glabrata, Candida tropicalis- C. tropicalis, Cellulosimicrobium cellulans- C. cellulans, Enterobacter aerogenes- E. aerogenes, Enterococcus cloacae- E. cloacae, Enterococcus faecium- E. faecium, Escherichia coli- E. coli, Fusobacterium nucleatum- F. nucleatum, Haemophilius paraphrohaemlyticus- H. paraphrohaemlyticus, Helicobacter pylori- H. pylori, Klebsiella pneumonia- K. pneumonia, Lactobacillus rhamnosus- L. rhamnosus, Lysteria monocytogenes- L. monocytogenes, Listonella anguillarum- L. anguillarum, Micrococcus luteus- M. luteus, MRSA- methicillin resistant *Staphylococcus aureus, Porphyromonas gingivalis- P. gingivalis, Propionibacterium acnes- P. acnes, Pseudomonas aeruginosa- P. aeruginosa, Pseudomonas syringe- P. syringe,Serratia marcescens- S. marcescens, Staphylococcus aureus- S. aureus, Stenotrophomonas maltophilia- S. maltophilia, Streptococcus mutans- S. mutans, Streptococcus viridans- S. viridans, Streptococcus pneumonia- S. pneumonia, Streptococcus sobrinus- S. sobrinus, Vibrio cholera- V. cholera.*

Notably, the antibacterial properties of honey have been widely searched, while limited information on other antimicrobial properties (*e.g.* effects of honey on viruses, microscopic fungi and parasites) is available. The antifungal activity of honey has been evaluated against several representatives of *Candida* species, *Sacharomyces cerevisiae* and filamentous fungi of genera *Aspergillus* [23]. Kačániová *et al.* [47, 48] evaluated the antibacterial activity of flower honeys from different regions of Slovakia and one sample from the island of Rhodes against microscopic filamentous fungi of *Penicillium* genus: *P. verrucosum, P. raistrickii, P. griseofulvum, P. expansum* and *P. crustosum* obtaining maximum activity with 50% honey solution. Important reported data on the antifungal activity of honey is also summarized in Table **1**.

Up until some years ago, honey antimicrobial activity was thought to rely exclusively in the osmotic pressure generated by the high concentration of its sugars [49 - 52]. But more recently, particularly in honey produced by bees belonging to the genus *Apis*, other components have been reported. The honey unique physicochemical features that are conditioning factors for its bacteriostatic and bactericidal activities include its high sugar content associated with a low water content, the low pH, the presence of hydrogen peroxide (H_2O_2), methylglyoxal, and the content in certain phenolic compounds (flavonoids), enzymes (glucose oxidase, proteases and amylases), proteins, organic acids (*e.g.* formic, benzoic and phenolic acid) and antimicrobial peptides [2, 24, 53, 54] (Table **2**).

Table 2. Main factors affecting the antimicrobial properties of honey.

High osmotic pressure, low water activity (a_w)
Low pH
Hydrogen peroxide
Low protein content
High ratio of carbon to nitrogen
Low redox potential (affects the high content of reducing sugars)
Viscosity (limiting dissolved oxygen)
Chemicals (Pinocembrin, Lysosym, Acids, Terpenes, Benzyl alcohol, Volatile substances (phyto compounds affected by bee enzymes), Catalase, Bee defensin-1, Glucose oxidase, Methylglyoxal)

The high concentration of sugars together with the low humidity content reduce a_w (water activity) and cause osmotic stress, inhibiting the growth of almost any microorganism and preventing microbial spoilage of honey [55]. Mild dilution of honey may allow yeast growth, but the antibacterial activity is retained in honey diluted to approximately 30–40%. Comparative studies using natural and synthetic honeys (solutions of sugars in the ratios typically found in honey) showed however that the antibacterial activity is not just a result of water elimination driven by osmolarity, and that additional antimicrobial factors must be involved as MIC values of the studied natural and synthetic honeys were distinct [21]. An antibacterial role is also claimed for the low pH value of honey [56], which generally ranges between 3.2 and 4.5. The acidic pH is caused by gluconic acid - an organic acid derived from the glucose oxidase reaction - and is unfavorable to bacteria, which normally prefer neutral or slightly alkaline environments [57]. Yet these topics are still controversial, with some researchers arguing that osmolarity and acidity are not antimicrobial factors of honey, since *in vitro* studies are normally performed with diluted honey, which has neutral pH and low osmolarity [8]. In a wound or ulcer however, when honey is applied directly, bacteria can come into contact with honey and thus acidity and osmolarity are important factors to hinder microbial growth [54].

When the inhibitory properties of honey against bacteria were first reported, at the beginning of the last century, the responsible factor was named "inhibitor", but later White *et al.* [49] recognized that factor as being hydrogen peroxide (H_2O_2). Although other antimicrobial compounds are present in honey (Table **2**), some authors still consider H_2O_2 as the main one [2]. Hydrogen peroxide is mainly produced during glucose oxidation, a reaction catalyzed under aerobic conditions by the enzyme glucose oxidase, which is found in bee´s salivary glands and is introduced into honey during nectar collecting. H_2O_2 concentration thus depends on its production rate by glucose oxidase but also on the rate of its destruction by catalases [58]. The higher the glucose oxidase level, the lower the catalase level and the higher the peroxide level. Variances in antimicrobial activity among honeys from assorted floral sources may, in part, reflect such variations. H_2O_2 accumulation can be influenced by heat or light inactivation of glucose oxidase [59, 60] or through degradation by catalase [61]. Catalase is found in microorganisms, pollen and nectar and may occur in honey. Remarkably, H_2O_2 is also a powerful antimicrobial defense system in nectar [62] and different levels of H_2O_2 are found among nectar samples [63]. Interestingly too, H_2O_2 levels in tobacco and petunia nectars are inversely related to the rate of peroxidase activity [63] suggesting that this nectar enzyme rather than honey catalase could influence H_2O_2 accumulation in honey. Differences in concentration and/or activity of glucose oxidase could also explain H_2O_2 variations and concomitant differences in antimicrobial activity [21, 49]. Bankar *et al.* [64] showed that glucose oxidase has bactericidal effects against acetic acid bacteria and lactic acid bacteria, as well as on *Streptococcus mutans*. However, no studies regarding the evaluation of this enzyme concentration or of its activity in different honeys are available.

The apparent function of H_2O_2 is the prevention of spoilage of unripe honey when sugar concentration has not reached levels able to efficiently prevent microbial growth [55]. Glucose oxidase is inactivated during ripening of honey but regains activity on dilution of honey. The highest H_2O_2 accumulation occurs in the range of 30–50% honey but declines rapidly below 30% honey due to the relatively low enzyme affinity for its substrate glucose. Glucose oxidase is also unstable at high temperatures [65].

While microscopic filamentous fungi and yeasts are not as sensitive as bacteria to hydrogen peroxide [49], the importance of H_2O_2 as a predictor of the antibacterial potential of honey has been proposed by distinct authors. As an example, the medium/high antibacterial activity of Canadian honeys against *Escherichia coli* and *Bacillus subtilis* has been shown to correlate with H_2O_2 content [66], suggesting a possible role as biomarker for hydrogen peroxide. More recently however, Bizerra *et al.* [67] re-examined the role of H_2O_2 evaluating several honey samples, both natural and artificial, against the same two bacterial reference strains. Antibacterial susceptibility tests with exogenous H_2O_2 showed inhibition of *Escherichia coli* and *Bacillus subtilis* growth dependent on concentration. Endogenous H_2O_2 also inhibited the growth of *Escherichia coli* in a concentration-dependent way, yet with a MIC_{90} two-fold higher. Surprisingly, honey H_2O_2 did not inhibit *Bacillus subtilis* growth and even stimulated it at high honey dilutions and in the presence of high levels of H_2O_2, suggesting that physical features, such as high osmolarity, or other honey compounds contributed to antimicrobial properties. Bizerra and co-workers [67] also reported DNA degradation of *Escherichia coli* cells exposed either to endogenous or to exogenous H_2O_2 and suggested that the H_2O_2 oxidizing effect was amplified by other honey bioactive components such as Fe or Cu, considering the lower concentrations of honeys′ H_2O_2 able to efficiently damage chromosomal DNA.

The loss of antibacterial activity after neutralization of H_2O_2 identifies hydrogen peroxide as a key antimicrobial factor, but various honeys display substantial antibacterial activity even after H_2O_2 neutralization [11, 68]. When testing the effects of 345 unifloral unheated honeys from New Zealand against bacteria of the genus *Staphylococcus*, Allen *et al.* [68] found that honeys from the 26 different floral sources displayed variable antibacterial activity. Authors [68] also observed that the antibacterial activity of Manuka honey – a dense dark honey from the nectar of the manuka tree (*Leptospermum scoparium*), native to New Zealand, which is a medical grade honey commercially available as a topical treatment for wound infections - has non-peroxide components. When studying the same samples, it was proved that Manuka honey had the strongest antibacterial activity against *Escherichia coli, Proteus mirabilis, Pseudomonas aeruginosa, Helicobacter pylori, Salmonella typhimurium, Serratia marcescens, Staphylococcus aureus* and *Streptococcus pyogenes* [22, 36, 68, 69]. Moreover, Willix *et al.* [70] showed that both the hydrogen peroxide and the non-peroxide antibacterial activity of honeys effectively prevent the growth of bacteria which infect wounds, despite differences in potency against the seven bacterial strains tested.

Several works [38, 44, 55, 71] identified methylglyoxal (MGO) as the major antibacterial component in Manuka honey. MGO is a highly reactive dicarbonyl inducing non-enzymatic modification of the free amino groups of arginine and lysine residues in peptides and proteins, resulting in the formation of ending products of glycation [72]. It is postulated that MGO may have negative effects on the function and on the structure of other proteins-peptides of honey, including glucose oxidase. MGO is normally produced during heat treatment or prolonged storage of carbohydrate-containing beverages and foods, having been identified in wine, beer [73], bread [74], soybean [75] and honey [76]. In honey, MGO is formed in the nectar by nonenzymatic transformation of dihydroxyacetone, existing at very high concentrations [76]. The freshly produced Manuka honey contains low levels of MGO but the content increases during storage at 37 °C. It is argued [71, 77, 78] that a concentration of MGO over 150 mg/Kg is directly accountable for the antibacterial properties characteristic of Manuka honey. In a screening of 106 different honeys of different plant origin, MGO was detected but in concentrations not higher than 24 mg/Kg [56, 71]. Activity of Manuka honey against *Staphylococcus aureus* and *Bacillus subtilis* was eliminated or substantially reduced upon neutralization of MGO but was not affected in the case of *Escherichia coli* and *Pseudomonas aeruginosa* [79], showing that MGO is not the only factor responsible for Manuka non-peroxide antimicrobial activity.

Recently, bee defensin-1, a peptide consisting of 51 amino acids belonging to the extensive family of antimicrobial peptides termed defensins, has been identified as the antibacterial compound in Revamil® (medihoney, another medical-grade honey with antibacterial action for topical therapeutic use) [56]. Bee defensing-1 has a potent activity against Gram-positive bacteria such as *Bacillus subtilis* and *Staphylococcus,* as well as against *Paenibacillus larvae*, the causative agent of the bee larval disease American Foulbrood [80]. Although the mechanism of action has not been fully described, this peptide secreted in variable amounts by the bee´s hypopharyngeal gland is actually recognized as an antimicrobial agent of honey [79]. Besides the evident direct antimicrobial effect of bee defensin-1, there is a possibility that the bee peptides can also activate a variety of immune and inflammatory cells through the stimulation of a cytokine [81] but direct evidence for these immunomodulatory properties is still missing. Also, the bee defensin-1 content and its relationship to honey antimicrobial activity have not yet been systematically investigated, requiring further investigation.

After neutralization of MGO, H_2O_2 and bee defensin-1, the activity of Medihoney is reduced to that of an equivalent sugar solution [56]. However, Mundo *et al.* [11] observed that several honeys retain part of its antibacterial activity after such inactivation, indicating that additional factors are implicated in the antibacterial activity. Phytochemicals such as flavonoids, aromatic acids and phenolic antioxidants, well-known bioactive compounds of diverse plant products, have been proposed for the nonperoxide antibacterial activity of such honeys. Several phenolic compounds with antibacterial properties, like chlorogenic acid, gallic acid, benzoic acid, aesculetin, caffeic acid, syringic acid, rutin, scopoletin, *p*-coumaric acid, salicylic acid and vanillic acid, quercetin and naringenin have been identified in bee products (nectar, pollen, propolis and honey) [82]. On the other hand, phenolic extracts from different honeys display *in vitro* inhibition of bacterial growth: for example the phenolic extracts obtained from monofloral honeys of *Quillaja saponar* showed to be active against *Pseudomonas aeruginosa*, *Salmonella typhi*, *Staphylococcus aureus* and *Streptococcus* β *hemolyticus* [83] and the phenolic extracts of coconut and gelam honeys inhibited *Escherichia coli*, *Staphylococcus aureus*, methicillin resistant *Staphylococcus aureus* (MRSA) and methicillin sensitive *Staphylococcus aureus* (MSSA) [84]. Moreover, extracts containing luteolin, quercetin, apigenin, kaemferol, acacetin, tamarixetin, chrysin and galangin delay the formation of germination tubes and hyphae in *Candida albicans* [85]. Additionally, differences in the antimicrobial activity among flower honeys with different phenolic contents have been proved. Chestnut (*Castanea sativa* Mill.) honey contained more polyphenolic compounds and displayed greater effect than canola (*Brassica napus* subsp. *napus* L.) honey against *Alternaria infectoria, Scopulariopsis brevicaulis, Trichophyton ajelloi* and *Saccharomyces cerevisiae*.

Few studies have been performed on the mechanism of action of flavonoids but different compounds seem target diverse functions and components in bacterial cells, like the inhibition of the cytoplasmic membrane functioning; inhibition of DNA gyrase activity and its carrier protein; or neutralization of the distribution of porines in Gram-negative microorganisms, giving access to the interior of the cell [86]. However, the activity of individual phenolics isolated from honey is too low to significantly contribute to its activity [87] and a combination of different compounds can rather contribute to the antimicrobial activity of honey, *via* additive or synergistic effects, a question that remains to be investigated.

2.1. Inhibitory Effect of Honey Against Biofilm Formation

Biofilm is a microbial community living inside a self-produced extracellular polysaccharide (EPS) matrix [88]. The EPS provides the community of bacteria or fungi with protection to antimicrobial and phagocytic onslaught. Bacterial biofilm is formed when surface-adherent bacteria in aqueous environment begins to secrete mucilage, an adhesive substance that anchor them to all kinds of material, in particular medical plastics, implant materials, metals and animal or human tissues. Bacterial colonists that formed initially interact with each other by Van der Waals forces. When cell adhesion occurs, cells anchor permanently on surfaces. The biofilm then undergoes maturation by providing more diverse adhesion sites, starting to spread [89]. Biofilms are ubiquitous. They form in streams and oceans, also on teeth, inside bodies, on natural surfaces continually wetted by dripping water. They also form inside of water pipes, toilets, drains, and everywhere where there is persistent water.

Chronic and badly healing wounds are a major worldwide problem in health care. Tissue of all chronic wounds is colonized by polymicrobial flora. The bacterial population of the chronic wounds is often arranged in a highly organized biofilm that protects the bacteria against antimicrobial therapy and the patient's immune system [90 - 92]. Enterococci have been associated with biofilms on several medical devices such as artificial hip prosthesis, intrauterine devices, urinary catheters, central venous catheters and prostetic heart valves [93]. The ability of Enterococci to form biofilm has become one of its virulence factors in nosocomial infections [94]. Among this bacterial group, the ability to form biofilm was greater for *Enterococcus faecalis* than in the case of *Enterococcus faecium* [95]. Anyhow, the ability of forming biofilm would probably depend on the cultivation conditions and the origin of strain [94].

Antibiotic resistance is the biggest stumbling among all the properties of the biofilm. Even high doses of medicines are not enough to overcome infection in medical practice. Furthermore, laboratory testing of antibiotic susceptibility does not provide proper results because it is performed under optimal laboratory conditions in a culture medium, and the tested bacteria are planktonic and do not show resistance. The bacteria in the biofilm are the ones that are resistant to antibiotics [96]. To the present, only a few studies have examined the effect of honey on chronic wounds and biofilms [20 - 22, 44] (Table **3**). Still, Alandejani *et al.* [97] have shown that Manuka honey is effective in inhibiting biofilms formed by *Pseudomonas aeruginosa* and *Staphylococcus aureus*, which play a major part in the pathophysiology of chronic inflammations of sinus and upper respiratory tract.

Table 3. Antibacterial and antifungal activity of propolis.

Type of propolis	**MIC**	**Microorganisms**	**Clinical importance**	**Reference**
Egyptian propolis	13.3%	*S. aureus, S. intermedius,* S. *saprophyticus,* S.*epidermidis*	bovine mastitis	[131]
Pakistan propolis	2.0 - 5.0 mg/ml	*A. baumannii*	bacteraemia, pneumonia, meningitis, UTI, wound infections	[132]
Turkish propolis	1.0 - 5.25 mg/ml	*E. faecalis, C. albicans,S. enteritidis, L. monocytogenes*	infected root canals superficial mycoses, food-born diseases	[133 - 135]
Iranian propolis	5.8 - 12.25%	*C. albicans, C. tropicalis, C. kefyr, C. krusei, M. slooffiae, M. globosa, M. pachydermatis*	Onychomycosis	[136]
Brazilian propolis	1.7 - 48.4%	*C. albicans, C. glabrata, C. tropicalis, C. guilliermondii, C. parapsilosis, Trichosporon sp. , S. cerevisiae, T. asahii, T. ovoides, T. cutaneum, G. candidum*	vulvovaginal candidiasis onychomycosis	[137]
Polish propolis	0.39 - 3.13 mg/ml	MSSA, MRSA	hospital infection	[12]
French propolis	100 - 250 µg/ml	*C. albicans, C. glabrata, A. fumigatus, S. aureus, E. coli*	food poisoning, skin infections, abscesses, pneumonia, meningitis, sepsis	[138]

(Table 3) contd.....

Type of propolis	MIC	Microorganisms	Clinical importance	Reference
Portuguese propolis	0.59 - 13.9 mg/ml	*S. aureus, P. aeruginosa, E. coli, C. albicans, Salmonella sp.*	nosocomial infections, surgical site wound infections, bacteraemia, respiratory pneumonia, gastrointestinal and skin infections, vulvovaginal candidiasis, UTI	[119,139, 140]
Australian propolis	2 - 12%	*S. aureus, S. epidermis, E. cloacae, E. faecium, Corynebacterium sp., P. aeruginosa*	Ulcer	[141]
Argentina propolis	31.2 - 250 μg/ml	*C. albicans, C. tropicalis, S. cerevisiae, C. neoformans, A. flavus, A. fumigatus, A. niger, T. rubrum, T. mentagrophytes, M. gypseum*	cold, cough, muscle aches and superficial mycoses	[142]
Maltese, Italian, Algerian, Portuguese, Spanish, French, Tunisian, Maroccan	31.2 - 250 μg/ml	*S. aureus, S. epidermidis, E. coli, E. cloacae, K. pneumoniae, P. aeruginosa, C. albicans, C. tropicalis, S. mutans, S. viridans, C. glabrata*	pharyngeal swab, expectorate, bronchus-aspirate, urinoculture, peritoneal catheter infection, foot wound, sacral decubitus ulcer, leg ulcer, aerobic hemoculture, auricular pus	[143]
Austrian, German, French propolis	1.275 μg/ml	*S. aureus, E. coli, C. albicans*	nosocomial infections, UTI, respiratory pneumonia, surgical site wound infections, bacteraemia	[144]
Bulgarian propolis	3.25 - 250 μg/ml	*H. pylori, E. coli, E. faecalis, Streptococcus β-haemolyticus, C. albicans*	chronic gastritis, peptic ulcer, gastric malignancies, UTI	[145, 146]
Czech propolis	16 - 64 μg/ml	*C. albicans, C. krusei, C. tropicalis, C. glabrata, T. asahii, A. fumigatus, Absidia corymbifera, T. mentagrophytes*	skin defects, infections, dental carries, support epithelisation in ulcers cures or burns	[147]
Slovakian propolis	128 - 256 μg/ml	*L. monocytogenes, P. aeruginosa, S. aureus, S. enterica, E. coli, A. fumigatus, A. flavus, A. niger, C. krusei, C. albicans, C. glabrata, C. parapsilosis, C. tropicalis, G. candidum, R. mucilaginosa*	skin defects, infections, dental carries, support epithelisation in ulcers cures or burns	[148, 149]

Absidia corymbifera- A. corymbifera, Acinotobacter baumannii - A. baumannii, Aspergillus flavus- A. flavus, Aspergillus fumigatus- A. fumigatus,, Aspergillus niger- A. niger, Candida albicans- C. albicans, Candida tropicalis- C. tropicalis, Candida kefyr-, C. kefyr, Candida krusei- C. krusei, Candida guilliermondii- C.

guilliermondii, Candida parapsilosis- C. parapsilosis, Cryptococcus neoformans- C. neoformans, Escherichia coli- E. coli, Enterococcus cloacae- E. cloacae, Enterococcus faecium- E. faecium,, Enterococcus faecalis- E. faecalis, Geotrichum candidum- G. candidum Lysteria monocytogenes- L. monocytogenes, Malassezia globose- M. globosa, Malassezia slooffiae- M. slooffiae, Malassezia pachydermatis- M. pachydermatis, Pseudomonas aeruginosa- P. aeruginosa, Rhodotorula mucilaginosa- R. mucilaginosa, Saccharomyces cerevisiae- S. cerevisiae , Salmonella enterica- S. enterica, Salmonella enteritidis- S. enteritidis, Staphylococcus aureus- S. aureus, Staphylococcus intermedius- S. intermedius, Staphylococcus saprophyticus- S. saprophyticus, Staphylococcus epidermidis- S. epidermidis, S. mutans- Streptococcus mutans, Streptococcus viridans- S. viridans, Trichophyton rubrum- T. rubrum, Trichophyton mentagrophytes- T. mentagrophytes, M Trichosporon asahii- T. asahii,, Trichosporon ovoides- T. ovoides, Trichosporon cutaneum- T. cutaneum

The presence of carbohydrates reduces the formation of bacterial biofilms [95]. Okhiria *et al.* [98] demonstrated that the biofilm of *Pseudomonas aeruginosa* was significantly inhibited by a 20% solution of Manuka honey, while the culture of plankton bacterial cells were sensitive to Manuka honey at concentrations less than 10%. Maximum reduction of the increase in biofilm was observed after 11 hours of incubation in 40% honey solution. In a recent study, 10% Manuka honey was able to influence the formation of biofilms in the oral pathogenic bacteria *Streptococcus* sp. and also repressed the adherence of Streptococci to the glass surface at sub-MIC concentrations [40]. These findings indicate that Manuka honey is able to diminish oral pathogens of dental plaque. Majtán *et al.* [44] demonstrated inhibition of biofilm formation in three microorganisms belonging to the group of very good biofilm producers. They found that 10% of honeydew honey can inhibit adherence of the biofilm of *Proteus mirabilis* to the surface of plastic culture vessel. Even more significant inhibition was demonstrated for *Escherichia coli*, when the active sub-inhibitory concentration of honeydew honey was 5-10%. Honeydew honey in the sub-MIC concentrations of 5 and 10% significantly allowed the formation of a bacterial biofilm, but at higher sub-MIC concentration (15%) inhibited its formation in *Enterobacter cloacae*.

Treatment of chronic wounds associated either with monomicrobial or polymicrobial biofilm is difficult. At present, antimicrobials in use are excellent for killing plankton bacteria, but they are not able to remove the infection in biofilm. In some cases, antibiotics may also further increase the biofilm formation in the sub-MIC concentrations. Therefore, the local and systemic antibiotic treatment of chronic wounds is not recommended [44]. Overall, honey is an attractive antimicrobial and antibiofilm agent, which could have an important role in the treatment of infectious wounds and biofilm. However, there is a need to gain more knowledge about the substances contained in honey with activity against the biofilm formation [44] as well as its effective concentrations.

3. ANTIMICROBIAL PROPERTIES OF PROPOLIS

Propolis, a Greek word derived from *pro* ("at the entrance to") and *polis* ("community"), reflects the importance of this waxy product to bees, which use it in their hives as protection against predators and microorganisms, as a thermal isolator, for sealing cracks and openings and smooth out the internal walls and to build aseptic locals for larvae [99 - 101]. Also known as bee glue, this hive sealant made by bees was well known by the Egyptians, who benefited from its anti-putrefactive properties in the mummification of corpses. It was also used by Greek and Roman physicians as a cicatrizing and antiseptic agent in wound treatment and as mouth disinfectant, and by Incas as an anti-pyretic agent [102]. Widely used between the 17th and 20th centuries, propolis come to be very popular in Europe and was listed as an official drug by the London pharmacopoeias [101]. Current applications of propolis include a diversity of bio and dermocosmetic products as well as functional foods and drinks [102, 103].

In the last decades, the chemical profiles and the pharmacological properties of propolis have been targeted by some reviews [99 - 104]. Raw propolis is composed of resins, essential oils, pollen, waxes, and other various organic and inorganic compounds [105, 106] but, as widely referred, its chemical composition is extremely complex and highly variable in samples from different zones of the globe, depending not only on the local flora but also on the collecting season and on the type of bees foraging [3, 99, 105, 106]. Propolis chemical heterogeneity is commonly linked to different bioactivities. On the other hand, and to make the scenario even more complex, there seems to be some consistency too in what concerns the independence of certain biological activities from propolis origin. Indeed, specific propolis biological activities like a similar spectra of antimicrobial activity can be associated with completely different phytochemical profiles in samples from different origins [107]. For instance, antimicrobial activity of European propolis is related with high concentrations of polyphenols -

flavonoids (chrysin, galangin, pinobanksin, pinocembrin), phenolic acids (caffeic acid, cinnamic acid, ferulic acid) and their esters [99, 108, 109] - while in Brazilian propolis that activity has mainly been correlated with prenylated phenylpropanoids, prenylated *p*-coumaric acids, caffeoyl quinic acids, acetophenones, diterpenic acids [99, 105], the flavonoids kaempferide, isosakuranetin and some amounts of kaempferol [110]. The most important antimicrobial components of propolis seem to be aromatic acids and phenolic compounds (in particular flavonoids and phenolic acids) [111], but, as a natural mixture of organic compounds, it may well be that the ratio of combined components is also important for its effects. It seems propolis has general pharmacological value rather like a natural mixture than a source of new powerful antimicrobial, antifungal and antiviral compounds [112]. Also, it is becoming more and more evident that it is not possible to ascribe a certain property solely to one individual component, so the active factor(s) responsible for many propolis biological properties still remains to be fully defined in many samples.

Recognized since antiquity, the antimicrobial activity is one of the best known propolis bioactivities, therein residing some of its applications and the development of the products that are actually marketed for human consumption [102, 113]. Antimicrobial properties are normally evaluated by disc diffusion assays or serial dilution methods in liquid or in solid media. Different methodologies have been avoiding comparison between results from several studies, as different parameters are determined. Minimum inhibitory concentration (MIC) determination seems to be the most widely accepted way of expressing propolis antimicrobial potential and even in this case values are often expressed in different units. As raw propolis cannot be used /tested directly, the usual procedure consists in extracting fractions rich in bioactive compounds (propolis balsam), discarding wax and debris. Common preparations are thus ethanol extracts (EE) as well as water extracts (WE) but also methanol (ME), *n*-hexan (HE) or dimethylsulfoxide (DE) extracts of propolis. The use of different solvents can change activity of propolis as different bioactive compounds can be differentially extracted.

Propolis can affect a wide spectrum of bacteria, fungi [114] as well as viruses [115] and parasites [116]. Propolis may act directly on microorganisms *in vitro*, and may stimulate the immune system *in vivo*, activating the mechanisms involved in the microorganisms killing [3]. Table **3** summarizes relevant reported data for the antimicrobial properties of propolis and its compounds.

3.1. Antibacterial Activity

Studies concerning the evaluation of propolis antimicrobial activity were performed *in vitro* by several researchers [117 - 119] using as reference strains many different bacterial species (Table **3**): Gram-positive and Gram-negative, aerobic and anaerobic, either from laboratory collections or isolated from samples. Data from such studies support the fact propolis is generally more active against Gram-positive bacteria than against Gram-negative bacteria. Kujumgiev *et al.* [112] assayed several propolis samples from different geographic regions (temperate and tropical zones) using *Staphylococcus aureus* and *Escherichia coli* as indicator strains. All the extracts were active against *Staphylococcus aureus* but none was active against *Escherichia coli*. Seidel *et al.* [120] described different susceptibility patterns of Gram-positive bacteria to propolis EE collected in different places, being the antibacterial effect greater for samples from a wet-tropical rain forest-type climate. In another study, 60–80 μg/cm^3 of propolis were necessary to suppress the growth of *Bacillus subtilis* and *Staphylococcus aureus,* while the concentration at least of 600–800 μg/cm^3 of propolis was needed to prevent the growth of *Escherichia coli*. Erkmen and Ozcan [121] reported MIC values of 0.02% propolis concentration against *Bacillus cereus* and *Bacillus subtilis*, 1.0% against *Staphylococcus aureus* and *Enterococcus faecalis*, but over than 0.2% against *Listeria monocytogenes*. Propolis extracts from different origins showed also to be active against *Bacillus cereus*, *Enterococcus faecalis*, *Enterobacter aerogenes, Listeria monocytogenes*, *Pseudomonas aeruginosa*, *Salmonella typhi*, *Staphylococcus aureus, Streptococcus pyogenes, Klebsiella pneumonia*, *Streptococcus mutans, Streptococcus sorbinus* [119, 122 - 124]. Wojtyczka *et al.* [125] tested a Polish propolis EE against MSSA and MRSA clinical isolates and reported different effectiveness levels against twelve *Staphylococcus aureus* strains, with MIC values extending from 0.39 to 0.78 mg/ml. *Paenibacillus larvae*, an important larval pathogen of the bee *Apis mellifera*, has become increasingly resistant to conventional antibiotics, and propolis extracts from several states of Brazil significantly inhibit this bacteria

[126].

Taking in consideration propolis antimicrobial effects on *Streptococcus mutans* group, Liberio *et al.* [127] suggested the use of propolis or of its compounds as potential cariostatic agents in the control of caries. Orsi *et al.* [128] pre-stimulated macrophages with Bulgarian or Brazilian propolis followed by a challenge with *Salmonella typhimurium* and observed that both samples improved the bactericidal activity of macrophages. A WE of Turkish propolis showed an inhibitory effect on the development of experimental tuberculosis infection caused by *Mycobacterium tuberculosis* in guinea-pigs [129]. Bulgarian propolis showed promising results against the enteric pathogen *Helicobacter pylori* and prevented growth of *Campylobacter jejuni* and *Campylobacter coli* [130].

Propolis and antimicrobial drugs may show synergistic effects, which potentiates drugs' effects and decreases the therapeutic doses. An increase of the effect of ampicillin, gentamycin and streptomycin against some clinically isolated Gram-positive bacteria was described by Scazzocchio *et al.* [117]. Instead, the action of chloramphenicol, ceftriaxone and vancomycin were moderately increased by the same propolis EE and no effect was observed in the case of erythromycin. Orsi *et al.* [150] showed a synergetic effect between EE of Brazilian and Bulgarian propolis and amoxicillin, ampicillin and cefalexin against *Salmonella typhi*, reporting that propolis diminished bacteria cell wall resistance to such antibiotics. Also, the action of different antistaphylococcal drugs against several MSSA and MRSA clinical isolates was potentiated by addition of a Polish propolis EE [125]. A synergistic activity was reported between ciprofloxacin and propolis in the treatment of experimental *Staphylococcus aureus* keratitis and propolis in combination with chlorhexidine inhibited the accumulation and adherence of cariogenic *Streptococcus mutans* on the tooth surface, suggesting its possible use in the treatment of dental caries [151].

About 300 chemical compounds have been identified in propolis but just a few exhibit antimicrobial activities. Bankova *et al.* [152] found that fractions of polar phenolic compounds from Brazilian propolis are active against *Staphylococcus aureus* and a link between flavonoids and the bacteriostatic activity was established in twelve propolis samples. Five terpenes isolated from a Cretan propolis showed antimicrobial activity against some Gram-positive and Gram-negative bacteria [153]. Hashimoto *et al.* [154] reported that a raw propolis EE and its active components *p*-coumaric acid, 3-prenyl-4-hydroxycinnamic acid and artepillin C are effective against *Helicobacter pylori*. Furthermore, caffeic acid phenethyl ester (CAPE), other propolis component, was shown to be a competitive inhibitor of *Helicobacter pylori* peptide deformylase, an enzyme essential for bacterial survival that catalyses the removal of formyl groups from the N-terminus of nascent polypeptide chains, and is considered a hopeful drug target for anti-*Helicobacter pylori* therapy [155]. Propolis volatiles have also been proved to be active against Gram-positive, as well as for some Gram-negative bacteria, non-pathogenic fungi and plant and human fungal pathogens [43, 156 - 158]. Propolis essential oils have considerable activity against both Gram-positive and Gram-negative bacteria too [158, 159].

The complex mechanism of propolis antimicrobial activity can be ascribed to a synergistic activity between phenolic and other compounds and follows multiple mechanisms of action: inhibition of cell division; disruption of the cytoplasmic membrane and of the cell wall [118]; inhibition of protein synthesis and RNA polymerase [160] and enzyme inhibition [86]. Propolis as well as its components cinnamic acid and flavonoids seem to affect the ion permeability of the inner bacterial membrane causing membrane potential dissipation and inhibiting bacterial motility [118]. Quercetin activity is partially detectable through DNA gyrase inhibition, while galangin causes a massive loss of potassium in the cells of *Staphylococcus aureus*, probably due to a direct effect on the cytoplasmic membrane or as an indirect consequence of bacterial cell wall weakening and resulting osmotic lysis [161]. At the membrane level, decoupling transduction of energy and consequent inhibition of bacterial motility were also observed [118]. It has been shown that propolis may limit the action of pathogenic *Staphylococcus aureus* by interfering with some virulence factors such as coagulase and lipase activities or the production of biofilm [117].

3.2. Antifungal Activity

Several *in vitro* studies have shown the antifungal effects of propolis from different geographic origin (Table **3**) [112, 137, 162 - 164]. Propolis seems to be the bee product with the highest antifungal activity, showing fungistatic activity at low concentrations and fungicidal effects at higher concentrations [165]. Different propolis samples show significant activity against *Candida albicans* although this yeast is one of the most resistant yeast species to propolis [111, 144, 166]. A Mexican propolis EE at a concentration of 0.8 mg/cm^3 inhibited 94.4% of the *Candida albicans* clinical isolates tested while the reference strain was inhibited at 0.6 mg/cm^3. In addition, several species of *Candida* genus (*C. albicans*, *C. tropicalis*, *C. krusei* and *C. guilliermondii*) showed to be sensitive to an EE of Brazilian propolis [163]. A Czech propolis dimethylsulfoxide extract was also assessed against *Candida albicans, Aspergillus fumigates, Microsporum gypseum, Microsporum canis* [124] and showed different effects: while the inhibitory action against *Candida albicans* and *Aspergillus fumigates* was strictly dose-dependent, the response of *Microsporum gypseum* displayed plateau across the concentrations range tested and for *Microsporum canis* decreased at the highest concentrations tested. Furthermore, propolis has been shown to exibit fungicide effects on juice spoilage fungi *Candida famata, Candida glabrata, Candida kefyr, Candida pelliculosa, Candida parapsilosis* and *Pichia ohmeri* [167] and the fungicidal effect was connected with the presence of flavonoids [168]. The susceptibility of yeast species isolated from patients with onychomycosis (mainly *Candida albicans, Candida tropicalis, Candida kefyr, Candida krusei, Malassezia globosa, Malassezia slooffiae,* and *Malassezia pachydermatis*) to propolis was studied *in vitro* [136] and the authors have concluded that the average MIC of propolis for Fluconazole-susceptible isolates was 5.8 μg/ml, whereas this value was 12.25 μg/ml for the Fluconazole-resistant ones.

Moreover, a fungicide action of several propolis extracts (EE, WE), propolis microparticles (PMs) and a propolis soluble dry extract against all *Candida albicans* morphotypes (yeast, pseudohyphae, and hyphae) was also reported [169]. A study on the effects of propolis on virulence factors of *Candida albicans* has shown a reduction in hyphal length, dose and time dependent inhibition of phospholipase activity and alteration of the structure the plasma membrane [170]. A more recent study with Brazilian propolis showed an induction of *Candida albicans* cell death mediated *via* metacaspase and Ras pathways [140]. An *in vitro* antifungal activity of EE and PMs of Brazilian propolis against yeast isolates from vulvovaginal candidiasis was described [137]. On the other hand, Castro *et al.* [171] demonstrated that propolis based gels and cream were partially able to control vulvovaginal candidiasis in a mouse model. Taken together, these results strongly indicate a therapeutic potential for propolis in the control of vulvovaginal candidiasis.

Murad *et al.* [162] registered an increase in the fungicidal activity of macrophages from BALB/C mice stimulated with Bulgarian and Brazilian propolis prior to be challenged with *Paracoccidioides brasiliensis,* the fungal agent of paracoccidioidomycosis, which is the most important human systemic mycosis in Latin America acquired through inhalation of mycelial fragments or airborne conidia. The *in vitro* antifungal activity of EE of Brazilian green and red propolis was tested against *Trichophyton* sp., being red propolis EE more efficient against this dermatophytosis causing fungus [164].

3.3. Antiviral Activity

Kujumgiev *et al.* [112] assessed the antiviral effects of propolis from different countries and most of the samples showed activity against the avian influenza virus. But propolis antiviral effects are extended to other viruses like herpes simplex types 1 (HSV-1) and 2 (HSV-2), adenovirus type 2, influenza virus or human immunodeficiency virus (HIV), vesicular stomatitis virus (VSV) and poliovirus type 2, among others [102, 115, 172 - 175]. The antiviral action of propolis WE and EE as well as of its constituents benzoic acid, caffeic acid, galangin, pinocembrin, *p*-coumaric acid and chrysin was tested against HSV-1 [173]: both extracts displayed high activity when cells were treated prior to viral infection, and galangin and chrysin showed to be the main bioactive compounds. Brazilian propolis proved also to be active against cutaneous HSV-1 infection in mice, reducing virus titers in skin and brain. Some propolis extracts displayed high antiviral activity when herpes viruses were treated prior to infection but no effects were detected when propolis extracts were added to uninfected cells prior to infection, suggesting an interference of propolis with viral compounds needed for adsorption or entry into host cells [176].

Brazilian propolis possesses anti-influenza virus activity, decreasing virus yields in the bronchoalveolar lavage fluids of lungs in infected mice [174]. When orally administered to BALB/C mice infected with influenza A virus, a Brazilian green propolis WE significantly extended treated mice lifetime [177]. Furthermore, therapeutic benefits of propolis were reported in the treatment of HIV, particularly for the aromatic compound moronic acid isolated from Brazilian propolis [178], while Gekker *et al.* [115] described an inhibition of HIV viral expression by propolis in a concentration-dependent manner and suggested an inhibition of viral entry as possible mechanism.

An anti HSV-1 activity of the major flavonoids of propolis, more specifically flavonols and flavones was observed [172]. A role of natural and synthetic flavonoids in picornavirus replication, through prevention of the decapsidation of viral particles and RNA release within cells or inhibition of viral RNA synthesis, was also described [179]. The mechanism of antiviral action of propolis is still far to be unveiled, although it is believed that the interaction of propolis with the cell membrane will block the penetration of the viral particles in the cell and/or induce changes in at intracellular level which, in turn, alter the cycle of viral replication [180]. Propolis flavonoids are thought to interfere with the intracellular redox processes induced by viral multiplication [172], scavenging free radicals that could mitigate the oxidative stress and, consequently, exerting antiviral action [181].

3.4. Inhibitory Effect of Propolis Against Biofilm Formation

Tooth decay is the most common oral disease associated with the formation of biofilm [182]. Tooth decay can progress to a series of complex diseases including oral and endodontic periapical periodontitis, with negative impacts in quality of life [183]. The application of natural substances with antimicrobial activity at tooth surfaces could lead to a reduction of biofilm formation [184]. Several studies have shown the *in vitro* inhibition of bacteria present in the dental plaque such as *Streptococcus sobrinus*, *Streptococcus mutans* and *Streptococcus cricetus* by propolis extracts [122, 123]. *In vitro* and *in vivo* experiments with Turkish propolis also showed positive effects against oral pathogens and bacteria in dental plaque [111, 121, 185]. Also, a flavonoids-free Brazilian propolis, with fatty acids (linoleic, oleic, palmitic and stearic) as the main compounds, revealed biological effects against *Streptococci* and inhibited the activity of glucosyltransferases. No significant effects on the viability of biofilms were observed but propolis samples expressively reduced acid production by the biofilms, inhibited the activity of F-ATPase by 60-65% and reduced considerably the incidence of smooth surface caries *in vivo*. The data suggest that cariostatic properties are related to propolis effects on acid production and acid tolerance of cariogenic Streptococci.

Chronic wounds provide ideal conditions for biofilm formation exposing proteins (collagen and fibronectin) and facilitating the bacterial attachment to damaged tissues. The ability of propolis and its individual components to inhibit biofilm formation or distortion of the already formed biofilm was only investigated in the case of oral or dental biofilms [186, 187] remaining the study of its role in chronic wounds to be performed [45]. Nevertheless, Scazzocchio *et al.* [117] confirmed the efficiency of a propolis ethanol extract in inhibiting the formation of biofilm and adherence of *Staphylococcus aureus* ATCC 6538P at the sub-MIC concentrations up to 40%.

4. ANTIMICROBIAL PROPERTIES OF POLLEN

Flower pollen represents the main food of bees with concentrations of phytochemicals and nutrients, rich in carotenoids, flavonoids and phytosterols [188]. Bee pollen is an agglomerate of pollen grains collected by bees from various botanical sources and mixed with nectar and secretions from the insect´s hypopharyngeal glands. This bee product has a complex chemical composition: it is constituted by aminoacids, carbohydrates, lipids, proteins, vitamins, carotenoids and polyphenolics such as flavonoids and minerals. Used since ancient times for its medical properties in conditions such as ageing, anemia, colds, colitis, flu and ulcers [189], pollen is greatly appreciated by the actual natural medicine because of its medical and nutritional potential, particularly due to the presence of phenolic compounds with antioxidant activity [190] and in general, this is increasingly being used as health food supplement, mainly to ameliorate the effects of ageing. Pollen phenolic composition mainly consists of flavonol glycosides and hydroxycinnamic acids [191]. This composition tends to be species-specific [191, 192] and has been associated with pollen therapeutic properties (antibiotic, antidiarrhoeic, antioxidant and antineoplasic) [190, 192,

193].

According to the literature, pollen extracts have been tested much more seldom for their antimicrobial activity than propolis or honey, which have been both extensively investigated for this property particularly against human pathogenic bacteria and fungi. One of the reported studies was performed by Basim *et al.* [188] who demonstrated the inhibitory effects of pollen against thirteen plant pathogenic bacteria. Of these, the most-susceptible was *Agrobacterium tumefaciens* while *Pseudomonas syringae* was the most resistant. This antibacterial activity was correlated with the presence of flavonoids, galangin, pinobaksin and pinocembrin, which were also demonstrated by the authors to possess antifungal activity. Additional active ingredients included esters of coumaric and caffeic acid, but also diterpenic acids, lignans of furfural or prenylated cumeric acid, which exhibited a cytotoxic effect, in addition to their inhibitory and antibacterial effects [188, 194, 195]. Researchers attributed these effects to the high content of quercetin and kaempferol glycosides, which are mainly known for its antimicrobial activity against the human pathogenic bacteria and fungi.

Similar results concerning antimicrobial effects of bee pollen were obtained by other researchers [149, 194 - 198] although selectivity and MIC values varied for individual microorganisms [197]. Different antimicrobial activity is displayed not only by pollen of different botanical origins but also by pollen from different geographical areas. Such variations are thought to mainly result from the amount of polyphenols, the nature of the compounds and possibly by the presence of non-volatile components [188, 199, 200]. Other authors claim that storage of pollen causes the loss of its beneficial properties and changes its chemical composition [192, 199]. Table **4** summarizes the most relevant reported data for the antimicrobial properties of pollen and of its compounds.

Different effects on microorganisms can not only be attributed to the chemical composition of pollen. The type of the tested organism is equally an important factor [197]. Graikou *et al.* [194] and Morais *et al.* [197] investigated the antibacterial properties of pollen on Gram-positive and Gram-negative bacteria. In both cases, microbial growth inhibition was observed. However, higher susceptibility was showed by Gram-positive bacteria. Barbosa *et al.* [201] observed that only the Gram-positive bacteria tested responded to pollen extracts. The resistance of Gram-negative bacteria is justified especially by flexible cell walls and their complex chemical structure. Additionally, these organisms have higher lipid content than Gram-positive bacteria. These features may be the key to their higher resistance [194, 197, 201].

Table 4. Antimicrobial activity of bee pollen.

Pollen	Microorganisms	Reference
Eucalyptus globulus *Ranunculus sardous* *Ulex europeans* bee pollen	*P. aeruginosa*	[207]
unknown nature	*S. viridans*	[208]
Turkish bee pollen	spoilage and pathogenic microorganisms, *S. aureus, Trichosporon sp., Candida spp. F. oxysporium, A. alternata, A. parasiticus,*	[121, 188, 202, 209]
Bred pollen	*S. epidermidis, S. aureus, B. subtilis,P. aeruginosa, Klebsiella pneumoniae*	[209]
Brazilian pollen	*L. monocytogenes, P. aeruginosa, S. aureus, S. enterica, E. coli,*	[199]
Slovak pollen	*A. fumigatus, A. flavus, A. niger C. albicans: C. glabrata, C. krusei, C. parapsilosis C. tropicalis, G. candidum R. mucilaginosa, L. monocytogenes, P. aeruginosa, S. aureus, S. entericaE. coli, C. butyricum, C. hystoliticum, C. intestinale, C. perfringens, C. ramosum.*	[149, 196, 204, 205]
Portuguese pollen	*S, aureus, B, cereus, E, coli, Z. mellis, Z. bailii, Z. rouxii, C. magnolia*	[197]
Maroccan pollen	*E. coli, S. enteritidis, P. aeruginosa, S. aureus, B. cereus*	[199]
Chilean pollen	*E. coli, P. aeruginosa, S. aureus,S. pyogenes, A. alternata, B. cinerea, F. oxysporum*	[210]
Greek pollen	*S. aureus, S. epidermidis, P. aeruginosa, E. cloacae, K. pneumoniae, E. coli, C. albicans, C. tropicalis, C. glabrata*	[194]
Egyptian pollen	*E. coli,L. monocytogenes, Salmonella enteritidis, Pseudomonas aeruginosa*	[211]
Saudi Arabian pollen	*S. aureus, B. cereus, B. subtillis, P. aeruginosa, K. pneumonia*	[212]
Indian pollen	*E. coli S. aureus P. aeruginosa B. subtilis, P. mirabilis S. typhi S. flexneri K. pneumonia*	[213]

(Table 4) contd.....

Pollen	Microorganisms	Reference
Cistaceae, Fabaceae, *Ericaceae*, *Boraginaceae*	*S. aureus*, *P. aeruginosa*, *E. coli*, *C. glabrata*	[214]
Egyptian pollen	*A. hydrophila*	[215]

Aeromonas hydrophila- A. hydrophila, Alternaria alternata- A. alternate, Aspergillus flavus- A. flavus, Aspergillus fumigatus- A. fumigatus,, Aspergillus niger- A. niger, Aspergillus parasiticus- A. parasiticus, Bacillus cereus- B. cereus, Bacillus subtilis- B. subtilis, Botrytis cinérea- B. cinérea, Candida albicans- C. albicans, Candida krusei- C. krusei, Candida parapsilosis- C. parapsilosis, Candida tropicalis- C. tropicalis, Candida glabrata - C. glabrata, Candida magmolia- C. magnolia, Clostridium butyricum- C. butyricum, Clostridium hystoliticum- C. hystoliticum, Clostridium intestinale- C. intestinale, Clostridium perfringens- C. perfringens, Clostridium ramosum- C. ramosum, Enterococcus cloacae- E. cloacae, Escherichia coli- E. coli, Fusarium oxysporum- F. oxysporum, Geotrichum candidum- G. candidum Klebsiella pneumonia- K. pneumonia, Lysteria monocytogenes- L. monocytogenes, Pseudomonas aeruginosa- P. aeruginosa, Proteus mirabilis- P. mirabilis, R. mucilaginosa- Rhodotorula mucilaginosa, Salmonella enterica- S. enterica, Salmonella enteritidis- S. enteritidis, Salmonella typhi- S. typhi, Shigella flexneri- S. flexneri, Staphylococcus aureus- S. aureus, Staphylococcus epidermidis- S. epidermidis, Streptococcus viridans- S. viridans, Streptococcus pyogenes- S. pyogenes, Zygosacharomyces mellis- Z. mellis, Zygosacharomyces bailii- Z. bailii, Zygosacharomyces rouxii- Z. rouxii

In addition to the antibacterial action, it was also reported an antifungal effect of pollen in some studies. One example is the monitoring of antifungal activity of bee products from different regions of Turkey against *Candida* spp. and *Trichosporon* spp. [202]. The results demonstrated the ability of the extracts to inhibit 40 kinds of yeast. The resultant antifungal activity decreased in the order: propolis - pollen - royal jelly - honey. A similar experiment was carried out by Özcan *et al.* [203]. However, the authors did not observe a complete inhibitory effect of pollen extracts on mycelia of *Aspergillus parasiticus* and the antifungal activity of pollen was again lower than that of propolis [202]. Erkmen and Özcan [121] tested the antimicrobial activities of pollen extract at concentrations from 0.02% to 2.5% (vol/vol) against bacteria (*Bacillus cereus, Bacillus subtilis, Enterococcus faecalis, Escherichia coli, Listeria monocytogenes, Salmonella typhimurium, Staphylococcus aureus* and *Yersinia enterocolitica*), yeasts (*Saccharomyces cerevisiae* and *Candida rugosa*), and molds (*Aspergillus niger* and *Rhizopus oryzae*) but no antimicrobial effects were detected.

Most of the authors who tested for antimicrobial properties of pollen have considered polyphenols as the main active ingredients and, in particular, documented a microbicidal activity for polyphenols against a number of pathogenic bacteria and microscopic filamentous fungi [149, 204, 205]. The current mechanism of polyphenol toxicity against microorganisms is related to the inhibition of hydrolytic enzymes (proteases, carbohydrolases) but also to other interactions that prevent microbial adhesion, the transfer of proteins in microbial cells or non-specific reactions with sugars [188].

Yang *et al.* [206] evaluated the inhibitory activity of bee pollen amylose against nine bacteria using pollen extracts with concentrations ranging from 2.1 to 33.6 mg/ml. The most effective were extracts with higher concentrations (16.8 to 33.6 mg/ml), able to inhibit *Salmonella, Pasteurella multocida, Erysipelothrix rhusiopathiae*, etc [206]. Barbosa *et al.* [201] devoted special attention to the lipid fraction of bee pollen due to the significance of fats as an energy source, but also because of their antimicrobial activity. Significant bacterial inhibitors were linoleic acid and linolenic acid. The authors evaluated the lipid fractions against the Gram-positive and also Gram-negative bacteria and concluded that the highest antimicrobial activity was shown by pollen from plant *Cistus landanifer* which had the highest concentration of linoleic acid and other fatty acids. In the case of *Rubus* pollen, this activity was attributed solely to linolenic acid [201].

CONCLUSION

Honey, propolis and pollen have been described to be effective in a number of human pathologies. Records of the use of these products date back to the earliest civilizations, either as food or for therapeutic applications. Several *in vitro* and a few clinical studies have confirmed the broad-spectrum antimicrobial properties of these bee products, reaching numerous microorganisms of many different genera. The number of studies demonstrating such bioactivity highlights the potential of these food sources as promising antimicrobial agents for diverse applications, namely in human and veterinary medicine, pharmacology, and cosmetics. The strong activity of honey and propolis, mainly against antibiotic-resistant bacteria has further enlarged the interest for application against microbial infections of medical interest. Furthermore, to date, there have been no reports documenting microbial resistance to honey, propolis or pollen.

The mechanism of antibacterial, antifungal and antiviral activities of pollen, propolis and honey are extremely complex due to the involvement of multiple compounds and to the discrepancies in their concentrations among bee products from different origins. This incomplete knowledge of the antimicrobial activity and the absence of studies supporting the systemic use of these natural products as antimicrobial agents represent a major obstacle for their clinical applicability. Future achievements, focusing on chemistry profiling and pharmacological traits, will certainly open the way to new therapeutic approaches and add considerable market value to honeybee plant-derived products.

CONFLICT OF INTEREST

The authors declare no conflict of interest regarding this publication.

ACKNOWLEDGEMENTS

Cristina Almeida-Aguiar is grateful to FEDER/COMPETE/POCI– Operacional Competitiveness and Internacionalization Programme, under Project POCI-0--0145-FEDER-006958 and to FCT - Portuguese Foundation for Science and Technology, under the project UID/AGR/04033/2013. Miroslava Kačániová thanks the European Community under project no 26220220180: Building Research Centre AgroBioTech", by grant of VEGA 1/0611/14 and Food and Agriculture COST Action FA1202.

REFERENCES

[1] Media, W.H. Media centre Antimicrobial resistance. *Development,* **2011**, 4-7.

[2] Mandal, M.D.; Mandal, S. Honey: its medicinal property and antibacterial activity. *Asian Pac. J. Trop. Biomed.,* **2011**, *1*(2), 154-160. [http://dx.doi.org/10.1016/S2221-1691(11)60016-6] [PMID: 23569748]

[3] Sforcin, J.M.; Bankova, V. Propolis: is there a potential for the development of new drugs? *J. Ethnopharmacol.,* **2011**, *133*(2), 253-260. [http://dx.doi.org/10.1016/j.jep.2010.10.032] [PMID: 20970490]

[4] Viuda-Martos, M.; Ruiz-Navajas, Y.; Fernández-López, J.; Pérez-Alvarez, J.A. Functional properties of honey, propolis, and royal jelly. *J. Food Sci.,* **2008**, *73*(9), R117-R124. [http://dx.doi.org/10.1111/j.1750-3841.2008.00966.x] [PMID: 19021816]

[5] Alvarez-Suarez, J.M.; Giampieri, F.; Battino, M. Honey as a source of dietary antioxidants: structures, bioavailability and evidence of protective effects against human chronic diseases. *Curr. Med. Chem.,* **2013**, *20*(5), 621-638.

[http://dx.doi.org/10.2174/092986713804999358] [PMID: 23298140]

[6] Molan, P.C. The evidence supporting the use of honey as a wound dressing. *Int. J. Low. Extrem. Wounds,* **2006**, *5*(1), 40-54. [http://dx.doi.org/10.1177/1534734605286014] [PMID: 16543212]

[7] Simon, A.; Traynor, K.; Santos, K.; Blaser, G.; Bode, U.; Molan, P. Medical honey for wound carestill the latest resort. *Evidence-based Complement. Altern. Med.,* **2009**, *6*(2), 165-173.

[8] Ortiz-Vázquez, E.; Cuevas-Glory, L.; Zapata-Baas, G.; Martínez-Guevara, J.; Ramón-Sierra, J. Which bee honey components contribute to its antimicrobial activity? A review. *Afr. J. Microbiol. Res.,* **2013**, *7*(51), 5758-5765.

[9] Carina, L.; Soledad, V.; Marina, B. Antibacterial activity of honey : A review of honey around the world. *J. Microbiol. Antimicrob.,* **2014**, *6*, 51-56. [http://dx.doi.org/10.5897/JMA2014.0308]

[10] AL-Waili, N.; Al Ghamdi, A.; Ansari, M.J.; Al-Attal, Y.; Al-Mubarak, A.; Salom, K. Differences in composition of honey samples and their impact on the antimicrobial activities against drug multiresistant bacteria and pathogenic fungi. *Arch. Med. Res.,* **2013**, *44*(4), 307-316. [http://dx.doi.org/10.1016/j.arcmed.2013.04.009] [PMID: 23684665]

[11] Mundo, M.A.; Padilla-Zakour, O.I.; Worobo, R.W. Growth inhibition of foodborne pathogens and food spoilage organisms by select raw honeys. *Int. J. Food Microbiol.,* **2004**, *97*(1), 1-8. [http://dx.doi.org/10.1016/j.ijfoodmicro.2004.03.025] [PMID: 15527912]

[12] Liu, J.R.; Ye, Y.L.; Lin, T.Y.; Wang, Y.W.; Peng, C.C. Effect of floral sources on the antioxidant, antimicrobial, and anti-inflammatory activities of honeys in Taiwan. *Food Chem.,* **2013**, *139*(1-4), 938-943. [http://dx.doi.org/10.1016/j.foodchem.2013.02.015] [PMID: 23561193]

[13] Smith, M.R.; McCaughey, W.F.; Kemmerer, A.R. Biological effects of honey. *J. Apic. Res.,* **1969**.

[14] French, V.M.; Cooper, R.A.; Molan, P.C. The antibacterial activity of honey against coagulase-negative staphylococci. *J. Antimicrob. Chemother.,* **2005**, *56*(1), 228-231. [http://dx.doi.org/10.1093/jac/dki193] [PMID: 15941774]

[15] Lusby, P.E.; Coombes, A.; Wilkinson, J.M. Honey: a potent agent for wound healing? *J. Wound Ostomy Continence Nurs.,* **2002**, *29*(6), 295-300. [PMID: 12439453]

[16] Wilkinson, J.M.; Cavanagh, H.M. Antibacterial activity of 13 honeys against *Escherichia coli* and *Pseudomonas aeruginosa*. *J. Med. Food,* **2005**, *8*(1), 100-103. [http://dx.doi.org/10.1089/jmf.2005.8.100] [PMID: 15857217]

[17] Maddocks, S.E.; Lopez, M.S.; Rowlands, R.S.; Cooper, R.A. Manuka honey inhibits the development of Streptococcus pyogenes biofilms and causes reduced expression of two fibronectin binding proteins. *Microbiology,* **2012**, *158*(Pt 3), 781-790. [http://dx.doi.org/10.1099/mic.0.053959-0] [PMID: 22294681]

[18] Badawy, O.F.; Shafii, S.S.; Tharwat, E.E.; Kamal, A.M. Antibacterial activity of bee honey and its therapeutic usefulness against *Escherichia coli* O157:H7 and *Salmonella typhimurium* infection. *Rev. - Off. Int. Epizoot.,* **2004**, *23*(3), 1011-1022.

[PMID: 15861897]

[19] Tan, H.T.; Rahman, R.A.; Gan, S.H.; Halim, A.S.; Hassan, S.A.; Sulaiman, S.A.; Kirnpal-Kaur, B. The antibacterial properties of Malaysian tualang honey against wound and enteric microorganisms in comparison to manuka honey. *BMC Complement. Altern. Med.,* **2009**, *9*, 34. [http://dx.doi.org/10.1186/1472-6882-9-34] [PMID: 19754926]

[20] Jervis-Bardy, J.; Foreman, A.; Bray, S.; Tan, L.; Wormald, P.J. Methylglyoxal-infused honey mimics the anti-*Staphylococcus aureus* biofilm activity of manuka honey: potential implication in chronic rhinosinusitis. *Laryngoscope,* **2011**, *121*(5), 1104-1107. [http://dx.doi.org/10.1002/lary.21717] [PMID: 21520131]

[21] Cooper, R.A.; Molan, P.C.; Harding, K.G. The sensitivity to honey of Gram-positive cocci of clinical significance isolated from wounds. *J. Appl. Microbiol.,* **2002**, *93*(5), 857-863. [http://dx.doi.org/10.1046/j.1365-2672.2002.01761.x] [PMID: 12392533]

[22] Cooper, R.A.; Molan, P.C.; Harding, K.G. Antibacterial activity of honey against strains of Staphylococcus aureus from infected wounds. *J. R. Soc. Med.,* **1999**, *92*(6), 283-285. [PMID: 10472280]

[23] Kuncič, M.K.; Jaklič, D.; Lapanje, A.; Gunde-Cimerman, N. Antibacterial and antimycotic activities of Slovenian honeys. *Br. J. Biomed. Sci.,* **2012**, *69*(4), 154-158. [PMID: 23304790]

[24] Alzahrani, H.A.; Alsabehi, R.; Boukraâ, L.; Abdellah, F.; Bellik, Y.; Bakhotmah, B.A. Antibacterial and antioxidant potency of floral honeys from different botanical and geographical origins. *Molecules,* **2012**, *17*(9), 10540-10549. [http://dx.doi.org/10.3390/molecules170910540] [PMID: 22948516]

[25] Taormina, P.J.; Niemira, B.A.; Beuchat, L.R. Inhibitory activity of honey against foodborne pathogens as influenced by the presence of hydrogen peroxide and level of antioxidant power. *Int. J. Food Microbiol.,* **2001**, *69*(3), 217-225. [http://dx.doi.org/10.1016/S0168-1605(01)00505-0] [PMID: 11603859]

[26] Mohapatra, D.P.; Thakur, V.; Brar, S.K. Antibacterial efficacy of raw and processed honey. *Biotechnol. Res. Int.,* **2011**, *2011*, 917505. [http://dx.doi.org/10.4061/2011/917505]

[27] Sherlock, O.; Dolan, A.; Athman, R.; Power, A.; Gethin, G.; Cowman, S.; Humphreys, H. Comparison of the antimicrobial activity of Ulmo honey from Chile and Manuka honey against methicillin-resistant *Staphylococcus aureus, Escherichia coli* and *Pseudomonas aeruginosa. BMC Complement. Altern. Med.,* **2010**, *10*, 47. [http://dx.doi.org/10.1186/1472-6882-10-47] [PMID: 20813024]

[28] Mullai, V.; Menon, T. Bactericidal activity of different types of honey against clinical and environmental isolates of *Pseudomonas aeruginosa. J. Altern. Complement. Med.,* **2007**, *13*(4), 439-441. [http://dx.doi.org/10.1089/acm.2007.6366] [PMID: 17532737]

[29] Chan-Rodríguez, D.; Ramón-Sierra, J.; Lope-Ayora, J.; Sauri-Duch, M.E.; Cuevas-Glory, L.; Ortiz-Vázquez, E. Antibacterial Properties of Honey Produced by *Melipona beecheii* and *Apis mellifera* against Foodborn Microorganisms. *Food Sci. Biotechnol.,* **2012**, *21*(3), 905-909.

[http://dx.doi.org/10.1007/s10068-012-0118-x]

[30] Dardón, M.J.; Enríquez, E. Caracterización fisicoquímica y antimicrobiana de la miel de nueve especies de abejas sin aguijón (meliponini) de Guatemala. *Interciencia,* **2008**, *33*(12), 916-922.

[31] Gaitanis, G.; Magiatis, P.; Hantschke, M.; Bassukas, I.D.; Velegraki, A. The Malassezia genus in skin and systemic diseases. *Clin. Microbiol. Rev.,* **2012**, *25*(1), 106-141. [http://dx.doi.org/10.1128/CMR.00021-11] [PMID: 22232373]

[32] Boorn, K.L.; Khor, Y.Y.; Sweetman, E.; Tan, F.; Heard, T.A.; Hammer, K.A. Antimicrobial activity of honey from the stingless bee Trigona carbonaria determined by agar diffusion, agar dilution, broth microdilution and time-kill methodology. *J. Appl. Microbiol.,* **2010**, *108*(5), 1534-1543. [http://dx.doi.org/10.1111/j.1365-2672.2009.04552.x] [PMID: 19811569]

[33] Dryden, M.; Lockyer, G.; Saeed, K.; Cooke, J. Engineered honey: *In vitro* antimicrobial activity of a novel topical wound care treatment. *J. Glob. Antimicrob. Resist.,* **2014**, *2*(3), 168-172. [http://dx.doi.org/10.1016/j.jgar.2014.03.006]

[34] Andualem, B. Combined antibacterial activity of stingless bee (Apis mellipodae) honey and garlic (Allium sativum) extracts against standard and clinical pathogenic bacteria. *Asian Pac. J. Trop. Biomed.,* **2013**, *3*(9), 725-731. [http://dx.doi.org/10.1016/S2221-1691(13)60146-X] [PMID: 23998014]

[35] Mulu, A.; Tessema, B.; Derbie, F. *In vitro* assessment of the antimicrobial potential of honey on common human pathogens. Ethiop. *J. Health Dev,* **2004**, *18*, 107-112.

[36] Molan, P. The antibacterial activity of honey: 2. Variation in the potency of the antibacterial activity. *Bee World,* **1992**, *73*, 59-76. [http://dx.doi.org/10.1080/0005772X.1992.11099118]

[37] Garedew, A.; Schmolz, E.; Lamprecht, I. Microcalorimetric investigation on the antimicrobial activity of honey of the stingless bee Trigona spp. and comparison of some parameters with those obtained with standard methods. *Thermochim. Acta,* **2004**, *415*(1–2), 99-106. [http://dx.doi.org/10.1016/j.tca.2003.06.004]

[38] Al-Habsi, N.A.; Niranjan, K. Effect of high hydrostatic pressure on antimicrobial activity and quality of Manuka honey. *Food Chem.,* **2012**, *135*(3), 1448-1454. [http://dx.doi.org/10.1016/j.foodchem.2012.06.012] [PMID: 22953879]

[39] Isla, M.I.; Craig, A.; Ordoñez, R.; Zampini, C.; Sayago, J.; Bedascarrasbure, E.; Alvarez, A.; Salomón, V.; Maldonado, L. Physico chemical and bioactive properties of honeys from Northwestern Argentina. *LWT -. Food Sci. Technol. (Campinas.),* **2011**, *44*(9), 1922-1930.

[40] Badet, C.; Quero, F. The *in vitro* effect of manuka honeys on growth and adherence of oral bacteria. *Anaerobe,* **2011**, *17*(1), 19-22. [http://dx.doi.org/10.1016/j.anaerobe.2010.12.007] [PMID: 21195787]

[41] Manyi-Loh, C.E.; Clarke, A.M.; Munzhelele, T.; Green, E.; Mkwetshana, N.F.; Ndip, R.N. Selected South African honeys and their extracts possess *in vitro* anti-Helicobacter pylori activity. *Arch. Med. Res.,* **2010**, *41*(5), 324-331. [http://dx.doi.org/10.1016/j.arcmed.2010.08.002] [PMID: 20851288]

[42] Ndip, R.N.; Malange Takang, A.E.; Echakachi, C.M.; Malongue, A.; Akoachere, J.F.; Ndip, L.M.;

Luma, H.N. In-vitro antimicrobial activity of selected honeys on clinical isolates of Helicobacter pylori. *Afr. Health Sci.,* **2007**, *7*(4), 228-232.
[PMID: 21499488]

[43] Melliou, E.; Stratis, E.; Chinou, I. Volatile constituents of propolis from various regions of Greece - Antimicrobial activity. *Food Chem.,* **2007**, *103*(2), 375-380.
[http://dx.doi.org/10.1016/j.foodchem.2006.07.033]

[44] Majtán, J.; Majtánová, L.; Xu, M.; Majtán, V. *In vitro* effect of subinhibitory concentrations of antibiotics on biofilm formation by clinical strains of Salmonella enterica serovar Typhimurium isolated in Slovakia. *J. Appl. Microbiol.,* **2008**, *104*(5), 1294-1301.
[http://dx.doi.org/10.1111/j.1365-2672.2007.03653.x] [PMID: 18028358]

[45] Moussa, A.; Noureddine, D.; Abdelmelek, M.; Saad, A. Antibacterial activity of various honey types of Algeria against Pathogenic Gram-Negative Bacilli: Escherichia coli and Pseudomonas aeruginosa. *Asian Pacific J. Trop. Dis.,* **2012**, *2*(3), 211-214.
[http://dx.doi.org/10.1016/S2222-1808(12)60048-6]

[46] Estevinho, L.; Pereira, A.P.; Moreira, L.; Dias, L.G.; Pereira, E. Antioxidant and antimicrobial effects of phenolic compounds extracts of Northeast Portugal honey. *Food Chem. Toxicol.,* **2008**, *46*(12), 3774-3779.
[http://dx.doi.org/10.1016/j.fct.2008.09.062] [PMID: 18940227]

[47] Kačániová, M.; Vukovic, N.; Bobková, A.; Fikselová, M.; Rovná, K.; Haščík, P.; Čuboň, J.; Hleba, L.; Bobko, M. Antimicrobial and antiradical activity of Slovakian honeydew honey samples. *J. Microb.Biotech. Food Sci.,* **2011**, *1*(3), 354-36.

[48] Kacániová, M.; Fatrcová-Sramková, K.; Nozková, J.; Melich, M.; Kadasi-Horáková, M.; Knazovická, V.; Felsöciová, S.; Kunová, S.; Máriássyová, M. Antiradical activity of natural honeys and antifungal effect against Penicillium genera. *J. Environ. Sci. Health B,* **2011**, *46*(1), 92-96.
[http://dx.doi.org/10.1080/03601234.2011.534416] [PMID: 21240710]

[49] White, J.W., Jr; Subers, M.H.; Schepartz, A.I. The identification of inhibine, the antibacterial factor in honey, as hydrogen peroxide and its origin in a honey glucose-oxidase system. *Biochim. Biophys. Acta,* **1963**, *73*, 57-70.
[http://dx.doi.org/10.1016/0926-6569(63)90108-1] [PMID: 14000328]

[50] Seymour, F.I.; West, K.S. Honey--its role in medicine. *Med. Times,* **1951**, *79*(2), 104-108.
[PMID: 14805702]

[51] Bose, B. Honey or sugar in treatment of infected wounds? *Lancet,* **1982**, *1*(8278), 963.
[http://dx.doi.org/10.1016/S0140-6736(82)91959-6] [PMID: 6122795]

[52] Tovey, F.I. Honey and healing. *J. R. Soc. Med.,* **1991**, *84*(7), 447.
[PMID: 1865464]

[53] Gul, W.; Farooq, N.; Khan, U.; Rehan, F.; Anees, D. Honey: a nectarous anti-infective agent. *World J Pharm Sci,* **2015**, *4*(4), 208-215.

[54] Molan, P.C. The antibacterial activity of honey 1. The nature of the antibacterial activity. *Bee World,* **1992**, *73*(1), 5-28.
[http://dx.doi.org/10.1080/0005772X.1992.11099109]

[55] Kwakman, P.H.; Zaat, S.A. Antibacterial components of honey. *IUBMB Life,* **2012**, *64*(1), 48-55. [http://dx.doi.org/10.1002/iub.578] [PMID: 22095907]

[56] Kwakman, P.H.; te Velde, A.A.; de Boer, L.; Speijer, D.; Vandenbroucke-Grauls, C.M.; Zaat, S.A. How honey kills bacteria. *FASEB J.,* **2010**, *24*(7), 2576-2582. [http://dx.doi.org/10.1096/fj.09-150789] [PMID: 20228250]

[57] Molan, P.C. Why honey is effective as a medicine. 2. The scientific explanation of its effects. *Bee World,* **2001**, *82*(1), 22-40. [http://dx.doi.org/10.1080/0005772X.2001.11099498]

[58] Weston, R.J. The contribution of catalase and other natural products to the antibacterial activity of honey: a review. *Food Chem.,* **2000**, *71*, 235-239. [http://dx.doi.org/10.1016/S0308-8146(00)00162-X]

[59] White, J.W.; Subers, M.H. Studies on honey inhibine. 3. Effect of heat. *J. Apic. Res.,* **1964**, *3*(1), 45-50.

[60] White, J.W.; Subers, M.H. Studies on honey inhibine. 4. Destruction of the peroxide accumulation system by light. *J. Food Sci.,* **1964**, *29*(6), 819-828. [http://dx.doi.org/10.1111/j.1365-2621.1964.tb00455.x]

[61] Schepartz, A.I. Honey Catalase: Occurrence and Some Kinetic Properties. *J. Apic. Res.,* **1966**, *5*(3), 167-176.

[62] Carter, C.; Thornburg, R.W. Is the nectar redox cycle a floral defense against microbial attack? *Trends Plant Sci.,* **2004**, *9*(7), 320-324. [http://dx.doi.org/10.1016/j.tplants.2004.05.008] [PMID: 15231276]

[63] Hillwig, M.S.; Kanobe, C.; Thornburg, R.W.; Macintosh, G.C. Identification of S-RNase and peroxidase in petunia nectar. *J. Plant Physiol.,* **2011**, *168*(7), 734-738. [http://dx.doi.org/10.1016/j.jplph.2010.10.002] [PMID: 21093100]

[64] Bankar, S.B.; Bule, M.V.; Singhal, R.S.; Ananthanarayan, L. Glucose oxidase--an overview. *Biotechnol. Adv.,* **2009**, *27*(4), 489-501. [http://dx.doi.org/10.1016/j.biotechadv.2009.04.003] [PMID: 19374943]

[65] Schepartz, A.I.; Subers, M.H. The glucose oxidase of honey. I. Purification and some general properties of the enzyme. *Biochim. Biophys. Acta,* **1964**, *85*, 228-237. [PMID: 14212969]

[66] Brudzynski, K. Effect of hydrogen peroxide on antibacterial activities of Canadian honeys. *Can. J. Microbiol.,* **2006**, *52*(12), 1228-1237. [http://dx.doi.org/10.1139/w06-086] [PMID: 17473892]

[67] Bizerra, F.C.; Da Silva, P.I., Jr; Hayashi, M.A. Exploring the antibacterial properties of honey and its potential. *Front. Microbiol.,* **2012**, *3*, 398. [http://dx.doi.org/10.3389/fmicb.2012.00398] [PMID: 23189076]

[68] Allen, K.L.; Molan, P.C.; Reid, G.M. A survey of the antibacterial activity of some New Zealand honeys. *J. Pharm. Pharmacol.,* **1991**, *43*(12), 817-822. [http://dx.doi.org/10.1111/j.2042-7158.1991.tb03186.x] [PMID: 1687577]

[69] Molan, P.C. *Honey as a topical antibacterial agent for treatment of infected wounds*; World Wide Wounds, **2001**, pp. 1-13.

[70] Willix, D.J.; Molan, P.C.; Harfoot, C.G. A comparison of the sensitivity of wound-infecting species of bacteria to the antibacterial activity of manuka honey and other honey. *J. Appl. Bacteriol.,* **1992**, *73*(5), 388-394.
[http://dx.doi.org/10.1111/j.1365-2672.1992.tb04993.x] [PMID: 1447054]

[71] Mavric, E.; Wittmann, S.; Barth, G.; Henle, T. Identification and quantification of methylglyoxal as the dominant antibacterial constituent of Manuka (Leptospermum scoparium) honeys from New Zealand. *Mol. Nutr. Food Res.,* **2008**, *52*(4), 483-489.
[http://dx.doi.org/10.1002/mnfr.200700282] [PMID: 18210383]

[72] Thornalley, P.J. Cell activation by glycated proteins. AGE receptors, receptor recognition factors and functional classification of AGEs. *Cell. Mol. Biol. (Noisy-le-grand),* **1998**, *44*(7), 1013-1023.
[PMID: 9846883]

[73] Barros, A.; Rodrigues, J.A.; Almeida, P.J.; Oliva-Teles, M.T. Determination of glyoxal, methylglyoxal, and diacetyl in selected beer and wine, by HPLC with UV spectrophotometric detection, after derivatization with *o*-phenylenediamine. *J. Liq. Chrom. Rel. Technol.,* **1999**, *22*(13), 2061-2069.
[http://dx.doi.org/10.1081/JLC-100101786]

[74] Hayashi, T.; Shibamot, T. Analysis of methyl glyoxal in foods and beverages. *J. Agric. Food Chem.,* **1985**, *33*(6), 1090-1093.
[http://dx.doi.org/10.1021/jf00066a018]

[75] Nagao, M.; Wakabayashi, K.; Fujita, Y.; Tahira, T.; Ochiai, M.; Sugimura, T. Mutagenic compounds in soy sauce, Chinese cabbage, coffee and herbal teas. *Prog. Clin. Biol. Res.,* **1986**, *206*, 55-62.
[PMID: 3952098]

[76] Adams, C.J.; Manley-Harris, M.; Molan, P.C. The origin of methylglyoxal in New Zealand manuka (*Leptospermum scoparium*) honey. *Carbohydr. Res.,* **2009**, *344*(8), 1050-1053.
[http://dx.doi.org/10.1016/j.carres.2009.03.020] [PMID: 19368902]

[77] Weigel, K.U.; Opitz, T.; Henle, T. Studies on the occurrence and formation of 1,2-dicarbonyls in honey. *Eur. Food Res. Technol.,* **2004**, *218*(2), 147-151.
[http://dx.doi.org/10.1007/s00217-003-0814-0]

[78] Atrott, J.; Henle, T. Methylglyoxal in manuka honey— correlation with antibacterial properties. *Czech J. Food Scie.,* **2009**, *27*, S163-S165.

[79] Kwakman, P.H.; Te Velde, A.A.; de Boer, L.; Vandenbroucke-Grauls, C.M.; Zaat, S.A. Two major medicinal honeys have different mechanisms of bactericidal activity. *PLoS One,* **2011**, *6*(3), e17709.
[http://dx.doi.org/10.1371/journal.pone.0017709] [PMID: 21394213]

[80] Fujiwara, S.; Imai, J.; Fujiwara, M.; Yaeshima, T.; Kawashima, T.; Kobayashi, K. A potent antibacterial protein in royal jelly. Purification and determination of the primary structure of royalisin. *J. Biol. Chem.,* **1990**, *265*(19), 11333-11337.
[PMID: 2358464]

[81] Niyonsaba, F.; Ushio, H.; Nakano, N.; Ng, W.; Sayama, K.; Hashimoto, K.; Nagaoka, I.; Okumura, K.;

Ogawa, H. Antimicrobial peptides human β-defensins stimulate epidermal keratinocyte migration, proliferation and production of proinflammatory cytokines and chemokines. *J. Invest. Dermatol.,* **2007**, *127*(3), 594-604.
[http://dx.doi.org/10.1038/sj.jid.5700599] [PMID: 17068477]

[82] Gómez-Caravaca, A.M.; Gómez-Romero, M.; Arráez-Román, D.; Segura-Carretero, A.; Fernández-Gutiérrez, A. Advances in the analysis of phenolic compounds in products derived from bees. *J. Pharm. Biomed. Anal.,* **2006**, *41*(4), 1220-1234.
[http://dx.doi.org/10.1016/j.jpba.2006.03.002] [PMID: 16621403]

[83] Ortiz-Vázquez, E.; Cuevas-Glory, L.; Zapata-Baas, G.; Martínez-Guevara, J.; Ramón-Sierra, J. Which bee honey components contribute to its antimicrobial activity? A review. *Afr. J. Microbiol. Res.,* **2013**, *7*(51), 5758-5765.

[84] Aljadi, A.M.; Kamaruddin, M.Y. Evaluation of the phenolic contents and antioxidant capacities of two Malaysian floral honeys. *J. Agric. Food Chem.,* **2004**, *85*, 513-518.
[http://dx.doi.org/10.1016/S0308-8146(02)00596-4]

[85] Candiracci, M.; Citterio, B.; Piatti, E. Antifungal activity of the honey flavonoid extract against Candida albicans. *Food Chem.,* **2012**, *131*(2), 493-499.
[http://dx.doi.org/10.1016/j.foodchem.2011.09.012]

[86] Cushnie, T.P.; Lamb, A.J. Antimicrobial activity of flavonoids. *Int. J. Antimicrob. Agents,* **2005**, *26*(5), 343-356.
[http://dx.doi.org/10.1016/j.ijantimicag.2005.09.002] [PMID: 16323269]

[87] Weston, R.J.; Brocklebank, L.K.; Lu, Y. Identification and quantitative levels of antibacterial components of some New Zealand honeys. *Food Chem.,* **2000**, *70*, 427-435.
[http://dx.doi.org/10.1016/S0308-8146(00)00127-8]

[88] Cutting, K.F. Honey and contemporary wound care: an overview. *Ostomy Wound Manage.,* **2007**, *53*(11), 49-54.
[PMID: 18057446]

[89] Proal, A. *Understanding biofilms. Available: http://bacteriality.com/2008/05/26/biofilm/,* **2015**. (2 April 2015)

[90] Bjarnsholt, T.; Kirketerp-Møller, K.; Jensen, P.O.; Madsen, K.G.; Phipps, R.; Krogfelt, K.; Høiby, N.; Givskov, M. Why chronic wounds will not heal: a novel hypothesis. *Wound Repair Regen.,* **2008**, *16*(1), 2-10.
[http://dx.doi.org/10.1111/j.1524-475X.2007.00283.x] [PMID: 18211573]

[91] **2007**.

[92] Davis, S.C.; Ricotti, C.; Cazzaniga, A.; Welsh, E.; Eaglstein, W.H.; Mertz, P.M. Microscopic and physiologic evidence for biofilm-associated wound colonization *in vivo. Wound Repair Regen.,* **2008**, *16*(1), 23-29.
[http://dx.doi.org/10.1111/j.1524-475X.2007.00303.x] [PMID: 18211576]

[93] Tendolkar, P.M.; Baghdayan, A.S.; Gilmore, M.S.; Shankar, N. Enterococcal surface protein, Esp, enhances biofilm formation by *Enterococcus faecalis. Infect. Immun.,* **2004**, *72*(10), 6032-6039.
[http://dx.doi.org/10.1128/IAI.72.10.6032-6039.2004] [PMID: 15385507]

[94] Necidová, L.; Janštová, B.; Karpíšková, S.; Cupáková, S.; Dušková, M.; Karpíšková, R. Importance of Enterococcus spp. for forming a biofilm. *Czech J. Food Sci., Praha,* **2009**, *27*(2), S354-S356.

[95] Mohamed, J.A.; Huang, D.B. Biofilm formation by enterococci. *J. Med. Microbiol.,* **2007**, *56*(Pt 12), 1581-1588.
[http://dx.doi.org/10.1099/jmm.0.47331-0] [PMID: 18033823]

[96] George, A. Microtiter dish biofilm formation assay. *J. Vis. Exp.,* **2011**, *30*(47), pii: 2437.

[97] Alandejani, T.; Marsan, J.; Ferris, W.; Slinger, R.; Chan, F. Effectiveness of honey on *Staphylococcus aureus* and *Pseudomonas aeruginosa* biofilms. *Otolaryngol. Head Neck Surg.,* **2009**, *141*(1), 114-118.
[http://dx.doi.org/10.1016/j.otohns.2009.01.005] [PMID: 19559969]

[98] Okhiria, O.A.; Henriques, A.F.; Burton, N.F.; Peters, A.; Cooper, R.A. Honey modulates biofilms of *Pseudomonas aeruginosa* in a time and dose dependent manner. *J. ApiProd. ApiMed. Sci.,* **2009**, *1*(1), 6-10.

[99] Bankova, V.; Castro, S.L.; Marcucci, M.C. Propolis: recent advances in chemistry and plant origin. *Apidologie (Celle),* **2000**, *31*, 3-15.
[http://dx.doi.org/10.1051/apido:2000102]

[100] Fokt, H.; Pereira, A.; Ferreira, A.M.; Cunha, A.; Aguiar, C. How do bees prevent hive infections? The antimicrobial properties of propolis. *Curr. Res. Technol. Educ. Top. Appl. Microbiol. Microb. Biotechnol. Formatex Research Center,* **2010**, *1*, 481-4493.

[101] Marcucci, M.C. Propolis: chemical composition, biological properties and therapeutic activity. *Apidologie (Celle),* **1995**, *26*(2), 83-99.
[http://dx.doi.org/10.1051/apido:19950202]

[102] Burdock, G.A. Review of the biological properties and toxicity of bee propolis (propolis). *Food Chem. Toxicol.,* **1998**, *36*(4), 347-363.
[http://dx.doi.org/10.1016/S0278-6915(97)00145-2] [PMID: 9651052]

[103] Banskota, A.H.; Tezuka, Y.; Kadota, S. Recent progress in pharmacological research of propolis. *Phytother. Res.,* **2001**, *15*(7), 561-571.
[http://dx.doi.org/10.1002/ptr.1029] [PMID: 11746834]

[104] Toreti, V.C.; Sato, H.H.; Pastore, G.M.; Park, Y.K. Recent progress of propolis for its biological and chemical compositions and its botanical origin. *Evid. Based Complement. Alternat. Med.,* **2013**, *2013*, 697390.
[http://dx.doi.org/10.1155/2013/697390] [PMID: 23737843]

[105] Bankova, V. Chemical diversity of propolis and the problem of standardization. *J. Ethnopharmacol.,* **2005**, *100*(1-2), 114-117.
[http://dx.doi.org/10.1016/j.jep.2005.05.004] [PMID: 15993016]

[106] Barlak, Y.; Değer, O.; Çolak, M.; Karataylı, S.C.; Bozdayı, A.M.; Yücesan, F. Effect of Turkish propolis extracts on proteome of prostate cancer cell line. *Proteome Sci.,* **2011**, *9*(1), 74.
[http://dx.doi.org/10.1186/1477-5956-9-74] [PMID: 22152088]

[107] Peña, R.C. Propolis standardization: a chemical and biological review. *Cien. Inv. Agr,* **2008**, *35*(1), 17-26.

[108] Falcão, S.I.; Vilas-Boas, M.; Estevinho, L.M.; Barros, C.; Domingues, M.R.; Cardoso, S.M. Phenolic characterization of Northeast Portuguese propolis: usual and unusual compounds. *Anal. Bioanal. Chem.,* **2010**, *396*(2), 887-897.
[http://dx.doi.org/10.1007/s00216-009-3232-8] [PMID: 19902191]

[109] Sun, L-P.; Chen, A-L.; Hung, H-C.; Chien, Y-H.; Huang, J-S.; Huang, C-Y.; Chen, Y-W.; Chen, C-N. Chrysin: a histone deacetylase 8 inhibitor with anticancer activity and a suitable candidate for the standardization of Chinese propolis. *J. Agric. Food Chem.,* **2012**, *60*(47), 11748-11758.
[http://dx.doi.org/10.1021/jf303261r] [PMID: 23134323]

[110] Park, Y.K.; Alencar, S.M.; Aguiar, C.L. Botanical origin and chemical composition of Brazilian propolis. *J. Agric. Food Chem.,* **2002**, *50*(9), 2502-2506.
[http://dx.doi.org/10.1021/jf011432b] [PMID: 11958612]

[111] Uzel, A.; Sorkun, K.; Onçağ, O.; Coğulu, D.; Gençay, O.; Salih, B. Chemical compositions and antimicrobial activities of four different Anatolian propolis samples. *Microbiol. Res.,* **2005**, *160*(2), 189-195.
[http://dx.doi.org/10.1016/j.micres.2005.01.002] [PMID: 15881836]

[112] Kujumgiev, A.; Tsvetkova, I.; Serkedjieva, Y.; Bankova, V.; Christov, R.; Popov, S. Antibacterial, antifungal and antiviral activity of propolis of different geographic origin. *J. Ethnopharmacol.,* **1999**, *64*(3), 235-240.
[http://dx.doi.org/10.1016/S0378-8741(98)00131-7] [PMID: 10363838]

[113] Castaldo, S.; Capasso, F. Propolis, an old remedy used in modern medicine. *Fitoterapia,* **2002**, *73*(1) Suppl. 1, S1-S6.
[http://dx.doi.org/10.1016/S0367-326X(02)00185-5] [PMID: 12495704]

[114] Sforcin, J.M.; Fernandes, A., Jr; Lopes, C.A.; Bankova, V.; Funari, S.R. Seasonal effect on Brazilian propolis antibacterial activity. *J. Ethnopharmacol.,* **2000**, *73*(1-2), 243-249.
[http://dx.doi.org/10.1016/S0378-8741(00)00320-2] [PMID: 11025162]

[115] Gekker, G.; Hu, S.; Spivak, M.; Lokensgard, J.R.; Peterson, P.K. Anti-HIV-1 activity of propolis in CD4(+) lymphocyte and microglial cell cultures. *J. Ethnopharmacol.,* **2005**, *102*(2), 158-163.
[http://dx.doi.org/10.1016/j.jep.2005.05.045] [PMID: 16046088]

[116] Freitas, S.F.; Shinohara, L.; Sforcin, J.M.; Guimarães, S. *In vitro* effects of propolis on Giardia duodenalis trophozoites. *Phytomedicine,* **2006**, *13*(3), 170-175.
[http://dx.doi.org/10.1016/j.phymed.2004.07.008] [PMID: 16428024]

[117] Scazzocchio, F.; D'Auria, F.D.; Alessandrini, D.; Pantanella, F. Multifactorial aspects of antimicrobial activity of propolis. *Microbiol. Res.,* **2006**, *161*(4), 327-333.
[http://dx.doi.org/10.1016/j.micres.2005.12.003] [PMID: 16427259]

[118] Mirzoeva, O.K.; Grishanin, R.N.; Calder, P.C. Antimicrobial action of propolis and some of its components: the effects on growth, membrane potential and motility of bacteria. *Microbiol. Res.,* **1997**, *152*(3), 239-246.
[http://dx.doi.org/10.1016/S0944-5013(97)80034-1] [PMID: 9352659]

[119] Silva, J.C.; Rodrigues, S.; Feás, X.; Estevinho, L.M. Antimicrobial activity, phenolic profile and role in the inflammation of propolis. *Food Chem. Toxicol.,* **2012**, *50*(5), 1790-1795.

[http://dx.doi.org/10.1016/j.fct.2012.02.097] [PMID: 22425940]

[120] Seidel, V.; Peyfoon, E.; Watson, D.G.; Fearnley, J. Comparative study of the antibacterial activity of propolis from different geographical and climatic zones. *Phytother. Res.,* **2008**, *22*(9), 1256-1263. [http://dx.doi.org/10.1002/ptr.2480] [PMID: 18570199]

[121] Erkmen, O.; Ozcan, M.M. Antimicrobial effects of Turkish propolis, pollen, and laurel on spoilage and pathogenic food-related microorganisms. *J. Med. Food,* **2008**, *11*(3), 587-592. [http://dx.doi.org/10.1089/jmf.2007.0038] [PMID: 18800911]

[122] Park, Y.K.; Ikegaki, M. Preparation of water and ethanolic extracts of propolis and evaluation of the preparations. *Biosci. Biotechnol. Biochem.,* **1998**, *62*, 2230-2232. [http://dx.doi.org/10.1271/bbb.62.2230]

[123] Park, Y.K.; Koo, M.H.; Abreu, J.A.; Ikegaki, M.; Cury, J.A.; Rosalen, P.L. Antimicrobial activity of propolis on oral microorganisms. *Curr. Microbiol.,* **1998**, *36*(1), 24-28. [http://dx.doi.org/10.1007/s002849900274] [PMID: 9405742]

[124] Netíková, L.; Bogusch, P.; Heneberg, P. Czech ethanol-free propolis extract displays inhibitory activity against a broad spectrum of bacterial and fungal pathogens. *J. Food Sci.,* **2013**, *78*(9), M1421-M1429. [http://dx.doi.org/10.1111/1750-3841.12230] [PMID: 23915150]

[125] Wojtyczka, R.D.; Dziedzic, A.; Idzik, D.; Kępa, M.; Kubina, R.; Kabała-Dzik, A.; Smoleń-Dzirba, J.; Stojko, J.; Sajewicz, M.; Wąsik, T.J. Susceptibility of Staphylococcus aureus clinical isolates to propolis extract alone or in combination with antimicrobial drugs. *Molecules,* **2013**, *18*(8), 9623-9640. [http://dx.doi.org/10.3390/molecules18089623] [PMID: 23941882]

[126] Bastos, E.M.; Simone, M.; Jorge, D.M.; Soares, A.E.; Spivak, M. *In vitro* study of the antimicrobial activity of Brazilian propolis against Paenibacillus larvae. *J. Invertebr. Pathol.,* **2008**, *97*(3), 273-281. [http://dx.doi.org/10.1016/j.jip.2007.10.007] [PMID: 18054037]

[127] Libério, S.A.; Pereira, A.L.; Araújo, M.J.; Dutra, R.P.; Nascimento, F.R.; Monteiro-Neto, V.; Ribeiro, M.N.; Gonçalves, A.G.; Guerra, R.N. The potential use of propolis as a cariostatic agent and its actions on mutans group streptococci. *J. Ethnopharmacol.,* **2009**, *125*(1), 1-9. [http://dx.doi.org/10.1016/j.jep.2009.04.047] [PMID: 19422903]

[128] Orsi, R.O.; Sforcin, J.M.; Funari, S.R.; Bankova, V. Effects of Brazilian and Bulgarian propolis on bactericidal activity of macrophages against *Salmonella Typhimurium. Int. Immunopharmacol.,* **2005**, *5*(2), 359-368. [http://dx.doi.org/10.1016/j.intimp.2004.10.003] [PMID: 15652765]

[129] Yildirim, Z.; Hacievliyagil, S.; Kutlu, N.O.; Aydin, N.E.; Kurkcuoglu, M.; Iraz, M.; Durmaz, R. Effect of water extract of Turkish propolis on tuberculosis infection in guinea-pigs. *Pharmacol. Res.,* **2004**, *49*(3), 287-292. [http://dx.doi.org/10.1016/j.phrs.2003.10.007] [PMID: 14726226]

[130] Boyanova, L.; Derejian, S.; Koumanova, R.; Katsarov, N.; Gergova, G.; Mitov, I.; Nikolov, R.; Krastev, Z. Inhibition of *Helicobacter pylori* growth *in vitro* by Bulgarian propolis: preliminary report. *J. Med. Microbiol.,* **2003**, *52*(Pt 5), 417-419. [http://dx.doi.org/10.1099/jmm.0.04895-0] [PMID: 12721318]

[131] Aamer, A.A.; Abdul-Hafeez, M.M.; Sayed, S.M. Minimum Inhibitory and Bactericidal Concentrations (MIC and MBC) of Honey and Bee Propolis against Multi-Drug Resistant (MDR) Staphylococcus sp. isolated from bovine clinical Mastitis. *Altern. Integr. Med.,* **2014**, *3*, 4.

[132] Hannan, A.; Batool, A.; Qamar, M.U.; Khalid, F. Propolis as an antibacterial agent against clinical isolates of mdr-acinetobacter baumannii. *J. Ayub Med. Coll. Abbottabad,* **2015**, *27*(1), 216-219. [PMID: 26182780]

[133] Arslan, S.; Ozbilge, H.; Kaya, E.G.; Er, O. *In vitro* antimicrobial activity of propolis, BioPure MTAD, sodium hypochlorite, and chlorhexidine on E. faecalis and C. Albicans. *Saudi Med. J.,* **2011**, *90*, 479-483.
[PMID: 21556468]

[134] Sílící, S.; Koc, A.; Mistik, S. Comparsion of*in vitro* activities of antifungal drugs and propolis against yeasts isolated from patients with superficial mycoses. *Ann. Microbiol.,* **2007**, *57*(2), 269-272. [http://dx.doi.org/10.1007/BF03175218]

[135] Temiz, A.; Şener, A.; Tuylu, A.Ö.; Sorkun, K.; Salih, B. Antibacterial activity of bee propolis samples from different geographical regions of Turkey against two foodborne pathogens, Salmonella Enteritidis and Listeria monocytogenes. *Turk. J. Biol.,* **2011**, *35*(4), 503-511.

[136] Khosravi, A.R.; Shokri, H.; Nikaein, D.; Mansouri, P.; Erfanmanesh, A.; Chalangari, R.; Katalin, M. Yeasts as important agents of onychomycosis: *in vitro* activity of propolis against yeasts isolated from patients with nail infection. *J. Altern. Complement. Med.,* **2013**, *19*(1), 57-62. [http://dx.doi.org/10.1089/acm.2011.0722] [PMID: 22954465]

[137] Dota, K.F.D.; Consolaro, M.E.L.; Svidzinski, T.I.E.; Bruschi, M.L. Antifungal activity of Brazilian propolis microparticles against yeasts isolated from vulvovaginal candidiasis. *Evidence-Based Complement. Altern. Med.,* **2011**, *eCAM 2011*, 201953.

[138] Boisard, S.; Le Ray, A.-M.; Landreau, A.; Kempf, M.; Cassisa, V.; Flurin, C.; Richomme, P. Antifungal and Antibacterial Metabolites from a French Poplar Type Propolis. *Evidence-Based Complement. Altern. Med.,* **2015**, *2015*, 319240.

[139] Dias, L.G.; Pereira, A.P.; Estevinho, L.M. Comparative study of different Portuguese samples of propolis: pollinic, sensorial, physicochemical, microbiological characterization and antibacterial activity. *Food Chem. Toxicol.,* **2012**, *50*(12), 4246-4253. [http://dx.doi.org/10.1016/j.fct.2012.08.056] [PMID: 22981908]

[140] de Castro, P.A.; Bom, V.L.; Brown, N.A.; de Almeida, R.S.; Ramalho, L.N.; Savoldi, M.; Goldman, M.H.; Berretta, A.A.; Goldman, G.H. Identification of the cell targets important for propolis-induced cell death in Candida albicans. *Fungal Genet. Biol.,* **2013**, *60*, 74-86. [http://dx.doi.org/10.1016/j.fgb.2013.07.001] [PMID: 23856128]

[141] Henshaw, F.R.; Bolton, T.; Nube, V.; Hood, A.; Veldhoen, D.; Pfrunder, L.; McKew, G.L.; Macleod, C.; McLennan, S.V.; Twigg, S.M. Topical application of the bee hive protectant propolis is well tolerated and improves human diabetic foot ulcer healing in a prospective feasibility study. *J. Diabetes Complications,* **2014**, *28*(6), 850-857. [http://dx.doi.org/10.1016/j.jdiacomp.2014.07.012] [PMID: 25239451]

[142] Agüero, M.B.; Svetaz, L.; Baroni, V.; Lima, B.; Luna, L.; Zacchino, S.; Saavedra, P.; Wunderlin, D.;

Feresin, G.E.; Tapia, A. Urban propolis from San Juan province (Argentina): Ethnopharmacological uses and antifungal activity against Candida and dermatophytes. *Ind. Crops Prod.,* **2014**, *57*, 166-173. [http://dx.doi.org/10.1016/j.indcrop.2014.03.009]

[143] Miguel, M.G.; Nunes, S.; Cruz, C.; Duarte, J.; Antunes, M.D.; Cavaco, A.M.; Mendes, M.D.; Lima, A.S.; Pedro, L.G.; Barroso, J.G.; Figueiredo, A.C. Propolis volatiles characterisation from acaricide-treated and -untreated beehives maintained at Algarve (Portugal). *Nat. Prod. Res.,* **2013**, *27*(8), 743-749. [http://dx.doi.org/10.1080/14786419.2012.696261] [PMID: 22724515]

[144] Hegazi, A.G.; Abd El Hady, F.K.; Abd Allah, F.A.; Abd Allah, F.A. Chemical composition and antimicrobial activity of European propolis. *Z. Naturforsch., C, J. Biosci.,* **2000**, *55*(1-2), 70-75. [http://dx.doi.org/10.1515/znc-2000-1-214] [PMID: 10739103]

[145] Gardjeva, P.A.; Dimitrova, S.Z.; Kostadinov, I.D.; Murdjeva, M.A.; Peyche, L.P.; Lukanov, L.K.; Stanimirova, I.V.; Alexandrov, A.S. A study of chemical composition and antimicrobial activity of Bulgarian propolis. *Folia Med. (Plovdiv),* **2007**, *49*(3-4), 63-69. [PMID: 18504937]

[146] Boyanova, L.; Gergova, G.; Nikolov, R.; Derejian, S.; Lazarova, E.; Katsarov, N.; Mitov, I.; Krastev, Z. Activity of Bulgarian propolis against 94 Helicobacter pylori strains *in vitro* by agar-well diffusion, agar dilution and disc diffusion methods. *J. Med. Microbiol.,* **2005**, *54*(Pt 5), 481-483. [http://dx.doi.org/10.1099/jmm.0.45880-0] [PMID: 15824428]

[147] Buchta, V.; Černý, J.; Opletalová, V. *In vitro* antifungal activity of propolis samples of Czech and Slovak origin. *Cent. Eur. J. Biol.,* **2011**, *6*(2), 160-166.

[148] Kacániová, M.; Melich, M.; Knazovická, V.; Felsöciová, S.; Sudzinova, J The Antimicrobial Activity of Honey and Propolis Against Yeasts Candida Species *Zootehnie şi Biotehnologii,* **2009**, *42*(2), 167-173.

[149] Kacániová, M.; Vuković, N.; Chlebo, R.; Haščík, P.; Rovná, K.; Cubon, J.; Dzugan, M.; Pasternakiewicz, A. The antimicrobial activity of honey, bee pollen loads and beeswax from Slovakia. *Arch. Biol. Sci.,* **2012**, *6*(3), 927-934. [http://dx.doi.org/10.2298/ABS1203927K]

[150] Orsi, R.O.; Sforcin, J.M.; Funari, S.R.; Fernandes Junior, A.; Bankova, V. Synergistic effect of propolis and antibiotics on the Salmonella typhi. *Braz. J. Microbiol.,* **2006**, *37*(2), 108-112. [http://dx.doi.org/10.1590/S1517-83822006000200002]

[151] Wagh, V.D. Propolis: A wonder bees product and its pharmacological potentials. *Adv. Pharmacol. Sci.,* **2013**, *2013*, 308249. [http://dx.doi.org/10.1155/2013/308249]

[152] Bankova, V.; Christov, R.; Kujumgiev, A.; Marcucci, M.C.; Popov, S. Chemical composition and antibacterial activity of Brazilian propolis. *Zeitschrift fur Naturforsch. C-Journal Biosci.,* **1995**, *50*, 167-172.

[153] Popova, M.P.; Chinou, I.B.; Marekov, I.N.; Bankova, V.S. Terpenes with antimicrobial activity from Cretan propolis. *Phytochemistry,* **2009**, *70*(10), 1262-1271. [http://dx.doi.org/10.1016/j.phytochem.2009.07.025] [PMID: 19698962]

[154] Hashimoto, T.; Aga, H.; Tabuchi, A. *Shibuya T.*; *Chaen H.*; *Fukuda* S.; Kurimoto M. Anti-*Helicobacter pylori* compounds in Brazilian propolis. *Nat. Med.,* **1998**, *52*, 518-520.

[155] Ayala, G.; Escobedo-Hinojosa, W.I.; de la Cruz-Herrera, C.F.; Romero, I. Exploring alternative treatments for Helicobacter pylori infection. *World J. Gastroenterol.,* **2014**, *20*(6), 1450-1469. [http://dx.doi.org/10.3748/wjg.v20.i6.1450] [PMID: 24587621]

[156] Oliveira, A.P.; França, H.S.; Kuster, R.M.; Teixeira, L.A.; Rocha, L.M. Chemical composition and antibacterial activity of Brazilian propolis essential oil. *J. Venom. Anim. Toxins Incl. Trop. Dis.,* **2010**, *16*, 121-130. [http://dx.doi.org/10.1590/S1678-91992010005000007]

[157] Hames-Kocabas, E.E.; Demirci, B.; Uzel, A.; Demirci, F. Volatile composition of anatolian propolis by headspace-solid-phase microextraction (HS-SPME), antimicrobial activity against food contaminants and antioxidant activity. *J Med Plants Res,* **2013**, *7*, 2140-2149.

[158] Bankova, V.; Popova, M.; Trusheva, B. Propolis volatile compounds: chemical diversity and biological activity: a review. *Chem. Cent. J.,* **2014**, *8*(1), 28. [http://dx.doi.org/10.1186/1752-153X-8-28] [PMID: 24812573]

[159] Bankova, V.; Christov, R.; Popov, S.; Marcucci, M.C.; Tsvetkova, I.; Kujumgiev, A. Antibacterial activity of essential oils from Brazilian propolis. *Fitoterapia,* **1999**, *70*(2), 190-193. [http://dx.doi.org/10.1016/S0367-326X(98)00045-8]

[160] Takaisi-Kikuni, N.B.; Schilcher, H. Electron microscopic and microcalorimetric investigations of the possible mechanism of the antibacterial action of a defined propolis provenance. *Planta Med.,* **1994**, *60*(3), 222-227. [http://dx.doi.org/10.1055/s-2006-959463] [PMID: 8073087]

[161] Cushnie, T.P.; Lamb, A.J. Detection of galangin-induced cytoplasmic membrane damage in *Staphylococcus aureus* by measuring potassium loss. *J. Ethnopharmacol.,* **2005**, *101*(1-3), 243-248. [http://dx.doi.org/10.1016/j.jep.2005.04.014] [PMID: 15985350]

[162] Murad, J.M.; Calvi, S.A.; Soares, A.M.; Bankova, V.; Sforcin, J.M. Effects of propolis from Brazil and Bulgaria on fungicidal activity of macrophages against Paracoccidioides brasiliensis. *J. Ethnopharmacol.,* **2002**, *79*(3), 331-334. [http://dx.doi.org/10.1016/S0378-8741(01)00404-4] [PMID: 11849837]

[163] Ota, C.; Unterkircher, C.; Fantinato, V.; Shimizu, M.T. Antifungal activity of propolis on different species of Candida. *Mycoses,* **2001**, *44*(9-10), 375-378. [http://dx.doi.org/10.1046/j.1439-0507.2001.00671.x] [PMID: 11766101]

[164] Siqueira, A.B.; Gomes, B.S.; Cambuim, I.; Maia, R.; Abreu, S.; Souza-Motta, C.M.; de Queiroz, L.A.; Porto, A.L. Trichophyton species susceptibility to green and red propolis from Brazil. *Lett. Appl. Microbiol.,* **2009**, *48*(1), 90-96. [http://dx.doi.org/10.1111/j.1472-765X.2008.02494.x] [PMID: 19018958]

[165] Quintero-Mora, M.L.; Londoño-Orozco, A.; Hernández-Hernández, F.; Manzano-Gayosso, P.; López-Martínez, R.; Soto-Zárate, C.I.; Carrillo-Miranda, L.; Penieres-Carrillo, G.; García-Tovar, C.G.; Cruz-Sánchez, T.A. [Effect of Mexican propolis extracts from *Apis mellifera* on *Candida albicans in vitro* growth]. *Rev. Iberoam. Micol.,* **2008**, *25*(1), 22-26.

[http://dx.doi.org/10.1016/S1130-1406(08)70006-8] [PMID: 18338923]

[166] Sosa, S.; Bornancin, A.; Tubaro, A.; Loggia, R.D. Topical antiinflammatory activity of an innovative aqueous formulation of actichelated propolis vs two commercial propolis formulations. *Phytother. Res.,* **2007**, *21*(9), 823-826.
[http://dx.doi.org/10.1002/ptr.2206] [PMID: 17605155]

[167] Koc, A.N.; Silici, S.; Mutlu-Sariguzel, F.; Sagdic, O. Antifungal activity of propolis in four different fruit juices. *Food Technol. Biotechnol.,* **2007**, *45*(1), 57-61.

[168] Farnesi, A.P.; Aquino-Ferreira, R.; De Jong, D.; Bastos, J.K.; Soares, A.E. Effects of stingless bee and honey bee propolis on four species of bacteria. *Genet. Mol. Res.,* **2009**, *8*(2), 635-640.
[http://dx.doi.org/10.4238/vol8-2kerr023] [PMID: 19554760]

[169] Berretta, A.A.; de Castro, P.A.; Cavalheiro, A.H.; Fortes, V.S.; Bom, V.P.; Nascimento, A.P.; Marquele-Oliveira, F.; Pedrazzi, V.; Ramalho, L.N.Z.; Goldman, G.H. Evaluation of mucoadhesive gels with propolis (EPP-AF) in preclinical treatment of candidiasis vulvovaginal infection. *Evidence-Based Complement. Altern. Med.,* **2013**, *2013*, 641480.

[170] D'Auria, F.D.; Tecca, M.; Scazzocchio, F.; Renzini, V.; Strippoli, V. Effect of propolis on virulence factors of Candida albicans. *J. Chemother.,* **2003**, *15*(5), 454-460.
[PMID: 14598937]

[171] de Castro, P.A.; Bom, V.L.; Brown, N.A.; de Almeida, R.S.; Ramalho, L.N.; Savoldi, M.; Goldman, M.H.; Berretta, A.A.; Goldman, G.H. Identification of the cell targets important for propolis-induced cell death in Candida albicans. *Fungal Genet. Biol.,* **2013**, *60*, 74-86.
[http://dx.doi.org/10.1016/j.fgb.2013.07.001] [PMID: 23856128]

[172] Amoros, M.; Simões, C.M.; Girre, L.; Sauvager, F.; Cormier, M. Synergistic effect of flavones and flavonols against herpes simplex virus type 1 in cell culture. Comparison with the antiviral activity of propolis. *J. Nat. Prod.,* **1992**, *55*(12), 1732-1740.
[http://dx.doi.org/10.1021/np50090a003] [PMID: 1338212]

[173] Schnitzler, P.; Neuner, A.; Nolkemper, S.; Zundel, C.; Nowack, H.; Sensch, K.H.; Reichling, J. Antiviral activity and mode of action of propolis extracts and selected compounds. *Phytother. Res.,* **2010**, *24*(S1) Suppl. 1, S20-S28.
[http://dx.doi.org/10.1002/ptr.2868] [PMID: 19472427]

[174] Shimizu, T.; Hino, A.; Tsutsumi, A.; Park, Y.K.; Watanabe, W.; Kurokawa, M. Anti-influenza virus activity of propolis *in vitro* and its efficacy against influenza infection in mice. *Antivir. Chem. Chemother.,* **2008**, *19*(1), 7-13.
[http://dx.doi.org/10.1177/095632020801900102] [PMID: 18610553]

[175] Shimizu, T.; Takeshita, Y.; Takamori, Y.; Kai, H.; Sawamura, R.; Yoshida, H.; Watanabe, W.; Tsutsumi, A.; Park, Y.K.; Yasukawa, K. Efficacy of Brazilian propolis against herpes simplex virus type 1 infection in mice and their modes of antiherpetic efficacies. *Evidence-Based Complement. Altern. Med.,* **2011**, *eCAM 2011*, 976196.

[176] Nolkemper, S.; Reichling, J.; Sensch, K.H.; Schnitzler, P. Mechanism of herpes simplex virus type 2 suppression by propolis extracts. *Phytomedicine,* **2010**, *17*(2), 132-138.
[http://dx.doi.org/10.1016/j.phymed.2009.07.006] [PMID: 19682876]

[177] Takemura, T.; Urushisaki, T.; Fukuoka, M.; Hosokawa-Muto, J.; Hata, T.; Okuda, Y.; Hori, S.; Tazawa, S.; Araki, Y.; Kuwata, K. 3, 4-dicaffeoylquinic acid, a major constituent of Brazilian propolis, increases TRAIL expression and extends the lifetimes of mice infected with the influenza A virus. *Evidence-Based Complement. Altern. Med.,* **2012**, *eCAM2012*, 946867.

[178] Ito, J.; Chang, F-R.; Wang, H-K.; Park, Y.K.; Ikegaki, M.; Kilgore, N.; Lee, K-H. Anti-AIDS agents. 48.(1) Anti-HIV activity of moronic acid derivatives and the new melliferone-related triterpenoid isolated from Brazilian propolis. *J. Nat. Prod.,* **2001**, *64*(10), 1278-1281. [http://dx.doi.org/10.1021/np010211x] [PMID: 11678650]

[179] Tait, S.; Salvati, A.L.; Desideri, N.; Fiore, L. Antiviral activity of substituted homoisoflavonoids on enteroviruses. *Antiviral Res.,* **2006**, *72*(3), 252-255. [http://dx.doi.org/10.1016/j.antiviral.2006.07.003] [PMID: 16934879]

[180] Huleihel, M.; Isanu, V. Anti-herpes simplex virus effect of an aqueous extract of propolis. *Isr. Med. Assoc. J.,* **2002**, *4*(11) Suppl., 923-927. [PMID: 12455181]

[181] Nakayama, T.; Yamada, M.; Osawa, T.; Kawakishi, S. Suppression of active oxygen-induced cytotoxicity by flavonoids. *Biochem. Pharmacol.,* **1993**, *45*(1), 265-267. [http://dx.doi.org/10.1016/0006-2952(93)90402-I] [PMID: 8381006]

[182] Robertson, A. Social inequalities and the burden of food-related ill-health. *Public Health Nutr.,* **2001**, *4*(6A), 1371-1373. [http://dx.doi.org/10.1079/PHN2001221] [PMID: 11918483]

[183] Taubman, M.A.; Nash, D.A. The scientific and public-health imperative for a vaccine against dental caries. *Nat. Rev. Immunol.,* **2006**, *6*(7), 555-563. [http://dx.doi.org/10.1038/nri1857] [PMID: 16729013]

[184] Selwitz, R.H.; Ismail, A.I.; Pitts, N.B. Dental caries. *Lancet,* **2007**, *369*(9555), 51-59. [http://dx.doi.org/10.1016/S0140-6736(07)60031-2] [PMID: 17208642]

[185] Özan, F.; Sümer, Z.; Polat, Z.A.; Er, K.; Özan, U.; Deger, O. Effect of mouthrinse containing propolis on oral microorganisms and human gingival fibroblasts. *Eur. J. Dent.,* **2007**, *1*(4), 195-201. [PMID: 19212467]

[186] Duarte, S.; Rosalen, P.L.; Hayacibara, M.F.; Cury, J.A.; Bowen, W.H.; Marquis, R.E.; Rehder, V.L.; Sartoratto, A.; Ikegaki, M.; Koo, H. The influence of a novel propolis on mutans streptococci biofilms and caries development in rats. *Arch. Oral Biol.,* **2006**, *51*(1), 15-22. [http://dx.doi.org/10.1016/j.archoralbio.2005.06.002] [PMID: 16054589]

[187] Koo, H.; Rosalen, P.L.; Cury, J.A.; Park, Y.K.; Bowen, W.H. Effects of compounds found in propolis on *Streptococcus mutans* growth and on glucosyltransferase activity. *Antimicrob. Agents Chemother.,* **2002**, *46*(5), 1302-1309. [http://dx.doi.org/10.1128/AAC.46.5.1302-1309.2002] [PMID: 11959560]

[188] Basim, E.; Basim, H.; Ozcan, M. Antibacterial activities of turkish pollen and propolis extracts against plant bacterial pathogens. *J. Food Eng.,* **2006**, *77*, 992-996. [http://dx.doi.org/10.1016/j.jfoodeng.2005.08.027]

[189] Hanssen, M. *The healing power of pollen and other products from the beehive, propolis, royal jelly,*

honey. Wellingborough; Thorsons Publishers Ltd.: Hammersmith, London, **1979**.

[190] Carpes, S.T.; Mourão, G.B.; Alencar, S.M.; Masson, M.L. Chemical composition and free radical scavenging activity of *Apis mellifera* bee pollen from Southern Brazil. *Braz. J. Food. Technol.,* **2009**, *12*, 220-229.
[http://dx.doi.org/10.4260/BJFT2009800900016]

[191] Campos, M.G.; Markham, K.R.; Mitchell, K.A.; Cunha, A.P. An approach to the characterization of bee pollens *via* their flavonoid/phenolic profiles. *Phytochem. Anal.,* **1997**, *8*, 181-185.
[http://dx.doi.org/10.1002/(SICI)1099-1565(199707)8:4<181::AID-PCA359>3.0.CO;2-A]

[192] Campos, M.G.; Webby, R.F.; Markham, K.R.; Mitchell, K.A.; Da Cunha, A.P. Age-induced diminution of free radical scavenging capacity in bee pollens and the contribution of constituent flavonoids. *J. Agric. Food Chem.,* **2003**, *51*(3), 742-745.
[http://dx.doi.org/10.1021/jf0206466] [PMID: 12537451]

[193] Almaraz-Abarca, N.; Campos, M.G.; Ávila-Reyes, J.A.; Naranjo-Jiménez, N.; Herrera-Corral, J.; González-Valdez, L.S. Variability of antioxidant activity among honeybee-collected pollen of different botanical origin. *Interciencia-Caracas,* **2004**, *29*, 574-578.

[194] Graikou, K.; Kapeta, S.; Aligiannis, N.; Sotiroudis, G.; Chondrogianni, N.; Gonos, E.; Chinou, I. Chemical analysis of Greek pollen - Antioxidant, antimicrobial and proteasome activation properties. *Chem. Cent. J.,* **2011**, *5*(1), 33.
[http://dx.doi.org/10.1186/1752-153X-5-33] [PMID: 21699688]

[195] Özkalp, B.; Özcan, M.M. Antibacterial activity of pollen and propolis extracts. *J. Food Agric. Environ.,* **2010**, *8*(2), 17-19.

[196] Kačániová, M.; Vatľák, A.; Vukovic, N.; Petrová, J.; Brindza, J.; Nôžková, J.; Fatrcová-Šrámková, K. Antimicrobial Activity of Bee Collected Pollen against Clostridia. *Animal Sci Biotech,* **2014**, *47*(2), 362-365.

[197] Morais, M.; Moreira, L.; Feás, X.; Estevinho, L.M. Honeybee-collected pollen from five Portuguese Natural Parks: palynological origin, phenolic content, antioxidant properties and antimicrobial activity. *Food Chem. Toxicol.,* **2011**, *49*(5), 1096-1101.
[http://dx.doi.org/10.1016/j.fct.2011.01.020] [PMID: 21291944]

[198] Carpes, S.T.; Begnini, R.; De Alencar, S.M.; Masson, M.L. Study of preparations of bee pollen extracts, antioxidant and antibacterial activity. *Cienc. Agrotec.,* **2007**, *31*(6), 1818-1825.
[http://dx.doi.org/10.1590/S1413-70542007000600032]

[199] Abouda, Z.; Zerdani, I.; Kalalo, I.; Faid, M.; Ahami, M.T. The antibacterial activity of Moroccan bee bread and bee-pollen (fresh and dried) against pathogenic bacteria. *Res. J. Microb,* **2011**, *6*, 376-384.
[http://dx.doi.org/10.3923/jm.2011.376.384]

[200] Almeida-Muradian, L.B.; Pamplona, L.C.; Coimbra, S.; Barth, O.M. Chemical composition and botanical evaluation of dried bee pollen pellets. *J. Food Compos. Anal.,* **2005**, *18*, 105-111.
[http://dx.doi.org/10.1016/j.jfca.2003.10.008]

[201] Barbosa, S.I.; Silvestre, A.J.; Simoes, M.M.; Estevinho, M.L. Composition and Antibacterial Activity of the Lipophilic Fraction of Honeybee Pollen from Native Species of Montesinho Natural Park. *Inter. J. Agricul. Res.,* **2006**, *1*(5), 471-479.

[http://dx.doi.org/10.3923/ijar.2006.471.479]

[202] Koc, A.N.; Silici, S.; Ercal, B.D.; Kasap, F.; Hörmet-Oz, H.T.; Mavus-Buldu, H. Antifungal activity of Turkish honey against Candida spp. and Trichosporon spp: an *in vitro* evaluation. *Med. Mycol.,* **2009**, *47*(7), 707-712. [http://dx.doi.org/10.3109/13693780802572554] [PMID: 19888803]

[203] Özcan, M.; Ünve, A.; Ceylan, D.A.; Yetisir, R. Inhibitory effect of pollen and propolis. *extracts Food/Nahrung,* **2004**, *48*(3), 188-194.

[204] Fatrcová-Šramková, K.; Nôžková, J.; Kačániová, M.; Máriássyová, M.; Rovná, K.; Stričík, M. Antioxidant and antimicrobial properties of monofloral bee pollen. *J. Environ. Sci. Health B,* **2013**, *48*(2), 133-138. [http://dx.doi.org/10.1080/03601234.2013.727664] [PMID: 23305281]

[205] Hleba, L. Antimicrobial effect of bee collected pollen extract to Enterobacteriaceae genera after application of bee collected pollen in their feeding. *Sci. Pap. Anim. Sci. Biotechnol.,* **2013**, *46*(2), 108-113.

[206] Yang, X.; Guo, D.; Zhang, J.; Wu, M. Characterization and anti-tumor activity of pollen polysaccharide. *Int. Immunopharmacol.,* **2007**, *7*(3), 401-408. [http://dx.doi.org/10.1016/j.intimp.2006.11.001] [PMID: 17276899]

[207] Campos, M.; Cunha, A.; Markham, K. Detection of antimicrobials in bee products with activity against viridans streptococci. *Polyphenol communications 98,* **1998**.LilleXIXth international conference on polyphenols

[208] Tichy, J.; Novak, J. Inibition of Virulence of Pseudomonas auruginosa cultures, by flavonoids isolated from bee-pollen: possible structure-activity relantionships. *J. Altern. Complement. Med.,* **2000**, *6*(5), 383-389. [http://dx.doi.org/10.1089/acm.2000.6.383] [PMID: 11059500]

[209] Baltrušaitytė, V.; Venskutonis, P.R.; Čeksterytė, V. Antibacterial activity of honey and beebread of different origin against S. aureus and S. epidermidis. *Food Technol. Biotechnol.,* **2007**, *45*(2), 201-208.

[210] Cabrera, C.; Montenegro, G. Pathogen control using a natural Chilean bee pollen extract of known botanical origin. *Cien. Inv. Agr.,* **2013**, *40*(1), 223-230. [http://dx.doi.org/10.4067/S0718-16202013000100020]

[211] Khider, M.; Elbanna, K.; Mahmoud, A.; Owayss, A.A. Egyptian honeybee pollen as antimicrobial, antioxidant agents, and dietary food supplements. *Food Sci. Biotechnol.,* **2013**, *22*(5), 1-9. [http://dx.doi.org/10.1007/s10068-013-0238-y]

[212] Basuny, A.; Arafat, S.; Soliman, H. Chemical analysis of olive and palm pollen: Antioxidant and antimicrobial activation properties. *Wudpecker J. Food Technol.,* **2013**, *1*(2), 14-21.

[213] Eswaran, V.U.; Bhargava, R.H. Chemical Analysis and Anti-Microbial Activity of Karnataka Bee Bread of Apis species. *World Appl. Sci. J.,* **2014**, *32*(3), 379-385.

[214] Pascoal, A.; Rodrigues, S.; Teixeira, A.; Feás, X.; Estevinho, L.M. Biological activities of commercial bee pollens: antimicrobial, antimutagenic, antioxidant and anti-inflammatory. *Food Chem. Toxicol.,* **2014**, *63*, 233-239.

[http://dx.doi.org/10.1016/j.fct.2013.11.010] [PMID: 24262487]

[215] El-Asely, A.M.; Abbass, A.A.; Austin, B. Honey bee pollen improves growth, immunity and protection of Nile tilapia (Oreochromis niloticus) against infection with Aeromonas hydrophila. *Fish Shellfish Immunol.,* **2014**, *40*(2), 500-506.
[http://dx.doi.org/10.1016/j.fsi.2014.07.017] [PMID: 25086230]

CHAPTER 12

Add Value Products of Honeybee Plant-Derived Origin

Silvia C.F. Iop[1,*], Elsa C.D. Ramalhosa[2], Dalva M.R. Dotto[1], Andreia Cirolini[1], Naiane Beltrami[1]

[1] *Multidisciplinary Department - Federal University of Santa Maria, Rua Francisco Guerino, 407 – Silveira Martins – RS, Brasil*

[2] *Mountain Research Centre (CIMO) - School of Agriculture, Polytechnic Institute of Bragança, Campus de Stª Apolónia, Apartado 1172, 5301-855 Bragança, Portugal*

Abstract: Although honey extraction is an ancient practice, its use as a source of income is more recent. Over the years several techniques have been developed to achieve greater amounts of honey and also to ensure the quality of the product. Furthermore, bees during their life cycle produce wax, royal jelly, pollen and propolis besides honey. Pollen and propolis are such as honey, plant derivatives, while the wax and royal jelly are products secreted by glands of the bees. This chapter aims to present ways to add value to bee products plant-derived, namely: honey, pollen and propolis. Uses and new applications are presented. Moreover, the production of honey-fermented products such as mead, honey beer and honey vinegar are also discussed.

Keywords: Antibacterial properties, Antioxidant properties, Applications, Chemical composition, Cosmetics, Drink, Fermentation, Food industry, Health, Honey, Honey beer, Honey vinegar, Mead, Medicine, Natural preservatives, Pollen, Propolis, Quality, Uses, Wort.

* **Address Correspondence to Silvia Iop:** Multidisciplinary Department - Federal University of Santa Maria, Rua Francisco Guerino, 407 – Silveira Martins – RS, CEP 97 195 000, Brasil; Tel: +55-55-3224 4700; Email: silviaiop@ufsm.br.

Susana M. Cardoso (Ed)

1. INTRODUCTION

During their life cycle, bees produce honey, propolis, pollen, wax and royal jelly for their maintenance. Each of these products has an important role in the development and routine of the hive members. Since the beginning of beekeeping as an economic activity, research has been developed to understand the best way of management the bees' hives in order to obtain higher productivity.

Over the years many techniques have been developed to optimize the production of honey, pollen, wax and royal jelly that had contributed to these honey derivatives become a source of income for many rural families. Although honey production has already sufficient supply to put itself into a central position in agribusiness, more research is necessary in order to other products originating from beekeeping reach the market.

This chapter aims to show how to add value to products of honeybee plant-derived origin such as honey, propolis (bee glue) and pollen, as well as through the development of mead (honey fermented drink) and other honey fermented products.

2. ADDING VALUE TO PRODUCTS OF BEEKEEPING

2.1. Honey

Honey is the "natural sweet substance, produced by *Apis mellifera* bees from the nectar of plants or from secretions of living parts of plants, or excretions of plant-sucking insects on the living parts of plants, which the bees collect, transform by combining with specific substances of their own, deposit, dehydrate, store and leave in honeycombs to ripen and mature" [1]. After harvesting, honey does not necessarily need further treating. However, honey must always be handled hygienically and the cleaning status of all equipment is of great importance. Moreover, as honey is hygroscopic, all equipment and bottles used in its processing must be dry in order to reduce the chances of fermentation. Nevertheless, honey has a long shelf-life if harvested with care and stored in containers with appropriate lids.

Since ancient times honey has a high cultural value as a food. Besides that, this is also used for anointing in traditional birth, marriage and funeral ceremonies. This cultural connection is evident in the term "honeymoon". In general terms and, as cited by Krell [1], honey has been used in many applications such as:

- For bees as a food supply for those times when the flowers are scarce;
- As food for humans. Honey can be eaten in its unprocessed state such as liquid or crystallized forms or in the comb;
- Food ingredient as natural sweetener for drinks or in other dishes. Honey has been used in several food products, for example: *i*) honey liqueurs; *ii*) spreads with dried fruits or milk, like "Dulce de Leche", a popular Argentinean spread; *iii*) fruits and nuts (whole, chopped or pureed) in honey; *iv*) honey with pollen and propolis; *v*) fruit marmalade or jams where a portion of the sugar is replaced by honey; *vi*) honey jelly; *vii*) syrups; *viii*) rose honey, which is produced by mixing honey with red rose petals and boiling water; *ix*) caramels and candies; *x*) honey gums; *xi*) gingerbread; *xii*) marzipan; *xiii*) bakery products; *xiv*) non-alcoholic beverages like strengthening or replenishing isotonic drinks; and *xv*) sauces and some distilled alcoholic beverages. Some of these have honey as a flavoring agent such as Drambuie in Scotland, Benedictine in France, Irish Mist in Ireland, Krupnik in Poland, Grappa al Miele in Italy and Barenfang in Germany. Moreover, in the breakfast cereal industry, mixtures of honey (liquid or dried form) with cereal flakes and dried fruits such as roasted peanuts are used, improving oxidative stability [2];
- Medicine or tonic due to its medicinal properties. Indeed, hot milk, tea or other infusions with honey are a very usual home remedy to treat colds and throat infections. Recently, Raeessi *et al.* [3] found that "honey with coffee" is an effective treatment for post-infectious cough. As reviewed by Ediriweera and Premarathna [4], fresh bee's honey may be used in the treatment of several diseases such as bronchial asthma, throat infections, hiccups and ulcers, while old bee's honey is more used to treat diarrhea, vomiting, rheumatoid arthritis, diabetes mellitus and obesity. Nowadays, several products with honey in their formulation have been produced such as wound dressings [5 - 8] and gels for topical treatments for reepithelialization [9]. Furthermore, carbon nanoparticles from honey have given good results in real-time photoacoustic imaging [10] and

honey-based films are showing to be a potential product to be used in tissue engineering applications and regenerative medicine [11];

- In cosmetics. Earlier civilizations already used honey for skin treatment. Nowadays, honey is used in cosmetic formulations due to its several properties such as keeps the skin young and delays wrinkle formation, it is emollient, humectant, soothing and has hair conditioning effects [12]. It also regulates the pH and prevents pathogen infections [12]. Lip ointments, hydrating creams, cleansing milks, tonic lotions, after sun, shampoos and conditioners are some examples of honey-based cosmetic products [1, 12];
- Other uses: honey is used to improve aroma and humidity of tobacco. Moreover, honey is used to attract insects for pollination of some agricultural crops by mixing honey with other substances in solution [1] and as an indicator for environmental pollution such as heavy metals [13].

Table **1** resumes some applications of honey in medicine, cosmetics, food and environment areas.

Table 1. Several applications of honey.

Area	**Product**	**Reference**
Medicine Wound care dressings Gel, balms New materials Treatment uses	Honey-based wound dressings	[5 - 7]
	MEDIHONEY® gel	[8]
	L-Mesitran Soft® (gel that contains 40% honey, hypoallergenic lanolin, vitamin C, vitamin E and polyethylene glycol 4000)	[9]
	Sepropol balms (contain extracts of propolis and honey)	[14]
	Carbon nanoparticles from honey	[10]
	Nanofibers[1]	[15]
	Honey-blended silk fibroin films	[11]
	Honeydew honey (Treatment of non-healing leg ulcers)	[16]
	Pure Natural Honey (Treatment of radiation mucositis)	[17]
Cosmetics	Gel-based lipstick	[18]

(Table 1) contd.....

Area	Product	Reference
Food	Natural honey as a nutraceutical agent	[19]
	Portuguese "água-mel" and Italian "abbamele"	[20]
	Honey coating on roasted peanuts	[2]
	Gels from whey protein concentrate and honey (to be utilized in desserts formulation)	[21]
	Functional products, *e.g.* pralines, honey-sweetened cashew apple juice, biscuits, baked products, savory snacks	[22, 26]
Marker of environmental pollution	Heavy metals	[13]

[1] – Prepared from honey produced by *Apis dorsata*.

2.2. Propolis

Propolis is a substance produced by bees from vegetable materials actively secreted by plants or from its wounds, including lipophilic materials on leaves and leaf buds, resins, mucilage, gums, lattices [27], which are modified by secretions of worker bees.

The collection of propolis is indicated for areas with favorable vegetation without pollution by agrotoxic agents because, in the absence of vegetation bees seek other sources such as ink residue, asphalt and other products that can be toxic to humans. Thus, for production of propolis as a source of income, it is worth to notice that the quality of the final product is dependent on the source of resin available to the bees. The presence of plant species suppliers of good quality raw materials, like pine, Brazilian pine, bracatinga, peach and plum, among others with good density and diversity directly influence the type and quantity of propolis produced [28].

Propolis classification, according to its physicochemical and biological characteristics is a way to facilitate the grouping and comparison available data among them. Currently in Brazil there are 15 types of propolis, differentiated by their chemical composition, 12 described by Park *et al.* [29], the 13 type by Daugsch *et al.* [30] and the 14 and 15 types by Righi [31]. The propolis of northeastern Brazil (group 13) has similarity with propolis of Cuba, considering its composition in flavonoid compounds [32]. More information about propolis chemical compositions can be found in the publication of Huang *et al.* [33].

Data on the world production of propolis are scarce. According to Miguel and Antunes [34], 10 to 15% of the worldwide production is from Brazil, being this country the third world producer. In Brazil, Minas Gerais is the largest producer of propolis with 35 tons/year [35].

The chemical composition of propolis directs its applications and the presence of phenolic, aromatic compounds, flavonoids (*e.g.* flavones and flavanones) makes this beekeeping product an important ally to health. Since ancient times there are reports which describe the medical use of propolis in general and febrile infections by the Incas, in mummification rituals by the Egyptian priests and, use as healing agent by the Greeks. The largest reported contribution of propolis use was in the Boer War in South Africa, where the alcoholic extract was used in inflamed wounds and as a healing agent [36].

Even though propolis is known for some time ago, nowadays it has gained popularity in cosmetics, medicine and food industries, as described in Table **2**.

Table 2. Relevant propolis applications.

Area	Propolis application	Reference
Cosmetics	Suntan creams	[37]
	Topical ointments	[38, 39]
Medicine	Antibacterial mouth-dissolving fibers	[40]
	Mucoadhesive propolis gel	[41]
	Membranes for regenerative medicine	[42]
	Topical products	[43, 44]
	Brown and green propolis for leishmaniasis treatment	[45, 46]
	Red propolis for cell carcinoma treatment	[47]

(Table 2) contd.....

Area	Propolis application	Reference
Food	Plant oils as a natural antioxidant	[48]
	Butter as a natural antioxidant	[49]
	Fish paste as a natural antioxidant and antimicrobial additive	[50]
	Pork sausages, fresh oriental sausage, Italian-type salami, beef patties, sausages and ground pork as natural antioxidant and antimicrobial additive	[51 - 56]
	"a la Piedra" Turron as functional ingredient	[57]
	Drinks such as sugarcane spirit and fruit juices as natural biocide to control bacterial contamination	[58 - 60]
	Vegetables sanitization	[61]
	Post-harvest of fruits	[62, 63]
	Edible coatings for table grapes (antimicrobial properties) and chilli (control post-harvest anthracnose)	[64 - 66]
	Packaging films such as chitosan-propolis coated polypropylene films (antimicrobial activity)	[67 - 69]
	Microencapsulation of propolis extract by complex coacervation and encapsulation for production of chitosan-propolis beads as examples of natural antimicrobial delivery systems used to prevent the growth of pathogenic/spoilage bacteria in food	[70, 71]
Agriculture	Foliar application, soil drench and against nematode infection	[72]
	Animal performance	[73, 74]
Textiles	Cotton textiles	[75]

In cosmetics propolis has been added to (*i*) suntan creams to shield essential oils from UV stress [37] and (*ii*) in topical ointments used against herpes *labialis* such as the "Propolis Extract ACF®" [38] or in the prevention of psoriasis [39].

As mentioned before and described in previous chapters, propolis has great potential therapeutic use, especially due to its antioxidant, antibacterial, antiviral and anti-inflammatory properties, as well as immunomodulatory action, beyond other ameliorative effects. Due to its antioxidant capacity, propolis or ethanolic extracts of propolis (EEP) have been used as an agent in ameliorating the toxicity resulted from the use of several anticancer drugs that can induce toxicity and oxidative stress [76 - 78]. Propolis has also shown great antibacterial activity, being added to antibacterial mouth-dissolving fibers [40], mucoadhesive propolis gel to prevent radiation-induced oral mucositis [41], and in membranes from latex for regenerative medicine [42]. Moreover, commercially available propolis has shown interesting properties, such as that of prevention of transthyretin amyloidosis (*i.e.* a rare disease resulting from accumulation of inappropriately folded proteins) [43] previously reported for Bio30™ from New Zealand or that of corneal wound healing and anti-inflammatory effects [44], which has been reported for topical application of the Brazilian stingless bee propolis (*Scaptotrigona* sp.). Brown and green propolis have also been used in the treatment of leishmaniasis [45,46] while red propolis has shown significant modulatory effect on the development of chemically-induced squamous cell carcinoma [47].

In food science and technology, propolis has also shown great potential as natural antioxidant and antimicrobial additive, as well as functional ingredient and food [79]. Propolis extracts may also be applied after harvesting of vegetable products. Significant extension of the storage life of some fruits such as dragon fruit [63] and citrus fruits [62] has been achieved after treatments with EEP and propolis ethyl acetate extracts, respectively. Moreover, nowadays there is a growing interest in active and biodegradable packaging materials, having EEP great potential to be added to these products. EEP application as a coating significantly decreased the incidence of illness and helped to retain moisture, firmness, peel colour and soluble solids content of some fruits such as chilli [65, 66], all important factors in the maintenance of fruits' quality and shelf life extension. Furthermore, propolis may be used to develop anti-microbial packaging films [67]. EEP incorporation to gelatin-based films, for example, may promote reduction in rupture tension and water vapor permeability, as well as activity against *S. aureus* [69].

Beyond the positive results reported for vegetable origin products, potential applications for propolis in meat products have also been found (*e.g.* as a preservative used as an alternative for nitrites) [55]. For example, the addition of propolis and the application of heat to ground pork may have a synergistic effect, causing an increased thermal injury to *Escherichia coli* O157:H7 [56]. Moreover, the use of natural biocides such as brown and green propolis instead synthetic antibacterial compounds in fermentation processes as well as the addition of propolis extracts as natural antioxidant and antimicrobial additives, can be of great importance to extend the shelf life of food products.

In agriculture, interesting results have also been obtained for propolis as foliar application or soil drench, as well as against nematode infection. Aqueous, ethanolic and acetonic propolis extracts increased shoot height, root dry weight, the total of branches and pods per plant, number of seeds/pod and seed index, as well as the carbohydrate, protein and phenolic compounds contents in faba bean plants [72]. Furthermore, propolis extracts may also be infused in synthetic adsorbents, being the volatiles released able to inhibit molds and post-inoculation bacterial colonization on brown rice during storage [80]. Furthermore, propolis added in the diet of bulls may improve animal performance and carcass weight, namely, the final weight, feed efficiency, average daily gain and hot carcass weight [73]. Similar results were obtained with fish, as rapid muscular growth and reduced mortality of fish eggs were observed [74]. New applications for propolis are now being reported such as its use in cotton textiles in order to produce fabrics with superior antibacterial activity, water repellent, UV protection and ease of care characteristics [75]. Consequently, all studies show the great potential of propolis in several distinctive areas.

2.3. Pollen

Pollen comes from the Latin word "pollenins" which means "very fine powder" [81]. The pollen grains or just pollen are microscopic structures that represent the male gamete of plants which are necessary for the fertilization and seeds generation to perpetuate the plant species.

The pollen grains are formed by haploid cells with two nuclei: one vegetative that has as function to form the pollen tube and other reproductive that is necessary to fertilize the ovule. Pollen grains have a hard outer wall (exine) that is protected with a layer of wax. This wax makes the pollen very difficult to digest. Moreover, it is also the cause why pollen can become fossilized and remains undamaged for many millions of years [82].

Bee pollen is a product obtained from the assemblage of different pollen grains from various vegetable sources harvested by bees, which are mixed with nectar and secretions from glands in hipofaringeans and stored in pollen basket (specialized bags) located in the rear legs of worker bees [83, 84].

Bee pollen is the main solid food source for bees and it is essential for a normal growth and development of all individuals of a colony, as well as for their reproduction [85].

The size of pollen grains ranges from 5.5 to 15 mg. According to Negrão [86], smaller pollen grains (less than 2 mg) have higher protein content than the bigger grains, being this fact considered by these searchers as a bees' strategy to meet their needs. When Nature is plenty of pollen, bees collect between 15 and 55 kg of pollen annually [87].

The harvest of bee pollen is done through installation of pollen collectors at hive entrance. Pollen collectors are constituted by a barrier (grid or retention barrier) with perforations which allow passage of bees but are sufficiently narrow to retain the trapped pollen loads. Thus, the pollen loads fall into a collecting tray, separated by a screen in order to prevent the collection by the bees. In general, pollen collectors are made of wood, with acrylic sheet retention barriers, with 2.0 to 3.0 mm thick, perforated with round holes 4.3 to 4.5 mm in diameter. There are three types of collectors: frontal, background and top. The difference is that in frontal collector the pollen loads are captured before the beehive entrance, staying outside, and this fact implies the need for daily harvest. For the other types of collectors (*i.e.* background and top collectors), pollen is protected and its harvest can be done every two days [88]. When the harvest is done in a daily basis, it is recommended to be done in the afternoon and overall, the collecting should not exceed two weeks, to avoid weakening of the hive. The collected material is presented as grain acorns with varying color, indicating the various botanical communities collected by the bees to form a mixture known as "mix" pollen [89].

After harvest, pollen grains should be cleaned, preferably by techniques using purified air in order to avoid contamination. The next step is to reduce the moisture content. The drying of the pollen can be made in an oven with temperature not exceeding 40°C, till 6% or less moisture. The obtained pollen can be stored for up to 15 months. Pollen grains can also be preserved by techniques such as freezing, freezing followed by storage in liquid nitrogen and freeze-drying [90].

In Brazil, the pollen is classified in two ways: 1) bee pollen, which is the product collected in its original form which should not exceed 30% moisture, and 2) dehydrated pollen *i.e.*, the product submitted to the dehydration process in temperatures exceeding 42°C and moisture content maximum of 4% [91].

The Brazilian production of pollen began modestly in the late 80s, however, it has grown due to consumers' demand for natural products [92].

Concerning applications, as pollen is the main protein and lipid source for honey bees [93], sometimes beekeepers use pure pollen, pollen supplements or pollen substitutes to feed their colonies when natural pollen sources are limited [1]. Pollen has also excelled in the scientific field due to its bioactive properties such as antibacterial, antifungal, anti-inflammatory, immunomodulator and anti-cariogenic activities, exerting antioxidant functions and inhibiting the destructive action of free radicals [94, 95].

Related to that, bee pollen has also been used in the treatment of diseases such as colds, flu, ulcers, aging, anemia, colitis, allergic reactions, enteritis, intra-abdominal postoperative adhesion, and reduction in cholesterol and blood lipids [96]. In more detail, pollen extracts have been injected subcutaneously in order to desensitize allergic patients [1] and bee pollen has been used in medicine in order to correct atherogenic dyslipidemia [97] that comprises a triad of increased blood of low-density lipoprotein (LDL) particles, decreased high-density lipoprotein (HDL) particles, and increased triglycerides [98]. Kasianenko *et al.* [97] had stated a significant hypolipidemic effect in patients taking honey in combination with pollen and bee bread that is a pollen stored by honeybees that undergoes a lactic acid fermentation, showing a characteristic sour taste and it is more digestible and enriched with new nutrients than pollen [1]. Moreover, bee pollen ethanol extract also showed anti-inflammatory properties, inhibiting strongly paw edema, *via* the inhibition of NO production and COX-2 activity (an enzyme responsible for inflammation) [99]. Bee pollen has also the ability to alleviate the negative effects caused by the carbamate pesticide, carbaryl, in what regards to several oxidative stress markers and serum biochemical parameters [100]. Carbaryl has been used in agriculture and house pest control, as well as against ectoparasites of animals, but it is mildly toxic for humans and domestic animals, and moderately toxic for fish [100], causing damage in theirs biological systems.

Regarding pollen addition to fish, pollen can be used in tilapia diet because its use improved the growth rate and increases non-specific immunity that also is beneficial in controlling infections by *Aeromonas hidrofila* [101]. Interesting results have also been obtained on the development of digestive organs in broiler chickens fed with a basic diet supplemented with pollen [102]. These results suggest that bee pollen can be used as a food supplement for certain health conditions such as short bowel syndrome [102]. So, these works show that bee pollen can be used as a food additive, administered for its alleviating or curative effect, or for medical protection against toxic compounds.

Few works have been published on pollen addition to food; however this product has been added to candy bars [1]. Furthermore, Turhan *et al.* [103] reported recently that the addition of pollen to meatballs enhanced their nutritional and storage quality through prevention of lipid oxidation and inhibiting the bacterial growth with minimal changes in composition and/or sensory properties, showing the possibility of new applications of bee pollen to food industry.

Pollen can be associated with other bee products such as honey, propolis and royal jelly, but these combinations should be described in their labeling, as stated in service DIPA nº002/04 [104]. Another way to commercialize pollen is in the form of pollen extract, a cream obtained from pollen granules without allergenic elements according to Silva [105], with compounds such as vitamin E, fatty acids, flavonoids, steroids and hormones.

Nevertheless, pollen inclusion in some cosmetic preparations should be considered with some care as there is some allergy risk for a high percentage of the population [1]. Nevertheless several pollen extracts may be prepared by changing the solvents used. Several types of alcohols are the most common solvents for extraction [1]; however, by varying the alcohol concentration, different extracts can be obtained. Krell [1] refers that a propylene glycol extract contains most water soluble material, being proteins left behind. In this way all or almost all allergenic material is eliminated, being such as extract suitable for external applications such as in cosmetics.

Bee pollen has also been used for mechanical or hand pollination, as pollen transfer efficiency is a factor that influences the number of flowers [106]. Some studies on pollination have been performed in food and sexually deceptive orchids [106], sorghum [107], watermelon flowers [108] and *Epimedium* species [109], beyond other works.

Honeybee products and in particular bee pollen may be useful in monitoring environmental contamination by metals [110] and inorganic anions such as fluoride [111]. Abnormalities during pollen germination [112] and development [111], as well as on shape and size of pollen grains [111 - 114], may occur in the presence of pollutants. Pukacki and Chalupka [112] stated that pollen from a contaminated area presented less total phospolipids, lower contents of soluble proteins and low molecular antioxidants (*e.g.* ascorbic acid and thiols). Nevertheless air pollution might induces flavonoids accumulation in pollen collected on polluted areas [113, 115], suggesting that plants try to adjust their metabolism to air pollutants in order to decrease damages [113].

2.4. Mead and Other Fermented Honey-Based Beverages

Fermentation is an easy, inexpensive and efficient way of preserving perishable raw materials [116]. Fermented foods are known since antiquity where findings from 6000 B.C. have denoted that ancient people fermented milk, meat, vegetables and honey [117 - 119]. Fermentation, from the Latin language signs "boil" and was defined by Pasteur as "life without air". It is a metabolic process of deriving energy from organic compounds without the involvement of an oxygenating exogenous agent [120]. This process involves the action of microorganisms on organic food components [121, 122].

During the fermentation process, carbohydrates are used by microorganisms for producing different types of metabolites and due to their characteristics such as volatility, acidity and capacity of water absorption, among others, they can contribute to the formation of a new product. Among the advantages of the fermentation process for the preparation of food, one can highlight an increase of nutritional quality (increase in the digestibility of some foods) and a reduced toxicity way of generating functional compounds (vitamins and antioxidants)

[117].

Several fermented products such as mead, beer and vinegar can be produced from honey. Following, a more detail description of these products can be found.

2.4.1. Mead

Mead is a beverage produced from ancient times, according to historical findings, since 7000 B.C. [123] and is considered as one of the oldest beverages in the world.

Mead contains 8 to 18% ethanol (v/v) which is obtained by the fermentative process of a honey solution by yeast [124]. It receives different designations such as *Pyment*, *Cyser*, *Melomel*, *Metheglin*, *Spiced pyment*, *Sack mead*, *Sack metheglin*, *Mulsum*, *Hypocras*, *Brochet* or *Bochet*, *Ogol*, *Tej* [125–127], depending on the ingredients used in their formulation.

The quality of mead is directly related to the sensory properties of honey [128]. Bahiru *et al.* [129] emphasized that the microbiota, physical and chemical environment and duration of fermentation and preparation processes result in variations in the physicochemical properties of the final product. These authors analyzed the composition of 200 samples of '*Tej*', a drink obtained from an indigenous Ethiopian honey, and found mean values of total carbohydrates from 1.49 to 3.73 mg/mL, total lipids less than 1.00 mg/mL, total protein between 0.33 to 4.66 mg/mL and reducing sugars values from 0.46 to 2.09 mg/mL.

Concerning the Brazilian law, the Ordinance 64/2008 of the Ministry of Agriculture, Livestock and Supply defines mead as a beverage that contains alcohol from 4-14% (v/v) at 20°C and total acidity between 50 to 130 mEq/L; fixed acidity at least equal to 30 mEq/L; volatile acidity, expressed as acetic acid can be up to 20 mEq/L and reduced dry extract as a minimum value of 7.0 g/L [130].

The nutritional value of mead can be affected by the formation of hydroxy-methylfurfural (HMF). This compound is a cyclic aldehyde and it is formed by the degradation of sugars and its formation causes reduction in the nutritional value of the product. The presence of high concentrations of hydroxymethylfurfural and the absence of phenolic compounds are the indicators of higher heat treatment during mead's production [131]. Wintersteen *et al.* [132] demonstrated that the antioxidant capacity of honey to provide protection against diseases associated with oxidative stress can be extended to mead. However, the study revealed that a drastic heat treatment altered the antioxidant capacity throught phenolic profile changes. Gupta and Sharma [133] also emphasize that mead is a nutritious alcoholic beverage, with an effect on digestion and metabolism and that it is recommended for individuals with anemia and chronic diseases of the gastro-intestinal tract.

2.4.1.1. Commercialization and Marketing

Even though some fermented honey-based products exist (*e.g.* mead, honey vinegar and honey beer), mead is probably the most known and consumed in several places of the world including Latin America, Argentina and Bolivia. Although Brazil had already provided significant production with several commercially available brands, mead is not yet a popular drink in Brazil. The low popularity of mead may be explained by the lack of knowledge about this product and also due to a low number of investigations on its technological processing [134]. According to Rivaldi *et al.* [135], the mead market consists of consumers of beverages and organic foods. The price of meads packaged in 750 mL fluctuates between U$ 10.9 and U$ 20.0, and products that are considered to be premium have prices reaching about U$ 70.0.

Other factors driving the consumption of mead are the interest in their therapeutic and nutritional properties and an increased demand for gourmet products [132, 136].

2.4.1.2. Production

Meads can be classified into dry, sweet and frothy, according to manufacturing technology. To obtain dry mead it is necessary to let the honey solution to ferment longer than in sweet mead. In dry mead the consumption of sugars is higher originating a less sweet drink and with a higher alcohol content. On contrary, in sweet mead a short fermentation is performed, being obtained a more sweet drink. Production of mead also depends on honey variety and yeast species used in the fermentation of the wort and of pH [137].

2.4.1.2.1. Ingredients

The quality of the raw material is the principal factor to guarantee the quality of the final product. It is impossible to have a product with high quality using an inadequate raw material. Mead, being a product of honey fermentation, encompasses the characteristics of honey that originated it. Therefore, the color, aroma and honey chemical composition has influence on the sensory characteristics of the mead obtained.

The taste of honey is given by the combination of its components, including sugar, acids, ash and volatile compounds. Studies on the aromatic quality of honey and their characterization are important to assist in the selection of honey for the manufacturing of mead. Strong *et al.* [138] described sensory characteristics of different commercial honeys that can be used to facilitate the choice of raw material for mead production (tasty characteristics as sweet, caramelized, pungent, waxy, creamy and buttery; and related to origin as citrusy, green leaves, woody and spicy).

The choice of yeast species is also a crucial issue when the art of fermentation is used as a lucrative activity, since it allows standardizing the sensory characteristics of the final product. For commercial production of mead, selected yeast species from bread making, wine and champagne, as well as wild yeasts (found in honey itself), are commonly used. According to Gupta and Sharma [133], meads made from wine yeast exhibited faster fermentations, higher alcohol contents and lower amounts of residual sugar than those produced with mead yeast at 20 or 30ºC.

Some yeast species do not produce more than 4% alcohol, while with others such as *Saccharomyces bayanus,* alcoholic contents up to 15% may be reached, yielding alcoholic beverages with very dry characteristics. Some of them also exhibit low attenuation alcohol, *i.e.*, low ratio of conversion of sugar into alcohol, which results in an increased amount of residual sugar, and thus a more sweet drink [139].

Juice or concentrated fruit, spices, herbs and grains can also be used as ingredients in meads. The use of fruit juice originates meads called *Melomels,* with variable characteristics. *Melomels* have distinct sensory characteristics and its standardization requires more efforts. Besides the common factors, the fruit ripeness, its variety and climatic conditions before harvest, must also be taken into account. In addition, as color, aroma and flavor of the fruits are transferred to mead, the producer has to work hard to guarantee a harmonious sensory quality, not failing to display the characteristics of honey and fruit from which the mead was originated [125]. Strong *et al.* [138] describe sensory characteristics of different fruits used in the preparation of meads that can help in the characterization of the product. Tart characteristics can be obtained using apple, black raspberry, cranberry, elderberry, key lime, lemon, lime, pineapple, cherry, passion fruit, raspberry and redcurrant. Passion fruit can be contribute with musky aroma and flavor, for example.

Spices and herbs when added to mead allow obtaining different products (*Metheglin, Hypocras*, among others). The herbs and spices can be added directly to the final product in an extract or in the raw form.

Meads can be prepared with the addition of unmalted grain, being barley the most common one and the resulting mead named as *Braggot*. Obviously, *Braggot* beverage must balance the characteristics of honey and grain. Hop can also be added, favoring tannins precipitation and the clarification of the final product [126].

Besides the above ingredients, other nutritional supplements are commonly used in the production of mead, including acids, sulfites, stabilizers and buffering agents. Indeed, small quantities of acid (such as malic, tartaric and citric) allow balancing the taste of mead. Some experts recommend an acid mixture consisting of citric acid (25%), malic acid (30%) and tartaric acid (45%) [125]. Sodium sulfite or potassium metabisulfite are commonly used in wines. The use of 25 to 50 mg/L of sulfur or 50-100 mg/L of potassium metabisulfite is recommended in mead manufacture [133]. Potassium sorbate or a wine stabilizer may be added in the bottling stage of mead in order to inhibit the occurrence of a secondary fermentation, thereby eliminating the remaining yeast cells. On the other hand, nitrogen deficiency can be compensated by adding 250 mg/L of diammonium phosphate and 250 mg/L of potassium bitartrate to the wort [133]. In addition, calcium and potassium carbonate are used as buffering agents, maintaining the pH suitable for the fermentation process; however, one should maintain the concentration used in order to avoid the appearance of bitter/salty taste [140].

2.4.1.2.2. Processing

The composition of the final product depends on the honey:water proportion used in the elaboration process that can vary between 1:0.5 and 1:3. Several variations of this drink can be obtained by adding fruits and juices, bitter herbs, grains, teas and spices, a practice attributed to the Romans. The honey floral source, the type of yeast used for fermentation and the use of additives and nutritional supplements during fermentation process also contribute in obtaining different products [1, 125, 133, 141, 142].

Wort is a term used to refer to a honey solution to be fermented. Its composition in fermentable sugars is a very important factor for the characteristics of the final fermented product. Sugars are very important in mead because the amount of residual sugar contributes to the body and sweetness of the final product. As stated before, the wort used to prepare mead can be a solution of honey in water in different proportions (traditional mead) or can result of adding honey to fruit juices or juice concentrate solutions prepared in water (*Melomels*).

When meads are prepared from wild yeast, the wort to be fermented does not undergo any specific treatment. In turn, meads obtained by selected yeast are often subjected to control treatments of heat or ultra-filtration, in order to prevent the action of competing microorganisms and also facilitate the production of mead. Meads obtained from heat-treated wort tend to have lower acidity [133]. The heat treatment of wort also precipitates proteins and other turbidity compounds; however, depending on the relationship between time/temperature used in the process, undesirable flavors can be produced [132]. A milder heat treatment (avoiding boiling) is a viable and attractive alternative since it facilitates the removal of coagulated proteins without any interference in the taste of the final product. The use of a temperature between 71-76 ºC for 15 minutes followed by cooling to 21°C is a good option [140]. On the other hand, ultra-filtration is a non-thermal process that can be used in the wort preparation stage and produces mead with desirable characteristics such as lightness, clarity, pleasant taste and no aftertaste.

Sulfitation, *i.e.* the use of sodium metabisulfite, is a common practice in wine-making that can also be used in mead-making. The addition of sulphurous anhydride or sodium metabisulfite has the advantage to prevent growth of lactic bacteria [133]. However, some people are allergic to metabisulfite and its use in mead making should be declared. Mead that has been sulfited needs to be clarified.

Moreover, the addition of nutrients and facilitators to the wort in order to increase yeast's activity is a common practice. Supplementation with diammonium phosphate, magnesium sulfate, folic acid, niacin, thiamine, and sodium pantothenate provide a better yield to the process [140]. Gupta and Sharma [133] reported a successful reduction of the fermentation time of mead wort by using nutritional supplementation for yeasts. The addition of potassium tartrate and ammonium phosphate at 0.04% results in a mead beverage in 6 weeks. Supplementation with ammonium chloride, potassium bicarbonate and sodium phosphate in concentrations from 0.04 to 0.08% allows the production of mead in 4 weeks. The addition of 0.2% cream of tartar, 0.1% ammonium phosphate, 0.5% citric acid, 0.025% magnesium chloride, and 0.025% calcium chloride resulted in mead with 12-13% alcohol in 25 days [133].

Pollen supplementation also influences the production process and the final characteristics of mead. When comparing sweet meads obtained from different honeys (wild, "angico" flowers and honeydew) using yeast for bread and wine making in a ratio of 1:4 (honey:water), Kempka and Mantovani [143] observed that the alcohol content at the end of fermentation differed between processes. Moreover, in the wort of "angico" mead supplemented with 1% pollen, a reduction on the fermentation time to 72 h was obtained with a final product with 10% alcohol. This alcoholic content was greater than those obtained for the "angico", wild and honeydew meads equal to 8.0, 9.0 and 6.5%, respectively, where the total fermentation time was 168 h. Roldán *et al.* [144] observed that supplementation with pollen improved the fermentation rate, alcohol yields and the characteristics of the meads obtained, regarding sensory profile as well as by increasing their volatile contents. The fermentation process and the choice of honey to be fermented are equally important factors that affect the quality of the final product. The fermentation planning includes: the choice of mead to be produced; use or not of nutrients; choice of the fermentation temperature and the safeguarding method of the end point of the process [133]. Note that along the fermentation process several problems may occur because yeast cells are subjected to several stressful conditions such as high osmotic pressure due to the high sugar content of the wort, ethanol production and the presence of antimicrobial compounds such as sulfur dioxide [145]. *E.g.* succinic acid production, depending on the yeast species and nitrogenous compounds, can induce a sudden drop in pH in the initial hours of the process, causing the stop of the fermentation [146].

After fermentation mead matures at temperatures between 2-4ºC. After maturation, the mead is left for rest up to two weeks [133].

2.4.2. Other fermented Honey-Based Beverages

Beer is an alcoholic beverage made by brewing from malted cereal grains such as barley and wheat, and flavored with hops; however, other ingredients may be added. It is supposed that the early Anglo-Saxons already drank beer prepared from a brew of water and honeycomb made in a clay pot [147]. To improve flavor, herbs were also used.

Ginger, cinnamon, cloves, orange peel and other herbs may be added to the herb beers, while molasses, brewer's caramel or chocolate among other products are added to specialty beers [147]. Honey may be added to herb beers and specialty beers. At present, several countries already produce and commercialize honey beers. *E.g.* in England the Fuller's sell the Organic Honey Dew; in Argentina the Buller brand sells a honey beer, while the Brazilian "Cervejaria Colorado" produces "Appia" and in Canada several honey beers can be found such as the Boréale Dorée, Granville Island Cypress Honey Lager and U Miel Pilsner.

Depending on the type of honey used, different end products may be obtained, varying in color, aroma, rounding effect and flavor [148]. The National Honey Board [148] refers that brewers tend to prefer mild honeys such as clover honey in lagers, while in porters, stouts and herb or spice beers honeys of other floral sources (*e.g.* sage or citrus) are more used. Nevertheless the addition of honey to beer must be performed with caution because wild yeasts and bacteria that are present in the honey might grow and proliferate when honey is diluted in water or wort [147]. Moreover, when honey is added to beer further problems for brewing may occur due to the addition to the wort of diastatic enzymes (α- and β-amylase) that are present in honey. These enzymes may degrade dextrins into fermentable sugars in an undesirable extension, causing loss of beer´s body and texture [147]. So, pasteurization of honey is recommended, although honey must not be exposed to high temperatures for a long time [148]. Another point to be considered is the possibility of increasing the alcohol content of the end product by increasing the proportion of fermentable sugars. Furthermore, the stage of the brewing process at which the honey is added, the type of beer that it is intended to obtain, and the type and quantity of honey used are important factors that will affect the strength of the honey flavor in the final product [148].

There are only a few scientific studies focusing honey beers. However, Brunelli *et al.* [149] produced recently several honey beers from worts with three concentrations of the original extract (11, 13 and 15 ºBrix) and three honey concentrations (0, 20 and 40%). In their study, the authors verified that honey addition enhanced carbonation, foam density and total foam, while the obtained beers were less bitter and less acid. These results indicate that this topic needs to be more studied in the future.

Another fermented honey-based beverage is honey vinegar. Vinegar is a condiment produced by the successive alcoholic and acetic fermentations of sugar solutions by certain yeasts and bacteria and it may also be produced from honey. In general terms this must be firstly fermented by yeast to produce ethanol and afterwards an acetic fermentation must be performed.

As explained by Fabian [150], to produce honey vinegar, it is first necessary to dilute the honey to a sugar concentration compatible with yeasts activity. At the same time some minerals such as potassium tartrate, ammonium phosphate, ammonium chloride, potassium bicarbonate or sodium phosphate should be added. Then, the solution must be heated in order to give a better color to the vinegar and kill all, or at least almost all the microorganisms since these could spoil the final product. The next step is to inoculate the honey solution with a starter obtained from a pure culture of yeast. Afterwards, a second fermentation by bacterium such as *Acetobacter aceti*, transforms the produced alcohol into acetic acid (acetic fermentation).

Few scientific studies have been performed in honey vinegar; however Ilha *et al.* [151] used bee (*Apis mellifera*) honey for vinegar production and they were able to obtain honey vinegar (5 L) with 9% acetic acid (w/v) and about 1% alcohol (v/v) from 1 kg of honey, showing this vinegar good consumer's acceptability. These results indicate that honey vinegar may have potential to be successfully marketed.

Other important factors that have to be considered in vinegar making enclose:

i. Temperature control. Indeed Fabian [150] already recommended a temperature between 18 and 24 ºC, since yeasts do not grow well in higher temperatures and alcohol evaporation occurs. In opposition, if temperature is too low, the yeast's grow is very slow and off-fermentations and vinegar disease may occur;
ii. Air. It is necessary to leave a considerable headspace above the liquid level in order to store the fermentation gas produced and because air is absolutely necessary for the acetic fermentation;
iii. Fining or clearing the vinegar. Gelatin and isinglass are examples of products that may be used;
iv. Pasteurization and storage. After the vinegar has been made, a pasteurization must be performed to prevent further fermentation. The pasteurized vinegar should be stored in air-tight bottles or in completely filled tightly bunged barrels.

Few studies have been performed on the impact of honey vinegar ingestion on health. Nevertheless Derakhshandeh□Rishehri *et al.* [152], when evaluating the intake of honey vinegar syrup (a mixture of these two ingredients) on glycemic parameters and lipid profiles in healthy individuals, found that honey vinegar syrup ingestion increased fasting insulin level and decreased total cholesterol; however, an unfavorable effect on high-density lipoprotein cholesterol was observed. Thus, further studies are needed in this subject.

CONCLUSION

Beekeeping activity is important not only because it helps in the maintenance of plant species since bees are the main pollinators for most plants but also because it serves as a source of income for small farmers. Furthermore, several products are being produced from honeybee plant-derived origin such as honey, pollen and propolis with added value, increasing farmers' incomes. Moreover due to their healthy chemical and biological properties, these products may be used in several applications such as food, cosmetics, medicine, as natural additives, beyond others. Furthermore, nowadays consumers look for healthy and/or "gourmet" foods, a field in which honey, pollen and propolis have great potential to be used as raw materials. In addition, the production of fermented honey-based beverages such as mead, honey beer and honey vinegar, is also a promising activity. Nevertheless more studies related to both standardization of the production process and on the understanding of the main factors that control and are important to guarantee the quality assurance of the product, whether chemical or sensory properties, are needed in order to increase their global production and popularization.

CONFLICT OF INTEREST

The authors confirm that they have no conflict of interest to declare for this publication.

ACKNOWLEDGEMENTS

Silvia Cristina Ferreira Iop gratefully acknowledge the Research Incentive Fund (FIPE) of Federal University of Santa Maria by funding the "Mead and sustainability in beekeeping" (Project Research – 031559) and the Professors Luis G. Dias and Susana M. Carvalho by invitation to participate in this work and the trust placed to us with respect to the contribution in this chapter to the work that tries to lighten the latest advances in beekeeping area Elsa C.D. Ramalhosa gratefully acknowledge the Portuguese Foundation for Science and Technology (FCT) for funding CIMO by the Project PEst-OE/AGR/UI0690/2014.

REFERENCES

[1] Krell, R. *Value-added products from beekeeping. FAO Agricultural Services Bulletin No. 124*; FAO, **1996**.

[2] Wambura, P.; Tegete, H.; Verghese, M. Application of high-Power ultrasound to improve adhesion of honey on roasted peanuts to improve oxidative stability. *Food Bioprocess Technol.,* **2012**, *5*, 2012-2016.
[http://dx.doi.org/10.1007/s11947-010-0467-5]

[3] Raeessi, M.A.; Aslani, J.; Raeessi, N.; Gharaie, H.; Zarchi, A.A.; Raeessi, F.; Ahmadi, M. "Persistent post-infectious cough" is better treated by which one? Prednisolone, Honey, Coffee, or Honey plus coffee: A meta-analysis. *Indian J. Traditional Knowledge,* **2014**, *13*, 453-460.

[4] Ediriweera, E.R.H.S.S.; Premarathna, N.Y.S. Medicinal and cosmetic uses of Bee's Honey - A review. *Ayu,* **2012**, *33*, 178-182.
[http://dx.doi.org/10.4103/0974-8520.105233]

[5] Fox, C. Honey as a dressing for chronic wounds in adults. *Br. J. Community Nurs.,* **2002**, *7*, 530-534.
[http://dx.doi.org/10.12968/bjcn.2002.7.10.10667]

[6] Molan, P.C.; Betts, J.A. Clinical usage of honey as a wound dressing: an update. *J. Wound Care,* **2004**, *13*, 953-956.
[http://dx.doi.org/10.12968/jowc.2004.13.9.26708]

[7] Pieper, B. Honey-based dressings and wound care: an option for care in the United States. *J. Wound Ostomy Continence Nurs.,* **2009**, *36*, 60-66.
[http://dx.doi.org/10.1097/01.WON.0000345177.58740.7d]

[8] Tirado, D.J.; Hudson, N.R.; Maldonado, C.J. Efficacy of medical grade honey against multidrug-resistant organisms of operational significance: part I. *J. Trauma Acute Care Surg.,* **2014**, *77*, S204-S207.
[http://dx.doi.org/10.1097/TA.0000000000000324]

[9] Boekema, B.K.H.L.; Pool, L.; Ulrich, M.M.W. The effect of a honey based gel and silver sulphadiazine on bacterial infections of *in vitro* burn wounds. *Burns,* **2013**, *39*, 754-759.
[http://dx.doi.org/10.1016/j.burns.2012.09.008]

[10] Wu, L.; Cai, S.; Nelson, K.; Xing, W.; Xia, J.; Zhang, R.; Stacy, A.; Luderer, M.; Lanza, G.M.; Wang, L.V. A green synthesis of carbon nanoparticles from honey and their use in real-time photoacoustic imaging. *Nano Res.,* **2013**, *6*, 312-325.
[http://dx.doi.org/10.1007/s12274-013-0308-8]

[11] Rajput, M.; Bhandaru, N.; Barui, A.; Chaudhary, A.; Paul, R.R.; Mukherjee, R.; Chatterjee, J. Nano-patterned honey incorporated silk fibroin membranes for improving cellular compatibility. *RSC Adv.,,* **2014**, *4*, 44674-44688.
[http://dx.doi.org/10.1039/C4RA05799F]

[12] Burlando, B.; Cornara, L. Honey in dermatology and skin care: a review. *J. Cosmet. Dermatol.,* **2013**, *12*, 306-313.
[http://dx.doi.org/10.1111/jocd.12058]

[13] Popa, M.; Bostan, R.; Popa, D. Honey – marker of environmental pollution. Case study – the transylvania region, romania. *J. Environ. Prot. Ecol.,* **2013**, *14*, 273-280.

[14] Jastrzębska-Stojko, Z.; Stojko, R.; Rzepecka-Stojko, A.; Kabała-Dzik, A.; Stojko, J. Biological activity of propolis-honey balm in the treatment of experimentally-evoked burn wounds. *Molecules,* **2013**, *18*, 14397-14413.
[http://dx.doi.org/10.3390/molecules181114397]

[15] Puttamayutanon, P.; Sutjarittangtham, K.; Eitssayeam, S.; Chantawannakul, P. Nanofiber production using polyvinylalcohol (PVA) and thai honey and its properties. *Adv. Mat. Res.,* **2012**, *506*, 541-544.
[http://dx.doi.org/10.4028/www.scientific.net/AMR.506.541]

[16] Mayer, A.; Slezak, V.; Takac, P.; Olejnik, J.; Majtan, J. Treatment of non-healing leg ulcers with honeydew honey. *J. Tissue Viability,* **2014**, *23*, 94-97.
[http://dx.doi.org/10.1016/j.jtv.2014.08.001]

[17] Jayachandran, S.; Balaji, N. Evaluating the effectiveness of topical application of natural honey and benzydamine hydrochloride in the management of radiation mucositis. *Indian J. Palliat. Care,* **2012**, *18*, 190-195.
[http://dx.doi.org/10.4103/0973-1075.105689]

[18] Wanjai, C.; Sirisa-ard, P.; Chantawannakul, P. Formulation development of gel-based lipstick containing longan honey and natural color extracted from sappan wood and roselle. *Adv. Mat. Res.,* **2012**, *506*, 375-378.
[http://dx.doi.org/10.4028/www.scientific.net/AMR.506.375]

[19] Ajibola, A.; Chamunorwa, J.P.; Erlwanger, K.H. Nutraceutical values of natural honey and its contribution to human health and wealth. *Nutr. Metab.,* **2012**, *9*(61), 1-12.

[20] Miguel, M.G.; Antunes, M.D.; Aazza, S.; Duarte, J.; Faleiro, M.L. Honey-based "água-mel" chemical characterization and microbiological quality. *Ital. J. Food Sci.,* **2013**, *25*, 275-282.

[21] Yamul, D.K.; Lupano, C.E. Properties of gels from whey protein concentrate and honey at different pHs. *Food Res. Int.,* **2003**, *36*, 25-33.
[http://dx.doi.org/10.1016/S0963-9969(02)00104-7]

[22] Popov-Raljić, J.V.; Laličić-Petronijević, J.G.; Georgijev, A.S.; Popov, V.S.; Mladenović, M.A. Sensory evaluation of pralines containing different honey products. *Sensors,* **2010**, *10*, 7913-7933.
[http://dx.doi.org/10.3390/s100907913]

[23] Silva, R.A.; Maia, G.A.; Costa, J.M.C.; Rodrigues, M.C.P.; Fonseca, A.V.V.; Sousa, P.H.M.; Carvalho, J.M. Honey-sweetened cashew apple juice: development and stability. *Ciênc. Tecnol. Aliment.,* **2008**, *28*, 348-354.
[http://dx.doi.org/10.1590/S0101-20612008000200012]

[24] Conforti, P.A.; Lupano, C.E. Functional properties of biscuits with whey protein concentrate and honey. *Int. J. Food Sci. Technol.,* **2004**, *39*, 745-753.
[http://dx.doi.org/10.1111/j.1365-2621.2004.00838.x]

[25] Ropa, D.J. The functional role of honey in baked goods. *Cereal Foods World,* **1999**, *44*, 140-142.

[26] Cardetti, M. Functional role of honey in savory snacks. *Cereal Foods World,* **1997**, *42*, 746-748.

[27] Crane, E. *Beekeeping: Science, Practice and World Resources*; Heinemann: London, **1988**.

[28] Tristão, P. *Produção e Processamento de Própolis e Cera. http://www.portalagropecuario.com.br/apicultura/producao-e-processamento-de-propolis-e-cera/,* **2015**. (Accessed at 11th February 2015).

[29] Park, Y.K.; Ikegaki, M.; Alencar, S.M. *Classificação das própolis brasileiras a partir de suas características físico-químicas e propriedades biológicas http://www.apacame.org.br/mensagemdoce/58/artigo.htm,* **2000**. (Accessed at 7th April 2015).

[30] Daugsch, A.; Moraes, C.S.; Fort, P.; Pacheco, E.; Lima, I.B.; Abreu, J.A.; Park, Y.K. *Própolis vermelha e sua origem botânica http://www.apacame.org.br/mensagemdoce/89/artigo.htm,* **2006**. (Accessed at 19th March 2015).

[31] Righi, A.A. *Perfil químico de amostras de própolis brasileiras. Dissertação (mestrado)*; Instituto de Biociências da Universidade de São Paulo. Departamento de Botânica: São Paulo, **2008**, p. 102.

[32] Daugsch, A. *A própolis vermelha do nordeste do Brasil e suas características químicas e biológicas*; Tese (doutorado) – Universidade Estadual de Campinas. Faculdade de Engenharia de Alimentos, **2007**, p. 144.

[33] Huang, S.; Zhang, C-P.; Wang, K.; Q.Li, G.; Hu, F-L. Review recent advances in the chemical composition of propolis. *Molecules,* **2014**, *19*, 19610-19632.
[http://dx.doi.org/10.3390/molecules191219610]

[34] Miguel, M.G.; Antunes, M.D. Is propolis safe as an alternative medicine? *J. Pharm. Bioallied Sci.,* **2011**, *3*, 479-495.
[http://dx.doi.org/10.4103/0975-7406.90101]

[35] Valverde, M. *Preço da própolis em alta no mercado. Agronegócio 20/03/2014. http://www.diariodocomercio.com.br/noticia.php?id=132227,* (Accessed at 23rd March 2015)

[36] Wiese, H. *Novo manual de apicultura*; Livraria e Editora Agropecuária: Guaíbac, **1995**.

[37] Angelo, G.; Lorena, C.; Marta, G.; Antonella, C. Biochemical composition and antioxidant properties of *Lavandula angustifolia* Miller essential oil are shielded by propolis against UV radiations. *Photochem. Photobiol.,* **2014**, *90*, 702-708.
[http://dx.doi.org/10.1111/php.12229]

[38] Bankova, V.; Galabov, A.S.; Antonova, D.; Vilhelmova, N.; Di Perri, B. Chemical composition of Propolis Extract ACF® and activity against herpes simplex virus. *Phytomedicine,* **2014**, *21*, 1432-1438.
[http://dx.doi.org/10.1016/j.phymed.2014.04.026]

[39] Oršolić, N.; Skurić, J.; Dikić, D.; Stanić, G. Inhibitory effect of a propolis on Di-n-Propyl Disulfide or n-Hexyl salycilate-induced skin irritation, oxidative stress and inflammatory responses in mice. *Fitoterapia,* **2014**, *93*, 18-30.
[http://dx.doi.org/10.1016/j.fitote.2013.12.007]

[40] Asawahame, C.; Sutjarittangtham, K.; Eitssayeam, S.; Tragoolpua, Y.; Sirithunyalug, B.; Sirithunyalug, J. Antibacterial Activity and Inhibition of Adherence of *Streptococcus mutans* by

Propolis Electrospun Fibers. *AAPS Pharm Sci Tech,* **2015**, *16*, 182-191.
[http://dx.doi.org/10.1208/s12249-014-0209-5]

[41] Noronha, V.R.A.S.; Araújo, G.S.; Gomes, R.T.; Iwanaga, S.H.; Barbosa, M.C.; Abdo, E.N.; Ferreira e Ferreira, E.; Campos, A.C.V.; Souza, A.A.; Abreu, S.R.L.; Santos, V.R. Mucoadhesive propolis gel for prevention of radiation-induced oral mucositis. *Curr. Clin. Pharmacol.,* **2014**, *9*, 359-364.
[http://dx.doi.org/10.2174/1574884709666140205210051]

[42] Silva, A.J.; Silva, J.R.; de Souza, N.C.; Souto, P.C.S Membranes from latex with propolis for biomedical applications. *Mater. Lett.,* **2014**, *116*, 235-238.
[http://dx.doi.org/10.1016/j.matlet.2013.11.045]

[43] Yokoyama, T.; Kosaka, Y.; Mizuguchi, M. Inhibitory activities of poropolis and its promising component, caffeic acid phenethyl ester, against amyloidogenesis of human transthyretin. *J. Med. Chem.,* **2014**, *57*, 8928-8935.
[http://dx.doi.org/10.1021/jm500997m]

[44] Martin, L.F.T.; Rocha, E.M.; Garcia, S.B.; Paula, J.S. Topical Brazilian propolis improves corneal wound healing and inflammation in rats following alkali burns. *BMC Complement. Altern. Med.,* **2013**, *13*(337), 1-7.

[45] Ferreira, F.M.; Castro, R.A.O; Batista, M.A.; Rossi, F.M.O; Silveira-Lemos, D.; Frézard, F.; Moura, S.A.L.; Rezende, S.A. Association of water extract of green propolis and liposomal meglumine antimoniate in the treatment of experimental visceral leishmaniasis. *Parasitol. Res.,* **2014**, *113*, 533-543.
[http://dx.doi.org/10.1007/s00436-013-3685-8]

[46] Santana, L.C.L.R; Carneiro, S.M.P.; Caland-Neto, L.B.; Arcanjo, D.D.R.; Moita-Neto, J.M.; Citó, A.M.G.L; Carvalho, F.A.A. Brazilian brown propolis elicits antileishmanial effect against promastigote and amastigote forms of *Leishmania amazonensis. Nat. Prod. Res.,* **2014**, *28*, 340-343.
[http://dx.doi.org/10.1080/14786419.2013.856904]

[47] Pinheiro, K.S.; Ribeiro, D.R.; Alves, A.V.; Pereira-Filho, R.N.; Oliveira, C.R.; Lima, S.O.; Reis, F.P.; Cardoso, J.C.; Albuquerque-Júnior, R.L. Modulatory activity of Brazilian red propolis on chemically induced dermal carcinogenesis. *Acta Cir. Bras.,* **2014**, *29*, 111-117.
[http://dx.doi.org/10.1590/S0102-86502014000200007]

[48] Ozcan, M. Use of propolis extract as a natural antioxidant for plant oils. *Grasas Aceites,* **2000**, *51*, 251-253.
[http://dx.doi.org/10.3989/gya.2000.v51.i4.420]

[49] Ozcan, M.; Ayar, A. Effect of propolis extracts on butter stability. *J. Food Qual.,* **2003**, *26*, 65-73.

[50] Kim, G.W.; Kim, G.H.; Kim, J.S.; An, H.Y.; Hu, G.W.; Park, I.S.; Kim, O.S.; Cho, S.Y. Quality characteristics of fried fish paste of Alaska Pollack meat paste added with propolis. *J. Korean Soc. Food Sci. Nutr.,* **2008**, *37*, 485-489.
[http://dx.doi.org/10.3746/jkfn.2008.37.4.485]

[51] Han, S.K.; Yamauchi, K.; Park, H.K. Effect of nitrite and propolis preservative on volatile basic nitrogen changes in meat products. *Microbios,* **2001**, *105*, 71-75.

[52] Ali, F.H.; Kassem, G.M.; Atta-Alla, O.A. Propolis as a natural decontaminant and antioxidant in fresh

oriental sausage. *Vet. Ital.,* **2010**, *46*, 167-172.

[53] Bernardi, S.; Favaro-Trindade, C.S.; Trindade, M.A.; Balieiro, J.C.C.; Cavenaghi, A.D.; Contreras-Castilo, C.J. Italian-type salami with propolis as antioxidant. *Ital. J. Food Sci.,* **2013**, *25*, 433-440.

[54] Vargas-Sanchez, R.D.; Torrescano-Urrutia, G.R.; Acedo-Félix, E.; Carvajal-Millán, E.; González-Córdova, A.F.; Vallejo-Galland, B.; Torres-Llanez, M.J.; Sánchez-Escalante, A. Antioxidant and antimicrobial activity of commercial propolis extract in beef patties. *J. Food Sci.,* **2014**, *79*, C1499-C1504.
[http://dx.doi.org/10.1111/1750-3841.12533]

[55] Gutiérrez-Cortés, C.; Mahecha, H.S. Antimicrobial activity of propolis and its effect on the physicochemical and sensorial characteristics in sausages. *Vitae,* **2014**, *21*, 90-96.

[56] Kim, Y-H.; Kim, S-A.; Chung, H-J. Synergistic effect of propolis and heat treatment leading to increased injury to *Escherichia coli* O157:H7 in ground pork. *J. Food Saf.,* **2014**, *34*, 1-8.
[http://dx.doi.org/10.1111/jfs.12088]

[57] Narbona, E.; García-García, E.; Vázquez-Araújo, L.; Carbonell-Barrachina, A.A. Volatile composition of functional "a la Piedra" turrón with propolis. *Int. J. Food Sci. Technol.,* **2010**, *45*, 569-577.
[http://dx.doi.org/10.1111/j.1365-2621.2009.02167.x]

[58] Mutton, M.J.R.; Oliveira Filho, J.H.; Costa, G.H.G.; Roviero, J.P.; Freita, L.A. Green and brown propolis: efficient natural biocides for the control of bacterial contamination of alcoholic fermentation of distilled beverage. *Food Sci. Technol.,* **2014**, *34*, 767-772.
[http://dx.doi.org/10.1590/1678-457X.6469]

[59] Koc, A.N.; Silici, S.; Mutlu-Sariguzel, F.; Sagdic, O. Antifungal activity of propolis in four different fruit juices. *Food Technol. Biotech.,* **2007**, *45*, 57-61.

[60] Silici, S.; Karaman, K. Inhibitory effect of propolis on patulin production of *Penicillium expansum* in apple juice. *J. Food Process. Pres.,* **2014**, *38*, 1129-1134.
[http://dx.doi.org/10.1111/jfpp.12072]

[61] Feás, X.; Pacheco, L.; Iglesias, A.; Estevinho, L.M. Use of propolis in the sanitization of lettuce. *Int. J. Mol. Sci.,* **2014**, *15*, 12243-12257.
[http://dx.doi.org/10.3390/ijms150712243]

[62] Yang, S.; Peng, L.; Cheng, Y.; Chen, F.; Pan, S. Control of citrus green and blue molds by Chinese propolis. *Food Sci. Biotechnol.,* **2010**, *19*, 1303-1308.
[http://dx.doi.org/10.1007/s10068-010-0186-8]

[63] Zahid, N.; Ali, A.; Siddiqui, Y.; Maqbool, M. Efficacy of ethanolic extract of propolis in maintaining postharvest quality of dragon fruit during storage. *Postharvest Biol. Tech.,* **2013**, *79*, 69-72.
[http://dx.doi.org/10.1016/j.postharvbio.2013.01.003]

[64] Pastor, C.; Sánchez-González, L.; Marcilla, A.; Chiralt, A.; Cháfer, M.; González-Martínez, C. Quality and safety of table grapes coated with hydroxypropylmethylcellulose edible coatings containing propolis extract. *Postharvest Biol. Tech.,* **2011**, *60*, 64-70.
[http://dx.doi.org/10.1016/j.postharvbio.2010.11.003]

[65] Ali, A.; Chow, W.L.; Zahid, N.; Ong, M.K. Efficacy of propolis and cinnamon oil coating in controlling post-harvest anthracnose and quality of chilli (*Capsicum annuum* L.) during cold storage.

Food Bioprocess Tech., **2014**, *7*, 2742-2748.
[http://dx.doi.org/10.1007/s11947-013-1237-y]

[66] Ali, A.; Wei, Y.Z.; Mustafa, M.A. Exploiting propolis as an antimicrobial edible coating to control post-harvest anthracnose of bell pepper. *Packag. Technol. Sci.,* **2015**, *28*, 173-179.
[http://dx.doi.org/10.1002/pts.2088]

[67] Mascheroni, E.; Guillard, V.; Nalin, F.; Mora, L.; Piergiovanni, L. Diffusivity of propolis compounds in polylactic acid polymer for the development of anti-microbial packaging films. *J. Food Eng.,* **2010**, *98*, 294-301.
[http://dx.doi.org/10.1016/j.jfoodeng.2009.12.028]

[68] Torlak, E.; Sert, D. Antibacterial effectiveness of chitosan-propolis coated polypropylene films against foodborne pathogens. *Int. J. Biol. Macromol.,* **2013**, *60*, 52-55.
[http://dx.doi.org/10.1016/j.ijbiomac.2013.05.013]

[69] Bodini, R.B.; Sobral, P.J.A.; Favaro-Trindade, C.S.; Carvalho, R.A. Properties of gelatin-based films with added ethanol–propolis extract. *LWT –. Food Sci. Technol.,* **2013**, *51*, 104-110.

[70] Nori, M.P.; Favaro-Trindade, C.S.; Matias de Alencar, S.; Thomazini, M.; Balieiro, J.C.C; Castillo, C.J.C Microencapsulation of propolis extract by complex coacervation. *LWT –. Food Sci. Technol.,* **2011**, *44*, 429-435.

[71] Mascheroni, E.; Figoli, A.; Musati, A.; Limbo, S.; Drioli, E.; Suevo, R.; Talarico, S.; Rollini, M. An alternative encapsulation approach for production of active chitosan–propolis beads. *Int. J. Food Sci. Technol.,* **2014**, *49*, 1401-1407.
[http://dx.doi.org/10.1111/ijfs.12442]

[72] Noweer, E.M.; Dawood, M.G. Efficiency of propolis extract on faba bean plants and its role against nematode infection. *Commun. Agric. Appl. Biol. Sci.,* **2009**, *74*, 593-603.

[73] Valero, M.V.; Martin do Prado, R.; Zawadzki, F.; Eiras, C.E.; Madrona, G.S.; Nunes do Prado, I. Propolis and essential oils additives in the diets improved animal performance and feed efficiency of bulls finished in feedlot. *Acta Sci. Anim. Sci.,* **2014**, *36*, 419-426.
[http://dx.doi.org/10.4025/actascianimsci.v36i4.23856]

[74] Velotto, S.; Vitale, C.; Varricchio, E.; Crasto, A. Effect of propolis on the fish muscular development and histomorphometrical characteristics. *Acta Vet. Brno,* **2010**, *79*, 543-550.
[http://dx.doi.org/10.2754/avb201079040543]

[75] Sharaf, S.; Higazy, A.; Hebeish, A. Propolis induced antibacterial activity and other technical properties of cotton textiles. *Int. J. Biol. Macromol.,* **2013**, *59*, 408-416.
[http://dx.doi.org/10.1016/j.ijbiomac.2013.04.030]

[76] El-Naggar, S.A.; Alm-Eldeen, A.A.; Germoush, M.O.; El-Boray, K.F.; Elgebaly, H.A. Ameliorative effect of propolis against cyclophosphamide-induced toxicity in mice. *Pharm. Biol.,* **2015**, *53*, 235-241.
[http://dx.doi.org/10.3109/13880209.2014.914230]

[77] Rizk, S.M.; Zaki, H.F.; Mina, M.A. Propolis attenuates doxorubicin-induced testicular toxicity in rats. *Food Chem. Toxicol.,* **2014**, *67*, 176-186.
[http://dx.doi.org/10.1016/j.fct.2014.02.031]

[78] Wided, K.; Hassiba, R.; Mesbah, L. Polyphenolic fraction of Algerian propolis reverses doxorubicin induced oxidative stress in liver cells and mitochondria. *Pak. J. Pharm. Sci.,* **2014**, *27*, 1891-1897.

[79] Luo, C.; Zou, X.; Li, Y.; Sun, C.; Jiang, Y.; Wu, Z. Determination of flavonoids in propolis-rich functional foods by reversed phase high performance liquid chromatography with diode array detection. *Food Chem.,* **2011**, *127*, 314-320. [http://dx.doi.org/10.1016/j.foodchem.2011.01.006]

[80] Griffiths, G.A.; Toshitaka, U.; Fumihiko, T.; Daisuke, H. Effect of vapors from fractionated samples of propolis on microbial and oxidation damage of rice during storage. *J. Food Eng.,* **2008**, *88*, 341-352. [http://dx.doi.org/10.1016/j.jfoodeng.2008.02.019]

[81] Coronel, B.B.; Grasso, D.; Pereira, S.C.; Fernadez, G. *Caracterización Bromatológica Del Pólen Apícola Argentino*; Ciência, Docência y Tecnologia, Universidade Nacional de Entre Rios: Concepción del Uruguay, Argentina, **2004**, pp. 145-181.

[82] Mutsaers, M.; Blitterswijk, H.; Leven, L.; Kerkvliet, J.; Waerdt, J. *Bee products: properties, processing and marketing. Agrodok 42*; Agromisa Foundation: Wageningen, **2005**.

[83] Milfont, M. de O.; Freitas, B.M.; Alves, J.E. *Pólen Apícola: Manejo para a Produção de Pólen no Brasil,* 1st ed; Aprenda Fácil Editora: Viçosa, MG, **2011**.

[84] BRASIL. Instrução Normativa no. 3, de 19 de Janeiro de 2001. Regulamento Técnico para Fixação de Identidade e Qualidade de Pólen Apícola do Ministério de Agricultura e do Abastecimento. *Brasília, Seção,* **2001**, *1*, 18-23.

[85] Moreti, A.C. de C.C. *Pólen: alimento protéico para as abelhas – Complemento alimentar para o homem. Instituto de Zootecnia de São Paulo,* **2006**. http://www.iz.sp.gov.br/pdfs/1178193451.pdf (Accessed at 14th February 2008).

[86] Negrão, A.F. *Efeito da sazonalidade no teor proteico e composição de aminoácidos no pólen apícola produzido em Botucatu, estado de São Paulo*; Dissertação-Programa de Pós-Graduação em Zootecnia: Botucatu, SP, **2014**.

[87] Winston, M.L. *The Biology of the Honey Bee*; Harvard University Press: Cambridge, **1987**.

[88] Alves, M.L.T.M.F. Produção de pólen apícola *Pesquisa & Tecnologia,* **2013**, *10, n. 2.*

[89] Buena, A.F. El pólen: su regulamentación como alimento dietético. *Vida Apícola,* **1991**, *46*, 25-28.

[90] Mărgăoan, R.; Al. Mărghitaş, L.; Dezmirean, D.; Mihai, C.M.; Bobiş, O. Bee Collected Pollen – General Aspects and Chemical Composition. *Bulletin UASVM Animal Science and Biotechnologies,* **2010**, *67*, 254-259.

[91] BRASIL. *Ministério da Agricultura – Secretaria de Defesa Agropecuária. Instrução Normativa n. 3 de 19 de janeiro de 2001*; ANEXO VI Regulamento Técnico para Fixação de Identidade e Qualidade de Própolis, **2001**.

[92] Barreto, L.M.R.C.; Funari, S.R.C; Orsi, R.O. Composição e Qualidade do Pólen Apícola Proveniente de Sete Estados Brasileiros e do Distrito Federal. *B. Indústr. Anim.,* **2005**, *62*, 167-175.

[93] Avni, D.; Hendriksma, H.P.; Dag, A.; Uni, Z.; Shafir, S. Nutritional aspects of honey bee-collected pollen and constraints on colony development in the eastern Mediterranean. *J. Insect Physiol.,* **2014**,

69, 65-73.
[http://dx.doi.org/10.1016/j.jinsphys.2014.07.001]

[94] Carpes, S.T.; Prado, A.; Moreno, I.A.M; Mourão, G.B.; Alencar, S.M.; Masson, M.L. Avaliação do potencial antioxidante do pólen apícola produzido na Região Sul do Brasil. *Quim. Nova,* **2008**, *31*, 1660-1664.
[http://dx.doi.org/10.1590/S0100-40422008000700011]

[95] Menezes, J.D.S.; Maciel, L.F.; Miranda, M.S.; Druzian, J.I. Compostos bioativos e potencial antioxidante do pólen apícola produzido por abelhas africanizadas (*Apis mellifera L.*). *Rev. Inst. Adolfo Lutz,* **2010**, *69*, 233-242.

[96] Ulusoy, E.; Kolayli, S. Phenolic composition and antioxidant properties of anzer bee pollen. *J. Food Biochem.,* **2014**, *38*, 73-82.
[http://dx.doi.org/10.1111/jfbc.12027]

[97] Kasianenko, V.I.; Komisarenko, I.A.; Dubtsova, E.A. Correction of atherogenic dyslipidemia with honey, pollen and bee bread in patients with different body mass. *Ter. Arkh.,* **2011**, *83*, 58-62.

[98] Musunuru, K. Atherogenic dyslipidemia: cardiovascular risk and dietary intervention. *Lipids,* **2010**, *45*, 907-914.
[http://dx.doi.org/10.1007/s11745-010-3408-1]

[99] Maruyama, H.; Sakamoto, T.; Araki, Y.; Hara, H. Anti-inflammatory effect of bee pollen ethanol extract from *Cistus* sp. of Spanish on carrageenan-induced rat hind paw edema. *BMC Complement. Altern. Med.,* **2010**, *10*(30), 1-11.

[100] Eraslan, G.; Kanbur, M.; Silici, S. Effect of carbaryl on some biochemical changes in rats: the ameliorative effect of bee pollen. *Food Chem. Toxicol.,* **2009**, *47*, 86-91.
[http://dx.doi.org/10.1016/j.fct.2008.10.013]

[101] El-Asely, A.M.; Abbass, A.A.; Austin, B. Honey bee pollen improves growth, immunity and protection of Nile tilapia (*Oreochromis niloticus*) against infection with Aeromonas hydrophila. *Fish Shellfish Immunol.,* **2014**, *40*, 50-506.
[http://dx.doi.org/10.1016/j.fsi.2014.07.017]

[102] Wang, J.; Li, S.; Wang, Q.; Xin, B.; Wang, H. Trophic effect of bee pollen on small intestine in broiler chickens. *J. Med. Food,* **2007**, *10*, 276-280.
[http://dx.doi.org/10.1089/jmf.2006.215]

[103] Turhan, S.; Yazici, F.; Saricaoglu, F.T.; Mortas, M.; Genccelep, H. Evaluation of the nutritional and storage quality of meatballs formulated with bee pollen. *Korean J. Food Sci. An.,* **2014**, *34*, 423-433.
[http://dx.doi.org/10.5851/kosfa.2014.34.4.423]

[104] Faria, R.N. *Instrução de serviço DIPA n°002/04, Instituto Mineiro de Agropecuária vinculada à Secretaria de Estado de Agricultura*; Pecuária e Abastecimento: Belo Horizonte, **2004**.

[105] Silva, C.H.M. *Novos Fatores contribuindo para longevidade humana*; , **1994**, p. 119.

[106] Scopece, G.; Schiestl, F.P.; Cozzolino, S. Pollen transfer efficiency and its effect on inflorescence size in deceptive pollination strategies. *Plant Biol. (Stuttg.),* **2015**, *17*, 545-550.
[http://dx.doi.org/10.1111/plb.12224]

[107] Gill, J.R.; Rooney, W.L.; Klein, P.E. Effect of humidity on intergeneric pollinations of *iap* (*Inhibition of Alien Pollen*) sorghum [*Sorghum bicolor* (L.) Moench]. *Euphytica,* **2014**, *198*, 381-387. [http://dx.doi.org/10.1007/s10681-014-1113-5]

[108] Sugiyama, K.; Kami, D.; Muro, T. Induction of parthenocarpic fruit set in watermelon by pollination with bottle gourd (*Lagenaria siceraria* (Molina) Standl.) pollen. *Sci. Hortic.,* **2014**, *171*, 1-5. [http://dx.doi.org/10.1016/j.scienta.2014.03.008]

[109] Li, X.X.; Wang, H.; Gituru, R.W.; Guo, Y.H.; Yang, C.F. Pollen packaging and dispensing: adaption of patterns of anther dehiscence and flowering traits to pollination in three *Epimedium* species. *Plant Biol. (Stuttg.),* **2014**, *16*, 227-233. [http://dx.doi.org/10.1111/plb.12040]

[110] Formicki, G.; Gren, A.; Stawarz, R.; Zysk, B.; Gal, A. Metal content in honey, propolis, wax, and bee pollen and implications for metal pollution monitoring. *Pol. J. Environ. Stud.,* **2013**, *22*, 99-106.

[111] Malayeri, B.E.; Noori, M.; Jafari, M. Using the pollen viability and morphology for fluoride pollution biomonitoring. *Biol. Trace Elem. Res.,* **2012**, *147*, 315-319. [http://dx.doi.org/10.1007/s12011-011-9290-8]

[112] Pukacki, P.M.; Chalupka, W. Environmental pollution changes in membrane lipids, antioxidants and vitality of Scots pine (*Pinus sylvestris* L.) pollen. *Acta Soc. Bot. Pol.,* **2003**, *72*, 99-104. [http://dx.doi.org/10.5586/asbp.2003.012]

[113] Rezanejad, F. Air pollution effects on structure, proteins and flavonoids in pollen grains of *Thuja orientalis* L. (Cupressaceae). *Grana,* **2009**, *48*, 205-213. [http://dx.doi.org/10.1080/00173130902949417]

[114] Rezanejad, F. The response of anther and pollen development, pollen cellular material release and pollen proteins to air pollution in *Petunia hybrida* Juss. (Solanaceae). *Iran. J. Sci. Technol.,* **2013**, *A1*, 63-68.

[115] Rezanejad, F. Air pollution effects on flavonoids in pollen grains of some ornamental plants. *Turk. J. Botany,* **2012**, *36*, 49-54.

[116] Marshall, E.; Mejia, D. *Traditional Fermented Food and Beverages for improved Livelihoods.* Diversification booklet number 21; Rural Infrastructure and Agro-Industries Division Food and Agriculture Organization of the United Nations: Rome, **2011**. www.fao.org/dorep/015/i2477e/i2477e00.pdf

[117] Caplice, E.; Fitzgerald, G.F. Food fermentation: role a microorganisms in food production and preservation. *Int. J. Food Microbiol.,* **1999**, *50*, 131-139. [http://dx.doi.org/10.1016/S0168-1605(99)00082-3]

[118] Blandino, A.; Al-Aseeri, M.E.; Pandiella, S.S.; Cantero, D.; Webb, C. Cereal-based fermented foods and beverages. *Food Res. Int.,* **2003**, *36*, 527-543. [http://dx.doi.org/10.1016/S0963-9969(03)00009-7]

[119] Farnworth, E.R., Ed. *Handbook of fermented functional foods,* 2nd ed; CRC Press: Boca Raton, **2008**.

[120] Bourdichon, F.; Casaregola, S.; Farrokh, C.; Frisvad, J.C.; Gerds, M.L.; Hammes, W.P.; Harnett, J.; Huys, G.; Laulund, S.; Ouwehand, A.; Powell, I.B.; Prajapati, J.B.; Seto, Y.; Schure, E.T.; Boven, A.V.; Vankerckhoven, V.; Zgoda, A.; Tuijtelaars, S.; Hansen, E.B. Food fermentations:

microorganisms with technological beneficial use. *Int. J. Food Microbiol.,* **2012**, *154*, 87-97. [http://dx.doi.org/10.1016/j.ijfoodmicro.2011.12.030]

[121] Quizani, N.; Mothershaw, A. Fermentation as a method for food preservation. In: *Handbook of Food Preservation,* 2nd ed; Rahman, M.S., Ed.; CRC Press: Boca Raton, **2007**; pp. 215-236.

[122] Aloys, N.; Angeline, N. Traditional fermented foods and beverages in Burundi. *Food Res. Int.,* **2009**, *42*, 588-594. [http://dx.doi.org/10.1016/j.foodres.2009.02.021]

[123] Hutkins, R.W. Wine fermentation. In: *Microbiology and Technology of Fermented Food*; Hutkins, R.W., Ed.; Blackwell Publishing: Iowa, **2006**. [http://dx.doi.org/10.1002/9780470277515.ch10]

[124] Pereira, A.P.; Mendes-Ferreira, A.; Oliveira, J.M.; Estevinho, L.M.; Mendes-Faria, A. Effect of *Saccharomyces cereviseae* cells immobilisation on mead production. *LWT –. Food Sci. Technol.,* **2014**, *56*, 21-30.

[125] NHB – National Honey Board. *Making Mead: the art and the science. www.bjcp.org/mead/makingmead.pdf,* (Accessed at 16th August 2014).

[126] Piatz, S. *The complete guide to making mead: the ingredients, equipment, processes, and recipes for crafting honey wine.* ; Voyageur Press: Minneapolis, **2014**.

[127] Bahiru, B.; Mehari, T.; Ashenafi, M. Yeast and lactic acid flora of *Tej*, an indigenous Ethiopian honey wine: variations with and between production units. *Food Microbiol.,* **2006**, *23*, 277-282. [http://dx.doi.org/10.1016/j.fm.2005.05.007]

[128] Barrios, C.; Principal, J.; Sánchez, J.; Guédez, J.C. Caracterización físico-química y análisis sensorial de un hidromiel elaborado de manera artesanal. *Zootecnia Trop.,* **2010**, *28*, 313-319.

[129] Bahiru, B.; Mehari, T.; Ashenafi, M. Chemical and nutritional properties of 'tej', an indigenous Ethiopian honey wine: variations within and between production units. *J. Food Technol. Africa,* **2001**, *6*, 104-108.

[130] BRASIL. Ministério da Agricultura, Pecuária e Abastecimento. Portaria no. 64, de 23/04/2008. Aprova o regulamento técnico para a fixação dos padrões de identidade e qualidade para as bebidas alcoólicas fermentadas: fermentado de fruta, sidra, hidromel, fermentado de cana, fermentado de fruta licoroso, fermentado de fruta composto e saquê. Diário Oficial [da] República Federativa do Brasil, Brasília, DF, 24 abr. **2008** *Seção, 1*, 9.

[131] Iglesias, A.; Pascoal, A.; Choupina, A.B.; Carvalho, C.A.; Feás, X.; Estevinho, L.M. Developments in the fermentation process and quality improvement strategies for mead production. *Molecules,* **2014**, *19*, 12577-12590. [http://dx.doi.org/10.3390/molecules190812577]

[132] Wintersteen, C.L.; Andrae, L.M.; Engeseth, N.J. Effect of heat treatment on antioxidant capacity and flavour volatiles of mead. *J. Food Sci.,* **2005**, *70*, 119-126. [http://dx.doi.org/10.1111/j.1365-2621.2005.tb07071.x]

[133] Gupta, J.K.; Sharma, R. Production technology and quality characteristics of mead and fruit honey wines: a review. *Nat. Prod. Radiance,* **2009**, *8*, 345-355.

[134] Mattieto, R.A.; Lima, F.; Venturieri, G.C.; Araújo, A.A. *Tecnologia para obtenção artesanal de hidromel do tipo doce.* ; Belém: Embrapa Amazônia Oriental, **2006**, Comunicado Técnico, 170, p. 5.

[135] Rivaldi, J.D.; Silva, M.M.; Coelho, T.C.; Oliveira, C.T.; Mancilha, I.M. Caracterização e perfil sensorial de hidromel produzido por *Saccharomyces cereviseae* IZ 888. *Braz. J. Food. Technol.,* **2009**, *11*, 58-63.

[136] Sroka, P.; Tuszynski, T. Changes in organic acid contents during mead wort fermentation. *Food Chem.,* **2007**, *104*, 1250-1257. [http://dx.doi.org/10.1016/j.foodchem.2007.01.046]

[137] Gomes, T.; Barradas, C.; Dias, T.; Verdial, J.; Morais, J.S.; Ramalhosa, E.; Estevinho, L.M. Optimization of mead production using response surface methodology. *Food Chem. Toxicol.,* **2013**, *59*, 680-686. [http://dx.doi.org/10.1016/j.fct.2013.06.034]

[138] Strong, G.; Ruud, S.; Kristen, E.; Schramm, K.; Stock, C.; Bakulić, P.; Zapolski, M. *BJCP Special Ingredient Descriptions. http://www.bjcp.org,* (Accessed at 1st March 2014).

[139] Ilha, E.C.; Bertoldi, F.C.; Reis, V.D.A.; Sant'Anna, E. *Rendimento e eficiência da fermentação alcoólica na produção de hidromel*; Corumbá. Embrapa Pantanal, **2008**, p. 14.

[140] McConnell, D.S.; Schramm, K.D.; Kent, G.W. *An analysis of mead, mead making and the role of its primary constituents. http://www.bjcp.org/mead/analysis.html,* (Accessed at 1st August 2014)

[141] Gogol, D.; Tuszynski, T. Miody pitne-tradycje surowce I technologie. *Przemysl Fermentacyjny-Warzywny,* **1996**, *8*, 25-27.

[142] Schramm, K. *The complete meadmaker: home production of honey wine from your first batch to award-winning fruit and herb variations*; Brewers Publications: Boulder, **2003**.

[143] Kempka, A.P.; Mantovani, G.Z. Produção de hidromel utilizando méis de diferentes qualidades. *Revista Brasileira de Produtos Agroindustriais,* **2013**, *15*, 273-281. [http://dx.doi.org/10.15871/1517-8595/rbpa.v15n3p273-281]

[144] Roldán, A.; van-Muiswingkel, G.C.; Lasanta, C.; Palacios, V.; Caro, I. Influence of pollen addition on mead elaboration: physicochemical and sensory characteristics. *Food Chem.,* **2011**, *126*, 574-582. [http://dx.doi.org/10.1016/j.foodchem.2010.11.045]

[145] Querol, A.; Fernández-Spinar, M.T.; Olmo, M.; Barrio, E. Adaptive evolution of wine yeast. *Int. J. Food Microbiol.,* **2003**, *86*, 3-10. [http://dx.doi.org/10.1016/S0168-1605(03)00244-7]

[146] Fleet, G.H. *Wine Microbiology and Biotechnology*; Hardwood Academic Publishers: Sidney, **1994**, pp. 166-179.

[147] NHB - National Honey Board. *Honey and Homebrewing: a sweet history. www.nhb.org,* (Accessed at 07th June 2015)

[148] NHB - National Honey Board. *Honey in Beer. www.nhb.org,* (Accessed at 07th June 2015)

[149] Brunelli, L.T.; Mansano, A.R.; Venturini, W.G. Caracterização físico-química de cervejas elaboradas com mel. *Braz. J. Food. Technol.,* **2014**, *17*, 19-27. [http://dx.doi.org/10.1590/bjft.2014.004]

[150] Fabian, F.W. *Honey Vinegar*, Extension Bulletin No. 149. Michigan State College of Agriculture and Applied Science, **1935**.

[151] Ilha, E.C.; Sant'Anna, E.; Torres, R.C.; Porto, A.C.; Meinert, E.M. Utilization of bee (*Apis mellifera*) honey for vinegar production at laboratory scale. *Acta Cient. Venez.,* **2000**, *51*, 231-235.

[152] Derakhshandeh-Rishehri, S.M.; Heidari-Beni, M.; Feizi, A.; Askari, G.R.; Entezari, M.H. Effect of Honey Vinegar Syrup on Blood Sugar and Lipid Profile in Healthy Subjects. *Int. J. Prev. Med.,* **2014**, *5*, 1608-1615.

SUBJECT INDEX

Made in the USA
Las Vegas, NV
04 November 2026

57436921R00293